DOUBLE–DOUBLE

Simplifying the Design and Manufacture of Composite Laminates

EDITORS
Stephen W. Tsai
Brian G. Falzon
Ali Aravand

SECOND EDITION

SECOND EDITION

DOUBLE–DOUBLE

Simplifying the Design and Manufacture of Composite Laminates

AUTHORS & EDITORS

Stephen W. Tsai
Brian G. Falzon
Ali Aravand

AUTHORS

Erik Kappel
Antonio Miravete
Aniello Riccio
Francesco Di Caprio
Scott Millen
Surajit Roy
Kirk Ming Yeoh
Jialong Liu
Gang Yang
Tong-Earn Tay
Vincent Tan
Dazhi Jiang
Jie Zhi
Dan Wang
Shuguang Li
Mingming Xu
Elena Sitnikova
Chuanxiang Zheng
Bo Jin
Albertino Arteiro
Naresh Sharma
Thierry Massard

Book designed by Alice Zheng (zhengalice7@gmail.com)

Library of Congress Control Number: 2023910809

ISBN 9778-0-9819143-5-0 for ebook
978-0-9819143-4-3 for hardbound
Printed in Belfast, Northern Ireland

Front cover is a set of superposed failure envelopes of 15 DD carbon/epoxy composites in [±75/±15], normalized by their respective areas. Master envelope for this group of DD laminates is simply the average of all 15 envelopes.

Department of Aeronautics & Astronautics
Stanford University, Stanford, CA 94305-4035

Preface: Double-double, Second Edition

A year has passed since the first edition of the book was published. Our global team members have made technical advances and reached out to several new members. It was decided to take advantage of the International Conference on Composites Materials (ICCM23), held in Belfast between 30th July 30 and 4th August 2023, so that this edition could be made available to all participants.

Chapter 1 lays the foundations of 'Double-Double' (DD) in laminate plate theory that are unique and underpin the work in the following chapters. They include the concepts of homogenization, trace (or Tsai's modulus), strength ratios, omni envelopes, unit circle criterion, and lamination parameter plots. Chapter 2, on master envelopes, is new. It shows that materials and laminates are equivalent and well-structured under the field of DD. A single parameter, represented by radius R, of a circle that is equal in area for any First Ply Failure (FPF) or Last Ply Failure (LPF) envelope, was found, which complements the trace for the stiffness of laminates. Consequently, rating and scaling of materials and laminates become simple. Manufacturing also becomes simpler and not burdened with rules on mid-plane symmetry, ten percent rule, ply-blocking limits on ply nesting, and others. Having a sub-laminate building block of only four plies, DD can be easily homogenized both across the thickness and throughout the entire component. Symmetry is naturally satisfied. Layup does not have to stop at the mid-plane to reverse the stacking sequence. Tapering to save weight is simple because the laminate properties do not change. Testing is simpler because the stiffness and strength of DD laminates are predictable and hence the level of testing required for material properties is significantly reduced. In fact, only tensile tests are needed because laminate compression and shear are independent of materials; i.e., the same value for all CFRP. Any additional testing is more for calibration and validation than a blind test of properties. Chapter 4 explores homogenization and the use of lamination parameter plots is expanded. Features of lamination plots on DD are further illustrated in a new Chapter 10. Chapter 6, also a new addition in this 2nd edition, provides Compression-After-Impact (CAI) residual strength data on DD laminates which shows that DD laminates are more than competitive when compared with the residual strength of comparable quad laminates currently in use. The notched response of DD laminates is discussed in an updated Chapter 11, and certification issues, presented in Chapter 13, further support DD as a viable technology. The new Chapter 12 applies the master envelopes concept in Chapter 2 to coupons with open holes. It opens the opportunity to extend such simplified approaches to stress raisers and damage beyond that of an open hole. The new Chapter 14 shows another novel application of DD in bending. Each application adds to the credibility of DD in its elegance, simplicity, and practicality.

In every instance of our discoveries, advances and new test data, DD has demonstrated to us its truly revolutionary nature. First of all, our goal for simplicity, as opposed to complexity, is being realised. DD is orderly, compact, and easy to manipulate. It is field-based, so existing rules of transformation and manipulation are correct and credible. Subjective choices in stacking sequence, blending, round-off plies, and many design rules are not only of questionable value but also inflict unnecessary complexity. Current design of components can call for dozens, if not hundreds of laminates that can be replaced by one DD laminate, with aggressive taper, single ply drops placed on the exterior surface, with no internal defects, no warpage, and easy to inspect and repair. DD can be described by one parameter in trace for stiffness and the effective radius of equal area of failure envelopes for strength. Rating and scaling are possible. Subjectivity is replaced by objectivity. Optimization becomes significantly more tractable for complex geometries and loading conditions.

It has been a busy year for DD. First, we conducted training sessions for 100 NASA engineers from many of their centers. We received support from Northrop-Grumman to promote DD. One regional jet manufacturer has decided to use DD for three components of their plane. We had more than 100 graduate students participating in our online workshop. Our goal is to make composite materials simpler and better understood and utilized through DD than the current practice of using quad-based laminates.

Design tools are available for those interested. The MicMac/Duplex is used to generate the master envelopes. The Lam search is an optimizer as described in Chapter 15. Both are available. MATLAB versions of these tools are also available. All for free.

Online training with Composites Design Workshop, held semiannually, is free for graduate students. Hundreds of students have participated in this workshop. Information and registration can be found on our webpage: compositesdesign.stanford.edu.

Steve Tsai
Stanford University

Brian G. Falzon
RMIT University

Ali Aravand
Queen's University Belfast

June 2023

Foreword: Double-double, Second Edition

Composite layup optimisation has made significant progress in recent years, driven by the innovative "DD" approach. This approach has opened up new opportunities for streamlining the design and manufacturing of structural composites, at the same time enhancing their strength. The limitations of the traditional "black aluminium" approach can only be overcome with a more advanced, flexible, and efficient alternative. Specifically, the necessity of adhering to the 10% rule and midplane symmetry in layup design associated with traditional quad-axial laminates has made them difficult to design and costly to manufacture and repair. The DD approach presents several notable advantages compared to the legacy quad laminates. It eliminates the need for midplane symmetry requirements, providing significant benefits in terms of laminate tapering and repair. This flexibility allows for more efficient and tailored modifications to the laminate structure. Additionally, the DD approach enables the achievement of laminate homogenisation with a minimum number of repeats. The latter is best demonstrated through a "Grid" representation of the minimum number of repeats required to achieve homogenisation for various DD staggers. Besides, the DD approach is analytically robust. Where applicable, closed-form solutions based on the in-plane and out-of-plane lamination parameters offer an instant and accurate means for converting the quad configurations into DD alternatives.

The DD conversion tools have been successfully integrated with the Finite Element Analysis solvers as part of a design optimisation workflow. Examples have been covered in this publication. These benefits will significantly facilitate the repair and layup of composite structures, making them more reliable and cost-effective in the long run. Since the introduction of the main concept by Prof. Stephen Tsai in 2017, several important developments have been made by the international DD community, which will greatly benefit the wider and faster adoption of the method in structural applications. The present publication includes some of the most recent advances in the field.

The Aerospace industry has shown a growing interest in the DD approach. For example, Spirit AeroSystems, which is currently involved in an academic collaboration in this area, has acknowledged and affirmed the potential benefits of the

DD design. Stephen McBurney, Head of the Stress & Mass Properties at Spirit AeroSystems Europe, recognises DD's potential in "tailoring the ply architecture for optimising components such as wing skin based on typical design load cases in different wing zones". He anticipates that "DD would prove advantageous in addressing the majority of the acreage of the planform and then accommodate the local high input zones separately". Margins in the spanwise direction are optimised, whilst those in the chordwise direction have high redundant capability. The question arises whether DD can remove this redundancy without contravening best practice rules.

The widespread adoption and use of double/double laminates, particularly in the aerospace industry where its potential advantages can be best exploited, would, however, require overcoming some key challenges. Resistance to depart from the long-standing traditional quad-axial layup and the embedded 10% rule in existing manufacturing processes remains a significant limiting factor. Additionally, sufficient additional testing and certification of optimised layups are needed, especially in aerospace applications where safety and reliability are paramount. The aerospace industry faces unique challenges due to stringent regulatory requirements and certification processes associated with composite materials in aircraft structures. Therefore, the adoption of the DD approach will require diligent scrutiny from regulatory bodies to ensure compliance with safety and reliability standards. Overcoming these challenges necessitates careful planning and implementation, supported by sufficient testing and experimentation to validate analytical and computational results. This will undoubtedly enable the composite manufacturers and users, in particular within the aerospace industry, to reap significant benefits from embracing the DD methodology.

Ali Aravand
Belfast, NI, UK,
May 2023

Foreword

Practicing composite structures design engineers, have initially relied on their brain to predict the final strength of laminates subjected to in-plane simple static loads; dynamic loads and buckling unleashed a plethora of rules, such as balanced and symmetric laminates, that only computer number crunching is able to estimate the strength and stiffness of such complex structures.

Industry has handled this increased complexity using Legacy ply angles, together with ever more complicated rules that cannot be all complied with, due to their contradictory and empirical nature. Weight reduction initiatives, further contribute to the relaxation of the rules, which results in significant deviations from the original objective, therefore the regulator reliance on testing for certification.

The established angle standards (quad) do not preclude, as it was initially assumed, high material qualification and aircraft certification costs. Perhaps, given this facts, it is now necessary for the composite professionals to look at this From a different perspective.

Professor Tsai's book looks at the problem with mathematical elegance, without ignoring the practical aspects.

The former, is clearly explained in this book by using "trace" an invariant that conveys the laminate material stiffness, also known as Tsai's modulus. Material choice can be done easily due to the "trace" scaling ability. It is also simply explained how the laminate stiffness can be understood by combining "trace" with the angle layup definition.

The simplicity extends to the latter: the "double-double" new angles proposal, offers an alternative to the legacy "quad"; layup is considerably easy and removes all the complexity imposed by the Legacy rules. Particularly, ply removal due to weight reduction will not impact as much the original design intent strength and stiffness. Though, it will be hard to relinquish the current practice to use zeros and nineties angles to accommodate varied load case structures.

Professor Tsai's work offers a credible challenge to well established, mostly empirical methods, that evolved little during the last 50 years. Remarkable engineering successes were possible, specifically in aerospace, even though at a considerable premium.

Perhaps, it is about time to give a chance to a different approach based on simpler methodology, based on his vast experience and rigorous scientific theories, focused as well in the simplification of the inherent manufacturing layup technology, demanded by industry.

Paulo Lage
Bristol, UK
April 2022

Foreword II

Industrializing the Game Changing ideas of a Global Composites Team

Philippe Sanial, Chomarat

People all around the world working in the Advanced Composites industry know that Stephen W. Tsai has helped guide our industry for countless years and is continuing to bring innovative and game changing ideas today. The size and scope of this work and the progression of these ideas into reality cannot be done by Steve alone, but with the tremendous global team that has been assembled and partially represented in this book…anything is possible…including Double-Double. The level of experience represented in these chapters cannot be measured in just years, but rather decades or maybe even centuries.

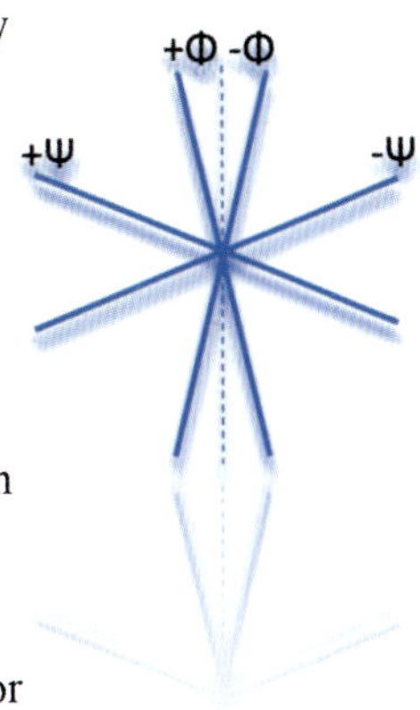

Double-Double (DD) is an innovative concept that utilizes an improved ply stacking sequence with the primary objective to simplify composites processing and predictability while also improving performance. This novel layup strategy incorporates 4-ply [$\pm\Phi/\pm\Psi$] building blocks, which are stacked upon each other. These building blocks are designed with angles and weights that are optimized for the application and then stacked repeatedly without concern mid-plane symmetry.

Game changing ideas like Double-Double are often stalled or disappear when they don't become industrialized. The key to success is the transition

From the concept or lab test to the manufacturing floor.

The ideas and concepts outlined in this book relative to DD can certainly be applied in production today depending on the application and process for manufacturing. For example, using the advanced manufacturing technologies of Automated Fiber Placement (AFP) and Automated Tape Laying (ATL) allows programming changes to incorporate revised DD layup strategies. These specific production systems are great for companies that build large and/or high-volume parts and where long-term programs can justify extremely expensive equipment. For the rest of the industry, industrialization needs to come from a different solution that provides the benefits of DD without the requirements of huge programs and capital investments.

Chomarat has been working with Steve and his global team since the mid-2000's and have collectively pioneered some innovative concepts and materials. The use of Chomarat's C-Ply™ product form is an important bridge that allows many of the design concepts to be scaled into processes that are usable by all parts manufacturers, not just those with large and automated layup equipment. C-Ply™ production is a fully industrialized technology for assembling plies of carbon together that can be specified and ordered today for use in almost any application. The base format of the material is carbon non-crimped fabric (NCF), but the capability range of C-Ply™ exceeds traditional carbon NCF. The full capability envelope for C-Ply™ relative to angles, ply weights, and widths can be seen in the table at the end of this forward.

What is Double-Double?

In simplest terms, DD is a revised ply orientation sequence that allows expanded design optimization while also adding many other advantages that can ultimately increase performance, reduce weight, and likely lower costs. More comprehensive technical details will follow in this book, but the charts below outline some basic concepts.

Angle Plies **(Double)**

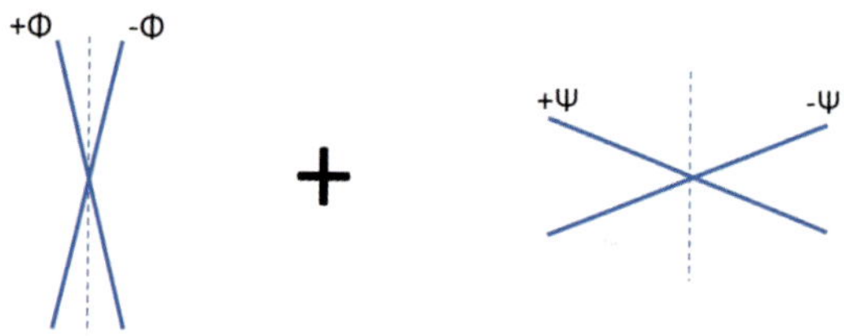

Sub Laminate (**Double-Double) ±Φ ±Ψ**

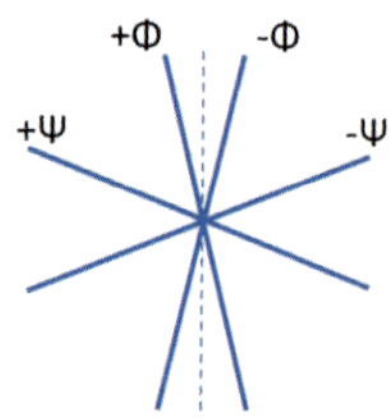

There is always a combination of Φ and Ψ to meet the required properties

Double-Double Performance Adjustment vs Quadriaxial

Quad performance managed by adding 0°, 90° or ± 45° plies (**Quad** is discrete)

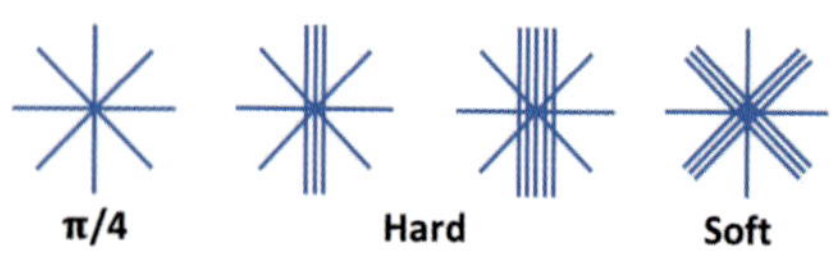

Increased number of plies: 4 to 6, 8, 10...

Double-Double performance managed by tuning Φ & Ψ (**DD** is continuous & adjustable)

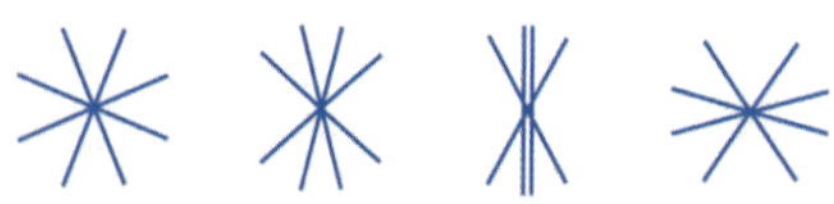

DD is always 4 plies, easier Homogenization

Ply drop strategy simple and flexible

DD can match all laminates: π/4, soft, hard, preferred direction

Key Benefits of Double-Double

- **DD leads to lighter parts**
  - DD optimization is continuous (quad is discrete)
  - 4-ply minimum gage allows fine thickness tuning
- **DD laminate is homogeneous and requires no mid-plane symmetry – if sufficient repeat is achieved**
- **Simple ply drop strategy**
  - Possible with minimum gage and no mid-plane symmetry requirement
  - Single ply drops are possible and can be positioned on inside or outside surfaces of laminate
- **Efficient manufacturing and easier repair**
  - Always the same simple 4-ply sub laminate $\pm\Phi/\pm\Psi$
  - DD can be tapered by additive layup or subtractive ply drop or machining
  - $\pm\Phi$ and $\pm\Psi$ can be prebuilt to speed up deposition rate using Non-Crimp Fabrics (i.e., C-Ply™)

DD overcomes most of the current composite limitations

What is C-Ply™?

The shortest answer is ***carbon reinforcements with UD properties and fabric handleability***. C-Ply™ fabrics alone are dry reinforcements that can be used within most composite part manufacturing processes, including: Prepreg, RTM, Vacuum Infusion, Wet Molding, Press Curing, Winding, Etc.

C-Ply™ is an enabling reinforcement material to be used where full layup automation (i.e., ATL or AFP) is not feasible or possible, but the producer wants to achieve UD properties with a more typical manufacturing process. The automation is done in the reinforcement/textile production process to create multi-ply carbon sub-laminates that generally have two angles and are stitched together for robust handling.

The ability to tailor the weight of the fabric allows the designer flexibility to achieve homogenization faster on relatively thin parts, while optimizing the number of plies for layup on thicker parts. In most cases, the pliability of the DD stacks made with C-Ply™ will be a vast improvement in quality and angle control when compared to handling typical UD materials.

"Sales" Overview of C-Ply™
Put the fibers where you need them and not where you don't. In this load case, it is best to have more reinforcement between the two big dogs.

Enabling DD Industrialization

Two separate biaxial C-Ply™ products are most often used to construct one DD configuration. The fully automated textile production process can usually be set up to allow both biaxial materials to be laid up in the same axis, but if the desired angles don't fall within the acceptable range below, it may be necessary to lay one of them up perpendicular. For example, if a ±15° is desired (outside of production range), it would be produced as a ±75° and then laid up perpendicular. In all cases, the optimized angles for the materials can be configured on the textile production line, so the layup process becomes much faster and simpler when compared to a quadriaxial layup.

An important point to consider is that the order of plies can vary if the designer prefers a different configuration within the DD stack. For example, one laminate may benefit from a stacking sequence of $\Phi/-\Phi/\Psi/-\Psi$, while a soft laminate may prefer $\Phi/-\Psi/-\Phi/\Psi$ or a hard laminate may prefer $\Phi/-\Psi/\Psi/-\Phi$. Any of these scenarios could be made from two separate C-Ply™ products that are then stacked together in the same layup axis to construct the DD stack without any real manufacturing consequences. In other words, the setup and production speed (both influencing cost) to produce a $\Phi/-\Phi$ is not much different than a $-\Phi/\Psi$, as long as the angles fall within the ranges shown in the table below. And...manufacturing an unbalanced biaxial product generally does not present any special challenges within the production process.

DD **Biaxial** **Biaxial**

$\Phi/-\Phi/\Psi/-\Psi = (\Phi/-\Phi) + (\Psi/-\Psi)$ ... *angles are balanced within each biaxial product*

$\Phi/-\Psi/-\Phi/\Psi = (\Phi/-\Psi) + (-\Phi/\Psi)$ ... *both biaxial products are unbalanced*

$\Phi/-\Psi/\Psi/-\Phi = (\Phi/-\Psi) + (\Psi/-\Phi)$ ... *both biaxial products are unbalanced*

One additional consideration is that 0° plies can also be incorporated into a ply stack depending on the application and desired performance. When including a 0° ply it can be coupled with a single Φ or even a $\pm\Phi$ (creating a triaxial), and the 0° ply can be a different areal weight, if preferred. There are very few manufacturing barriers for producing DD with C-Ply™. Range of C-Ply products are shown below:

Areal Weight Ranges:

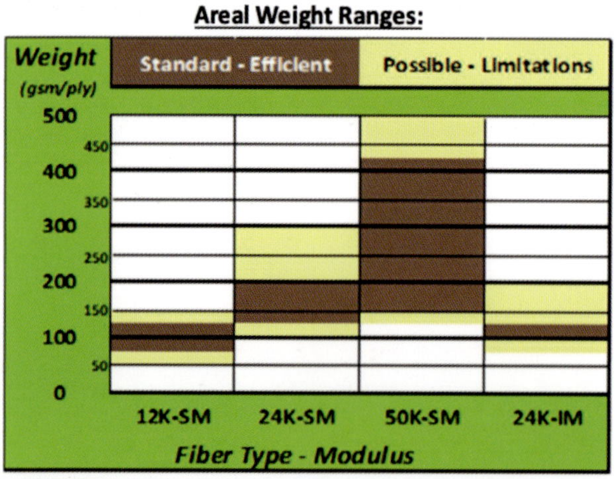

Ply Angle Ranges:

- Standard – Efficient:
  +30° to -30°
- Possible – Limitations:
  +20° to -20°

Overall Product Width:

- 2.54 meter (100 inch)
- Slit to desired widths

Preface

Double-double (DD) is a new class of composite laminates, replacing the conventional quad that limits ply angles to 0, ±45, 90, and many other laminates called unconventional. In contrast, DD consists of two pairs of ply angles: [$\pm\theta_1$,$\pm\theta_2$]. The goal is to have laminates that will be easier and faster to produce with lower weight and lower cost. Many manufacturing innovations can be achieved with DD and are not otherwise possible with the conventional quad and unconventional laminates.

For some time, we have been advocating the use of one parameter for stiffness in trace, one parameter for failure criterion by leveraging Nettles circle, indiv- id- ual ply drops instead of symmetric pairs, one-axis layup with no cross-plying, and one continuous layup without stopping for mid-plane symmetry. We are now advocating one test for interlaminar stress, one laminate for design allowable generation, and one of two stacking sequences for DD to reach homogenization the fastest, and more simplifications to come. All these steps that will make composites easier and lower cost to produce are enabled because of the homogenization of DD laminates. They are like orthotropic metals, but better than real metals because DD can have directional properties; DD can be tapered to save weight with which neither conventional and unconventional laminates, nor metals, can compete.

At the beginning of composite materials and structures for aerospace applications in the 1960s design rules were made without much challenge. Many were more philosophical than rational. The most prominent rules include the following:

1. Laminates must have mid-plane symmetry in order to avoid warpage
2. Laminates must be balanced; i.e., to be orthotropic so it is easier to model
3. Laminates must have 10 percent in each of quad angles to guard against secondary loads
4. Ply grouping to be 4 or less layers so interlaminar stresses can be minimized
5. Inter-ply angle should be 45 degree or less also to minimize interlaminar stresses

These rules impose severe penalties in laminates design and cause many self-in- flicted complexities in manufacturing. As we will see in this book, many of these rules are not necessary, often incur a weight penalty, and their intent can be satisfied by other means. For example, mid-plane symmetry can be replaced by through-the-thickness homogenization which is naturally symmetric. A direct benefit of this rule change beyond the faster and less prone to error in layup is to allow ply drops to be in singles rather than in symmetric pairs. Single ply drops can be located at the exterior surfaces. The laminate interior can now be void of discontinuities and wrinkles. When a component is homogenized, warpage and twist are prevented. Many benefits in manufacturing can be derived directly from homogenization.

There are other practices which have made composites difficult to understand and produce. One major deterrent is the design allowable generation process. It calls for testing multiple laminates, multiple batches, and multiple environments. In this book, we recognize the existence of master ply and modeling of failure modes that can lead to testing of one laminate, one batch, and one set of environmental conditions. A [±23/±75], for example, can be one such laminate, as derived in Chapter 2. Vincent Tan in Chapter 7 found that the interlaminar stress is the same for all homogenized DD laminates. Testing any one laminate will be sufficient to know what all the other laminates will do. Albertino Arteiro in Chapter 11 explores the same one-laminate concept to his design allowable generation for laminates with and without hole. Orders of magnitude in the number of test coupons can be systematically reduced. New materials and processes would be more easily exploited in making advances in production possible.

There are also many practices which have never been tested for their scientific foundation. Fiber and matrix dominated property, behavior and/or quality is an example. It is at best a vague description of the failure of a uniaxial tensile test. It is not an appropriate description of a laminate which may be subjected to different loading conditions. Their responses can be very complex and not sorted out into two categories of fiber- and matrix-dominated. DD has been unfairly tagged to be matrix dominated, not sufficient to be fiber dominated. Such simplistic judgement is subjective and has no evidence that it is true.

The 10 percent rule is another fuzzy rule. It is hard to prove that the laminate would be stronger or tougher with this rule. It only outlaws many laminates which might have performed more than adequately otherwise. The equivalent of a 10 percent rule for double-double would be to limit laminate options to those of quasi-isotropic angles; i.e., all inter-ply angles are to be between 40 and 50 degrees. In fact a more rational approach to what the inter-ply angles should be is covered by the work of Vincent Tan, et al. Single-double can be the optimum and presents no intrinsic fear. A good example would be [±45] as the best buckling laminate for many boundary conditions with constant and tapered laminates. This work is done by Dan Wang in Chapter 8.

Homogenization is the most important quality of laminates in our view of the composites world. It will supersede the need for mid-plane symmetry and open up ply drops in singles and more flexible locations. Better quality laminates with no interior defects and no gross warpage become possible. There are only 2 optimum stackings for double-double laminates, and they can be found by Antonio Miravete in Chapter 4. Conventional quad and unconventional laminates are heterogenous, just the opposite of homogeneous. Such laminates can have millions if not billions anisotropic flexural stiffness components making optimization impossible. Stacking sequence and blending challenges are self-inflicted. They would simply go away if laminates are homogenized. It is not surprising that only DD with 4-ply sub-laminate can be homogenized. There is a recommended practice to put plies to carry principal load in the center of a laminate. This practice is just the opposite of homogenization. Another practice of putting [±45] fabrics on the exterior surface to increase surface toughness. It assumes that an incoming object can distinguish between [±45] from [0/90].

On the fundamental side of laminates, carpet plots can now be replaced by laminate parameter plots. Carpet plot was invented to facilitate laminate design. It turned out to be a restricting factor that stymied growth in laminates. The [0] and [90] plies were locked in, and the only easy extension would be the change of [±45] to [±θ]. Also being able to only adjust the percentage of plies is excessively confining because plies are locked in discrete numbers: one cannot have 1.2 plies; there has to be 1 or 2 plies, and nothing in between. Thus, the percentage plies is not a continuous variable. One cannot have 37 percent [0], for example. Worse for quad is that the building block sub-laminates keep getting larger and larger, from 4 to 5 to 10 to 16 and even 32. Homogenization would be out of the question. Ply drops to save weight would not work because the laminate properties would not remain the same. As we will see in Chapter 2, lamination parameter plots can handle any laminate from quad to the unconventional to the DD. Direct comparison among all laminates can be made. It also drives laminate design based on field theory, and no longer remains a collection of discrete laminates.

Even more fundamental is the uncertainty of the interaction term in the Tsai-Wu failure criterion. We are happy to have Shuguang Li in Chapter 9 to show that -1/2 is in fact the correct value for the interaction term if it is assumed that the failure envelope is a paraboloid (with very large resistance to triaxial compression, the same assumption made for the von Mises criterion for isotropic materials). It is important to have removed one critical uncertainty of the Tsai-Wu criterion.

Tapering of composite laminates in lightly loaded areas is one way to save weight. With DD that is homogenized, tapering is achieved through the reduction in the number of repeats. Laminate properties remain constant while its thickness is reduced. For quad laminates that have thick sub-laminates, ply drops will change laminate properties. Optimization becomes so complex if possible at all.

The effect of the inter-ply angle on laminate toughness is also presented. To keep it at 45 or less degrees, the number of allowable quad laminates is reduced from 27 to 11. The reason for this drastic reduction in allowable quad is that the [0] must not be next to [90] plies, [+45] must be separated from [-45] plies. In these cases, the inter-ply angle is 90 degree. For DD, the only time when there is a 90 degree inter-ply angle for DD is when the laminate is [0/90]. In all other cases, the angle is always less than 90. Any constraint imposed by the inter-ply angle must also consider the effect of stacking and how DD may be produced as a non scrimp fabric.

Weight reduction by the ANASYS program by Aniello Riccio and Francesco Di Caprio in Chapter 5 made a very significant demonstration of what DD can offer. The program first considers the most feasible weight reduction by assuming that each element of the structure can have its own thickness or number of repeats. The new ply drops would be included in the next iteration in the stress analysis to determine a more refined stress at each element. In the end, all elements have as close to unity strength ratio R as possible. To make weight reduction feasible in production, at least 3 options are offered. One is with constant thickness for the entire component. The resulting weight savings reflects the advantage of DD over quasi-isotropic laminate assuming that is the starting laminate. Two 1-axis layup will also be considered; one with layup along the x-axis, the other along the y-axis. In all 3 cases, the maximum weight savings were reduced to facilitate layup. This conversion program is available for all interested parties free of charge. Surajit also showed that a homogenized laminate will improve its resistance to transverse impact.

A similar conversion program in ABAQUS is available from Surajit Roy in Chapter 6. More detail of his program can be found in his chapter. His program is also available free for interested parties.

In Chapter 13 Thierry Massard showed how Lam search, an optimizer, can select the best laminate from either the DD or quad family of laminates. Unique features of this Excel-based program include an easy to use material data file, load definition including auto-filled random loading after primary load sets, strength ratios for each load determined from 4 failure criteria (i.e., Tsai-Wu, max strain, unit circle and Nettles circle). Built-in load sets include wide band, fuselage, upper and lower wings, bulkheads, submarines, and shaft. A free download of a copy of this is also available. Lam search is also useful to rate different materials, as described in Chapters 1 and 2.

The greatest impact of double-double is on manufacturing. Erik Kappel in Chapter 3 and Bo Jin in Chapter 1 have shown how quality tapered stringers and panels can be produced with the card sliding technique. Ply drops are placed on the exterior surfaces; consequently, in the interior of laminates there are no defects from discontinuities, wrinkles and voids. The components do not warp, and easy to add plies for hard points and repair. Layup speed in 1- axis layup without cross-plying can be 6X faster, and less prone to error because layup is continuous and need not stop at the mid-plane to change stacking sequence.

Double-double is more than a replacement of quad laminates. It offers manufacturing and design that has no counterpart with quad and other laminates. The grid/skin structure is a double-double that takes advantage of the water jet and wire cutting on metals for the grid. This effort was done by Chuangxiang Zheng as one of the authors of Chapter 10. Assembly of grid/skin was done be using [±60] CFRP skin which has negative thermal expansion. This was done by chilling the components for an exterior skin (or by heating them for an interior skin) by Erik Kappel and Bo Jin. The pattern of the grid is totally flexible. Ribs can be straight or curved. They can different cross sections and in non repeating patterns. This assembly can be done in matter of hours, with no glue or fasteners. External skin will be pre-tensioned so buckling can be avoided. Dan Wang was the analyst in the design of grid and skin for their unusual properties.

A new path for certification for DD is described in Chapter 12 by Naresh Sharma. With many new ideas proposed by the authors of this book, certification can be a smoother and more expedient process.

In the end, double-double is conceptually straightforward but needs some effort to learn what it is. Casual observation to dismiss it is not fair. DD laminates are not matrix dominated. They are not delamination prone just because inter-ply angles greater than 45 degree is always present; i.e., if one inter-ply angle is less than 45 degree, its complement is always more than 45. This rule is not enforceable. DD opens up the world of optimization. With DD, optimization becomes more feasible and is readily implemented. New measures of laminate quality have emerged. Homogenization is the key. Optimum ply thickness is a function of laminate thickness, not an absolute measurement of 125 GSM, for example. For thick laminates, ply thickness should be 1/16 of the total thickness. In other words, all laminated panels can be done with 16 layers if homogenization is the criterion. Reaching out to manufacturing engineers is our most important challenge. It is our fondest hope that we can achieve this through this book.

Double-double is a new field. It is easy to reach the leading edge with effort and moral commitment. In nearly every case, the authors joined the DD team and within a matter of months they would have discovered new solutions that are patentable and/or publishable. This is not so for many same old, same old topics which offer only incremental advances at best. We welcome new believers in DD, and can help them to switch to DD and see immediate realization of surprising benefits. The train is at the station ready to leave. Each new believer can be the go-to, not the me-too, person in his/her company. Members of our team will do our best to help him/her.

Support of this book by JEC is hereby acknowledged. Michel Cognet, Anne-Carole Barbarin and Andrew Mafeld have made this book project possible. We also thank Alice Zheng and Thierry Massard for their design and compilation of this book.

Stephen W. Tsai
Honolulu
April 30, 2022

About Authors

Stephen W. Tsai
BE 1952, DEng 1961, Yale. Ford Motor, Washington University, US Air Force Materials Laboratory, Stanford University. Known for his Tsai-Hill failure criterion; lamination parameters with Pagano; Tsai-Wu failue criterion and strength ratios; Halpin-Tsai micromechanics formula; trace, master ply, omni envelopes, and unit circle failue criterion with Melo; double- double laminates; Lam search with Massard and Shah; metallic grid/[±60] CFRP skin for glueless assembly with Zheng, Wang, Kappel, and Jin. He is a member of the US Academy of Engineering since 1995.

Erik Kappel
Research Associate at German Aerospace Center's (DLR) Institute of Composite Structures and Adaptive Systems in Braunschweig. Rescarch Interest:
• Composite design, analysis and manufacturing with a strong focus on process-induced distortions (PID) of CFRP components in aerospace applications, efficient numerical PID prediction and compensation-measure definition.
• Doctoral degree in Mechanical Engineering from Otto v. Guericke University Magdeburg in 2013 and a Diploma degree in Mechanical Engineering from the Leibniz University of Hannover in 2009. Currently compiling his Habilitation at the Technical University of Braunschweig.
• Visiting researcher at the Materials-Engineering Department of the University of British Columbia (UBC) in 2013.
• Lecturer for "Composites in Consumer Product" at the Technical University of Braunschweig.

Antonio Miravete
He has a Ph.D. in Mechanical Engineering and 40+ years of experience in analysis, design and manufacturing of organic matrix composite materials. He started his collaboration with Prof. Tsai in the 1980's. He has authored or co-authored several books published in the US and Europe. Associated with academic institutions since the beginning of his career, he is a research affiliate at the Department

of Aeronautics and Astronautics, MIT. His major interests are double – double, nanoscale materials to enhance performance of advanced composite materials and continuous carbon fiber additive manufacturing. He is also a US chess master.

Aniello Riccio

Graduated in Aeronautical Engineering at the University of Naples in 1996. He got his PhD in 1999 at the Second University of Naples (SUN). In 2000, his PhD thesis was awarded with the first prize in the frame of "the Pratt & Whitney-EREAAward". Between 2000 and 2010, he was engaged by the Italian Aerospace Research Centre (CIRA). In 2003, he became member (chairman in 2008-2011) of the Structures and Materials Group of Responsibles (GoR) within the GARTEUR organization. Since 2004, he has been reviewer for many scientific journals. Between 2006 and 2012, he acted as chairman of European Projects and Mod funded projects. From 2010 he joined the SUN (then University of Campania "Luigi Vanvitelli"). From 2018, he is full professor in aerospace Structures. Prof. Riccio is author of several Publications on advanced materials and structures in international Journals and Conference proceeding. He also acts as associate editor for several international scientific Journals. Main topics of Research are fracture mechanics, design with composite materials, development of Finite Elements based numerical tools for Design and Optimization of Double-Double Composites.

Francesco Di Caprio

Graduated cum laude in Aerospace Engineering and completed the university education by achieving the PhD in Aerospace Science and Technology at Second University of Naples (currently University Campania "Luigi. Vanvitelli") with a thesis on damage tolerant composite structures. He has been working with CIRA (Italian Aerospace Research Centre) since 2011 in Airframe Design and Dynamic laboratory. He is author of several scientific works published on international peer-reviewed journals and Conference proceedings. His main research topics are: numerical methods for structural analysis; multi-objective optimization; design for additive manufacturing; composite materials; crashworthiness; development of numerical methods for Double-Double composites.

Distinguished Professor Brian G. Falzon holds the Chair of Advanced Composite Materials and Aerospace Structures at RMIT University and is the Director of the RMIT University's Space Industry Hub, the Victorian Node of the Smart Satellite Cooperative Research Centre (SmartSat CRC), and of the Education and Training Program for the Sovereign Manufacturing Automation of Composites (SoMAC) CRC. He was Head of School of Mechanical and Aerospace Engineering at Queen's University Belfast, from 2015 to 2021, and was appointed to the prestigious UK Royal Academy of Engineering – Bombardier Chair in Aerospace Composites in 2013. Between 2008 and 2012 Prof Falzon held the Foundation Chair in Aerospace Engineering at Monash University in Melbourne, Australia, where he was also Director of Research and Head of Aerospace Engineering within the Department of Mechanical and Aerospace Engineering. Between 1996 and 2008, Prof Falzon was at Imperial College London where he joined as a postdoctoral research fellow before becoming an academic staff member. He holds Adjunct Professorships at Queen's University Belfast, Monash University and the University of New South Wales and was a Visiting Researcher at the Japanese Aerospace Exploration Agency (JAXA). Prof Falzon is a passionate engineer and is internationally renowned for his industry-driven work on the computational analysis, design, manufacture and testing of advanced composite structures.

Dr Ali Aravand is a lecturer (assistant Professor) at the School of Mechanical Engineering, Queen's University Belfast. His PhD was focused on hierarchically structured fibre-reinforced nano-composites. He specialises in designing and manufacturing structural composites for aerospace and automotive applications, composite injection overmoulding, nano-engineered composite laminates, and fracture behaviour and toughness of polymer composites. He has led several academic/industry collaborations focusing on the design and automated manufacture of structural composite components. This includes a current collaboration with Spirit AeroSystems on DD-optimised composite laminates and collaborations on the design and manufacture of hybrid injection overmoulded thermoplastic composites involving Collins Aerospace.

Dr Scott Millen is a Lecturer in Manufacturing in the School of Mechanical and Aerospace Engineering at Queen's University Belfast. Scott joined the staff of Queen's in 2019 and again in 2021. He has a master's degree in Mechanical Engineering and a PhD in computational modelling of composite materials. After completing his PhD, Scott took up a research fellow role working with local industrial collaborators from the aerospace industry (Collins Aerospace and Spirit AeroSystems, Belfast). Scott has also worked within the Belfast Maritime Consortium UKRI Strength in Places project led by Artemis Technologies Ltd. Scott has significant experience in Finite Element Analysis (FEA) in particular developing high-fidelity tools using ABAQUS, Python and Fortran user-subroutines.

Surajit Roy Assistant Professor in the Department of Mechanical and Aerospace Engineering at California State University, Long Beach. He obtained his Ph.D. degree from the Department of Aeronautics and Astronautics at Stanford University in 2014. He teaches senior design capstone course in mechanical engineering at CSULB. He has mentored several industry sponsored projects in the past from SoCal gas, Boeing, Aerospace Corporation, and some local companies in the region. His current research interests are damage tolerance of novel composite materials, development of prognostic models for material ageing under harsh environments, non-destructive testing techniques, and data-driven algorithms for structural health monitoring.

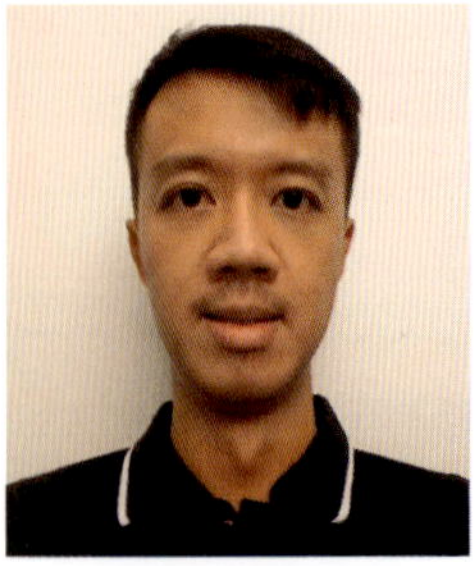

Kirk Ming Yeoh is a Ph.D. candidate at the Department of Mechanical Engineering, National University of Singapore. He obtained his B.Eng. in Mechanical Engineering from Nanyang Technological University, Singapore. His doctoral thesis is on the development of a multiscale and multiphysics numerical model for heterogeneous structures, focusing on its application towards fibre-reinforced composites. His research interests are in computational homogenization, multiscale modelling and composite materials.

Dr. Liu Jialong is a professor at the School of Materials, Sun Yat-Sen University. Prior to this, he worked as a research fellow at the Department of Mechanical Engineering, National University of Singapore, and he obtained bachelor's and Ph.D. degree from the same department. Currently, his research interests are in the damage mechanism and commercial application of Double-Double laminates, bio-inspired composite structures, and advanced composite materials. His specialization is composite structure design. It focuses on improving the mechanical performance of composite materials through microscale and mesoscale structure design. The composite materials include healable composites, natural fiber reinforced composites, 3D printed composites, fiber reinforced thermoplastic composites, impact resistance composites, and hybrid composites.

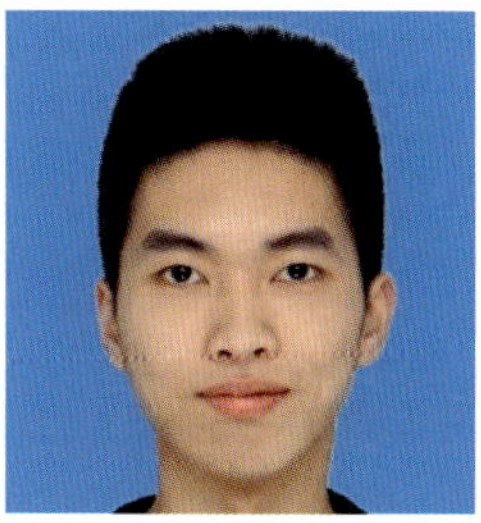

Yang Gang is a Ph.D. candidate at the Department of Materials, Sun Yat-sen University. He received his bachelor's degree from Sichuan University and his master's degree from China University of Geosciences. His research interests arc in the design and optimization of mesoscale level structure of thermoplastic and thermosetting composites.

Dazhi Jiang received the B.S. degree in Applied Chemistry and the M.S. degree in Composite Materials from National University of Defense Technology (NUDT), Changsha, China in 1987 and 1990, and the Ph.D in Engineering Mechanics from Huazhong University of Science and Technology, Wuhan, China in 1997. As a post-doctor and research fellow, he worked in University of Science and Technology of China, Hefei, China during 1998-1999 and Nanyang Technological University, Singapore during 2002-2004. He joined NUDT as an academic faculty from 1990 until 20018. He is currently working in the School of Materials, Sun Yat-Sen University, Shenzhen campus, China. His research interests include mechanics of composite structures and nano-phased polymer composites.

Jie Zhi is a professor at the School of Aerospace Engineering and Applied Mechanics, Tongji University. Prior to this, he worked as a research fellow at the Department of Mechanical Engineering, National University of Singapore, and he obtained Ph.D. degree from the same department in

2020. He also holds M.Eng. and B.Eng. degrees in Aerospace Engineering with a minor in Applied Mathematics from Beihang University. He is an expert on computational mechanics with an emphasis on composite structures. His current research interests include structural analysis and design of fiber-reinforced composites, numerical methodologies on damage and fracture, and multiscale/multiphysics modeling techniques. He has published more than 20 papers in reputable international journals, conference proceedings, etc.

T.E. (Tong-Earn) Tay is full Professor at the Department of Mechanical Engineering, National University of Singapore (NUS). He earned his PhD from the University of Melbourne, Australia in 1989. His research interests are in progressive damage, failure, fracture, impact, joining and multi-scale computational analysis of fiber-reinforced composite structures. He is a recipient of JEC Life Achievement Award, a registered Professional Engineer (PE), Chartered Engineer (CEng), Founding Fellow of the Singapore Academy of Engineering (FSAE). He was previously Head of Department of Mechanical Engineering, NUS (2011 to 2015), and Vice-Dean for Research for Faculty of Engineering, NUS (2009 to 2011).

Vincent Tan is an Associate Professor at the National University of Singapore. After obtaining his BEng and MEng in Mechanical Engineering from NUS, he pursued his PhD at Northwestern University, USA. His research interests are in computational homogenization, multiscale modelling, and fibre reinforced composites. He serves on the editorial boards of several journals, including Composites Part B, Mechanics of Materials, and International Journal of Applied Mechanics. He is a Vice President and Fellow of the Association for Computational Mechanics (Singapore) and a General Council Member of both the International Association of Computational Mechanics and Asian Pacific Association for Computational Mechanics.

Dan Wang is currently a Scientist working in the department of Engineering Mechanics (EM) at Agency for Science, Technology and Research (A*STAR Research Entities), Singapore. She is an expert on structural optimization with the emphasis on composite structures. Her research interest is in the development of optimization methodologies in order to explore the large design space and high tailorability of novel composite materials and hierarchical structures, such as Double-Double (DD) laminate layup, curved fiber paths, and non-periodic stiffener layout. She received a Master and a PhD (Combined program) in 2013 from Northwestern Polytechnical University (NWPU), China in Mechanical Engineering and did a Postdoc at the group of Aerospace Structures and Computational Mechanics in Delft University of Technology (TU-Delft), the Netherlands. She was awarded the National Outstanding PhD thesis of China on the subject of Aeronautical and Astronautical Science and technology, 2016 (only 15 award winners nationwide that year). Dan Wang currently works closely with Prof Steven Tsai on numerical simulation and buckling optimization of DD laminates including thickness tapering and grid stiffeners.

Shuguang Li
Professor of Aerospace Composites, is a part of the Institute for Aerospace Technology, Faculty of Engineering, University of Nottingham, UK. He obtained his PhD from University of Manchester in 1993 and was appointed to his present position in 2012. He was one of the two advisors to the 2nd World Wide Failure Exercise for polymer composites (on 3D failure theories). He is one of the organisers for the 3rd Exercise of the same (on damage theories). The outcomes have been published in Journal of Composite Materials. His recent research on the subject of composites failure criteria has led a number of publications aiming at the rationalisation of existing criteria, so that users will be able to apply them to the limit of their applicability with confidence. He is on the editorial board of International Journal of Mechanical Sciences. He is a visiting professor at Nanjing University of Aeronautics and Astronautics and Zhejiang University, China. He has published over 100 academic articles, most of them in highly reputable international journals. His main research interest is in the area of analysis of composite

materials and structures; in particular, on subjects of damage and failure, micromechanics and characterisation. As an outcome of his research on micromechanical modelling of composites, a monograph entitled 'Representative Volume Elements and Unit Cells© Concepts, Theory, Applications and Implementation' was published by Elsevier in 2020. An open-source software named as UnitCells© has also been made available by Elsevier which offers material scientists and structural designers a useful tool for the characterisation of modern materials in terms of effective properties of elasticity and thermal and electric conductivity.

Elena Sitnikova
Research Fellow in the department of Mechanical, Materials and Manufacturing Engineering at the University of Nottingham, which she joined in 2013. Prior to this, she took position of Research Associate at the University of Liverpool. She has PhD in Engineering from the University of Aberdeen, which was awarded for research into the problem of vibration reduction and control in impact systems using shape memory alloys. She obtained a BSc and an MSc in Applied Mathematics and Mechanics from Saint Petersburg State University. Elena's research interests and expertise span a wide range of subjects, with her main published research outputs being in the fields of composite materials, applied dynamics and dynamic failure of materials. These were formalised as 20 journal publications. Among her most substantial achievements is publication of a monograph 'Representative Volume Elements and Unit Cells — Concepts, Theory, Application and Implementation", to which she contributed as co-author. It advances the subject of formulation and application of RVEs and UCs, which are essential tools for the analysis of complex materials and their multiscale modelling. Elena's present research efforts are focused on the design and application of textile composites. Another subject of special interest is failure of materials under high intensity loading. Elena is developing the generalised dynamic failure analysis framework that would be applicable to both isotropic and anisotropic materials.

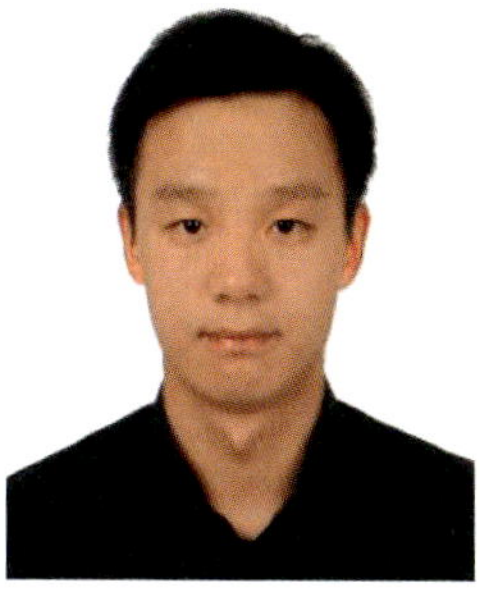

Mingming Xu
PhD student (2017-2022) in the Composites Research Group of Faculty of Engineering, University of Nottingham, UK. He obtained his master's and bachelor's degrees from the Beijing Institute of Technology in 2016 and 2014, respectively. His main research interest is in the multiscale analysis of textile composites and lightweight structures, nonlinear and dynamic behaviour of composites, in particular, blast and impact resistance.

Chuanxiang Zheng
Professor of Zhejiang University, China. Specialization in composite pressure vessels. He obtained his Bachelor's degree, Master's degree, and Doctor's degree from Zhejiang University in Chemical Process Machinery (Pressure Vessels). He started teaching at Zhejiang University from 1998, and was appointed as a professor in 2009. He visited the Aeronautic & Astronautic Department of Stanford University as a Visiting Scholar in 2008, and from 2011 to 2014. His research interests include composite pressure vessels, carbon fiber structure design, and strength analysis of carbon fiber tubes.

Bo Jin
A faculty at the University of Southern California. He teaches graduate and undergraduate courses at USC's Aerospace and Mechanical Engineering, and conducts research at USC's M.C. Gill Composites Center. He specializes in composites design, composites manufacturing and characterization methods, computational mechanics, and Finite Element Analysis. He was an assistant research scientist at the Spanish National Lab IMDEA Materials, and a composites FEA Engineer at MSC NASTRAN.
Bo is the recipient of several awards, including an ACM-ICPC Gold Medal, an SME Outstanding Young Manufacturing Engineer Award, a SAMPE National Leadership Award, a 1st Place of NASTRAN Simulating Reality Contest, and several best paper awards in major composites conferences including CAMX/SAMPE. He serves on the editorial board of the Journal of Composites Science, and as the conference chair and technical committee member at SAMPE/CAMX. Bo has published ~60 peer-reviewed journal and conference papers including a few featured articles in top journals in the composites field.

Albertino Arteiro is currently Assistant Professor at the Materials and Technological Processes Group of the Department of Mechanical Engineering (DEMec) of the Faculty of Engineering – University of Porto. His research interests include material development and numerical modelling at different length-scales, design and analysis of composite materials based on new concepts, and composites manufacturing. He received a Master's Degree in Mechanical Engineering - Specialization in Structural Engineering and Machine Design - from the Faculty of Engineering - University of Porto in 2012, and a Doctoral Degree in Mechanical Engineering (cum Laude) from the same University in 2016, during which he was a Visiting Student Researcher at the Department of Aeronautics and Astronautics of Stanford University. He was awarded the Best PhD Thesis by the Portuguese Association of Theoretical, Applied and Computational Mechanics (APMTAC).

Naresh Sharma is the CEO and Founder of NASHERO, a light aircraft manufacturer out of North Italy. NASHERO designs, develops, produces and sells carbon composite aircraft. With the impending AS9100 certification, they plan on supplying larger OEMs with technology, composite and complex precision machined metal parts. Since 2008, Naresh is an ASTM member and an active contributor to the ASTM's aviation standards including those for General Aviation.

Naresh earned a Ph.D. in Aerospace Engineering and Computer Sciences from the Delft University of Technology in 1997, a Master's in Aeronautical Engineering from the Indian Institute of Technology (IIT), Kanpur in 1989, and a Bachelor's in Mechanical Engineering from IIT, Kanpur in 1986.

Thierry Massard

Graduates from Ecole Centrale in material engineering (1975) and he holds a PhD from Ecole Centrale in mechanics of composite materials (1979).

Thierry Massard is an expert for innovative materials, especially in the field of composite materials. He was assistant professor on composite materials at ENSTA ParisTech. Engineering School. He is a World Fellow of the International Committee on Composite Materials.

He and Prof Tsai have been working together for more than 40 years. Together they have contributed to many topics in the field of composite materials. Thierry Massard is the head of Think Composites Europe a no-for-profit organization for the promotion of composite materials in the industry (https://www.think-composites.com/) . He developed many software for the optimum design of composite laminates. Recently he contributed with Prof Tsai to Lam search, a powerful tool for the design of optimum Double Double laminates a cutting edge technique that will revolutionize the way to design and build composites structures.

Steve's parents, Stephen and Lily Tsai,
passed away when they were 92 and 95, respectively

Celebrating birthday at his favorite restaurant China Stix in Santa Clara, CA, in 2019

TABLE OF CONTENTS

With Carlos Cimini, taken on Stanford Campus, taken June 2016

CHAPTER 1
Foundation of Double-Double

Stephen W. Tsai, Stanford University
stsai@stanford.edu

Unique topics in the development of double-double (DD) over and beyond the basics of laminate plate theory and failure criteria will be presented. Two goals are: first, DD can make manufacture and design robust, fast, and at lower weight and cost; secondly, DD is new and different from, not an extension of the traditional quad laminates, and leading to innovations not possible with quad. Homogenization provides multiple departures from the prevailing lamination practice will be described in section 1.1. Trace as an invariant can cover laminate stiffness with one parameter instead of 4 in section 1.2. Many simplifications from use of DD can be shown. Traditional carpet plots and revived lamination parameter plots are compared in their representation of quad, unconventional and DD laminates in section 1.3. The latter plot shows how well DD can not only match traditional laminates but also provide benefits not realized hitherto. Section 1.4 covers failure envelopes of laminates including the omni envelope for strength. It also provides a way for rating CFRP materials useful to guide materials and processing improvements. A summary of why DD is as simple to understand and use as metals, and why dispersed laminates should be recognized as the technology of the last century.

1.1 Homogenization

Homogenization as a laminate character has never received much respect. DD has changed all that. When a laminate is homogenized it means that its properties is the same across the thickness along the z-axis, and it can also be the same across a component in the x- and y-axis. With or without respect homogenization has made many of the rules governing laminate manufacture and design unnecessary. It is the beginning of simplification that we seek.

Homogenization is made simple by DD because its building block has only 4 plies in [$\pm\Phi/\pm\Psi$]. Traditional quad laminates in 0, ±45, 90 are based on 4 to 10 plies but required mid-plane symmetry. So their building blocks are 8 to 20 plies. Homogenization cannot be realized.

Quad is a very complicate collection of discrete laminates. In the traditional carpet plot of laminate properties, the x-axis is in percentage of [±45]. That percentage varies depending on the number of plies in the building block. It can be based on 4 to 10 plies. For 10-ply laminates, the percentages of [±45] are limited to 20, 40, 60, and 80. The 10 percent rule prohibits the use of 0 and 100 percent [±45]. For 8-ply laminates, the permissible [±45] percentages are 25, 50 and 75. These two building blocks have no connection between them. Two separate families.

DD on the other hand has no such limits. It is a field-based family. All members are continuously connected. Each family can be defined by the angle increments Δ that is chosen. It can be 15, 10, 7.5, … degrees depending how large the members of the family are desired.

In Figure 1.1, the stiffness component A_{11} is constructed for a 10-ply quad family of laminates. First, the entire field of 10 percent variations for the x-axis for the percentage of [±45], and the y-axis, of [0]. For the 10-ply building block, permissible [±45] are 20, 40, 60 and 80 percent. Then there is a 10 percent rule that prohibits the use of the 0 and 100 percent of each boxed columns in the figure. So there are only 16 out of 66 of the field that can be used, less than 1/4. That is not very efficient. DD on the other hand can use 100 percent of the field.

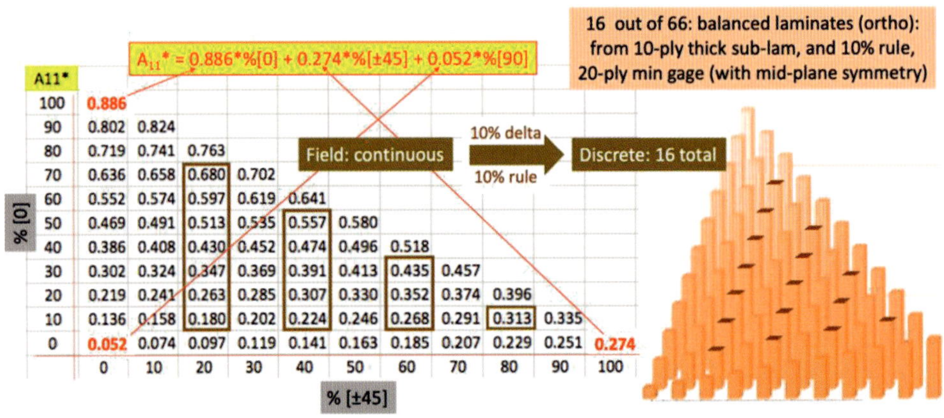

Figure 1.1 Construction of 10-ply quad laminates restricted to 10-ply and 10 percent rule

The same process use in the last figure can be applied to other quad families like 6-to 9-ply building blocks, and they are shown in the bottom row of Figure 1.2. The member laminates in Figure 1.1 can be matched by DD and the resulting DD laminates are shown in Figure 1.2. One complexity of quad is seen in the nonlinear spacing of the colored building blocks with different ply thicknesses. An example of this matching is given: for (40/20/40) the matching DD is [±13/±77]. The procedure for finding the matching DD will be given later in this chapter. .

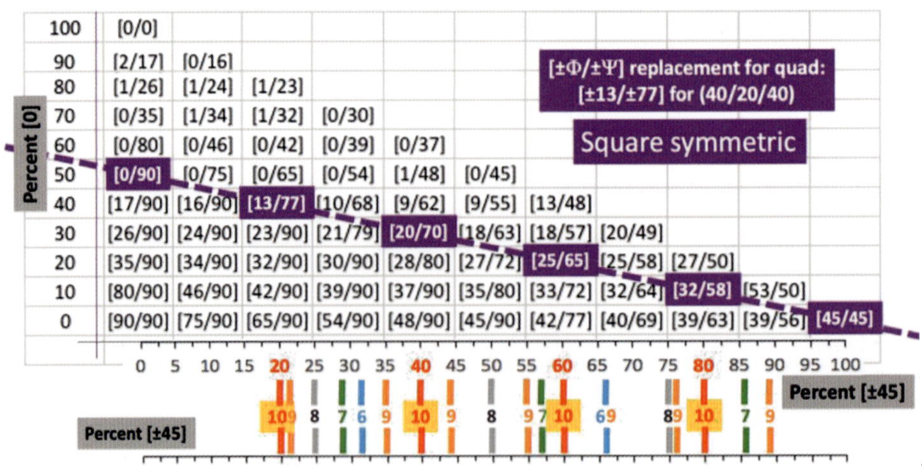

Figure 1.2 Matching DD of the 10 percent increments of quad in Figure 1.1

If a carpet plot must include every building block plies shown in the last figure, the same nonlinear spacing for the y-axis must also be done. Large number of vacant cells will be created because many laminates with the nonlinear [±45] and [0] do not exist. That would be another source of self-inflicted complexity. It is a bad way of assembling laminates, conceived in 1960's, and has not changed. As we will see, DD laminates do not have such complexities.

Another comparison between quad and DD can be seen in Figure 1.3. A collection with uneven spacing on the left, and continuous, orderly field on the right shows the contrast.

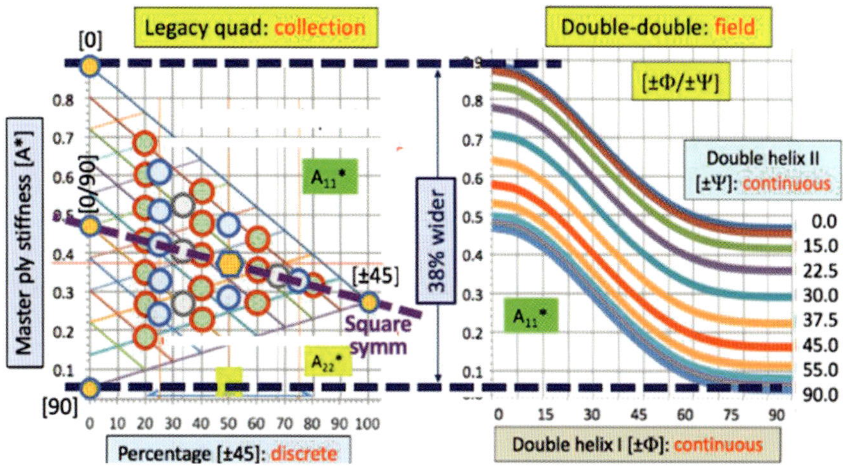

Figure 1.3 Difference between a collection of quad versus a field of DD

Mid-plane symmetry is automatically satisfied when the laminate across the thickness is homogenized. So laminate layup is simpler when there is no need to stop at the mid-plane to change the stacking sequence to mirror the laminate in the lower half. Stacking for DD can be continuous. Layup is so much simpler, faster and less prone to error.

When the laminate for a component is the same, this is a homogeneous component. It will not warp or twist. It is best to repair or reinforce local hard points with the same laminate for the patch. There is no mismatch when the laminate is the same; i.e., all laminates are an "in family" variety. A component made with quad laminates can have hundreds of laminates. Issues involve stacking sequence, ply drops, and blending. DD on the other hand can offer one DD to cover the entire component. The resulting simplicity in design, manufacturing and testing is unmatched.

With homogenization, ply drops can be in singles rather than in symmetric pairs. This is a game changer because single ply drops can be located at the exterior surfaces of the laminates leaving the interior free of discontinuities, voids, and wrinkles. A better quality laminate can be expected. When resin infusion is followed, homogenized laminates offer uniform resistance to flow and would expect smoother infiltration and better resulting laminates. A comparison between the mid-plane symmetric, interior ply drops in pairs of quad versus the exterior single ply drops of DD (by Erik Kappel of DLR) is shown in Figure 1.4. Again the difference is dramatic. Quality of products can be so dramatically different.

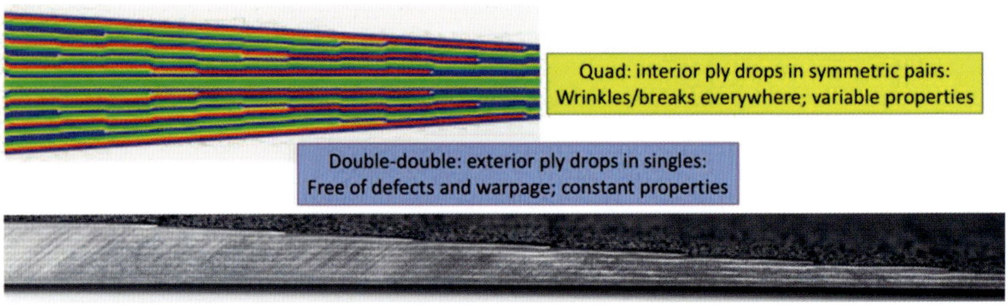

Figure 1.4 Ply drops between symmetric interior pairs of quad vs single exterior drops in DD

In design, homogenization means equal in-plane and flexural stiffness; i.e., [A] = [D], and [B] = 0. This is so different from heterogenous laminates where for a given laminate in [A] can have hundreds and thousands of [D] making laminate optimization so complex if not impossible. Stacking sequence and blending will no longer issues to be considered.

There will be more benefits from homogenization in manufacturing and design as we proceed with our chapters in this book. Suffice to say that the best laminate to reach homogenization is DD. It is easy to explain why it is so simple.

DD is a sub-laminate building block based on two ply angles in [Φ/Ψ]. There are two stacking sequences fastest to reach homogenization as described in Chapter 4: Staggered 1 in $[\Phi/-\Psi/-\Phi/\Psi]_{rT}$, and Staggered 3 in $[\Phi/-\Psi/\Psi/-\Phi]_{rT}$ where r is the number of repeats, T for total laminate (as opposed to S for mid-plane symmetry). Conventional quad has hundreds and thousands of stacking sequences and corresponding flexural rigidities. Finding the best laminates for buckling and vibration is a serious problem in having too many choices and also too many clusters where uniqueness cannot be assured; i.e., different designers are likely to select different laminates as their best. Most components are designed with subjectively selected stacking sequence and blending, it is one of a kind, not ever seen before, nor to be seen again. The structure is simply too complex and design with little objectivity. The component cannot be scaled or duplicated, and have no idea how optimum it is.

Double-double is straightforward and easy to learn and use. It is much simpler than the traditional quad and nearly all the unconventional laminates that we have seen. Students will have to study to master it. Just listening to presentations and seeing viewgraphs will not be sufficient. On the other hand if they are proficient in laminated plate theory, double-double is just one application. It is that simple but requires a curious mind to motivate learning how better, lighter and lower cost composites can really be. How simple the design process and validation tests can be.

Quad are governed by at least 5 rules in mid-plane symmetry, balanced laminates, 10 percent rule, ply nesting in 4 or fewer plies, and inter-ply angles no greater than 45 degree. They were essentially philosophies. They have made composites difficult to manufacture and design. DD on the other hand is much easier to use and can save weight that the conventional and unconventional laminates cannot do (for their inability to reach homogenization). Quad has an inherent bias against high shear laminates. There are 3 or higher number of laminates among the low shear laminates than the high shear ones. More explanation will be given later in this chapter. Suffice to say, DD is far more efficient, flexible and fair minded. The selection of the angle increment is an important decision. There can be angle increments like 7.5, 15, 22.5, and 30 degree that will yield a variety of combinations of stiffness and strength properties.

Yes, homogenization is the key to enable weight reduction of lightly loaded regions of a component. The laminate is inherently resistant to delamination, and easy to calculate interlaminar shear stress, buckling and natural frequencies. DD has only two stacking sequences to consider. Naturally symmetric and needs only 1 axis layup with no need of cross-plying. Examples of heterogeneous versus homogeneous laminates can be seen in Figure 1.5. Weight savings can be done by tapering with no change in properties. That is simple for DD, but not possible for quad.

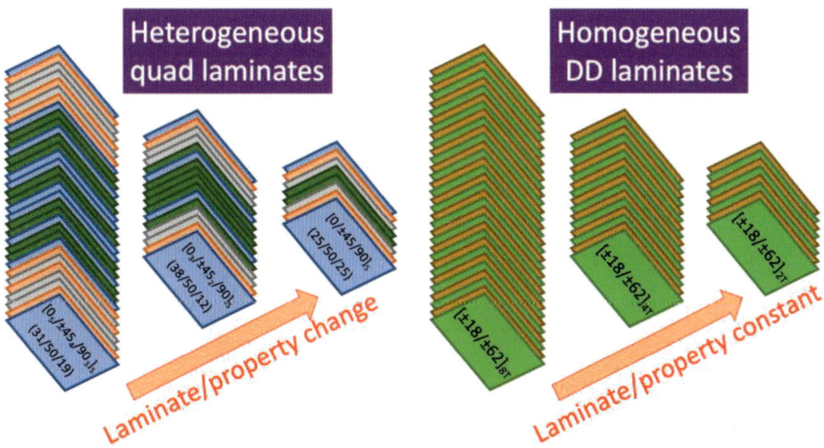

Figure 1.5 Tapering to save weight is simple for DD but very difficult for quad

1.2 One parameter stiffness: trace

Four elastic constants are usually used for stiffness of plies from which laminate stiffness is derived. It is now recognized only one constant that is needed is the trace, or also known as the Tsai's modulus. As seen in Figure 1.6, 15 representative CFRP laminate stiffness in engineering constants first, followed by plane stress components [Q]. Trace is then computed using Trace = $Q_{xx} + Q_{yy} + 2Q_{ss}$. Then trace-normalized [Q*] follows. At the bottom rows of the table, master ply of CFRP found is based on the average of [Q*] with its standard deviation and coefficient of variations also listed. The practical importance of master ply is that it represents all CFRP. So when composites are analyzed, designed and tested, the best starting material would be the master ply. The resulting general solution covers all CFRP and for specific material the scaling factor is its trace. If a different material is desired, just use a different value for trace, no recalculation is necessary. If a new process is introduced, it is only necessary to find the new value for trace. If it is 10 percent higher, the stiffness of the component will increase by 10 percent, deformation and strain decrease 10 percent, and stress increase 10 percent.

CFRP	Ex	Ey	Es	nu	Qxx	Qyy	Qxy	Qss	Trace	Qxx*	Qyy*	Qxy*	Qss*	Trace*
IM6/epoxy	203	11.20	8.40	0.32	204	11.26	3.60	8.40	232	0.879	0.049	0.016	0.036	1.00
IM7/977-3	191	9.94	7.79	0.35	192	10.00	3.50	7.79	218	0.883	0.046	0.016	0.036	1.00
T300/5208	181	10.30	7.17	0.28	182	10.35	2.90	7.17	206	0.880	0.050	0.014	0.035	1.00
IM7/MTM45	175	8.20	5.50	0.33	176	8.24	2.72	5.50	195	0.901	0.042	0.014	0.028	1.00
T800/Cytec	162	9.00	5.00	0.40	163	9.08	3.63	5.00	183	0.895	0.050	0.020	0.027	1.00
IM7/8552	159	8.96	5.50	0.32	160	9.01	2.88	5.50	180	0.889	0.050	0.016	0.031	1.00
T800S/3900	151	8.20	4.00	0.33	152	8.25	2.72	4.00	168	0.903	0.049	0.016	0.024	1.00
T300/F934	148	9.65	4.55	0.30	149	9.71	2.91	4.55	168	0.888	0.058	0.017	0.027	1.00
T700 C-Ply 6	141	9.30	5.80	0.30	142	9.36	2.81	5.80	163	0.871	0.057	0.017	0.036	1.00
AS4/H3501	138	8.96	7.10	0.30	139	9.01	2.70	7.10	162	0.857	0.056	0.017	0.044	1.00
T650/epoxy	139	9.40	5.50	0.32	140	9.47	3.03	5.50	160	0.872	0.059	0.019	0.034	1.00
T4708/MR60	142	7.72	3.80	0.34	143	7.77	2.64	3.80	158	0.903	0.049	0.017	0.024	1.00
T700/2510	126	8.40	4.20	0.31	127	8.45	2.62	4.20	144	0.883	0.059	0.018	0.029	1.00
AS4/MTM45	128	7.93	3.65	0.30	129	7.97	2.39	3.65	144	0.894	0.055	0.017	0.025	1.00
T700 C-Ply 5	121	8.00	4.70	0.30	122	8.05	2.41	4.70	139	0.875	0.058	0.017	0.034	1.00
									Average: Master ply	0.885	0.051	0.016	0.032	1.00
									cv %	1.5%	10.4%	9.9%	17.4%	0.00
									Stdev	0.013	0.005	0.002	0.005	0.00

Figure 1.6 Derivation of master ply partitioning from engineering constants [1]

The stiffness of a laminate is determined by the partitioning of the value of trace. It turns out that the strength of laminates can also have one parameter for the master strength envelope, to be covered in Chapter 2. This same parameter is valid for all CFRP.

Another source of the master ply stiffness is seen in Figure 1.7 where the database had over 200 CFRP. The master-ply stiffness is nearly the same as that shown in Figure 1.6.

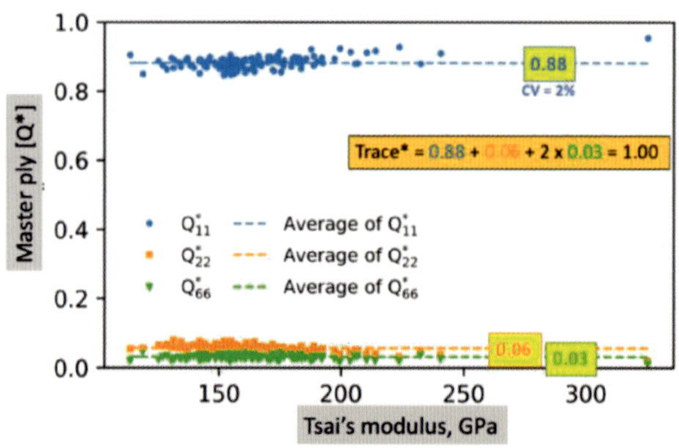

Figure 1.7 Partitioning of [Q] of CFRP [2]

Shear modulus A_{66} is 3 percent of trace. That is probably more accurate than any experimental measurement. Shear tests are difficult and also have extensive nonlinearity. The use of [90/0] is recommended because the Poisson's ratio is near zero to deduce shear modulus, to be explained later in the partitioning of trace for this laminate in 47 to 3 percent. For this coupon test, gripping would be easy, and the A_{11} and E_1 are equal. The load required for failure is ½ of all [0] coupons. The data scatter is also expected to be lower than when coupon is all [0].

As seen in Figure 1.8, laminate stiffness has two parts: 1) material constant in its trace listed on the left; 2) partition parameter as function of ply orientations like those of [Hard] and [QI] laminate shown here. Trace is the one and only one constant that will be needed, not 4 in E_x, E_y, E_s, ν_x as listed in Figure 1.6. Laminate stiffness is determined by the partitioning of trace; i.e., for the top hard laminate it is 57% for A_{11}; 22% for A_{22}, and 10.5% for A_{66}. For a given laminate this partitioning is the same for all CFRP. Similarly, for QI laminate, the partitioning is different but remains the same for all carbon composites at 37% for A_{11} and A_{22}; 13% for A_{66}.

One parameter

Ply material	Trace
IM6/epoxy	232
IM7/977-3	218
T300/5208	206
IM7/MTM45	195
T800/Cytec	183
IM7/8552	180
T800S/3900	168
T300/F934	168
T700 C-Ply 64	163
AS4/H3501	162
T650/epoxy	160
T4708/MR60H	158
T700/2510	144
AS4/MTM45	143
T700 C-Ply 55	139

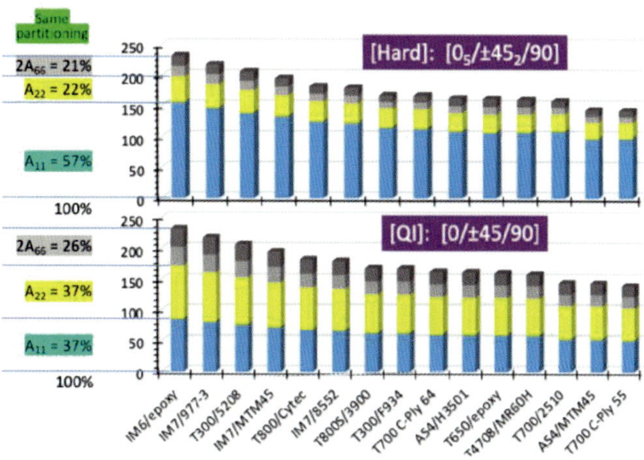

Figure 1.8 With fixed partitioning for each laminate, trace is the only constant needed.

Practical significance: For design allowable, only one constant is needed for each material making testing much simpler (no need for shear test, for example). There is one master ply that represents all CFRP. In design, no specific material needs to be selected until the end of optimization. Whatever material is picked it is defined by its trace. So different material can be scaled by its trace. Trace represents the total elastic stiffness potential. Different laminates have different partitioning of the trace, as seen in Figure 1.8. The sum remains the same for each material. The partitioning is unique for each laminate. The factor of 2 in the shear component came from the use of engineering strain which is twice as much as the tensorial strain.

Before trace was discovered, composite mechanics was always perceived to be more complicated than the isotropic material. Not only there are more elastic constants but also new ones in shear coupling. With trace design, testing and manufacturing are not only simplified but also provided with methods and short cuts not possible with the traditional laminates. It is more than a substitution of one laminate for another. For those who have learned DD they see a sea change like the switching from DC to AC for electricity, or from particles to waves for light. They often wished that they made the switch earlier.

The governing equation of bending of a homogenized laminate is shown in Figure 1.9. The coefficients of this equation in terms of trace for a hard laminate are the same as seen in the horizontal lines on the right chart. One solution is applicable to all CFRP that differ only by their respective trace. From the value of trace in Figure 1.8, IM7/977 is 218, and T800/Cyt, 183, their ratio is 218/183 = 1.2 or 20 percent stiffer. That superior stiffness is maintained for all laminates. For bending solutions, the same 20 percent will remain valid with no need for re-calculation. Buckling will be 20 percent higher, and natural frequency would be square root of 1.2 = 1.1 or 10 percent higher.

$$D_{11}\frac{\partial^4 w}{\partial x^4} + 2(D_{12}+2D_{66})\frac{\partial^4 w}{\partial x^2 \partial y^2} + D_{22}\frac{\partial^4 w}{\partial y^4} = q$$

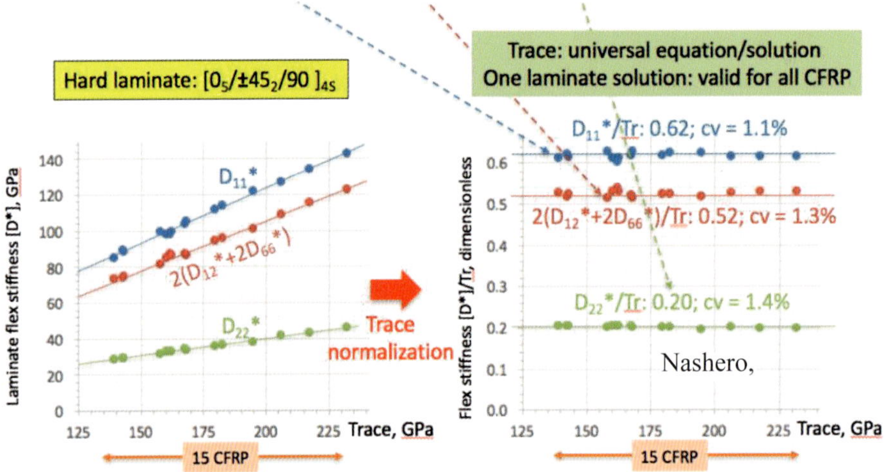

Figure 1.9 One laminate bending equation for all trace-normalized CFRP

For plane elastic equation of laminates, Figure 1.10 shows similar relations as the previous figure for bending. The compliance components are to be multiplied by the respective value of trace, rather than the usual normalization by division for the flexural stiffness earlier. Again, one laminate solution would be applicable to all CFRP. The difference between materials is by division by their respective trace. The 20 percent advantage in trace between IM7/977 and T800/Cyt in bending would be 1/1.2 = 0.83; i.e., the stress concentration would be 83 percent less for IM7/977 than T800/Cyt. This would be the case of an open hole laminate where the stress intensity of IM7/977 would be 83 percent of that of T800/Cyt.

With DD, composite laminates are as simple as metals. As mentioned earlier that Chapter 2 will show that one parameter for strength has also been found. Scaling for strength like that for stiffness can be applied in similar process, like switching aluminum from different heat treatments or that between aluminum and titanium. No re-calculation is necessary. For composites with DD it has additional advantage of being a field-based. Interpolation or zooming is possible in order to find better laminates. Between any 2 DD laminates that us separated by a fixed angle increment, say, 15 degrees, there are many possible laminates between them. This can be done by angles less than 15. Such feature is not possible with the quad laminates.

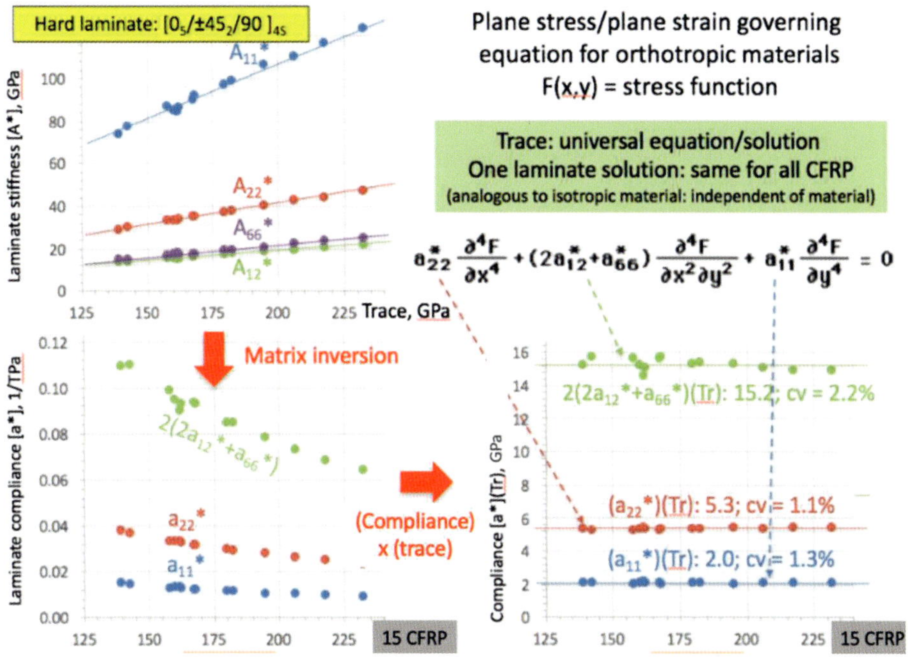

Figure 1.10 Solution of plane elasticity is the same for all CFRP for a given laminate

Invariants in coordinate transformation can lead to simplification as well as better understanding of composite properties. We have shown that trace was the one and only constant needed to capture the stiffness of a composite system. This is made possible by changing the transformation relations from power of a transformation from the 4th power of cosines and sines in Figure 1.11 to linear combinations of U's in Figure 1.12.

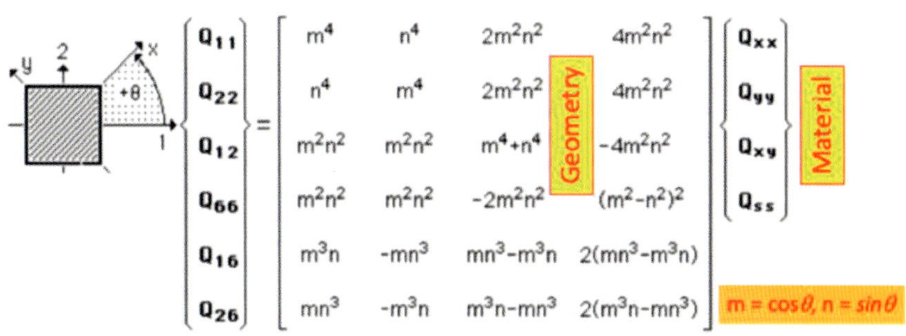

Figure 1.11 Traditional transformation of ply stiffness

Trigonometric identities between power and multiple angles are used as shown in the top part of Figure 1.12, invariants of stiffness can be seen in the bottom part of the figure. To go from power of sines and cosines to multiple angles, integration and differentiation become simpler and can maintain linear combinations of U's and [Q].

Geometry

$$m^4 = \frac{3 + 4\cos 2\theta + \cos 4\theta}{8}, \quad m^3 n = \frac{2\sin 2\theta + \sin 4\theta}{8},$$

$$m^2 n^2 = \frac{1 - \cos 4\theta}{8}, \quad mn^3 = \frac{2\sin 2\theta - \sin 4\theta}{8},$$

$$n^4 = \frac{3 - 4\cos 2\theta + \cos 4\theta}{8}$$

Table 1.2. Linear combinations of on-axis stiffness moduli.

Material

	Q_{xx}	Q_{yy}	Q_{xy}	Q_{ss}	Invariant?
$U_1 = U_4 + 2U_5$	3/8	3/8	1/4	1/2	Yes
U_2	1/2	-1/2	0	0	No
U_3	1/8	1/8	-1/4	-1/2	No
$U_4 = U_1 - 2U_5$	1/8	1/8	3/4	-1/2	Yes
$U_5 = (U_1 - U_4) / 2$	1/8	1/8	-1/4	1/2	Yes

Figure 1.12 Derivation of invariants of composite stiffness components

In particular, the physical significance of U_1 and U_5 invariants are shown as the average value of Q_{11} and Q_{66}, respectively in Figure 1.13. The asterisks in U_1* and U_5* mean trace normalized U_1 and U_5, respectively. They are the master ply values for composite materials as well as for isotropic metals. But for isotropic materials $U_2 = U_3 = 0$.

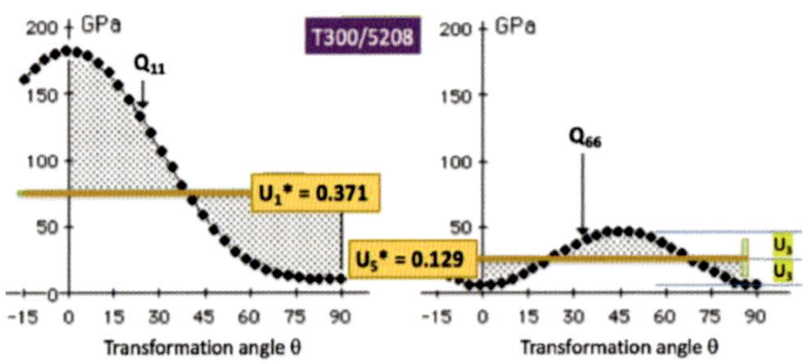

Figure 1.13 Physical of U_1 and U_5 invariants, where U_1*, U_5* are master ply invariants

These invariants are the master ply and universal constants for all composites and isotropic metals as seen in Figure 1.13. Not just for CFRP. There is only one constant because

$$U_1^* + U_5^* = 0.371 + 0.129 = 1/2$$

can be seen in the table in Figure 1.14. Such relation transcends materials beyond CFRP, like

$$U_1 + U_5 = \text{Trace}/2$$

Power of invariants cannot be overstated. Components of stiffness A_{11}, for example, belongs to a tensor which has fixed transformation and invariants. Others like E_1 which is not a component of tensor, and has not invariants. Failure criteria are subject to the same rules. Distinctions must be made and that is how simplicity can be realized as we will see in many chapters of this book.

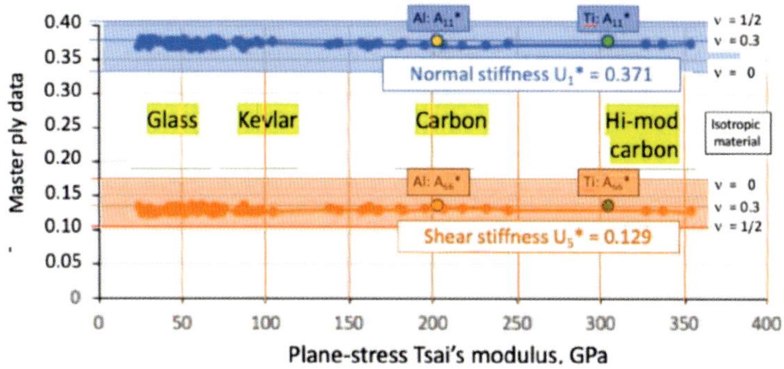

Figure 1.14 Universal master constants for composites and metals [3]

The traditional transformation relation shown in Figure 1.11 can be changed to that in Figure 1.15 with invariants shown explicitly. Again, the material and geometric portions are identified.

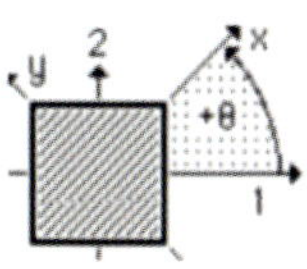

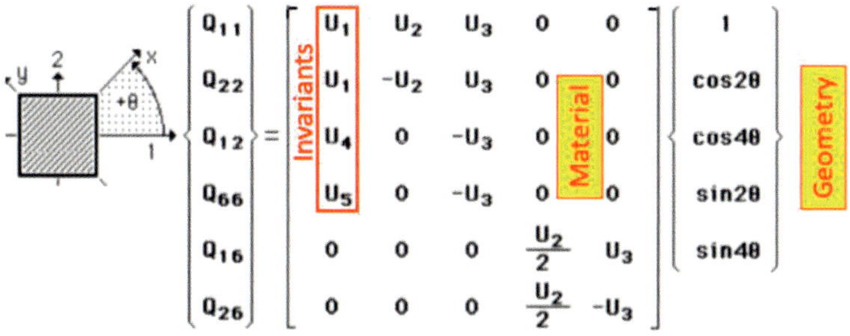

$$
\begin{Bmatrix} Q_{11} \\ Q_{22} \\ Q_{12} \\ Q_{66} \\ Q_{16} \\ Q_{26} \end{Bmatrix} =
\begin{bmatrix}
U_1 & U_2 & U_3 & 0 & 0 \\
U_1 & -U_2 & U_3 & 0 & 0 \\
U_4 & 0 & -U_3 & 0 & 0 \\
U_5 & 0 & -U_3 & 0 & 0 \\
0 & 0 & 0 & \frac{U_2}{2} & U_3 \\
0 & 0 & 0 & \frac{U_2}{2} & -U_3
\end{bmatrix}
\begin{Bmatrix} 1 \\ \cos 2\theta \\ \cos 4\theta \\ \sin 2\theta \\ \sin 4\theta \end{Bmatrix}
$$

(Labels in figure: Invariants — first column; Material — matrix; Geometry — right vector.)

Figure 1.15 Invariant transformation of ply stiffness

We have also discovered that there is a dependency between U_2 and U_3 of various materials as seen in Figure 1.16. For CFRP, the ratio is 4.26.

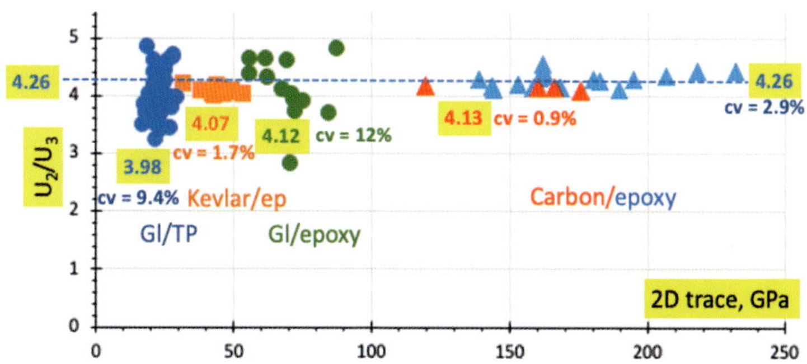

Figure 1.16 Empirical relation between U_2 and U_3 for various composite materials

It is an empirical but useful relation because it reduced the number of constant for laminate stiffness from two to one. The relation shown in Figure 1.15 can be rewritten in the following form for CFRP where for CFRP the U_2/U_3 is 4.26 as seen in Figure 1.17. It is just as easy to make U_2* as the only constant, also shown in Figure 1.17

$$\begin{bmatrix} A_{11}^* - U_1^* \\ A_{22}^* - U_1^* \\ A_{21}^* - U_4^* \\ A_{66}^* - U_5^* \\ A_{61}^* \\ A_{62}^* \end{bmatrix} = U_3^* \begin{bmatrix} 4.26V_1^* + V_2^* \\ -4.26V_1^* + V_2^* \\ -V_2^* \\ -V_2^* \\ 2.13V_3^* + V_4^* \\ 2.13V_3^* - V_4^* \end{bmatrix} \boxed{U_3/U_2 = 4.26} = U_2^* \begin{bmatrix} V_1^* + 0.224V_2^* \\ -V_1^* + 0.234V_2^* \\ -0.234V_2^* \\ -0.234V_2^* \\ V_3^* + 0.468V_4^* \\ V_3^* - 0.468V_4^* \end{bmatrix} \boxed{U_2/U_3 = 0.234}$$

Figure 1.17 One parameter of CFRP laminate stiffness based on U_3* or U_2*

The master ply values of U's are listed in the table in Figure 1.11. Note that U_1* and U_5* are universal constants for all materials, as seen in Figure 1.4. U_4* is dependent on U_1* and U_5*, as seen in Figure 1.12. Thus U_2* and U_3* are related, as seen in Figure 1.16, there remain only one independent constant in U_2* or U_3* as illustrated in Figure 1.17. Their master ply values and coefficients of variation are shown in the bottom of Figure 1.18. The variations in U_1* and U_5* are of particular universal importance and their coefficients of variations are 0.4 and 1.2 percent, respectively.

CFRP	U1	U2	U3	U4	U5	Trace	Ratio	U1*	U2*	U3*	U4*	U5*	Trace*
IM6/epoxy	85.88	96.44	21.83	25.43	30.23	232.2	1.000	0.370	0.415	0.094	0.110	0.130	1.000
IM7/977-3	80.61	91.11	20.51	24.01	28.30	217.8	1.000	0.370	0.418	0.094	0.110	0.130	1.000
T300/5208	76.37	85.73	19.71	22.61	26.88	206.5	1.000	0.370	0.415	0.095	0.109	0.130	1.000
IM7/MTM45	72.48	83.83	19.59	22.31	25.09	195.1	1.000	0.371	0.430	0.100	0.114	0.129	1.000
T800/Cytec	68.11	77.19	18.16	21.79	23.16	182.5	1.000	0.373	0.423	0.099	0.119	0.127	1.000
IM7/8552	66.82	75.46	17.65	20.53	23.15	179.9	1.000	0.371	0.419	0.098	0.114	0.129	1.000
T800S/3900	62.74	71.82	17.34	20.06	21.34	168.1	1.000	0.373	0.427	0.103	0.119	0.127	1.000
T300/F934	62.47	69.58	16.82	19.73	21.37	167.7	1.000	0.373	0.415	0.100	0.118	0.127	1.000
T700 C-Ply 64	60.30	66.24	15.30	18.10	21.10	162.8	1.000	0.370	0.407	0.094	0.111	0.130	1.000
AS4/H3501	59.66	64.90	14.25	16.96	21.35	162.0	1.000	0.368	0.401	0.088	0.105	0.132	1.000
T650/epoxy	59.55	65.25	15.17	18.20	20.67	160.4	1.000	0.371	0.407	0.095	0.113	0.129	1.000
T4708/MR60	59.06	67.56	16.27	18.91	20.07	158.3	1.000	0.373	0.427	0.103	0.120	0.127	1.000
T700/2510	53.48	59.18	14.15	16.77	18.35	143.7	1.000	0.372	0.412	0.099	0.117	0.128	1.000
AS4/MTM45	53.68	60.37	14.66	17.06	18.31	144.0	1.000	0.373	0.419	0.102	0.118	0.127	1.000
T700 C-Ply 55	51.62	56.84	13.27	15.68	17.97	139.2	1.000	0.371	0.408	0.095	0.113	0.129	1.000
							INTACT Average	0.371	0.416	0.097	0.114	0.129	1.000
							cv %	0.4%	2.0%	4.3%	3.9%	1.2%	0.0%

Figure 1.18 Master ply values for U's for CFRP [1]

A good example how the stiffness components of master ply CFRP in DD can be condensed in tables shown in Figure 1.19. Preliminary design without committing to any specific material is possible. That option is good because material selection will simply be the master ply value multiplied by the trace of a given material. Each table represent the entire field of DD with angle increment of 22.5 degrees. There are 15 independent components, as compared to 16 in quad laminates with 10 percent ply fractions to be shown in Figure 2.14.

U_1^*	0	23	45	68	90
0	0.37				
23	0.37	0.37			
45	0.37	0.37	0.37		
68	0.37	0.37	0.37	0.37	
90	0.37	0.37	0.37	0.37	0.37

U_5^*	0	23	45	68	90
0	0.13				
23	0.13	0.13			
45	0.13	0.13	0.13		
68	0.13	0.13	0.13	0.13	
90	0.13	0.13	0.13	0.13	0.13

U_4^*	0	23	45	68	90
0	0.11				
23	0.11	0.11			
45	0.11	0.11	0.11		
68	0.11	0.11	0.11	0.11	
90	0.11	0.11	0.11	0.11	0.11

U_2^*	0	23	45	68	90
0	0.42				
23	0.36	0.29			
45	0.21	0.15	0.00		
68	0.06	0.00	-0.15	-0.29	
90	0.00	-0.06	-0.21	-0.36	-0.42

U_3^*	0	23	45	68	90
0	0.10				
23	0.05	0.00			
45	0.00	-0.05	-0.10		
68	0.05	0.00	-0.05	0.00	
90	0.10	0.05	0.00	0.05	0.10

A_{11}^*	0	23	45	68	90
0	0.89				
23	0.78	0.67			
45	0.58	0.47	0.27		
68	0.48	0.37	0.18	0.08	
90	0.47	0.36	0.16	0.06	0.05

A_{22}^*	0	23	45	68	90
0	0.05				
23	0.06	0.08			
45	0.16	0.18	0.27		
68	0.36	0.37	0.47	0.67	
90	0.47	0.48	0.58	0.78	0.89

A_{66}^*	0	23	45	68	90
0	0.03				
23	0.08	0.13			
45	0.13	0.18	0.23		
68	0.08	0.13	0.18	0.13	
90	0.03	0.08	0.13	0.08	0.03

Figure 1.19 Master ply stiffness of 15 DD laminates as functions of U's

DD being a field-based laminate family it is very easy to present the stiffness in its entirety. Its member laminates can change as the angle increment changes; i.e., to use higher or lower angle increments than 22.5. The angle increments in this figure can change from 22.5 to whatever value that would be appropriate. They can be more or less than 22.5. They can also be variable rather than being constant. Such flexibility is possible because DD is field based and with continuous connection among all laminates. These factors apply to all CFRP materials. In an earlier example that IM7/977 is 20 percent stiffer than T800/Cyt (based on their trace values), it is 20 percent stiffer for all laminates. If a component has 200 laminates of IM7/977, it is 20 percent stiffer than the same component with the same 200 laminates of T800/Cyt. Such scaling is possible because trace is the one and only parameter that can determine the stiffness of laminates. Ability to scale gives design and manufacturing fast, accurate and confident.

In Figure 1.19, the top row of this figure shows the 3 invariants. They have constant values. They. apply to all materials including metals as seen in Figure 1.14. In the middle row, there are U_2* and U_3*. They are not invariants. For U_2* the square symmetric components are zero as seen in the figures in red. This $A_{11} = A_{22}$ is confirmed in the first 2 tables in the 3rd row. In the table for U_3*, there are 5 components with zero values, also shown in red. The shear modulus of these laminates is equal to U_5*, shown in the right table in the 3rd row.

Square symmetric laminates are very important and are anchored between [0/90] and [±45] with [QI] in the middle. They are clearly shown in A_{11} and A_{22} tables in the last row in Figure 1.19. They form the backbone of laminates that divides laminates into two groups. In the A_{11} table, higher stiffness values are those located above the square symmetric laminates. For A_{22} table, higher stiffness values are below the square symmetric laminates. In the A_{66} table, the highest shear modulus would be [±45], and have steady decline in values as laminates radiate from this maximum value. As we will see in Dan Wang's Chapter 8 on buckling, [±45] can be the best for many boundary conditions. For optimization, it will be the balance between the normal stiffness in A_{11} or A_{22}, and A_{66}. In many cases, square symmetric laminates may be the best compromise.

1.3 Carpet and lamination parameter plots

Full-field coverage on laminate stiffness in Figure 1.19 is unique for field-based laminates like DD. It is not possible with quad because its member laminates are discrete and have multiple building blocks in 6-, 8-, and 10-ply sub-laminates. The master ply A_{11}^* is shown in 3 charts in Figure 1.20. When 10 percent rule is applied to the 10-ply sub-laminate on the left, only 16 laminates are valid. The square symmetric laminates are seen to be the backbone and shown in red figures with grey shading. In the center, 8-ply sub-laminates are shown with 1/8 fraction rule applied. There are only 9 valid laminates. Finally, on the right is the 6-ply sub-laminates with 1/6 fraction rule and 4 valid laminates.

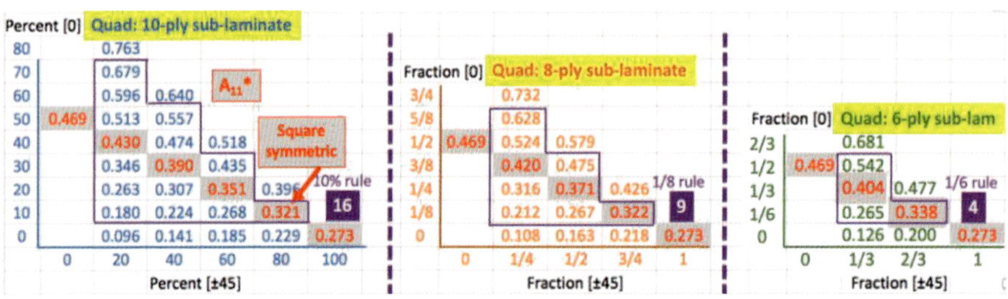

Figure 1.20 Valid laminates in 10-, 8- and 6-ply quad laminates

The 3 sub-laminates shown in Figure 1.20 can be presented in carpet plots each with the percent [±45] in the horizontal axis, and percent [0] in the vertical axis. The carpet plot for 10-ply sub-laminates is shown on the left of Figure 1.21, 8-ply in the middle and 6-ply on the right.

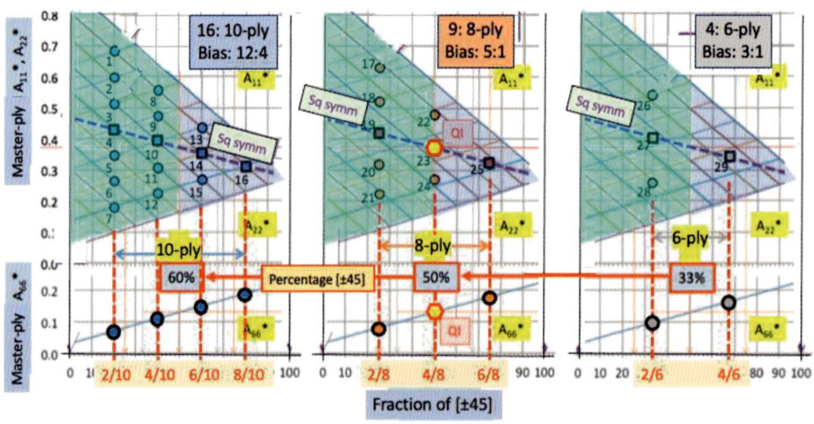

Figure 1.21 Biased partition in carpet plots based on 50 percent of [±45] in a laminate

Since [±45] must appear together (twice as thick as 0 and 90), there are gaps in each carpet plot. If 50 percent [±45] is the dividing lines, those less than 50 percent is shown with aqua background is called hard, and those with more than 50 percent, in blue called soft. There is a built-in bias in favor in the number of hard laminates with ratios shown in 12:4, 5:1 and 3:1, respectively. Such bias however does not exist for square symmetric laminates as their ratios of hard and soft are 2:2, 1:1 and 1:1, respectively. The same equity is found in the laminate shear modulus in A_{66}* in the bottom of the carpet plots.

If these 3 families of laminates plus 7-ply sub-laminates are shown in one table comparable to what was done for DD in Figure 1.19, the resulting table is shown in Figure 1.22. The coordinates are shown in both fractions and percentages of [±45] and [0], there are 71/106 cells for invalid laminates as compared with 35/106 valid laminates; i.e., twice as many. Shown in each valid cell are the angles of matching DD. There are no connections among the valid laminates because they are made with discrete number of plies. We can have 4 or 5 plies but not 4-1/2 plies. For DD which is field based, the ply angle can assume any value; i.e., 30, 30.5 or 30.54 if necessary. Plies are continuous that would make optimization simpler and more practical. The compact DD laminates are shown in the lower right as a comparison in their respective sizes. Again, there is a bias in favor of hard laminates in 24:8, or the same factor of 3.

35 Quad (Serial number shown in dark blue cells)

		%(±45)	1	2	3	4	5	6	7	8	9	10	11
	%[0]	Frat [0]	20%	25%	29%	33%	40%	50%	57%	60%	67%	75%	80%
Line	70%	7/10	32/1	8									
1	63%	5/8		38/0									
2	60%	6/10	42/0		13								
3	57%	4/7			43/0	17	20						
4	50%	4/8	65/0	59/0		52/0	49/1						
5	43%	3/7			67/6								
6	40%	4/10	77/13				62/9	25					
7	38%	3/8		75/15				57/12					
8	33%	2/6				73/18				30			
9	30%	3/10	90/23				70/20		28	57/18			
10	29%	2/7			84/23				58/19				
11	25%	2/8		90/27				68/22					
12	20%	2/10	90/32				80/28			65/25	33		
13	20%	1/6				90/32					62/27	34	
14	17%	1/7			90/35				71/30				
15	13%	1/8		90/38				78/33				60/30	35
16	10%	1/10	90/42				90/37			72/33			58/32
		Fract [±45]	2/10	2/8	2/7	2/6	4/10	4/8	4/7	6/10	4/6	6/8	8/10

26 DD

[±Φ] \ [±Ψ] Radius	0	15	30	45	60	75
15		243		Master		
30	363	306	208			
45	595	524	341	191		
60	856	783	586	341	208	
75	1053	979	783	524	306	243
90	1125	1053	856	595	363	

Figure 1.22 Valid quad laminates in carpet plot with twice as many empty cells vs DD

Laminate properties are integrated quantities of unidirectional plies. It was first conceived from discussion with Phillip Mast of the Naval Research Laboratory, and later with Nick Pagano [4] and Tom Hahn [5]. It was further promoted by Miki [6]. In-plane stiffness [A] is defined in the first line of Figure 1.23. Lamination parameters V_i* are then defined in line 4. For DD, simple relations can be derived and shown in the lower left. For all laminates, relations between laminate stiffness and lamination parameters are shown on the right side and are very easy to use. More detail derivation and explanation of lamination parameters will be covered in Chapter 4 and 10. Their plots give much better presentation of the properties of laminates of all types (quad, unconventional and DD) and the zones where they excel. They give visual appreciation why laminates are best used. For DD it shows the unique flexibility and simplicity of lamination.

$$A_{11}=\int_{-h/2}^{h/2} Q_{11}\,dz=\int_{-h/2}^{h/2}[U_1+U_2\cos 2\theta+U_3\cos 4\theta]dz$$

$$A_{11}=U_1h+U_2V_1+U_3V_2$$

...

$$A_{16}=\int_{-h/2}^{h/2} Q_{16}\,dz=\int_{-h/2}^{h/2}\left[\frac{U_2}{2}\sin 2\theta+U_3\sin 4\theta\right]dz=\frac{U_2}{2}V_3+U_3V_4$$

$$V_i^*=\frac{1}{h}V_i=V_{[1,2,3,4]}^*=\frac{1}{h}\int_{-h/2}^{h/2}[\cos 2\theta,\cos 4\theta,\sin 2\theta,\sin 4\theta]\,dz$$

For all laminates:

$$A_{11}^*=U_1+U_2V_1^*+U_3V_2^*$$
$$A_{22}^*=U_1-U_2V_1^*+U_3V_2^*$$
$$A_{61}^*=(U_2/2)V_3^*+U_3V_4^*$$
$$A_{62}^*=(U_2/2)V_3^*-U_3V_4^*$$

Solution

For double-double:

$$V_1^*=[\cos 2\Phi+\cos(-2\Phi)+\cos 2\Psi+\cos(-2\Psi)]/4=(\cos 2\Phi+\cos 2\Psi)/2;$$
$$V_2^*=[\cos 4\Phi+\cos(-4\Phi)+\cos 4\Psi+\cos(-4\Psi)]/4=(\cos 4\Phi+\cos 4\Psi)/2;$$
$$V_3^*=\sin 2\Phi+\sin(-2\Phi)+\sin 2\Psi+\sin(-2\Psi)=0;$$
$$V_4^*=\sin 4\Phi+\sin(-4\Phi)+\sin 4\Psi+\sin(-4\Psi)=0;$$

For all laminates:

$$V_1^*=(A_{11}^*-A_{22}^*)/2U_2$$
$$V_2^*=(A_{11}^*+A_{22}^*-2U_1)/2U_3$$
$$V_3^*=(A_{61}^*+A_{62}^*)/U_2$$
$$V_4^*=(A_{61}^*-A_{62}^*)/2U_3$$

Figure 1.23 Lamination parameters and their relations with master laminate [A*]

A new relation between lamination parameters and laminate stiffness is shown in Figure 1.24 which has the same structure as that for unidirectional ply is Figure 1.15.

$$\begin{Bmatrix}A_{11}\\A_{22}\\A_{12}\\A_{66}\\A_{16}\\A_{26}\end{Bmatrix}=\begin{bmatrix}U_1&U_2&U_3&0&0\\U_1&-U_2&U_3&0&0\\U_4&0&-U_3&0&0\\U_5&0&-U_3&0&0\\0&0&0&\frac{U_2}{2}&U_3\\0&0&0&\frac{U_2}{2}&-U_3\end{bmatrix}\begin{Bmatrix}1\\V_1^*\\V_2^*\\V_3^*\\V_4^*\end{Bmatrix}$$

Material

Geometry

Lamination parameters

Figure 1.24 In-plane stiffness [A] in terms of lamination parameters

The best use of DD is in its homogenized state. In Chapters 4 and 10, homogenization is reached in 3 or 4 repeats; i.e., 12 to 16 plies total. It is then possible to approximate [A] = [D], and coupling [B] = 0. There is no need to calculate [D] if DD is used. But for other laminates, conventional and unconventional, the thickness of the building block is too large. Laminates will be very thick before homogenization is reached.[1] It is therefore necessary to calculate the flexural stiffness for each of those non DD laminates. There are thousands of stacking sequences. As mentioned earlier heterogeneous laminates become a diificult issue for not only design but worse for manufacturing. They should be avoided. Guidelines promoting moving the "principal load carrying plies" to the center of a laminate in Hdbk17. Such recommendation runs in the opposite direction as DD's preference for homogeneity.

A critical relation between DD angles with the lamination parameters of any laminate is derived in Figure 1.25. If a laminate is within the domain of DD, an exact conversion can be found. The solution of a quadratic equation in Eq 1.6 would be real. So it is possible that a real solution may not be available. This happens when a laminate has more than 50 or higher percent [0] or [90]. In DD each ply angle can assume values up to and including 50 percent. No single ply angle can exceed that 50 percent limit. This explains why some laminates have imaginary roots in the quadratic equation solution. Then approximate match can be found.

For double-double: $\cos 2\Phi + \cos 2\Psi = 2V_1^*$ — Eq. 1.1

$\cos 4\Phi + \cos 4\Psi = 2V_2^*$ — Eq. 1.2

From Eq. 2.1: $\cos 2\Phi = 2V_1^* - \cos 2\Psi$ — Eq. 1.3

From trignometric idendities: $\cos 4\Phi = 2\cos^2 2\Phi - 1$; $\cos 4\Psi = 2\cos^2 2\Psi - 1$

Substituting into Eq. 2.2: $2\cos^2 2\Phi - 1 + 2\cos^2 2\Psi - 1 = 2V_2^*$ — Eq. 1.4

Substituting Eq. 2.3: $2(2V_1^* - \cos 2\Psi)^2 - 1 + 2\cos^2 2\Psi - 1 = 2V_2^*$

$$(2V_1^* - \cos 2\Psi)^2 + \cos^2 2\Psi - 1 - V_2^* = 0$$

Quadratic equation: $\cos^2 2\Psi - (2V_1^*)\cos 2\Psi + (2V_1^{*2} - V_2^*/2 - 1/2) = 0$ — Eq. 1.5

Thus: $\cos 2\Psi = V_1^* + [V_1^{*2} - 2V_1^{*2} + V_2^*/2 + 1/2]^{1/2} = V_1^* + [-V_1^{*2} + V_2^*/2 + 1/2]^{1/2}$ — Eq. 1.6

From Eq. 2.3: $\cos 2\Phi = 2V_1^* - \cos 2\Psi$ — Eq. 1.7

Figure 1.25 To convert any laminate to DD angles through lamination parameters

Question may be raised why DD cannot have ply percentage higher the 50 percent. This is possible if one of the double angles is thicker than the other angle. For example if one angle has twice the thickness of the other, then one angle can have 66 percent while the other angle 33 percent of the total laminates. But this unequal DD laminates have a drawback; i.e., the thickness of the building block will be 6 plies rather than 4 plies when the two DD angles have equal thickness. When the building block is thicker in 6 to 4 ratio, the number of repeats to reach homogenization will increase. That may not be an advantage of going beyond 50 percent limit on all ply angles. It is possible to go to the limit of single-double where there is only one angle that is 100 percent of the laminate. Homogenization is such critical feature in DD it should not be changed unless there is an overwhelming reason to do so. Examples of other extensions of DD can be seen in Figure 1.32. DD remains to be the best balanced family of laminates.

A better comparison between the carpet and lamination parameter plots for quad laminates can be seen in Figure 1.26 where the square symmetric laminate line is made horizontal for both plots. Each plot is divided into 4 regions: Hard1, Hard2, Neutral and Soft. They are defined by the composition of the laminates listed in the lower right corner of the figure. Their areas are equal.

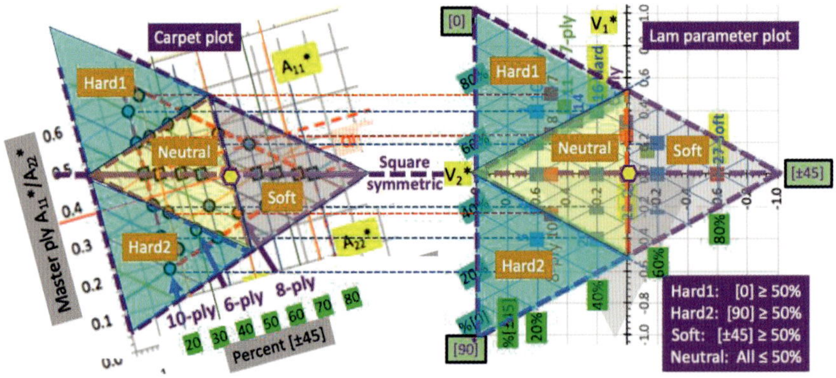

Figure 1.26 Matching of carpet and lamination parameter plots for quad laminates

It is interesting to compare the conventional carpet plot with the lamination parameter plot that we are advocating. Lamination parameter plot can be applied to conventional and unconventional laminates and DD so direct comparisons can be made. Carpet plot on the other hand is suitable only for the conventional quad laminates. DD cannot be shown in carpet plot.

The entire domain of DD is seen in Figure 1.27. On the left, the small figure identifies the boundaries as defined by the single-double as the bottom and on the top two boundaries defined by [0/Ψ] and [90/Ψ]. We also define hard laminates when $V_2* \geq 0$; soft laminates, $V_2* \leq 0$. Hard laminate resists normal stresses well while soft laminates, shear stress. Single-double like [±22.5] or smaller angle will behave like [0] under uniaxial load along the x-axis.

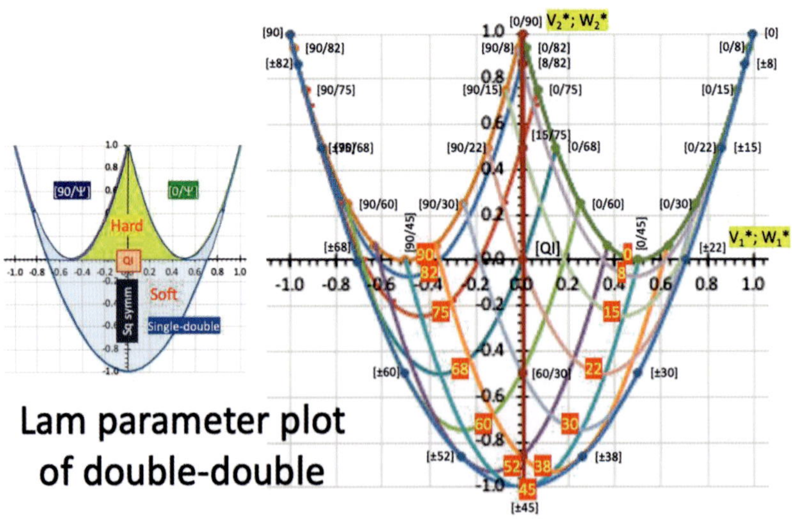

Figure 1.27 Laminate parameter plot of DD with hard and soft laminates identified

The angle increment Δ determines how large a DD family would be. In Figure 1.28, 3 values in 30, 22.5 and 15 are shown, in both graph and tabular forms. Dashed blue lines are the square symmetric laminates while the dashed red lines denote 50 percent [±45]. Blue circles with blue background are hard laminates while those with yellow background, soft laminates. Several circles with light green background are deemed neutral; i.e., not designated as either hard or soft. On the left, there are 10 DD laminates when angle increment is 30. Of the 10, 3 are in the hard domain, and 3 in the soft domain. For angle increments are 22.5 and 15, the ratios of hard to soft are 3 to 3 and 6 to 8, respectively. The 3:1 bias against soft laminates in quad laminates shown in Figure 1.21 has been essentially equalized in DD.

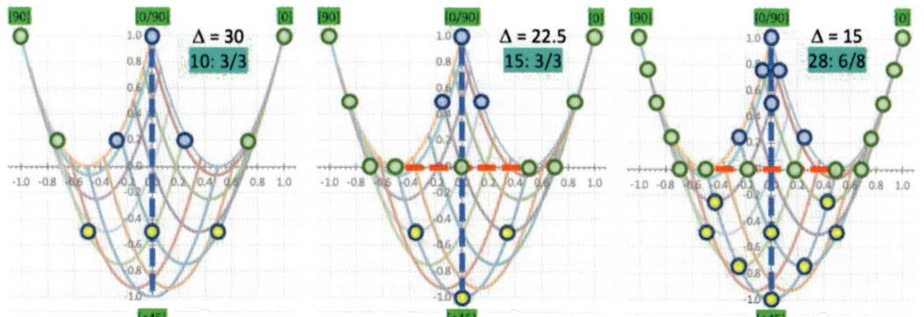

A11*	90	60	30	0 [±Ψ]
0				0.89
30			0.53	0.71
60		0.11	0.32	0.50
90	0.05	0.08	0.29	0.47

A11*	90	68	45	23	0 [±Ψ]
0					0.89
23				0.67	0.78
45			0.27	0.47	0.58
68		0.08	0.18	0.37	0.48
90	0.05	0.06	0.16	0.36	0.47

A11*	90	75	60	45	30	15	0 [±Ψ]
0							0.89
15						0.78	0.83
30					0.53	0.66	0.71
45				0.27	0.40	0.53	0.58
60			0.11	0.19	0.32	0.45	0.50
75		0.06	0.09	0.17	0.30	0.42	0.47
90	0.05	0.06	0.98	0.16	0.29	0.42	0.47

Figure 1.28 DD laminates as function of angle increment Δ

The 3:1 bias in favor of hard laminate in quad is shown again in Figure 1.29. The ratios are shown between squares with brown and yellow core. Those with olive green core are neutral laminates. The finite number of laminates for each angle increment in this figure in 9, 16, and 29 are different from 6- to 10-ply quad laminates. DD is continuous and can have more laminates with lower angle increments whereas quad laminates have reached dead ends. Cannot increase unless higher number of plies in sub-laminates increase beyond 10.

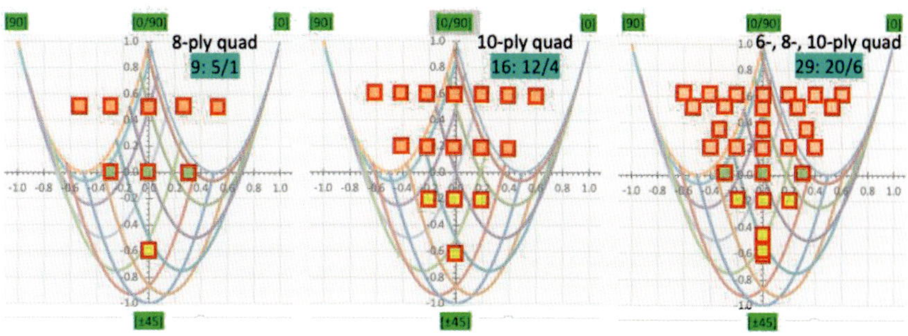

Figure 1.29 Distribution of quad in lamination parameter plot

In Figure 1.30 samples of unconventional laminates were able to be matched exactly (except for #3). Such exact matching is shown in columns U and Y when their values are unity for the in-plane [A] and flexure [D], respectively. These matching laminates can be plotted on the same lamination parameters as for quad and DD laminates in Figure 1.31.

	A	B	C	D	E	F	G	H	I	J	K	L	M	N	O	P	Q	R	S	T	U	V	W	X	Y
1	Sprengholz		1	2	3	4	5	6	7	8	9	10	11	12	1	[A]	[D]	Vi*	DD/A	[A]/DD	Ratio	Wi*	DD/D	[D]/DD	Ratio
2		Angle	45	-45	-45	45	45	0	0	-45	-45	45	-45	45	11	0.376	0.351	0.167	28.7	0.376	1.00	0.126	31.37	0.351	1.00
3		# ply	1	1	1	1	1	1	1	1	1	1	1	1	22	0.237	0.246	-0.667	50.9	0.237	1.00	-0.748	50.94	0.246	1.00
4															12	0.179	0.187	0.126		0.179	1.00			0.187	1.00
5															66	0.194	0.201	-0.748		0.194	1.00			0.202	1.00
6	Sprengholz		1	2	3	4	5	6	7	8	9	10	11	12	2	[A]	[D]	Vi*	DD/A	[A]/DD	Ratio	Wi*	DD/D	[D]/DD	Ratio
7		Angle	73.8	-34.4	42.8	-54.3	-43.7	-34.8	32.5	83.0	-9.4	-9.1	7.7	32.6	11	0.440	0.288	0.200	17.4	0.440	1.00	-0.100	32.37	0.288	1.00
8		# ply	1	1	1	1	1	1	1	1	1	1	1	1	22	0.273	0.371	-0.150	57.4	0.273	1.00	-0.425	64.39	0.372	1.00
9															12	0.129	0.156	-0.100		0.129	1.00			0.156	1.00
10															66	0.143	0.170	-0.425		0.143	1.00			0.170	1.00
11		Hagel	1	2	3	4	5	6	7	8	9	10	11	12	3	[A]	[D]	Vi*	DD/A	[A]/DD	Ratio	Wi*	DD/D	[D]/DD	Ratio
12		Angle	28.0	-24.4	18.4	-49.2	-2.4	-0.4	3.2	28.8					11	0.669	0.610	0.6748	0.0	0.655	0.98	0.597	12.68	0.610	1.00
13		# ply	1	1	1	1	1	1	1	1	0	0	0	0	22	0.107	0.113	0.1691	35.81	0.107	1.00	-0.099	36.58	0.113	1.00
14															12	0.098	0.124	0.5966		0.105	1.07			0.124	1.00
15															66	0.112	0.138	-0.099		0.119	1.06			0.138	1.00
16		Irisarri	1	2	3	4	5	6	7	8	9	10	11	12	4	[A]	[D]	Vi*	DD/A	[A]/DD	Ratio	Wi*	DD/D	[D]/DD	Ratio
17		Angle	-45	45	90	-45	45	-45	45	90	0				11	0.423	0.241	0.125	17.41	0.423	1.00	-0.203	37.2	0.241	1.00
18		# ply	2	1	2	1	2	1	1	1	5	0	0	0	22	0.319	0.410	0.000	62.41	0.319	1.00	-0.472	66.24	0.410	1.00
19															12	0.114	0.160	-0.203		0.114	1.00			0.160	1.00
20															66	0.129	0.175	-0.472		0.129	1.00			0.175	1.00
21		Tuan	1	2	3	4	5	6	7	8	9	10	11	12	5	[A]	[D]	Vi*	DD/A	[A]/DD	Ratio	Wi*	DD/D	[D]/DD	Ratio
22		Angle	-45	50	20	90									11	0.310	0.298	-0.102	29.14	0.311	1.00	0.017	35.85	0.298	1.00
23		# ply	1	1	1	1	0	0	0	0	0	0	0	0	22	0.395	0.284	-0.192	68.42	0.395	1.00	-0.822	53.16	0.284	1.00
24															12	0.133	0.194	0.0166		0.133	1.00			0.194	1.00
25															66	0.147	0.209	-0.822		0.147	1.00			0.209	1.00
26	Meddaikar		1	2	3	4	5	6	7	8	9	10	11	12	6	[A]	[D]	Vi*	DD/A	[A]/DD	Ratio	Wi*	DD/D	[D]/DD	Ratio
27		Angle	-45	-15	15	45	90								11	0.369	0.496	-0.045	22.1	0.369	1.00	0.356	15.5	0.496	1.00
28		# ply	1	1	1	1	2	0	0	0	0	0	0	0	22	0.406	0.199	0.167	71.9	0.406	1.00	-0.245	49.2	0.199	1.00
29															12	0.098	0.138	0.356		0.098	1.00			0.138	1.00
30															66	0.112	0.152	-0.245		0.112	1.00			0.152	1.00
31	Meddaikar		1	2	3	4	5	6	7	8	9	10	11	12	7	[A]	[D]	Vi*	DD/A	[A]/DD	Ratio	Wi*	DD/D	[D]/DD	Ratio
32		Angle	-45	-15	15	45	90								11	0.345	0.421	-0.033	25.6	0.345	1.00	0.241	24.13	0.421	1.00
33		# ply	2	1	1	2	2	0	0	0	0	0	0	0	22	0.373	0.220	-0.125	67.0	0.373	1.00	-0.523	50.27	0.220	1.00
34															12	0.126	0.165	0.2415		0.126	1.00			0.165	1.00
35															66	0.141	0.180	-0.523		0.141	1.00			0.179	1.00
36	Meddaikar		1	2	3	4	5	6	7	8	9	10	11	12	8	[A]	[D]	Vi*	DD/A	[A]/DD	Ratio	Wi*	DD/D	[D]/DD	Ratio
37		Angle	-45	-15	15	45	90								11	0.331	0.383	-0.027	27.83	0.331	1.00	0.181	28.2	0.383	1.00
38		# ply	3	1	1	3	2	0	0	0	0	0	0	0	22	0.353	0.232	-0.300	64.08	0.353	1.00	-0.657	50.5	0.232	1.00
39															12	0.144	0.178	0.181		0.143	1.00			0.178	1.00
40															66	0.158	0.193	-0.657		0.158	1.00			0.193	1.00
41		Almeida	1	2	3	4	5	6	7	8	9	10	11	12	9	[A]	[D]	Vi*	DD/A	[A]/DD	Ratio	Wi*	DD/D	[D]/DD	Ratio
42		Angle	45	-45	0	45	-45	45	-45	0	45	-45	45	-45	11	0.427	0.470	0.250	23.46	0.426	1.00	0.320	19.05	0.469	1.00
43		# ply	1	1	2	1	1	1	1	2	1	1	2	2	22	0.218	0.203	-0.500	50.27	0.219	1.00	-0.359	49.21	0.203	1.00
44															12	0.163	0.149	0.320		0.163	1.00			0.149	1.00
45															66	0.177	0.164	-0.359		0.177	1.00			0.164	1.00

Figure 1.30 DD match of unconventional laminates for [A] and [D]; [7-12]

It is interesting that all laminates, #1 through #9, have their in-plane and flexural stiffness components all fall in the soft region (having negative V_2^* except [A] for #3). The matching DD and original unconventional laminates have precisely the same stiffness and strength. DD for quad can be exact and approximate. The approximate ones are those shown as red squares with light blue core.

The principal message is that lamination parameter plots are suitable for finding the matching DD for the conventional quad and unconventional laminates. (Much more profound use can be found in Chapters 4 and 10.) Only DD is a field-based laminate family for having continuous laminates so it can match what the other laminates may be, some exactly and others approximately. The reverse cannot be done; i.e., there is no straightforward way for a DD to be matched by any conventional or unconventional laminates. DD as a conversion or a freely chosen laminate offers the simplest manufacturing and design. It can be homogenized and made better and lower cost laminated components.

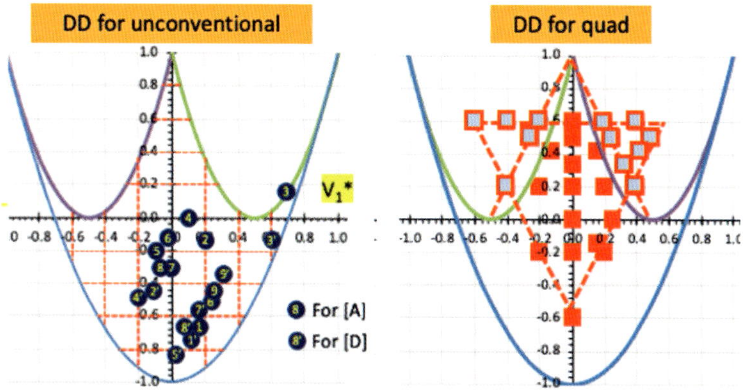

Figure 1.31 DD for unconventional and quad laminates

Antonio Miravete showed in Chapter 4 possible extensions of DD to go beyond its boundary as seen in the left of Figure 1.32. While such extensions are possible there are limitations as well. Areas gains are offset by areas lost. This can be seen in the options shown on the right.

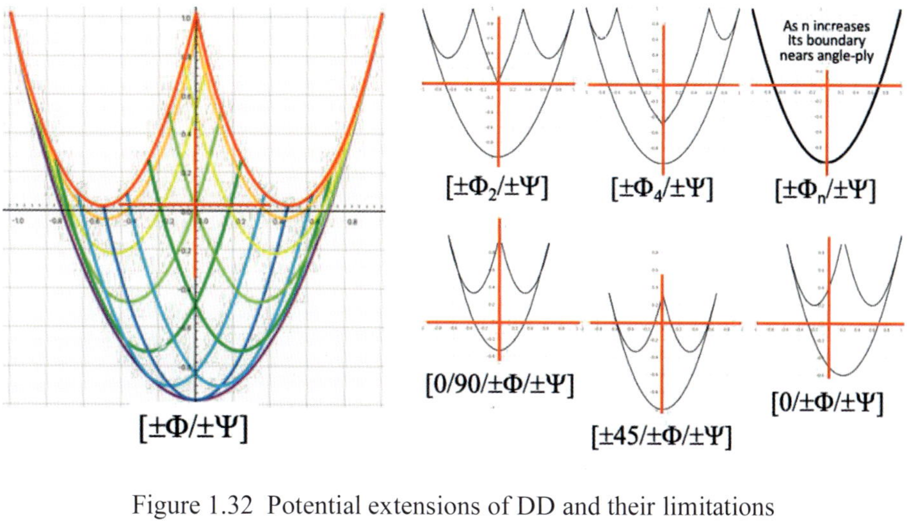

Figure 1.32 Potential extensions of DD and their limitations

In all cases, the building block thickness would be higher than 4-ply of the starting DD. Homogenization will be more difficult to reach when the building block sub-laminates are thicker. It seems to us that DD offers a good balance and should be retained.

It is appropriate to show at this time that when the stiffness of laminates is matched their strength will also be matched. This is an assurance that the switching to DD from quad or unconventional laminates does not result in loss of strength.

1.4 Failure envelopes

There are dozens of failure criteria that can be divided into two groups: interactive or non interactive. Tsai-Wu criterion is interactive while max stress, max strain, and Hashin are non interactive. In addition, Tsai-Wu is invariant so the criterion can be applied to any coordinate axes. Most other criteria have to be applied to the material axes only. Tsai-Wu criterion is a scalar product consisting of linear $\{F\}$ and quadratic $[F]$ terms only. Higher terms are ignored.

$F_{ij}\sigma_i\sigma_j + F_{ij}\sigma_i = 1$, **when expanded in the symmetry axes:**

$$F_{xx}\sigma_x^2 + 2F_{xy}\sigma_x\sigma_y + F_{yy}\sigma_y^2 + F_{ss}\sigma_s^2 + F_x\sigma_x + F_y\sigma_y = 1$$

The plane stress criterion can be represented in strain space by a straightforward substitution of the stress-strain relation.

$$F_{ij}\sigma_i\sigma_j = F_{ij}[Q_{ik}\epsilon_k][Q_{jl}\epsilon_l] = [F_{ij}Q_{ik}Q_{jl}]\epsilon_k\epsilon_l = G_{kl}\epsilon_k\epsilon_l,$$

Therefore $\boxed{G_{kl} = Q_{ki}Q_{lj}F_{ij}}$

$F_{ij}\sigma_i = F_i[Q_{ij}\epsilon_j] = [F_iQ_{ij}]\epsilon_j = G_j\epsilon_j$ **, therefore** $\boxed{G_l = Q_{jl}F_l}$

For an orthotropic material, the scalar equation in strains can be expanded as follows:

$G_{ij}\epsilon_i\epsilon_j + G_i\epsilon_i = 1$ **when expanded:**

$$\boxed{G_{xx}\epsilon_x^2 + 2G_{xy}\epsilon_x\epsilon_y + G_{yy}\epsilon_y^2 + G_{ss}\epsilon_s^2 + G_x\sigma_x + G_y\epsilon_y = 1}$$

where

$$G_{xx} = F_{xx}Q_{xx}^2 + 2F_{xy}Q_{xx}Q_{xy} + F_{yy}Q_{xy}^2 \ ,$$

$$G_{yy} = F_{xx}Q_{xy}^2 + 2F_{xy}Q_{xy}Q_{yy} + F_{yy}Q_{yy}^2 \ ,$$

$$G_{xy} = F_{xx}Q_{xx}Q_{xy} + F_{xy}(Q_{xx}Q_{yy} + Q_{xy}^2) + F_{yy}Q_{xy}Q_{yy} \ ,$$

$$G_{ss} = F_{ss}Q_{ss}^2 \ , \ G_x = F_xQ_{xx} + F_yQ_{xy} \ , \ G_y = F_xQ_{xy} + F_yQ_{yy}$$

Strength ratio R is defined in Figure 1.33 which is similar to the safety factor. It simplified the use of failure envelope with just one parameter. It determines more than failure or no failure like most failure criteria do. It gives numerical values how close the applied load is near failure, and what corrective actions can be taken.

For example, if R = 0.5, it means that the load is one half way from failure. Load can be doubled or thickness reduce to one half before failure occurs.

If R = 2, it means that the load is twice as high for the laminate to sustain. Load must be reduced to one half or thickness be double so the load can be carried.

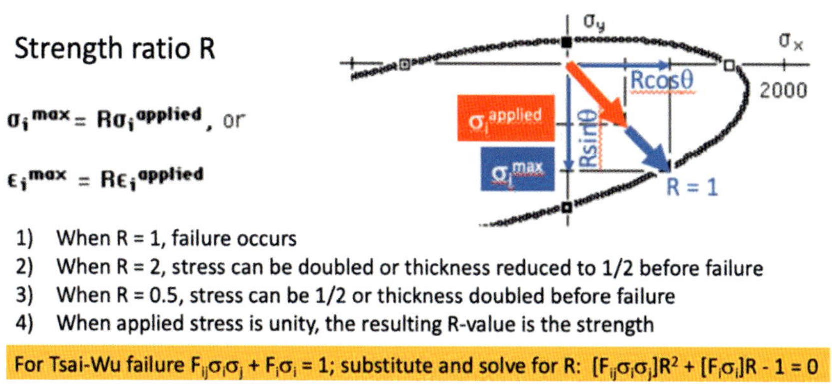

Figure 1.33 Strength ratio as a measure of safety for failure criteria

In Figure 1.34, at least 3 options are available in load set input. On the left is the normalized unit vectors. They are bounded by unit vectors between -1 and +1. The advantage of this input is that the resulting strength ratio is the working stress, as seen in value of 1079 MPa. This unit vector input is our recommendation for finding the working stress or strength of a laminate with multiple load set. It applies to running load, Pa/m, just as easily as having stress in Pa. Other load set inputs can certainly be used to serve different purposes.

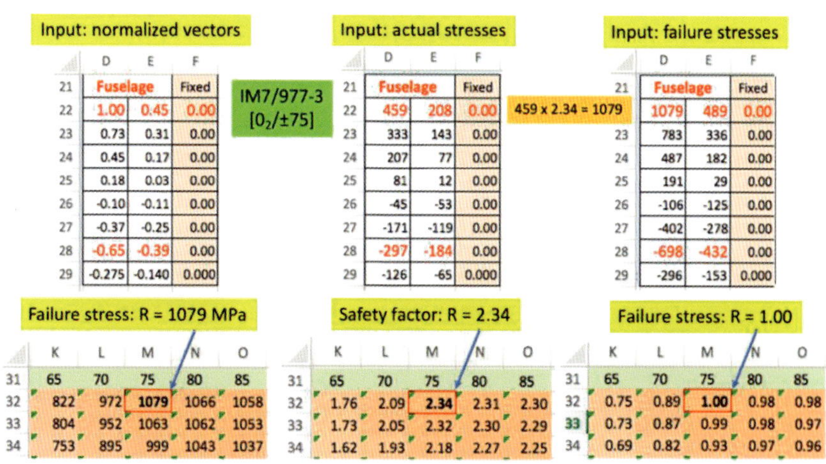

Input: normalized vectors

	D	E	F
21	Fuselage		Fixed
22	1.00	0.45	0.00
23	0.73	0.31	0.00
24	0.45	0.17	0.00
25	0.18	0.03	0.00
26	-0.10	-0.11	0.00
27	-0.37	-0.25	0.00
28	-0.65	-0.39	0.00
29	-0.275	-0.140	0.000

Input: actual stresses

	D	E	F
21	Fuselage		Fixed
22	459	208	0.00
23	333	143	0.00
24	207	77	0.00
25	81	12	0.00
26	-45	-53	0.00
27	-171	-119	0.00
28	-297	-184	0.00
29	-126	-65	0.000

Input: failure stresses

	D	E	F
21	Fuselage		Fixed
22	1079	489	0.00
23	783	336	0.00
24	487	182	0.00
25	191	29	0.00
26	-106	-125	0.00
27	-402	-278	0.00
28	-698	-432	0.00
29	-296	-153	0.000

Failure stress: R = 1079 MPa

	K	L	M	N	O
31	65	70	75	80	85
32	822	972	1079	1066	1058
33	804	952	1063	1062	1053
34	753	895	999	1043	1037

Safety factor: R = 2.34

	K	L	M	N	O
31	65	70	75	80	85
32	1.76	2.09	2.34	2.31	2.30
33	1.73	2.05	2.32	2.30	2.29
34	1.62	1.93	2.18	2.27	2.25

Failure stress: R = 1.00

	K	L	M	N	O
31	65	70	75	80	85
32	0.75	0.89	1.00	0.98	0.98
33	0.73	0.87	0.99	0.98	0.97
34	0.69	0.82	0.93	0.97	0.96

Figure 1.34 Options in load vectors for input format

Strength ratio R can be incorporated into failure criterion by substituting the maximum stress at failure leading to a quadratic equation in R that can be easily solved. This is not always done with application of failure criterion. Strength ratio R is a linear function of load or stress, and gives numerical values of safety. The derivation of R is seen in the following.

Letting σ_i reach maximum values when $F_{ij}\sigma_i^{max}\sigma_j^{max} + F_i\sigma_i^{max} = 1$

substitute $R\sigma_i^{applied}$ for σ_i^{max} $[F_{ij}\sigma_i\sigma_j]R^2 + [F_i\sigma_i]R - 1 = 0$

Solving quadratic equation: $aR^2 + bR - 1 = 0,$

$$a = F_{ij}\sigma_i\sigma_j,\ b = F_i\sigma_i$$

Positive quadratic root = strength ratio

$$R = -\left(\frac{b}{2a}\right) + \sqrt{\left(\frac{b}{2a}\right)^2 + \frac{1}{a}}$$

Strength ratio can be introduced into Tsai-Wu failure criterion in strain space that leads to a different quadratic equation. The solution yields the R value for strain space criterion:

$$[G_{ij}\epsilon_i\epsilon_j]R^2 + [G_i\epsilon_i]R = 1$$

$$aR^2 + bR - 1 = 0, \text{where } a = G_{ij}\epsilon_i\epsilon_j = F_{ij}\sigma_i\sigma_j,\ b = G_i\epsilon_i = F_i\sigma_i$$

Another feature of our approach to failure criterion of laminates is the omni envelopes. The derivation of the FPF and LPF envelopes are shown in Figure 1.35, on the left and right, respectively. In the middle are the respectively envelopes.

Ply orientation → FPF

Unit strain vector ↓

Angle	0	±15	±30	±45	±60	±75	90	Omni
0	18.0	14.0	9.7	7.5	6.6	6.3	6.3	6.3
15	14.6	12.8	9.9	8.1	7.2	6.8	6.6	6.6
30	11.2	10.7	9.7	8.7	8.0	7.6	7.5	7.5
45	8.9	8.9	8.9	8.9	8.9	8.9	8.9	8.9
60	7.5	7.6	8.0	8.7	9.7	10.7	11.2	7.5
75	6.6	6.8	7.2	8.1	9.9	12.8	14.6	6.6
90	6.3	6.3	6.6	7.5	9.7	14.0	18.0	6.3
105	6.2	6.2	6.3	7.1	9.2	13.8	18.5	6.2
120	6.4	6.3	6.3	6.8	8.6	12.6	16.3	6.3
135	7.0	6.7	6.4	6.8	8.3	11.3	13.7	6.4
150	8.0	7.4	6.8	7.0	8.1	10.3	11.8	6.8
165	9.3	8.4	7.5	7.4	8.2	9.7	10.5	7.4
180	10.4	9.5	8.4	8.1	8.5	9.4	9.8	8.1
195	10.9	10.3	9.3	8.9	9.0	9.3	9.5	8.9
210	10.5	10.3	10.0	9.7	9.6	9.6	9.6	9.6
225	10.0	10.0	10.0	10.0	10.0	10.0	10.0	10.0
240	9.6	9.6	9.6	9.7	10.0	10.3	10.5	9.6
255	9.5	9.3	9.0	8.9	9.3	10.3	10.9	8.9
270	9.8	9.4	8.5	8.1	8.4	9.5	10.4	8.1
285	10.5	9.7	8.2	7.4	7.5	8.4	9.3	7.4
300	11.8	10.3	8.1	7.0	6.8	7.4	8.0	6.8
315	13.7	11.3	8.3	6.8	6.4	6.7	7.0	6.4
330	16.3	12.6	8.6	6.8	6.3	6.3	6.4	6.3
345	18.5	13.8	9.2	7.1	6.3	6.2	6.2	6.2
360	18.0	14.0	9.7	7.5	6.6	6.3	6.3	6.3

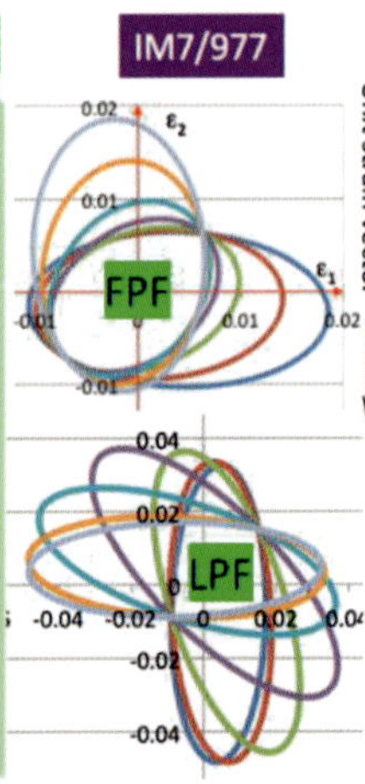

Ply orientation → LPF

Unit strain vector ↓

Angle	0	±15	±30	±45	±60	±75	90	Omni
0	17.0	18.0	21.4	27.9	34.7	32.6	29.6	17.0
15	17.3	18.0	20.2	24.1	28.6	31.5	32.2	17.3
30	18.7	19.1	20.3	22.1	24.1	25.7	26.3	18.7
45	21.5	21.5	21.5	21.5	21.5	21.5	21.5	21.5
60	26.3	25.7	24.1	22.1	20.3	19.1	18.7	18.7
75	32.2	31.5	28.6	24.1	20.2	18.0	17.3	17.3
90	29.6	32.6	34.7	27.9	21.4	18.0	17.0	17.0
105	19.6	23.3	36.0	34.4	24.1	19.1	17.7	17.7
120	13.6	15.9	27.1	42.8	29.0	21.6	19.7	13.6
135	10.7	12.2	19.1	43.2	37.4	26.3	23.5	10.7
150	9.2	10.2	14.7	31.2	47.9	34.8	30.6	9.2
165	8.5	9.3	12.4	21.8	44.6	46.6	42.4	8.5
180	8.4	9.0	11.3	16.8	28.9	43.2	46.8	8.4
195	8.9	9.4	11.0	14.2	19.5	25.8	28.8	8.9
210	10.1	10.4	11.3	12.9	14.8	16.7	17.4	10.1
225	12.5	12.5	12.5	12.5	12.5	12.5	12.5	12.5
240	17.4	16.7	14.8	12.9	11.3	10.4	10.1	10.1
255	28.8	25.8	19.5	14.2	11.0	9.4	8.9	8.9
270	46.8	43.2	28.9	16.8	11.3	9.0	8.4	8.4
285	42.4	46.6	44.6	21.8	12.4	9.3	8.5	8.5
300	30.6	34.8	47.9	31.2	14.7	10.2	9.2	9.2
315	23.5	26.3	37.4	43.2	19.1	12.2	10.7	10.7
330	19.7	21.6	29.0	42.8	27.1	15.9	13.6	13.6
345	17.7	19.1	24.1	34.4	36.0	23.3	19.6	17.7
360	17.0	18.0	21.4	27.9	34.7	32.6	29.6	17.0

Figure 1.35 Omni envelope for FPF and LPF for IM7/799

The omni envelopes are the inner most, controlling envelope of all possible ply angles from 0 and 360 degree. For FPF on the left, the controlling plies, shown with yellow shading, vary as ply orientations change. For LPF, matrix is degraded and only fiber remain as the primary load-carrying phase, the controlling plies are [0] and [90]. Only when load vector at 45 and 225 degree, all plies fail at the same load level. Since omni envelopes are no longer ply orientation dependent they become isotropic. The general strain can then be converted to principal strain. Then the envelope in strain space is invariant, same for each ply regardless the presence or absence of other plies. It becomes a material property. Exceptions include laminate subjected to bending (not in-plane only), and is not applicable for failure envelope in stress space (laminate dependent).

Thus, our LPF is a netting analysis that has been used for filament wound pressure vessels for more than 60 years. Netting analysis was limited to tensile failure from the internal pressure. The generalized version now must include compressive failure. It is however much simpler with only X and X', the uniaxial tensile and compressive strengths. Strength in the transverse direction, in Y and Y', and shear in S are not needed for the prediction of LPF. This is another significant simplification for DD. More detail will be given in Chapter 2.

Conversion from strain space failure envelopes to stress space ones is shown in Figure 1.36. The omni envelopes of T8S/3900 on the left and be converted to stress space envelope for quasi-isotropic laminate by multiplying the quasi-isotropic stiffness in U_1* and U_4* from the top row in Figure 1.19, with values in 0.37 and 0.11, respectively. The resulting envelopes in stress space is shown. The LPF is shown in red line. The unit circle failure criterion is the red shaded area that inscribe the LPF shown on the right. [13]

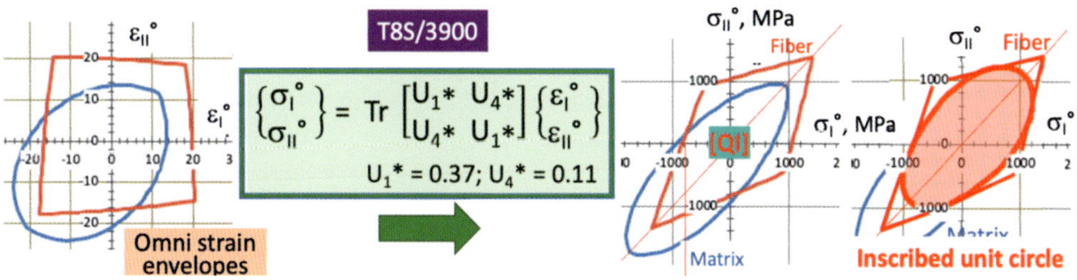

Figure 1.36 Conversion of failure envelope from strain to stress space, and unit circle criterion

As stated earlier, the strain space envelope, on the left in the last figure is a material property and covers FPF (in blue) and LPF (in red) for all ply orientations of this material. For stress space envelopes of different laminates only the appropriate stiffness of the laminates need to be used, instead of the QI constants in U_1* and U_4* shown above. For DD, their values in fraction of trace can be found in Figure 1.19. This is so simple because trace is the one and only parameter.

A summary of FPF and LPF envelopes in strain space can be seen in Figure 1.37, for T800/Cyt. On the left is the FPF omni envelope which is more of the matrix failure, and controlled by many plies as described in Figure 1.35. In the middle is the LPF omni envelope which is the fiber failure, and controlled by X and X' only, shown in circles with red core. Matrix controlled properties such Y, Y' and S are degraded or softened so they no longer matter in the LPF. The final omni FPF and LPF are seen in the right. To be conservative, the compressive strength is determined by the lower of Y' and X'. In most CFRP, Y' is much larger than X'.

We use failure criteria that are invariant from the scalar product of Tsai-Wu and facilitated with strength ratio R. They are much simpler to use and analytically correct than many failure criteria not being invariant. They have not transformation so failure criterion must be applied to the material axes. Cannot do it in laminate axes.

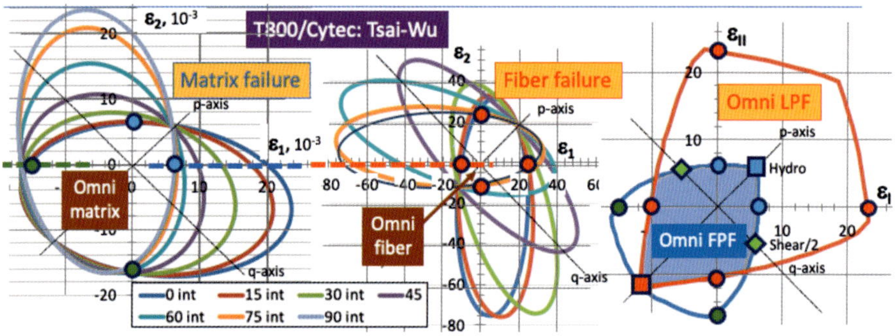

Figure 1.38 Summary of omni FPF and LPF for T800/Cyt

Shear strength for FPF are shown in the right figure in diamonds with olive green core. The extension of the q-axis to intercept of the omni LPF will be the ultimate shear failure strain. The FPF and LPF shear strengths of QI laminate will be the strain multiplied by the QI constants in U_1* and U_4*, shown in Figure 1.36. This shear strength determinations are for QI laminates only.

For quasi-isotropic [π/4] laminates in T700 C-Ply from Chomarat laid up and cured by Hexcel and tested by Alan Nettles of NASA Marshall Space Flight Center are shown in Figure 2.28. When coupons were cut with 22.5 degree rigid body rotation to form a DD in [±22.5/67.5], they had some strength as quad except the uniaxial tensile data shown in the upper right. It was 83 percent of the quad. As explained in reference 1 the failure mechanisms of quad and DD were different. Quad was a fiber tensile failure while the DD was a shear failure emendating from the edge of coupons. They cannot be compared directly. In other tests such as OHT and CAI, quad and DD have the same strength.

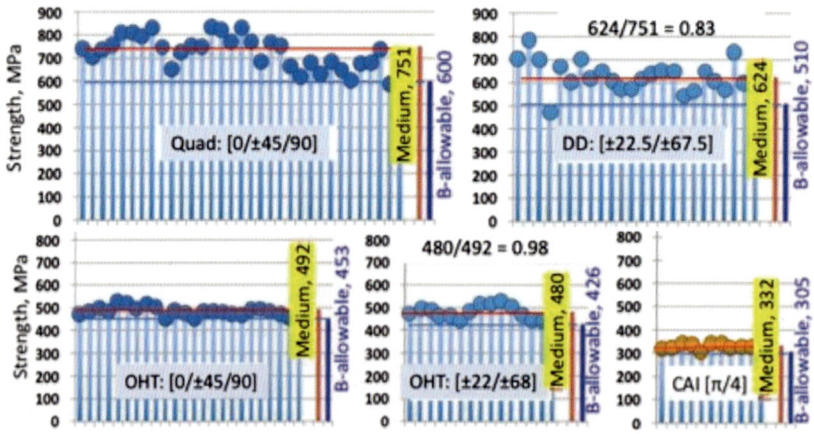

Figure 1.39 Comparison between quasi-isotropic quad and DD [3]

Additional data to compare hard and [±0/±50] are shown in Figure 1.40. They show nearly identical data for no-hole and open hole data between the two matching quad and DD. Waruna Seneviratne of NIAR provided the data. Note DD has 5 repeats while quad has none (with mid-plane symmetry only).

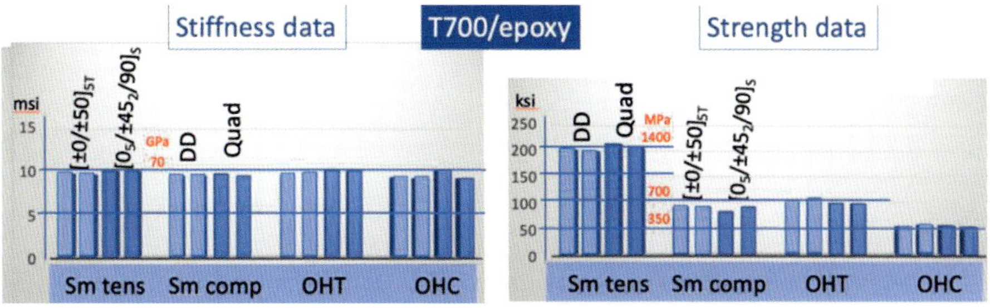

Figure 1.40 Comparable test data on hard quad and [±0/±50] [3]

Although we have covered this topic thorough previous publications and presentations, it is still important to bring out our approaches again that will simplify the process of design consistent with the theme of this book. The key features of our approach include:

1. Omni envelopes are the net domains for all ply orientations. It is thus independent of ply angles and becomes isotropic by definition. The principal strains can be used in place of regular strain. A 2-dimensional presentation is possible to cover all prevailing strains. No need for 3D. It is much simpler to work in a 2D space.
2. The degree of orthotropy of failure envelopes in strain space is much less than that in stress space. Do not have to deal with highly elongated failure envelopes that exist in stress space envelopes.
3. In strain space there is a symmetry axis along the hydrostatic tension and compression focal points where all ply angles must pass through.
4. For first-ply-failure, the omni envelope is controlled by all plies depending on the material; for ultimate failure it is by longitudinal tensile and compressive failure strains. Strains in the transverse directions and shear no longer matter after matrix degradation. Such simplification makes ultimate strength prediction simple.
5. We can convert omni principal strain envelopes to omni principal stress envelopes by using master ply isotropic constants in U_1, and U_4 for QI laminates, or the appropriate [A] for each laminate.

Omini envelopes are shown in Figure 1.38. They are the domains shared by all ply orientations from 0 to 90. FPF is based on plies with intact, pristine matrix is shown on the left. LPF is based on fiber tensile failure where matrix modulus are lowered to account for micro cracks is shown in the center. Fiber compressive failure is based on buckling. Its value is usually lower than the transverse compressive failure of [0] plies. It is therefore more conservative to use fiber buckling than transverse compression for the determination of the first-ply-failure. There are 4 anchors for each omni envelope shown in circles. They are the intercepts of the strain axes. Note the symmetry about the p-axis (along the 45-degree axis), as mentioned earlier.

1.5 Conclusions

The foundation of DD as we have presented can lead to a renaissance how composites are manufactured and designed. If the Handbook 17 is an indication of the conventional practice, many distinctions can be drawn.

First, the conventional quad of 0, ±45, 90 has been cited to be relaxed to 0, ±θ, 90. Is this a half way to DD? The new quad is still a collection of discrete laminates in 0 and 90. The number of plies to cover a wide range of properties will still have to be large. Making homogenization of this new quad will have the same degree of difficulty. A half way change does not reach a field-based family of laminates that DD offers.

Secondly, many recommended practices are philosophies with little or no scientific support. The stacking sequence of laminates consumed many paragraphs none of which had any model or data to show. To put ply orientations along the principal load direction was cited many times but difficult to do in practice. To place these loading bearing plies near the center of the laminate makes laminate more heterogeneous. In fact homogenization was mentioned but never realized for its ability to eliminate not only the stacking sequence but also the blending issues. The best solution to reach homogenization would be the use of DD. To treat laminates like parallel springs is just as unreliable.

Thirdly, allowable generation calls for many laminates, and many batches. There is no attempt to use trace to reduce the number of tests, and to recognize that homogenization can reduce interlaminar strength to test of one laminate. Variations caused by batches, environmental factors and stress raisers like open and filled holes can also be reduced to one laminate. For stiffness of laminates, there is only one constant in trace. Only one test is needed. Select the easiest one. Not the difficult ones like the shear and compression tests. For strength, there is one test for FPF that can be the transverse tensile in Y, and two for LPF in longitudinal tensile and compressive tests in X and X'. For interlaminar shear, there is one test (like short beam shear) on any homogenized laminate. Modeling helps to eliminate difficult tests. Bi-axial tests for example are critical but rarely used because of their complexity and lack of confidence. In Chapter 2 these issues will be discussed.

Finally carpet plots can now be replaced by Lam search or lamination parameter plots. In fact Lam search can supplement pages of derivations of laminated plate theory in the Handbook 17 and make it a search engine for composites engineers. The handbook does not recognized the potential of scaling to simplify design allowable generation. Scaling is made possible when laminate stiffness is recognized as the product of trace and its partitioning for given laminates. Similar scaling can be applied to the omni strength envelopes where its multipliers are not sensitive to ply materials. That is comparable to the master ply partitioning of laminate stiffness.

Manufacture and design of composite materials and structures can be simplified and strengthened by innovative solutions. Only DD offers such opportunities. The good news is that all laminates can readily converted to DD thereby receiving the benefits of better manufacturing and design. Not the same old, same old. The old quad is ready to be retired. Several cases have been presented here and other chapters of this book. We are in the beginning of a new venture. Many new discoveries can be expected. So will substantiation of claims to be forthcoming by test data or new solutions. It is with excitement that new approaches for the benefit of composites technology are emerging. It is hoped that engineers with wills to improve can find DD a good choice to learn, to understand, and to use.

References

[1] S.W. Tsai, S.D.D. Melo, Composite Materials Design and Testing (2015) Composites Design Group, Department of Aeronautics & Astronautics, Stanford University

[2] L. Jia, C. Zeng, J. Li, L Yao, C. Tang, "Validation and development of trace-based approach for composite laminates," *Composites Science and Technology* 221 (2022) 109348

[3] S.W. Tsai, "Double-double – a new family of laminates," *AIAA Journal*, Vol 39, No. 11, November 2021, 4293-4305

[4] S.W. Tsai, N.J. Pagano, "Invariants properties of composite materials," Composite Materials Workshop (1968) 223-253, Technomic

[5] S.W. Tsai, H.T. Hahn, *Introduction to composite materials* (1980) Technomic

[6] M. Miki, "A graphical method of design of laminated fibre composite plates with required in-plane stiffness," *Trans JSCM* 9, 2, (1983) 51-55

[7] M. Sprengholz, H. Traub, M. Sinapius, "Rapid transformation of lamination parameters into stacking sequences," *Composite Structures* (2021) 114514

[8] K. von Hagel, S. Joglekar, M. Pankow and S. Ferguson, "Explroing robust composite panels using multi-objective optimization and varying load conditions," *AIAA Aviation*, 15th AIAA/ISSMO Multidisciplinary Analysis and Optimization Conference, June 2014, Atlanta, GA.

[9] F-X Irisarri, D.H. Bassir, N. Carrere, J-F Maire, "Multiobjective stacking sequence optimization for laminated composite structures," *Composites Science and Technology* 69 (2009) 983-990,

[10] L-M Tuan, J. Lee, "Stacking sequence optimization for maximum strengths of laminated composite plates using generic algorithm and isogeometric analysis," *Compsoite Structures* (2014) COST 5692

[11] Y.M. Meddaikar, F-X Irisarri, M.M. Abdlia, "Laminate optimization of blended composite structures using a modified Shepard's method and stacking sequence tables," 11th World Congress on Structural and Multidisciplinary Optimization, Sydney, June 2016, Springer-Verlag

[12] P.S. de Almeida, "stacking sequence optimization for maximum buckling load of composite plates using harmony search algorithm," *Composite Structures* 143 (2016) 287-299

[13]. S. W. Tsai, JDD Melo, "A unit circle failure criterion for fiber reinforced polymer composites," *Composite Science Technology* 123 (2016) 71-78

With Ali Aravand at Paris Air Show, June 2019

CHAPTER 2

Master Failure Envelopes for Composite Laminates

Stephen W. Tsai, Stanford University
stsai@stanford.edu

It is commonly assumed that each composite material with different laminates will have unique failure envelopes. With omni failure envelopes it is simple to discover master envelopes to predict failure of each carbon composite material with all their laminates. To our surprise the rating of laminates from master envelopes remains the same for all materials. Each laminate can be made from multiple materials, and their rating is the same for all laminates. Materials and laminates are equivalent, interchangeable, and the same but from different viewpoints. For the ultimate strength of laminates, only uniaxial tension and compression of [0] are needed. Such fixed relations make design and testing of composites as simple as those for metals. With master envelopes there is no need to test every material from A to Z. The simplicity derived from master envelopes is analogous to one unique parameter in trace for each composite to define their stiffness of all laminates. Stiffness and strength measured from one [0/90] laminate can establish its material to be rated against other materials and all their laminates. Being field-based double-double laminate family is unique in capturing all these simplifying features, not possible by the quad laminates conceived at the dawn of structural composites more than 60 years ago.

2.1 Omni Envelope

Design of composite materials has long been assumed to be more complicated than metals. We have believed for some time that much of the complication is self-inflicted. Not enough effort has been explicitly directed to simplify the design, manufacturing and testing. Trace, double-double laminates and many steps that followed were our effort for this simplification. Trace reduced the number of parameters for stiffness to one from the current practice of four. Having one parameter is critically important because it can compare and rate stiffness between materials and laminates. With more than one parameter direct comparison is no longer simple. Double-double (DD) laminates bring another level of simplicity for being a field-based family as opposed to the quad and other laminates which are dispersed and unconnected. The master failure envelopes to be presented herein are another step in our effort to simplification.

Failure envelopes are commonly defined by the weakest ply in a laminate. In Figure 2.1 the first-ply failure (FPF) envelopes are controlled by either [0] or [90] plies as seen in the first envelope on the left. When others are added they change the laminate stiffness but not the FPF. In the last envelope, it is the omni envelope that cover all the plies in double-double (DD) plies between [0] and [90] at 15-degree angle increments such as [±15], [±30], [±45], [±60], [±75]. This last envelope is called omni envelope because it is the inner most of all ply orientations. It is no longer ply angle dependent. It is therefore isotropic. General strain can then be converted to principal strain, and shear strain is equal to equal tension and compression. The envelope can be viewed as a material property, independent of ply orientations and ply composition.

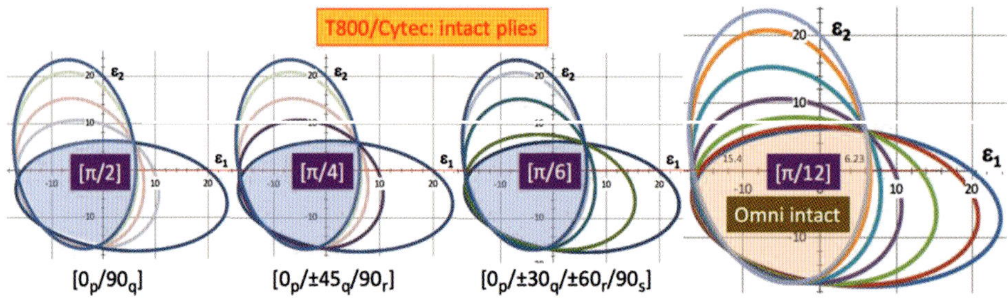

Figure 2.1 First-ply-failure envelopes of quad and double-double laminates

Figure 2.2 shows the small difference between quad and DD. Thus if the two laminates have the same stiffness, their FPF will be identical because the resulting strain from any given applied stress would be the same.

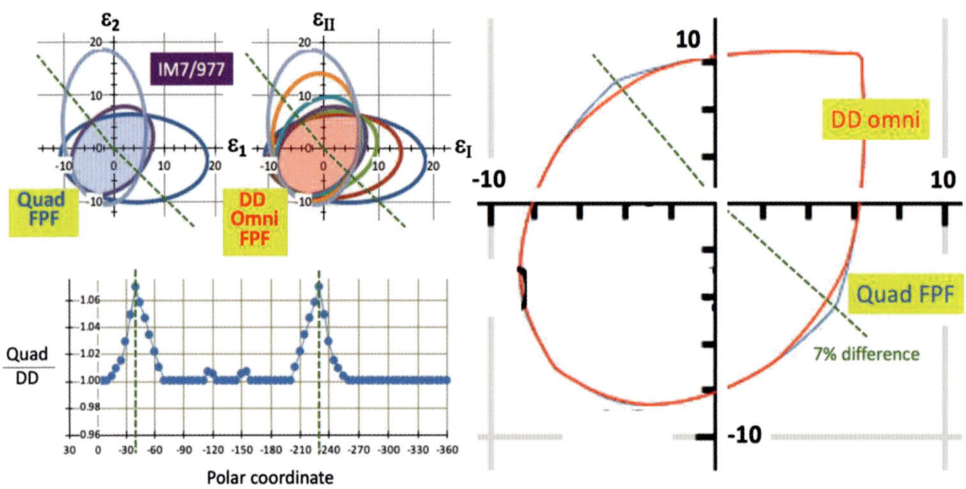

Figure 2.2 Close comparison of FPF between quad and DD for IM7/977

A more detail comparison between quad and DD in Figure 2.3 shows that DD can be more conservative as seen in the last failure envelope on the right. The example shows the actual FPF for [±30/±45] is larger (higher strength) than the omni FPF in the middle. As mentioned earlier, the omni FPF is kept because omni laminates are ply angle independent, therefore, isotropic. The general strain can then be converted principal strain. A feature of DD that can be described by omni envelope. The legacy quad laminates have an inner envelope almost identical to that of the omni envelope but is isotropic only when the number of plies in 0, ±45 and 90 are equal. When they are not equal which is the normal case, the laminate is not isotropic. Conversion from general to principal strain will no longer be possible. Simplification requires not only a giant step in trace but also small steps in the unique omni envelope for DD.

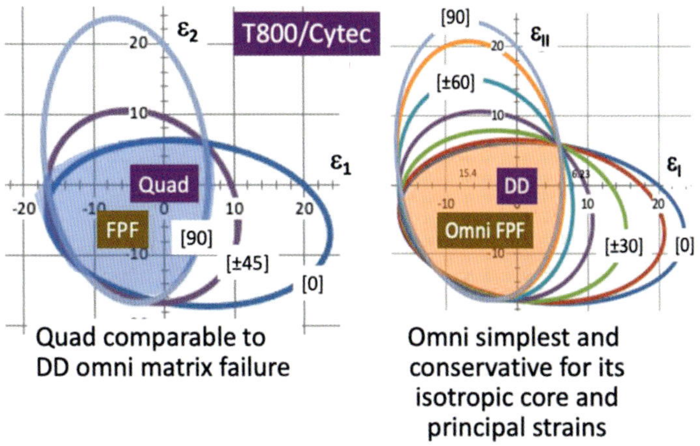

Figure 2.3 Comparison between quad, DD omni, and DD actual envelopes

The same approach can be applied to last-ply-failure (LPF). This is commonly done by reducing effective stiffness in the transverse direction of [0] to simulate the presence of micro cracks. Matrix becomes a rubber-like material with large failure strain. It forces failure from matrix to that of fiber. Figure 2.4 shows the case of LPF that is analogous to Figure 2.3 for FPF. The controlling plies are [0] and [90], like that for FPF. The difference is that for LPF X and X' (uniaxial longitudinal tensile and compressive strengths in blue and green circles, respectively) are the only strength data needed. For FPF, Y and Y' (transverse tensile, and compressive strength or X' whichever is lower) are needed. Having an isotropic core, shear strength (in circle with yellow core) can be estimated for DD laminates.

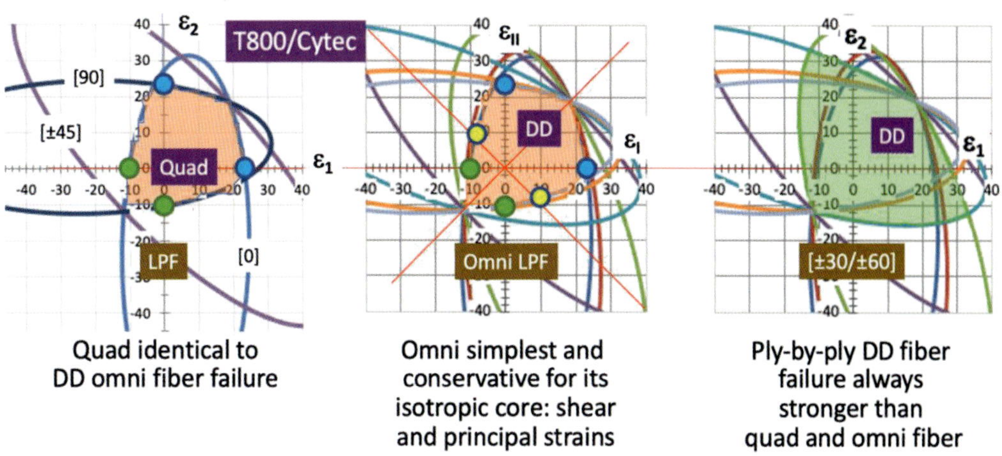

Figure 2.4 Last-ply-failure envelopes of quad, DD omni, and actual [±30/±60]

This fiber failure model is like the netting analysis for filament-wound pressure vessels to a laminate in general. With master envelopes that we are to propose in this work they establish relations between envelopes of all laminates with [0/90]. Testing of all remaining laminates (after [0/90]) should be for validation and calibration, not for data generation (from A to Z) as it is practiced now. Such tests are much simpler to do and should be sufficiently reliable if their values are predictable from the master envelopes. As more and more data become available, fewer and fewer tests should follow. This is so if master envelopes are realized and considered. Comparison among the FPF and LPF omni envelopes for some common CFRP's can be seen in Figure 2.5. The striking similarity among them provides the path to master envelopes. The FPF omni envelope can be controlled by any plies while the LPF is by [0] and [90] only. Thus, LPF is simpler to determine analytically as well as experimentally. It can be validated by uniaxial testing. FPF on the other hand is difficult to pinpoint experimentally because it is a measure of matrix failure which is difficult to observe. Shear strengths can be estimated with circles with yellow core in LPF envelopes because the omni envelopes are isotropic. In fact, it can be possible to use tensile tests only and avoid the difficult compression and shear tests.

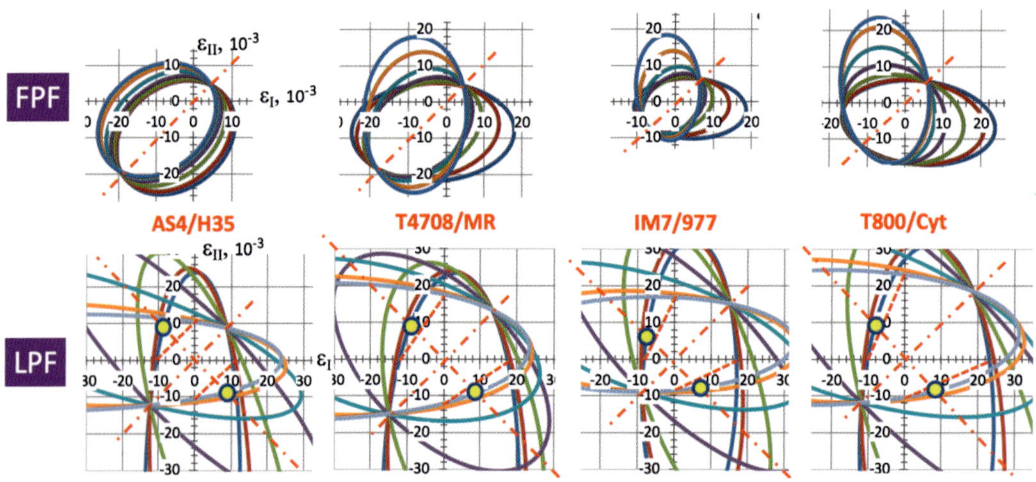

Figure 2.5 FPF and LPF of omni envelopes in strain space

More examples of omni LPF envelopes for representative CFRP are shown in Figure 2.6 which gives the LPF dimensions and shapes of various CFRP materials. It also shows the range of failure envelopes from the smallest to the largest among the most common CFRP materials. The maroon lines are Tsai-Wu criterion, while the blue lines are max strain criterion.

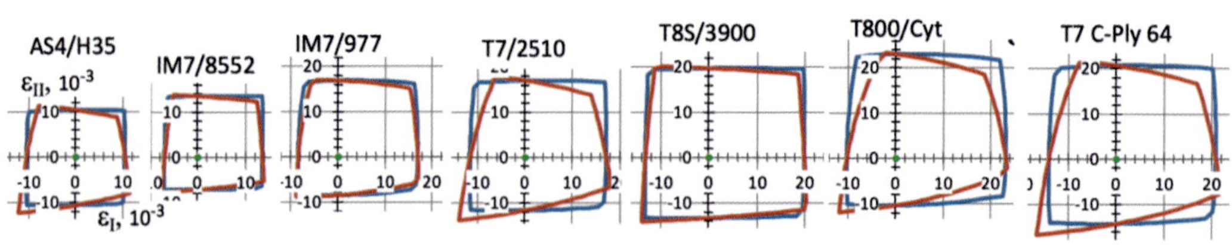

Figure 2.6 Representative LPF envelopes based on Tsai-Wu (maroon) and max strain (blue)

2.2 Radius of equivalent circles

We have found that the area of failure envelopes to be an important parameter that not only can represent the strength of a material and/or laminate, can also provide a parameter that can produce the master envelopes. It is a parameter or the average or total strength analogous to trace for stiffness that is the one and only variable that can represent the stiffness of a composite material.

One way to measure the area of the failure envelope is shown in Figure 2.7 for both stress- and strain-space envelopes. The radius of an equivalent circle having the same area as the failure envelope is shown in red in the lower left of this figure. This work was suggested by Antonio Miravete.

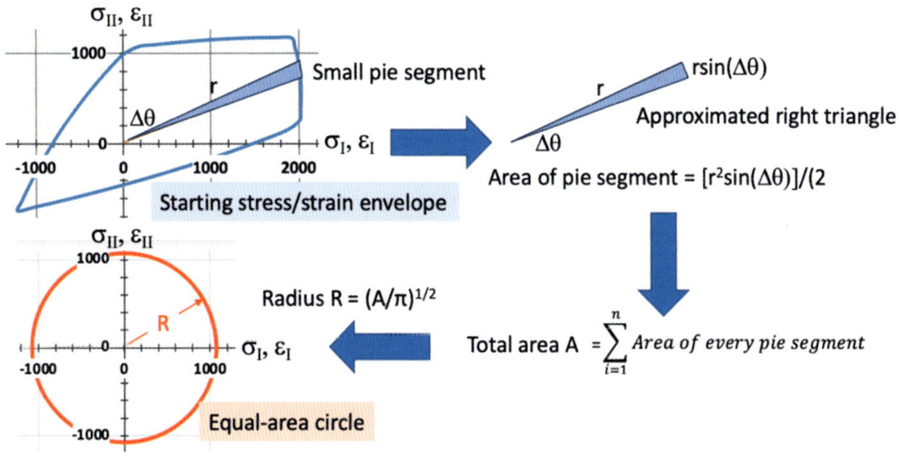

Figure 2.7 Determination of the radius of equivalent circle of failure envelopes

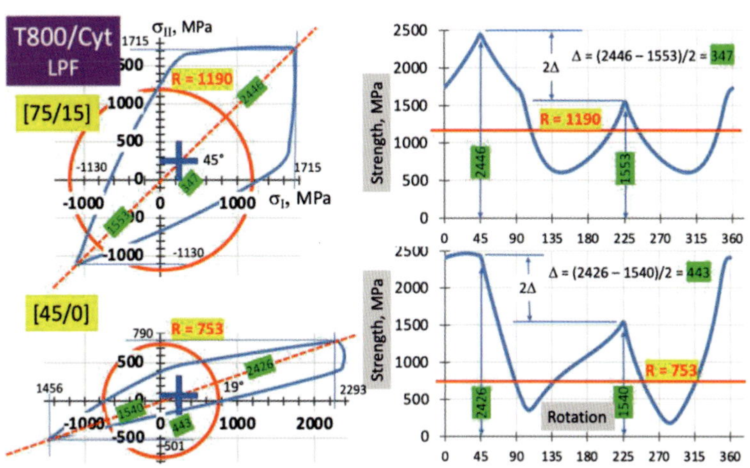

Figure 2.8 Geometric dimensions of failure envelopes

Each failure envelope, like those shown in Figure 2.5, can be normalized by its radius R in Figure 2.8. The equivalent circle with the same area as the failure envelope is in red, as it is in Figure 2.7. This radius is the average strength as well as a measure of total strength potential of the laminate. Other useful dimensions of failure envelope related to the radius include the length and inclination of the major diameter, and the center of the failure envelopes (shown in crosses) can also be seen in this figure.

In Figure 2.9, the evolution of failure envelopes from their original omni shapes to the master envelopes are shown. Note that the random appearance of the original omni envelopes in the first column for both FPF and LPF become more orderly as seen in the second column. This is done by the normalization of the original envelope by the radius. Suddenly the random appearance in the first column can be averaged to create the master envelopes, shown in the 3rd and 4th columns with dimensionless and 10^{-3} units, respectively. This master envelope in strain space can be used to recover all 15 original CFRP envelopes and all the laminates for each material. The implication is that all failure envelopes are connected and have definite structures. They are not random variables. Their shapes and sizes are all related. The generation of design allowable can be simplified and be selected with an understanding of how strength of laminates is controlled by material and geometric parameters. In fact, it can be shown that the only strength data needed are X and X' (uniaxial tensile and compressive strengths). After matrix degradation Y, Y' and S have no effect in the LPF strength of laminates.

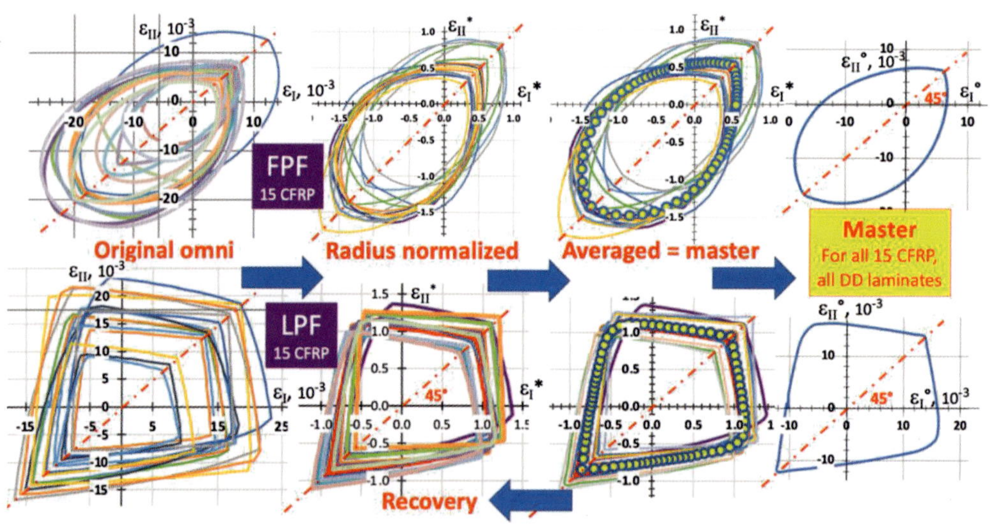

Figure 2.9 Master envelopes derived from the average of radius normalized envelopes

Normalization of the radius or another parameter like the major diameter can facilitate the process of finding master envelopes. The rigid structure of strength of material and their laminates can be found. Simplification of measuring strength of composites is now possible because their values can be predicted. They are not totally unknown. This is like what trace has simplified the stiffness, as seen in Figure 2.10. It is a single parameter for stiffness, no need for 4 for traditional orthotropic plies. Typical values are shown in the table on the left. Laminate components can be viewed as how trace is partitioned. It is separate and distinct from the material stiffness property and can be seen on how pies are cut for various laminates on the right. The percentages of trace are shown on each pie. For [0/90] on the left $A_{11} = A_{22} = 47$ percent of trace while A_{66} is 3 percent. That partitioning is [0/90] for all materials. So, trace is a master constant for stiffness, while radius is one measure of the masters for strength. They are analogous. With one parameter, it is easy to compare relative stiffness and strength of materials and laminates. They can be rated and scaled. It is worth repeating that if there is more than one parameter for a property, such simple rating and scaling eould no longer be possible.

Ply material	Trace
IM6/epoxy	232
IM7/977-3	218
T300/5208	206
IM7/MTM45	195
T800/Cytec	183
IM7/8552	180
T800S/3900	168
T300/F934	168
T700 C-Ply 64	163
AS4/H3501	162
T650/epoxy	160
T4708/MR60H	158
T700/2510	144
AS4/MTM45	143
T700 C-Ply 55	139

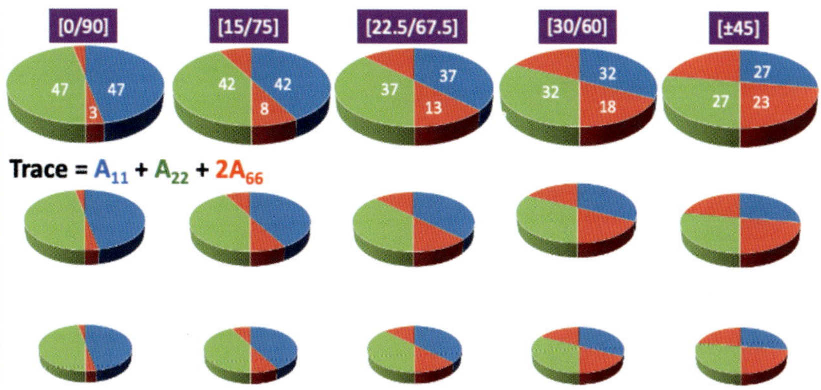

Figure 2.10 Trace on the left, and its fixed partitioning to create laminates on the right

2.3 Master envelopes

Master envelopes offer many features that make design, manufacturing, and testing of composites easier to understand and simpler to use. They will be explained as this chapter proceeds. These envelopes for different CFRP and laminates have been viewed as random variables having their own characteristics. Each combination of material/laminate is thought to need its allowable with no exception. Such prevailing view has in fact prevented many materials and processes of manufacturing from use because of the lack of design allowable from high cost and long period of time for its generation.

Master envelope can change such restrictive views. All envelopes are members of a family that is structured and rigidly connected. There are two critical features used in the development of master envelopes: 1) the omni envelopes in Figures 2.5; 2) the radius of the equivalent circle in Figure 2.8. The omni envelope for LPF depends on X and X' only. Other ply strengths like Y, Y' and S do not matter when the effective matrix stiffness is reduced to a small fraction of the intact state. As mentioned earlier, [0/90] will be the best laminate to measure X and X'. Quad and DD will have the same strength if their values of trace are equal. Then their laminate stiffness will be the same. Omni envelopes are ply orientation independent, thus is isotropic. General strain and stress can be converted to principal strain and stress. This is an important simplification for DD.

The second critical feature is the radius as a measure of strength. With it to normalize the original envelope as illustrated between the first two columns on the left in Figure 2.9, the randomness appearance in the first column is converted to more orderly stack of envelopes representing different CFRP. Their geometric shapes and sizes are now more similar. It makes finding their average to be used as the master look reasonable and easily accomplished.

In Figure 2.11, the radius-normalized failure envelopes for 15 CFRP in strain space are shown. As expected, their shapes and sizes are similar and comparable. The only difference is their x-axis intercepts in positive and negative directions (for tensile and compressive strengths) as listed in the top row table in this figure, in ascending order. The 15 CFRP can be grouped into 3 for low, 8 for medium and 4 for high categories. These 3 master envelopes are shown superposed in the lower right corner of this figure. Splitting into 3 masters is intended to increase accuracy of modeling strength of composites. The original envelopes in the top row are recovered, as seen in olive green dots. They are almost with 100 percent matching. The only difference between the original omni and the recovered envelopes are at the corners of the envelopes in the 2nd and 4th quadrants. The original has sharp corners while the recovered ones from master envelopes are rounded.

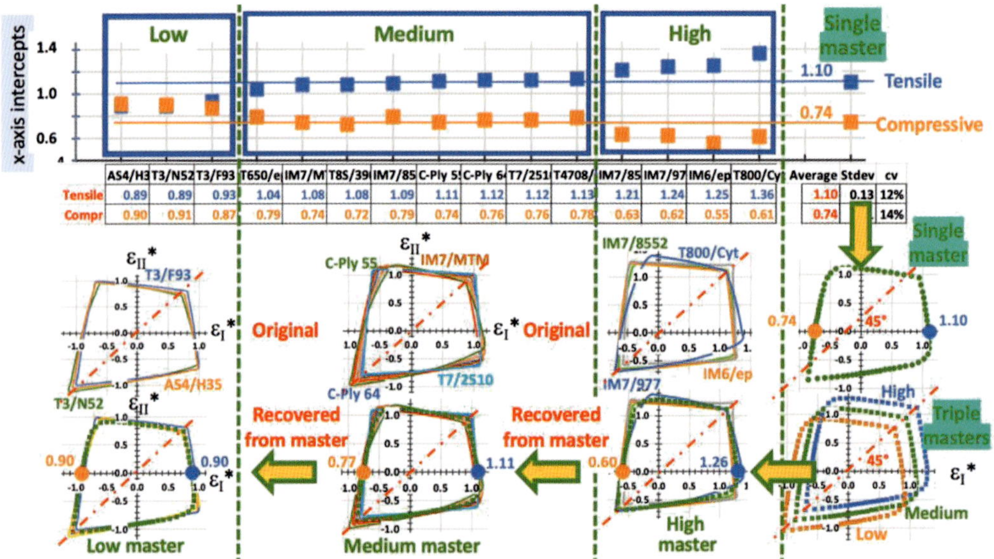

	AS4/H3	T3/N52	T3/F93	T650/e	IM7/M	T8S/39	IM7/85	C-Ply 5	C-Ply 6	T7/251	T4708/	IM7/85	IM7/97	IM6/ep	T800/Cy	Average	Stdev	cv
Tensile	0.89	0.89	0.93	1.04	1.08	1.08	1.09	1.11	1.12	1.12	1.13	1.21	1.24	1.25	1.36	1.10	0.13	12%
Compr	0.90	0.91	0.87	0.79	0.74	0.72	0.79	0.74	0.76	0.76	0.78	0.63	0.62	0.55	0.61	0.74		14%

Figure 2.11 Recovering original omni envelopes from master envelopes

It is remarkable that the envelopes in strain space do not vary much among the 15 CFRP. Master envelope or their 3 sub-groups are easy to determine, credit goes to the discovery of the radius used in the normalization. Credit also goes to the strain space failure envelopes, each of which contain all possible ply orientations from for all CFRP materials. These are omni envelopes as explained in Figure 1.35, and shown again in Figures 2.5 and 2.6. As we will see, such simplicity does not exist for failure envelopes in stress space. Each laminate must be evaluated individually.

As a reconfirmation of the triple masters in Figure 2.12, three materials in T800/Cyt as the high master, T4708/MR as the medium master, and T3/F97 as the low master. With 3 laminates in [60/30]. [60/0] and [90/0], the envelopes of original materials can be recovered from their respective master sub-groups. The original is shown in solid lines while the recovered are shown in red dots. Their respective agreements are necessary so the generation of master envelopes, and recovery from them to the original baseline laminates are reliable, and accurate. More examples of the recovery process will be seen in Figures 2.28, 2.32 and 2.33. The envelopes are in stress-space in MPa.

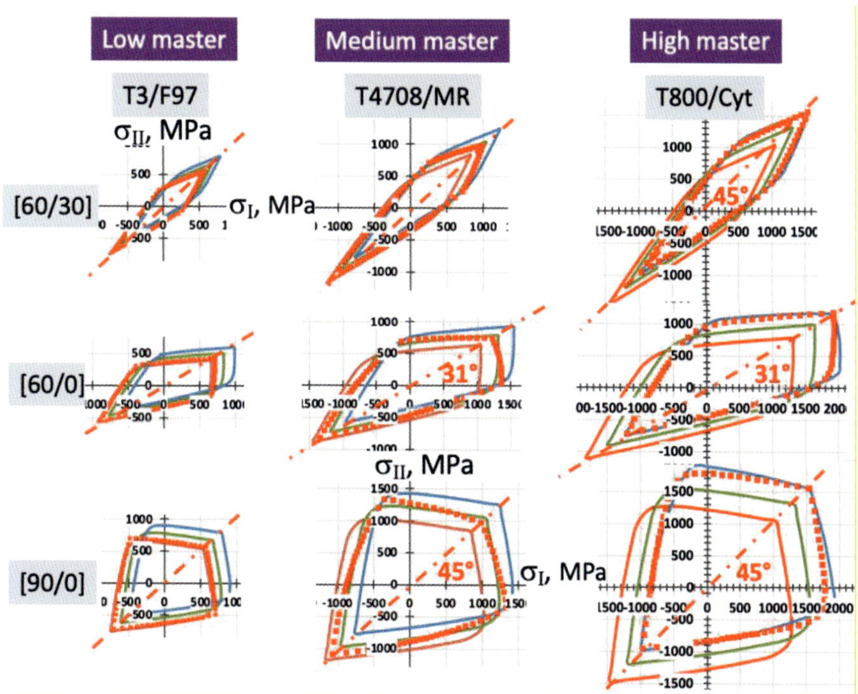

Figure 2.12 Recovery of respective envelopes from 3 representative materials and laminates

The use of [0/90] for the measurement of X and X' and failure strains e/x and e/x' is the simplest because the Poisson's ratio of this laminate is essentially zero, as seen in the horizontal line in Figure 2.13. Then $A_{11} = E_1$, and uniaxial stress along the x-axis in stress space resulting in uniaxial strain in strain space. Any other laminates will be more complicated because their Poisson's ratios are not zero. Lower load is required to reach the ultimate of [0/90] than the one for [0]. High confidence in back calculate the ply stress of [0] at failure from [0/90] is well accepted. Having near zero coupon makes gripping simpler and easier to perform the test. Quad laminates in 0, ±45 and 90 are complicated so many simplifications to be cited herein are not applicable. The fundamental difference is that quad is a collection of discrete laminates while double-double like [±75/±15] is field-based that can operate and interpolate following established rules of transformation, invariants, symmetry, and the number of independent laminates. With DD they are rational and well established. They provide simplification, utility, and confidence. (Our Poisson ratio index 21 is different from many others who use 12. Our notation is consistent with the matrix convention with index 21 to mean strain in the 2-axis from load applied in the 1-axis.)

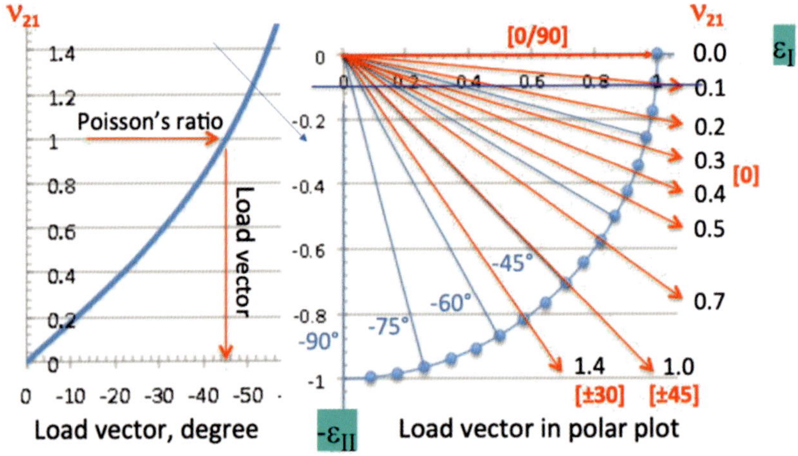

Figure 13. Poisson's strain from uniaxial tensile loading

2.4 Conversion to stress space

The emphasis of this paper is on the LPF (last-ply-failure) which can be validated experimentally. FPF (first-ply-failure) is not easy to observe or measure and will be cited in this work occasionally to gain some perspective of the failure process. The important message here is the number of strength parameters that is needed to generalize to the entire stress space: X and X'.

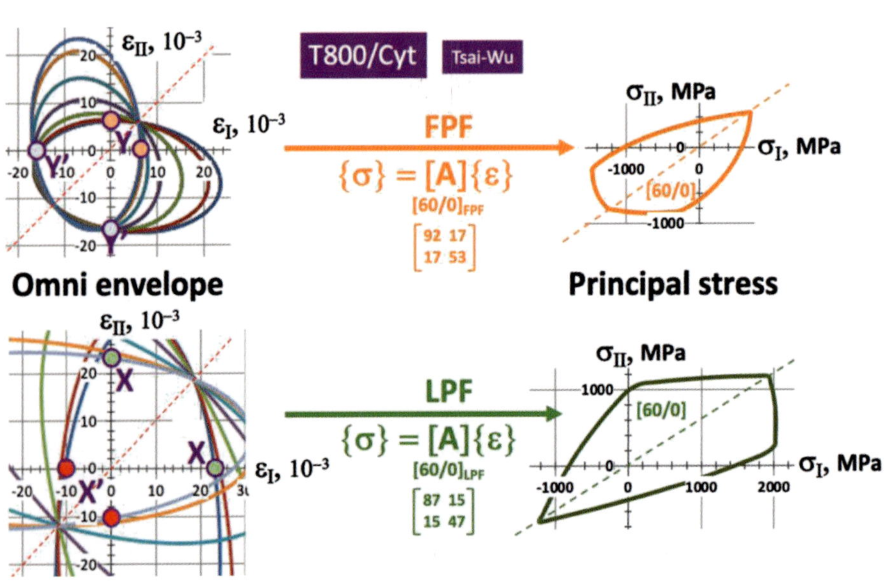

Figure 2.14 Conversion of strain to stress space envelopes

It is straightforward to find any failure envelope including the master envelope in stress space. It is simply a stiffness-multiplied strain envelope. This is explained in Figure 1.36, and shown in Figure 2.14 for T800/Cyt and [60/0] for both FPF on top, and LPF at the bottom. On the left is the omni envelopes with the failure strains of uniaxial test data listed. The process shown in the arrows are the stress-strain relation with the stiffness components A_{11}, A_{22} and A_{12} in GPa shown in brackets. For FPF, the stiffness is for intact matrix; for LPF, the degraded matrix simulating micro cracks in all plies. The resulting envelopes in stress space, in MPa, are shown on the right. This is a generic process that converts all envelopes from strain space to stress space. Notice the difference in the laminate stiffness of the intact and degraded laminates that are associated with the FPF and LPF, in [95, 53, 17], and [87, 47, 15], respectively.

2.5 Rating for stress space envelopes

The radius of the same-area circle in stress space envelopes can be obtained by the same method shown in Figure 2.14. Their values in MPa for 15 materials and 14 DD laminates are shown in Figure 2.15. The average values of each column and row would be good criteria for the respective rating for material and laminate. Their standard deviation and coefficients of variations are also shown with values of 56 and 19 percent, respectively.

		1	2	3	4	5	6	7	8	9	10	11	12	13	14	15	Laminate		
	Radius	AS4/H3	IM6/ep	T3/F93	T3/N5	C-Ply 5	C-Ply6	IM7/9	T800/	IM7/8	IM7/N	T7/25	T650/	IM7/8	T4708	T8S/3900	rating		cv
1	[75/15]	708	1158	623	753	1007	1168	1155	1226	849	1021	861	934	1015	984	1222	979	190	19%
2	[75/0]	760	1246	670	809	1083	1255	1243	1319	913	1098	926	1004	1092	1058	1314	1053	205	19%
3	[60/0]	619	1016	544	658	881	1023	1011	1076	742	891	754	816	886	861	1066	856	167	20%
4	[60/15]	567	929	497	602	806	936	924	984	678	814	690	746	810	787	974	783	153	19%
5	[60/30]	429	693	373	453	604	704	691	735	505	608	517	560	606	588	729	586	113	19%
6	[45/0]	434	708	377	458	613	716	702	753	513	614	526	567	613	598	737	595	116	20%
7	[45/15]	383	623	332	404	540	632	618	663	450	539	464	499	539	527	648	524	102	20%
8	[90/0]	812	1331	717	864	1158	1341	1328	1409	977	1174	990	1073	1167	1131	1405	1125	219	19%
9	[45/30]	256	403	215	267	354	419	400	430	288	346	304	326	349	341	416	341	66	19%
10	[±45]	160	219	121	160	202	249	217	239	149	184	175	186	190	187	225	191	35	18%
11	[30/0]	274	427	230	284	377	447	422	461	304	365	326	348	370	364	444	363	70	19%
12	[30/15]	235	359	194	241	319	381	353	390	253	304	276	294	309	306	371	306	59	19%
13	[±30]	171	242	131	170	220	271	237	266	163	197	193	202	206	206	245	208	40	19%
14	[±15]	196	279	155	197	258	315	273	312	191	230	227	237	241	243	290	243	46	19%
	Material															Avergae	582		
	rating	429	688	370	451	602	704	684	733	498	599	516	557	600	584	720	582		
	Stdev	228.1	388.3	207.9	247	333	381	389	407	289	348	283	310	342	329	412			
	cv	53%	56%	56%	55%	55%	54%	57%	56%	58%	58%	55%	56%	57%	56%	57%			

Figure 2.15. Rating of failure envelopes for 15 CFRP and 14 DD laminates

The material/laminate combinations that are highlighted with yellow background in 3 materials and 4 laminates in Figure 2.15 are examples used in Figure 2.17.

The profiles of CFRP materials in columns and DD laminates in rows are shown in Figure 2.16, and their repeated patterns are clearly visible. They give rise to master envelopes that will be the theme of this work. As a remarkable indication it can be seen that the structure of materials columns on the left is as repetitive and similar as that of laminate columns on the right. Each has its own variations. Within each material, the variations of laminates are the same. Thus, independent of materials. Conversely the variation of materials within each laminate is also the same, thus independent of laminates.

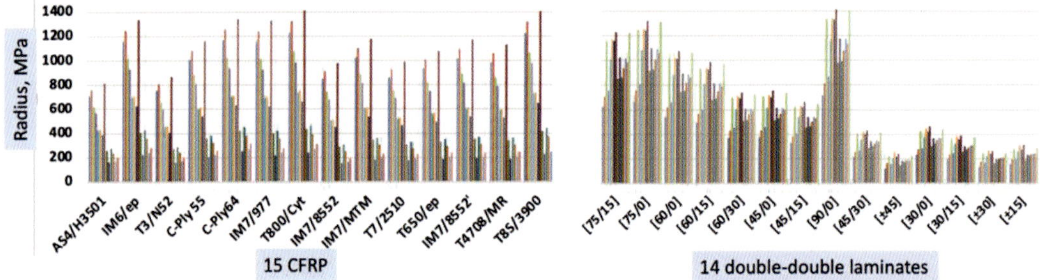

Figure 2.16 Profiles of materials and laminates of radii in columns and rows in Figure 2.15

Failure envelopes and their radii of equal-area circles are plotted in Figure 2.17, with radius values shown next to each envelope. Three ply materials are the lower bound T3/N52, medium or master T4708/MR, and upper bound T800/Cyt. Four DD laminates are [30/15], [60/30], [60/0] and [90/0]. Note materials and laminates are rated by single parameters (radii of same-area circles) highly configured. They are not random. This is the new concept that makes master envelopes possible, and useful. The red circles increase their radius linearly across materials vertically, from top to bottom, and across laminates horizontally, from left to right. The failure envelopes also follow definite patterns. Materials change proportionally along their vertical lines. The uniaxial tensile and compressive strength of DD laminates would follow definite patterns for each laminate, to be seen in Figure 2.23.

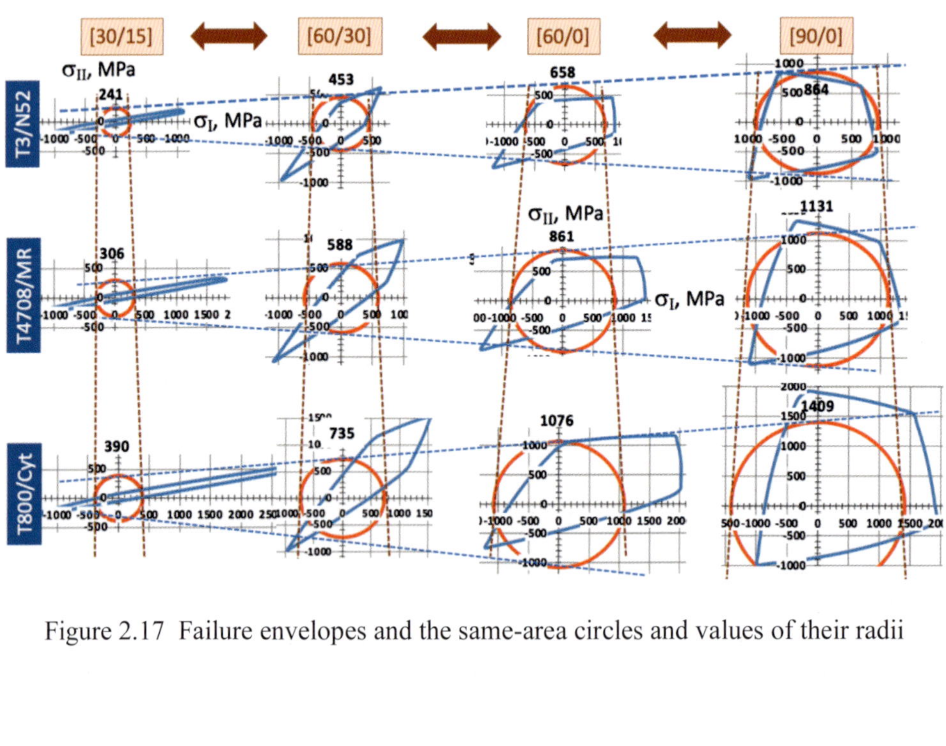

Figure 2.17 Failure envelopes and the same-area circles and values of their radii

Such simple relations make design data generation much simpler to implement and with higher confidence as cross checking between laminates and materials are totally predictable. Data generation for materials and laminates are based on calibration, not some exercise of blind testing with not much expectation what the data would be. With master envelopes, the ratios of strength data between two materials, say, 2:1, remains the same for all laminates. By the same token, the ratio between the strength data of 2 laminates, say, 3:1, remains the same for all materials. That is why allowable generation can change to that of validation/calibration from blind, point testing.

The radii in Figure 2.15 can be normalized by their master values. In Figure 2.18, normalization by row for each material is presented. The first material in AS4/H53 in the first row with value of 708 is normalized by the laminate rating shown in the column on the right in 979 led to the value of 708/979 = 0.72. in the top of the second column in Figure 2.18. The reason for the change in the column number is that the columns in this table are ranked in ascending order based on the material rating shown in the bottom row. These are material ratings/rankings for each CFRP. Note that with this normalization, the coefficients of variations are reduced from 56 percent in the bottom row of Figure 2.15 to at most 5.4 percent, shown in the last bottom row of Figure 2.18. They can be used for scaling as an important function of master envelopes. Note also their values are independent of laminates; i.e., having essentially the same value for all laminates.

		3	1	4	9	11	12	14	10	13	5	7	2	15	6	8
	Radius*	T3/F9:	AS4/H	T3/N5	IM7/8	T7/25	T650/	T4708	IM7/N	IM7/8	C-Ply 5	IM7/9	IM6/e	T8S/3	C-Ply6	T800/Cyt
1	[75/15]	0.64	0.72	0.77	0.87	0.88	0.95	1.01	1.04	1.04	1.03	1.18	1.18	1.25	1.19	1.25
2	[75/0]	0.64	0.72	0.77	0.87	0.88	0.95	1.01	1.04	1.04	1.03	1.18	1.18	1.25	1.19	1.25
3	[60/0]	0.64	0.72	0.77	0.87	0.88	0.95	1.01	1.04	1.04	1.03	1.18	1.19	1.24	1.19	1.26
4	[60/15]	0.63	0.72	0.77	0.87	0.88	0.95	1.01	1.04	1.03	1.03	1.18	1.19	1.24	1.20	1.26
5	[60/30]	0.64	0.73	0.77	0.86	0.88	0.96	1.00	1.04	1.03	1.03	1.18	1.18	1.24	1.20	1.25
6	[45/0]	0.63	0.73	0.77	0.86	0.88	0.95	1.01	1.03	1.03	1.03	1.18	1.19	1.24	1.20	1.26
7	[45/15]	0.63	0.73	0.77	0.86	0.88	0.95	1.01	1.03	1.03	1.03	1.18	1.19	1.24	1.21	1.27
8	[90/0]	0.64	0.72	0.77	0.87	0.88	0.95	1.01	1.04	1.04	1.03	1.18	1.18	1.25	1.19	1.25
9	[45/30]	0.63	0.75	0.78	0.85	0.89	0.96	1.00	1.01	1.02	1.04	1.17	1.18	1.22	1.23	1.26
10	[±45]	0.64	0.84	0.84	0.78	0.92	0.98	0.98	0.96	1.00	1.06	1.14	1.15	1.18	1.30	1.25
11	[30/0]	0.63	0.75	0.78	0.84	0.90	0.96	1.00	1.01	1.02	1.04	1.16	1.18	1.22	1.23	1.27
12	[30/15]	0.63	0.77	0.79	0.83	0.90	0.96	1.00	0.99	1.01	1.04	1.16	1.17	1.21	1.25	1.28
13	[±30]	0.63	0.82	0.82	0.78	0.93	0.97	0.99	0.95	0.99	1.06	1.14	1.16	1.18	1.30	1.28
14	[±15]	0.64	0.81	0.81	0.79	0.93	0.98	1.00	0.95	0.99	1.06	1.12	1.15	1.19	1.30	1.28
	Material rating	0.63	0.75	0.78	0.84	0.89	0.96	1.00	1.01	1.02	1.04	1.17	1.18	1.23	1.23	1.26
	Stdev	0.00	0.04	0.02	0.03	0.02	0.01	0.01	0.04	0.02	0.01	0.02	0.01	0.03	0.04	0.01
	cv	0.4%	5.4%	2.8%	4.0%	2.1%	0.9%	0.8%	3.6%	1.7%	1.2%	1.7%	1.2%	2.1%	3.5%	0.9%

Figure 2.18 Rating of normalized radii R* for each material

The material/laminate combinations that are highlighted with yellow background in 3 materials and 4 laminates in Figure 2.18 are examples used as material rating in Figure 2.23.

Normalization of radii in Figure 2.19 converted the starting values shown on the left of Figure 2.16, repeated herein, revealed remarkable relation between material and laminate. Relative strength of 15 CFRP are rated and ranked and are not sensitive to lamination; all normalized laminates are nearly equal. The highest strength material in T800/Cyt is 1.26 higher than the master while the lowest strength in T3/F93 is 0.63 of the master. If a structure has one material and can change to another these material factor/rating gives the precise change in strength. If T800/Cyt is to replace T4708/MR, the strength will increase by a factor of 1.26. No more, no less. There is no need for recalculation. The same ratio applied to all laminates; i.e., whatever laminate the original material has, the strength ratio is 1.26 when the material is changed. This is good because the first component may have more than one laminate, when it is changed to the second material, the strength change is 1.26 for all laminates. Again, no recalculation is needed. There is more to this material/laminate equivalence as we will see presently.

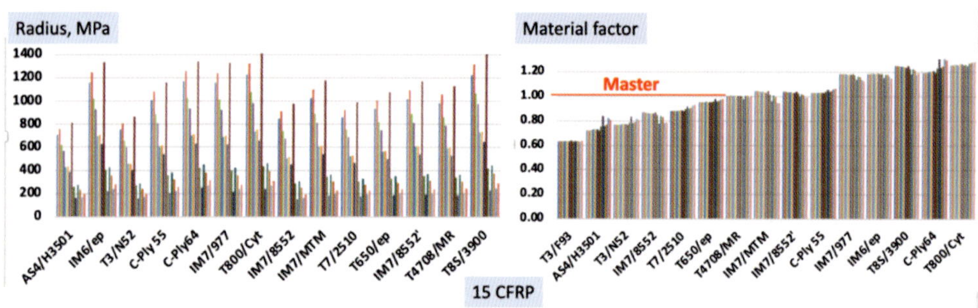

Figure 2.19 Comparison between absolute and normalized radii for 15 CFRP with 14 DD

Similarly to the material rating in Figure 2.18, it is possible to derive the laminate rating from normalized radii. This is seen in Figure 2.20 where radii of envelopes are normalized; e.g., in the top row and column the radius value is 708 in Figure 2.15. When it is normalized by the material factor in the bottom row with value 429: 708/429 = 1.65. This normalized radius is found in the second column in the table in Figure 20 for laminate [75/15]. On the right of this figure are 4 columns. The first column is the laminate rating, in ascending order. That is followed by the standard deviation, coefficient of variations, and relative radius based on [90/0] laminate.

The rating of materials and laminates (to be seen in Figure 2.21) is not currently considered. It has been assumed that each material/laminate combination is unique and independent from other combinations. It is an important realization in our work that material and laminate interact and connected by fixed rules of a field. Our new view is that material and laminate are totally connected and they are in fact two faces of the same entity. Each material and each laminate have a rating that is fixed. DD makes them easy to calculate and controlled by single parameters for stiffness and strength. DD family of laminates are orderly, compact, and fully populated. They are so different and simple as compared with the traditional quad. Above all, DD is simpler to understand and use. No man-made rules that quad is subjected to. Naturally evolved rules that make DD unique include no mid-plane symmetry, single ply drops placed on the exterior surface, one DD for entire highly tapered component, one-axis multi-axial layup with no cross-plying, … They lead to light weight, low cost, fast layup, less prone to error, easy inspection and repair, no voids, no wrinkles, no warpage, tensile tests only, . . .

Just a mid-course forecast what DD can offer.

		3	1	4	9	11	12	14	10	13	5	7	2	15	6	8	Laminate			
	Radius*	T3/F!	AS4/	T3/N	IM7/:	T7/2!	T650,	T470:	IM7/	IM7/:	C-Ply	IM7/!	IM6/	T8S/:	C-Ply	T800/Cyt	rating	Stdev	cv	[90/0]
10	[±45]	0.33	0.37	0.35	0.30	0.34	0.33	0.32	0.31	0.32	0.34	0.32	0.32	0.31	0.35	0.33	0.33	0.02	6.0%	0.17
13	[±30]	0.35	0.40	0.38	0.33	0.37	0.36	0.35	0.33	0.34	0.37	0.35	0.35	0.34	0.39	0.36	0.36	0.02	5.6%	0.19
14	[±15]	0.42	0.46	0.44	0.38	0.44	0.43	0.42	0.38	0.40	0.43	0.40	0.40	0.40	0.45	0.42	0.42	0.02	5.2%	0.22
12	[30/15]	0.52	0.55	0.53	0.51	0.54	0.53	0.52	0.51	0.52	0.53	0.52	0.52	0.52	0.54	0.53	0.53	0.01	2.2%	0.27
9	[45/30]	0.58	0.60	0.59	0.58	0.59	0.59	0.58	0.58	0.58	0.59	0.58	0.59	0.58	0.59	0.59	0.59	0.01	1.0%	0.30
11	[30/0]	0.62	0.64	0.63	0.61	0.63	0.63	0.62	0.61	0.62	0.63	0.62	0.62	0.62	0.63	0.63	0.62	0.01	1.4%	0.32
7	[45/15]	0.90	0.89	0.90	0.90	0.90	0.90	0.90	0.90	0.90	0.90	0.90	0.91	0.90	0.90	0.90	0.90	0.00	0.4%	0.47
5	[60/30]	1.01	1.00	1.00	1.01	1.00	1.01	1.01	1.02	1.01	1.00	1.01	1.01	1.01	1.00	1.00	1.01	0.00	0.5%	0.52
6	[45/0]	1.02	1.01	1.02	1.03	1.02	1.02	1.02	1.03	1.02	1.02	1.03	1.03	1.02	1.02	1.03	1.02	0.01	0.5%	0.53
4	[60/15]	1.34	1.32	1.33	1.36	1.34	1.34	1.35	1.36	1.35	1.34	1.35	1.35	1.35	1.33	1.34	1.34	0.01	0.8%	0.70
3	[60/0]	1.47	1.44	1.46	1.49	1.46	1.47	1.47	1.49	1.48	1.46	1.48	1.48	1.48	1.45	1.47	1.47	0.01	0.9%	0.76
1	[75/15]	1.69	1.65	1.67	1.70	1.67	1.68	1.68	1.71	1.69	1.67	1.69	1.68	1.70	1.66	1.67	1.68	0.02	0.9%	0.87
2	[75/0]	1.81	1.77	1.79	1.83	1.79	1.80	1.81	1.83	1.82	1.80	1.82	1.81	1.82	1.78	1.80	1.81	0.02	1.0%	0.94
8	[90/0]	1.94	1.89	1.91	1.96	1.92	1.93	1.94	1.96	1.95	1.92	1.94	1.93	1.95	1.90	1.92	1.93	0.02	1.0%	1.00
	Material rating	0.63	0.75	0.78	0.84	0.89	0.96	1.00	1.01	1.02	1.04	1.17	1.18	1.23	1.23	1.26				

Figure 2.20 Laminate rating by normalization of radii

The material/laminate combinations that are highlighted with yellow background in 3 materials and 4 laminates in Figure 2.20 are examples used as laminate rating in in the last column on the right in Figure 2.23. That is based on [90/0] as the baseline.

As seen earlier that normalization equalized material ratings in Figure 2.18, a similar effect of normalization in Figure 2.20 equalized laminate ratings. Both material and laminate can be rated and ranked based on their respective values of radii. They allow scaling among materials and laminates. An important feature of master envelopes. In Figure 2.21 the equalization of laminates, similar to that of materials in Figure 19, is shown. The normalized materials in each laminate remain the same.

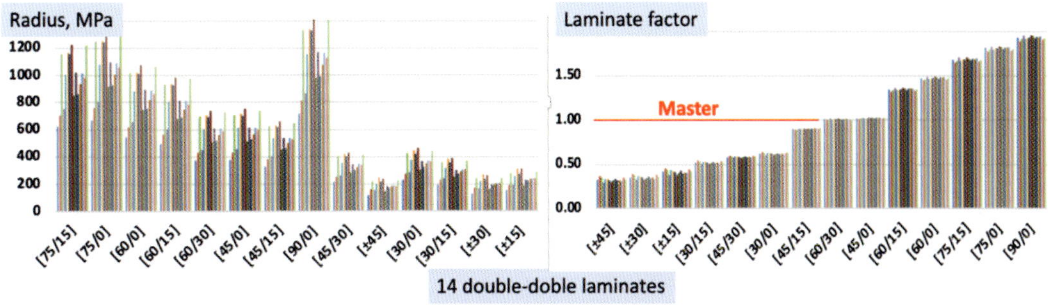

Figure 2.21 Comparison between absolute and normalized radii for 14 DD with 15 CFRP

The laminate rating can be shown in tabular forms as shown in Figure 2.22. The first table on the left is the value in the right column for laminate rating in Figure 2.20. The middle table is the same master but normalized to the radius of [90/0] with an R value of 1.93. That is the last column in Figure 2.20. The right table is the radius of DD laminates with T800/Cyt material.

This is obtained by multiplying the master radius by the material rating of the material. In this case its value is 1.26 as seen in the right of the bottom row of Figures 2.18 and 2.20. For each material their laminate rating is simply the corrected radii by the material rating, as seen in the example for T800/Cyt.

[±Φ] \ [±Ψ]	0	15	30	45	60	75
15		0.42		Master		
30	0.62	0.53	0.36			
45	1.02	0.9	0.59	0.33		
60	1.47	1.34	1.01	0.59	0.36	
75	1.81	1.68	1.34	0.9	0.53	0.42
90	1.93	1.81	1.47	1.02	0.62	

[±Φ] \ [±Ψ]	0	15	30	45	60	75
15		0.22		Norm: [90/0]		
30	0.32	0.27	0.19	Master/1.93		
45	0.53	0.47	0.30	0.17		
60	0.76	0.70	0.52	0.30	0.19	
75	0.94	0.87	0.70	0.47	0.27	0.22
90	1.00	0.94	0.76	0.53	0.32	

[±Φ] \ [±Ψ]	0	15	30	45	60	75
15		0.53		T800/Cyt		
30	0.79	0.66	0.45	1.26*master		
45	1.29	1.13	0.74	0.41		
60	1.85	1.69	1.27	0.74	0.45	
75	2.28	2.12	1.69	1.13	0.66	0.53
90	2.43	2.28	1.85	1.29	0.79	

Figure 2.22 Laminate rating of radius in tabular forms

The importance of the material and laminate ratings can be seen in their relations in Figure 2.23, which is Figure 2.17 with material and laminate ratings added. Note that the ratings in this figure are also marked in Figures 2.18 and 2.20. The material ratings are highlighted with green background, while laminate ratings, in aqua. They further reinforce the concept of master envelopes. All shape and size of envelopes are locked in and connected. For double-double laminate interpolation or zoom is possible and can be carried out to any desired degree of resolution. That is not possible for quad or other discrete laminates because they are not members of a field.

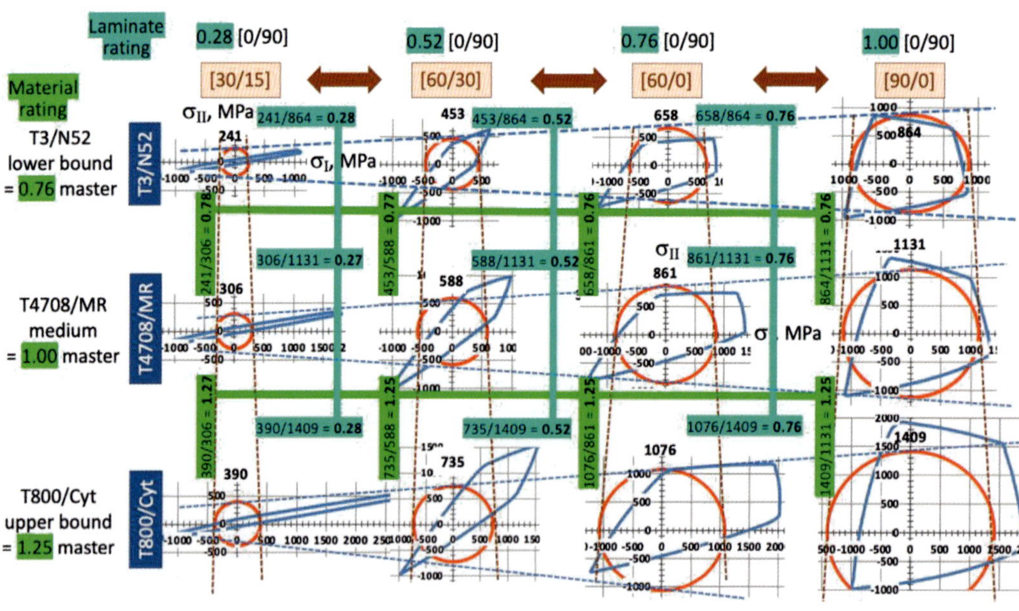

Figure 2.23 Material and laminate rating based on radii of failure envelopes

Double-double is unique not only because it is field-based, but also small thickness in their 4-ply building block so homogenization is easier to achieve (while 10-ply quad, for example, can not be homogenized without requiring multiples of 20 plies). Each material/laminate entity is controlled by its respective rating. Their values are fixed. As mentioned earlier, generation of material/laminate data is simplified considerably if we know what these rating values are. Testing is for validation. This is different from current practice that implicitly assume that each material and laminate is unique and independent. Not connected to any other material and laminate. Each material is tested the same way from A to Z. Huge number of data will be needed to generate allowable that costs millions of dollars and years of effort.

With an one-parameter of trace for stiffness and radius R or master envelopes for strength, ratings and rankings can be generated for all materials and all DD laminates. There is no longer any need to generate allowable with thousands of coupons. Just a few to validate what their predicted values are. Every material is related to any other material, and every laminate is related to any other laminate. A happy family! Data from any material can help those from others, just like those from any laminate can help equally those from others. As database for DD increases, it will soon only necessary to test for validation of the as built material and laminate. It may take just one week.

The material ratings can be correlated to the basic material data in X and X' as seen in Figure 2.24. Note that both radius R and diameter D correlate well with the average values of X and X'. For finding the LPF envelope of a new matcrial, it will be based on the measured X and X', and the trace of the material. From Figure 2.20, it is easy to find the normalized radius and diameter as described in Figure 2.8 have the same rating as the radius. Discussion of the major diameter will be in the next section

CFRP LPF	R*	D*	X	X'	(X+X')/2
T3/F93	0.63	0.64	1314	1220	1267
AS4/H3501	0.75	0.72	1447	1447	1447
T3/N52	0.78	0.78	1500	1500	1500
IM7/8552	0.84	0.86	2326	1200	1763
T7/2510	0.89	0.89	2172	1450	1811
T650/ep	0.96	0.96	2194	1653	1924
T4708/MR	1.00	1.04	2524	1700	2112
IM7/8552'	1.02	1.03	2501	1700	2101
IM7/MTM	1.03	1.03	2500	1700	2100
C-Ply 55	1.04	1.03	2530	1669	2100
IM7/977	1.17	1.15	3250	1600	2425
IM6/ep	1.18	1.18	3500	1540	2520
C-Ply 64	1.23	1.20	2944	1983	2464
T8S/3900	1.23	1.23	3000	2000	2500
T800/Cyt	1.26	1.26	3768	1656	2712

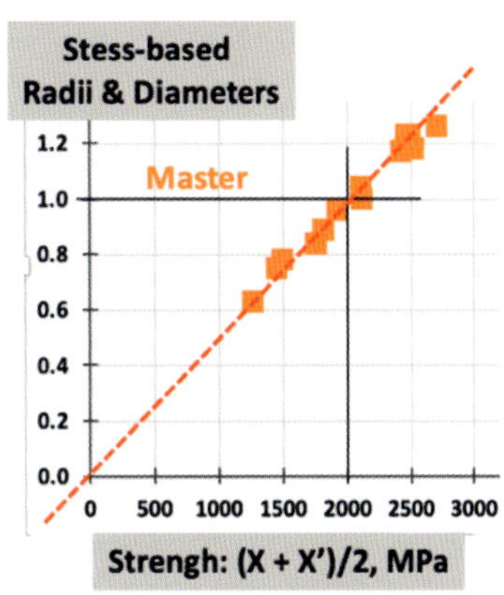

Figure 2.24 Correlation between material ratings with single parameter of (X+X')/2

2.6 Major diameters of master envelopes

There are other parameters that determines the master envelopes beyond the radius of the same-area circles. Three such dimensions are the major diameters, center locations, and slope, as shown in Figure 2.8. In Figure 2.25, the major diameters are rated in the same process as the radii in Figures 2.18 and 2.20. The results are shown with yellow background in the bottom row for materials, and right column for laminates. Their coefficients of variations are also shown with remarkably low values.

	Major	3	1	4	9	11	12	5	10	13	14	7	2	6	15	8		Laminate			Laminate	
	diameter	T3/F93	AS4/H3	T3/N52	IM7/8	T7/25	T650/	C-Ply 5	IM7/N	IM7/8	T4708	IM7/9	IM6/e	C-Ply6	T8S/3	T800/Cyt		master		cv	rating	cv
10	[±45]	1820	2070	2228	2470	2533	2737	2939	2961	2960	2921	3299	3369	3415	3535	3566		2855	543	19%	0.92	0.6%
5	[60/30]	1820	2070	2228	2470	2533	2737	2939	2961	2960	2921	3299	3369	3415	3535	3566		2855	543	19%	0.92	0.6%
1	[75/15]	1820	2070	2228	2470	2533	2737	2939	2961	2960	2921	3299	3369	3415	3535	3566		2855	543	19%	0.92	0.6%
8	[90/0]	1820	2070	2228	2470	2533	2737	2939	2961	2960	2921	3299	3369	3415	3535	3566		2855	543	19%	0.92	0.6%
2	[75/0]	1824	2074	2232	2475	2539	2743	2945	2968	2966	2927	3306	3376	3422	3542	3574		2861	544	19%	0.92	0.6%
4	[60/15]	1849	2102	2263	2509	2573	2780	2985	3008	3007	2967	3351	3422	3468	3591	3622		2900	551	19%	0.93	0.6%
3	[60/0]	1873	2130	2292	2542	2607	2817	3025	3049	3047	3006	3396	3468	3513	3638	3670		2938	559	19%	0.94	0.6%
9	[45/30]	1873	2130	2292	2542	2607	2817	3025	3049	3047	3006	3396	3468	3513	3639	3670		2938	559	19%	0.95	0.6%
7	[45/15]	1975	2244	2417	2682	2748	2970	3189	3216	3213	3187	3582	3657	3703	3837	3873		3100	590	19%	1.00	0.5%
13	[±30]	2024	2299	2476	2749	2816	3043	3268	3297	3294	3266	3671	3749	3794	3933	3970		3177	605	19%	1.02	0.5%
6	[45/0]	2024	2299	2483	2749	2824	3043	3268	3297	3293	3327	3671	3750	3814	3933	4010		3186	610	19%	1.02	0.3%
12	[30/15]	2186	2497	2729	2970	3103	3298	3539	3562	3557	3685	3967	4094	4191	4248	4420		3470	670	19%	1.12	0.9%
11	[30/0]	2281	2622	2868	3063	3276	3464	3708	3673	3702	3891	4120	4308	4424	4381	4678		3631	705	19%	1.17	1.6%
14	[±15]	2462	2826	3089	3278	3548	3747	4019	3932	4009	4212	4430	4690	4790	4639	5093		3918	765	20%	1.26	2.2%
	Material																Avergae	3110				
	master	1975	2250	2432	2674	2769	2976	3195	3207	3212	3226	3577	3675	3735	3823	3917	###					
	Stdev	203	240	276	261	322	318	339	313	330	417	361	414	437	363	484						
	Material	10%	11%	11%	10%	12%	11%	11%	10%	10%	13%	10%	11%	12%	9%	12%						
	rating	0.64	0.72	0.78	0.86	0.89	0.96	1.03	1.03	1.03	1.04	1.15	1.18	1.2	1.23	1.26						
	cv	0.6%	0.2%	0.5%	1.1%	0.7%	0.3%	0.3%	1.1%	0.6%	1.9%	0.7%	0.4%	0.8%	1.3%	1.3%						

Figure 2.25 Material and laminate ratings of major diameters of failure envelopes

The compositions of each material and laminate are shown in Figure 2.26. On the left is the listing of 15 CFRP. For each material, the rating of 14 laminates is the same. Conversely, for each laminate on the right, the rating of 15 CFRP is the same for each laminate. It is clear that materials and laminates are equivalent and interchangeable. This is true because the DD laminates are used and they are field-based. Quad and other individual laminates are most likely not field-based, and cannot be packaged in one simple table as seen in Figure 2.26 and earlier figures.

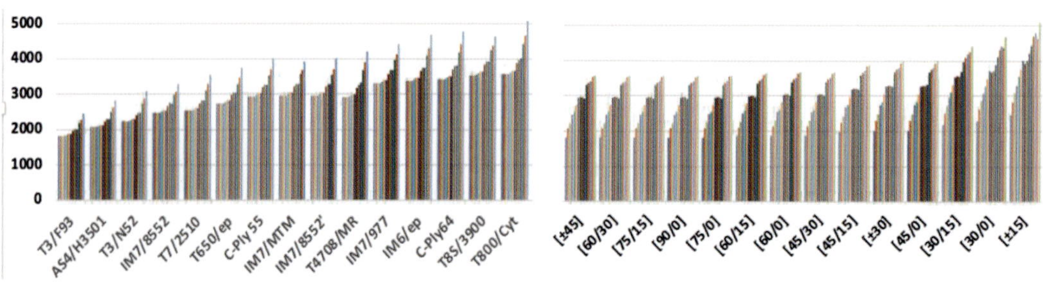

Figure 2.26 Major diameters of DD rated in materials on the left and laminates on the right

The laminate rating in Figure 2.25 can be shown in tabular form similar to those for radius in Figure 2.22 is shown in Figure 2.27. On the left in the top row is that for T4708/MR that is practically that of the master envelope. In the middle is the upper bound diameter in T800/Cyt, and on the right one of the lowest diameters in T3/N52. In the top row are the major diameters in MPa, and in the bottom row, the rating of normalized diameters for the same materials.

T4708/MR (Master*1)

[±Φ] \ [±Ψ]	0	15	30	45	60	75
15		3919		T4708/MR		
30	3631	3470	3177	Master*1		
45	3186	3100	2938	2855		
60	2938	2900	2855	2938	3177	
75	2861	2855	2900	3100	3470	3918
90	2855	2861	2938	3186	3631	

T800/Cyt (Master*1.26)

[±Φ] \ [±Ψ]	0	15	30	45	60	75
15		4938		T800/Cyt		
30	4575	4372	4003	Master*1.26		
45	4014	3906	3702	3597		
60	3702	3654	3597	3702	4003	
75	3605	3597	3654	3906	4372	4936
90	3597	3605	3702	4014	4575	

T3/N52 (Master*0.78)

[±Φ] \ [±Ψ]	0	15	30	45	60	75
15		3057		T3/N52		
30	2832	2706	2478	Master*0.78		
45	2485	2418	2292	4532		
60	2292	2262	2227	4665	5043	
75	2231	2227	2262	4921	5509	6220
90	2227	2231	2292	5058	5764	

T4708/MR (Master*1)

[±Φ] \ [±Ψ]	0	15	30	45	60	75
15		1.26		T4708/MR		
30	1.17	1.12	1.02	Master*1		
45	1.02	1.00	0.95	0.92		
60	0.94	0.93	0.92	0.95	1.02	
75	0.92	0.92	0.93	1.00	1.12	1.26
90	0.92	0.92	0.94	1.02	1.17	

T800/Cyt (Master*1.26)

[±Φ] \ [±Ψ]	0	15	30	45	60	75
15		1.59		T800/Cyt		
30	1.47	1.41	1.29	Master*1.26		
45	1.29	1.26	1.19	1.16		
60	1.19	1.18	1.16	1.19	1.29	
75	1.16	1.16	1.18	1.26	1.41	1.59
90	1.16	1.16	1.19	1.29	1.47	

T3/N52 (Master*0.78)

[±Φ] \ [±Ψ]	0	15	30	45	60	75
15		0.98		T3/N52		
30	0.91	0.87	0.80	Master*0.78		
45	0.80	0.78	0.74	0.72		
60	0.74	0.73	0.72	0.74	0.80	
75	0.72	0.72	0.73	0.78	0.87	0.98
90	0.72	0.72	0.74	0.80	0.91	

Figure 2.27 Absolute and relative major diameters for the master, upper and lower bounds

2.7 Center locations of master envelopes

In Figure 2.8, the center locations of failure envelopes were shown. They are also important geometric parameters that establish the relations between the envelopes of individual materials and their master envelopes. Such relations provide forward process going from original envelope to the master, as well as the recovery process going from master to original ply material. In Figure 2.28, such processes are shown with definition of the center locations in normalized displacements. Material on the left is T3/N52, the lower bound, and on the right, IM7/977, the upper bound. Two illustrations in [75/15] on top, and [60/0] at the bottom.

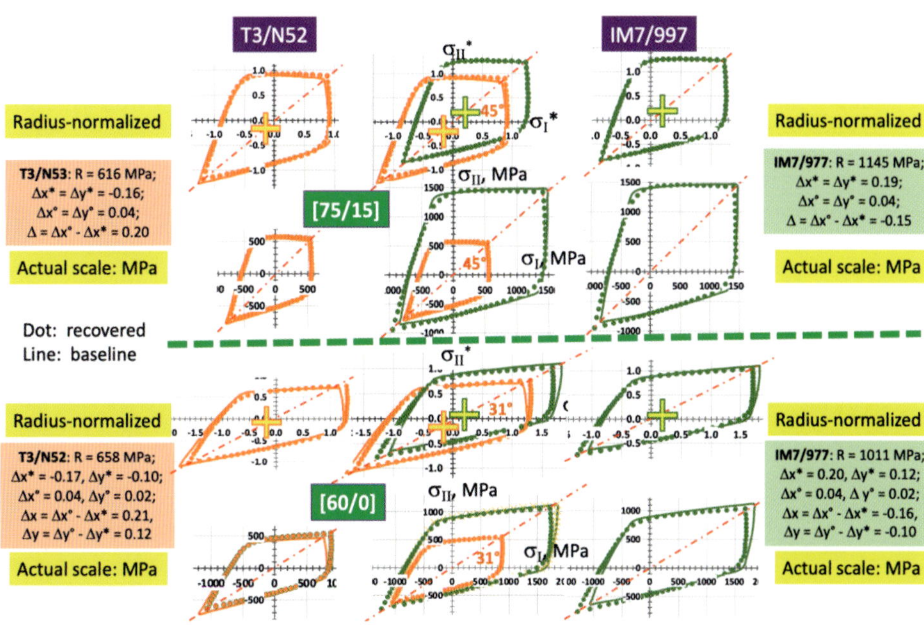

Figure 2.28 Displacements of center locations of T3/N52 and IM7/977 in [75/15] and [60/0]

For each of the 4 material/laminate combinations, the center locations are indicated in absolute and normalized envelopes. In the center column, the envelopes from the 2 materials are superposed to show the range of variation of all CFRP materials each laminate can have. The solid line represents the forward prediction of the master envelope from the original, and the dots represent the recovered envelopes from the master.

The center locations for the lower bound material in T3/N52 have negative values. They appear in the 3rd quadrants of each failure envelope. Those for the upper bound in IM7/977 have positive values.

The average value of the center location would be that of the master location, designated in Δx° and Δy°. Their values are so indicated in the boxes in the left and right columns in Figure 2.28.

The values for other CFRP materials and DD laminates than those shown above are listed in the tables in Figure 2.29. Note the Δx* values for the cases in Figure 2.28 are confirmed (except for Δy* to be confirmed presently). The laminate master values in Δx° shown (in the last column of Figure 29 (in 0.04 for both laminates) also agree with those in Figure 2.28.

		4	1	3	14	12	11	6	5	13	10	15	8	9	2	7	Lm master
	Δx*	T3/N5	AS4/H	T3/F9	T470	T650,	T7/2	C-Ply	C-Ply	IM7/	IM7/I	T8S/	T800,	IM7/	IM6/	IM7/977	Δx°
1	[75/15	-0.16	-0.12	-0.08	-0.04	-0.02	0.00	0.00	0.04	0.04	0.07	0.10	0.16	0.17	0.19	0.19	0.04
2	[75/0]	-0.15	-0.11	-0.07	-0.04	-0.02	0.00	0.00	0.04	0.04	0.07	0.09	0.15	0.16	0.17	0.18	0.03
3	[60/0]	-0.17	-0.13	-0.08	-0.04	-0.02	0.00	0.00	0.04	0.05	0.08	0.10	0.17	0.18	0.19	0.20	0.04
4	[60/15	-0.19	-0.15	-0.09	-0.05	-0.02	0.00	0.00	0.05	0.05	0.09	0.12	0.19	0.20	0.22	0.23	0.04
5	[60/30	-0.26	-0.20	-0.13	-0.06	-0.03	0.00	0.01	0.07	0.07	0.12	0.16	0.27	0.28	0.31	0.32	0.06
6	[45/0]	-0.17	-0.13	-0.08	-0.02	-0.02	0.00	0.01	0.04	0.05	0.08	0.10	0.18	0.18	0.19	0.20	0.04
7	[45/15	-0.22	-0.17	-0.11	-0.05	-0.02	0.00	0.00	0.06	0.06	0.10	0.14	0.23	0.24	0.26	0.27	0.05
8	[90/0]	-0.14	-0.11	-0.07	-0.03	-0.01	0.00	0.00	0.04	0.04	0.06	0.09	0.14	0.15	0.16	0.17	0.03
9	[45/30	-0.41	-0.31	-0.21	-0.10	-0.04	0.00	0.01	0.11	0.11	0.20	0.26	0.42	0.45	0.49	0.51	0.10
10	[±45]	-0.75	-0.55	-0.40	-0.20	-0.08	0.01	0.01	0.21	0.23	0.40	0.53	0.83	0.96	0.98	1.02	0.21
11	[30/0]	-0.11	-0.08	-0.06	0.01	0.00	0.03	0.03	0.05	0.04	0.06	0.08	0.17	0.14	0.18	0.17	0.05
12	[30/15	-0.19	-0.16	-0.11	0.00	-0.02	0.02	0.03	0.06	0.06	0.11	0.14	0.25	0.24	0.27	0.27	0.07
13	[±30]	-0.48	-0.35	-0.25	-0.11	-0.05	0.01	0.01	0.13	0.15	0.26	0.34	0.52	0.60	0.61	0.64	0.13
14	[±15]	-0.08	-0.06	-0.04	0.03	0.01	0.04	0.04	0.06	0.05	0.07	0.08	0.17	0.15	0.18	0.17	0.06
	Material																
	master	-0.25	-0.19	-0.13	-0.05	-0.02	0.01	0.01	0.07	0.07	0.13	0.17	0.28	0.29	0.31	0.32	

Figure 2.29 Displacement of center location along the x-axis for 15 CFRP and DD laminates

The displacements of the center along the y-axis are shown in Figure 2.30, with their laminate master values in the right columns

Center locations for 2 laminates in [75/15] and [45/30] are shown in Figure 2.31. On the left, the highest and lowest locations are IM7/977 and T3/N52, respectively, for both laminates. Their values are shown in Figures 2.28-2.30 and in crosses in Figure 2.31. The second group of envelopes are for all 15 CFRP envelopes normalized by their respective radius, from which their average values become the master.

		4	1	3	6	5	10	11	15	12	14	13	8	9	2	7	Lm master
	Δy*	T3/N5	AS4/H	T3/F9	C-Ply	C-Ply	IM7/I	T7/2!	T8S/3	T650,	T470	IM7/!	T800,	IM7/!	IM6/	IM7/977	Δy°
1	[75/15]	-0.16	-0.12	-0.08	0.00	0.04	0.07	0.17	0.10	0.20	0.24	0.22	0.16	0.17	0.19	0.19	0.04
2	[75/0]	-0.13	-0.10	-0.06	0.00	0.03	0.06	0.14	0.08	0.16	0.20	0.18	0.13	0.14	0.15	0.16	0.03
3	[60/0]	-0.10	-0.08	-0.05	0.00	0.03	0.05	0.11	0.06	0.13	0.16	0.14	0.10	0.11	0.12	0.12	0.02
4	[60/15]	-0.13	-0.10	-0.07	0.00	0.04	0.06	0.15	0.08	0.16	0.20	0.18	0.13	0.14	0.15	0.16	0.03
5	[60/30]	-0.26	-0.20	-0.13	0.01	0.07	0.12	0.29	0.16	0.33	0.41	0.37	0.27	0.28	0.31	0.32	0.06
6	[45/0]	-0.05	-0.04	-0.03	0.00	0.02	0.03	0.06	0.04	0.07	0.09	0.08	0.05	0.06	0.06	0.07	0.01
7	[45/15]	-0.09	-0.07	-0.04	0.00	0.02	0.04	0.10	0.06	0.11	0.14	0.13	0.09	0.10	0.11	0.11	0.02
8	[90/0]	-0.14	-0.11	-0.07	0.00	0.04	0.06	0.15	0.09	0.17	0.21	0.19	0.14	0.15	0.16	0.17	0.03
9	[45/30]	-0.25	-0.19	-0.13	0.00	0.07	0.12	0.28	0.16	0.32	0.39	0.36	0.26	0.27	0.30	0.31	0.06
10	[±45]	-0.75	-0.55	-0.40	0.01	0.21	0.40	0.01	0.53	-0.08	-0.20	0.23	0.83	0.96	0.98	1.02	0.21
11	[30/0]	-0.01	-0.01	-0.01	0.00	0.01	0.01	0.00	0.01	0.00	0.00	0.01	0.02	0.02	0.02	0.02	0.01
12	[30/15]	-0.03	-0.03	-0.02	0.00	0.01	0.02	0.00	0.03	0.00	0.00	0.01	0.05	0.05	0.05	0.05	0.01
13	[±30]	-0.17	-0.12	-0.09	0.00	0.05	0.09	0.00	0.12	-0.02	-0.04	0.05	0.18	0.21	0.21	0.22	0.05
14	[±15]	-0.01	0.00	0.00	0.00	0.00	0.00	0.00	0.01	0.00	0.00	0.00	0.01	0.01	0.01	0.01	0.00
	Material master	-0.16	-0.12	-0.08	0.00	0.05	0.08	0.11	0.11	0.11	0.13	0.15	0.17	0.19	0.20	0.21	

Figure 2.30 Displacement of center location along the y-axis for 15 CFRP and DD laminate

The center location for the master envelope is also the average of the locations of individual materials. Their values can be found in the right column in Figures 2.29 and 2.30. For the top laminate in [75/15] in Figure 2.31, their values are 0.04 for Δx° and Δy°. In the lower laminate in [45/30], their values are 0.10 and 0.06, respectively. The center location is important in the recovery process from master envelopes to those of individual baseline material. That will be presented presently.

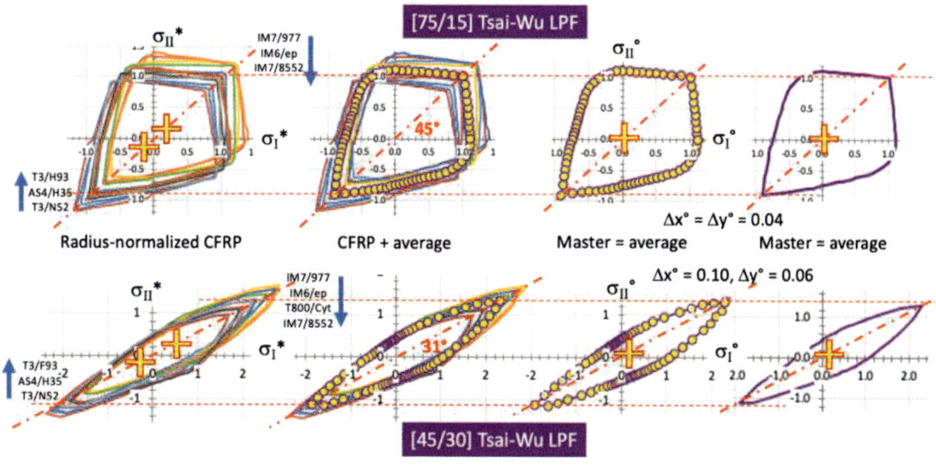

Figure 2.31 Center locations of individual and master envelopes

The complete cycle of the generation process of master envelope and the recovery of the original envelope from the master can be seen in two examples in Figure 2.32 for [75/15], and in Figure 2.33 for [60/0]. In the top row, the generation of master from baseline IM7/977 envelopes; in the bottom row, the recovery of the baseline from the master. As seen in Figure 2.28, the recovered envelopes shown in dots are very close to the baseline in continuous lines. In most cases, their differences are not distinguishable. Demonstrations of these reversible processes are necessary to validate the value of master envelopes. Without the reversable processes, the master envelopes will be of little value.

It is important to emphasize again the importance of the radius of the same-area circle of the failure envelope. The radius makes it possible to convert seemingly random sizes and shapes of failure envelopes to a collection of envelopes with similar sizes and shapes with differing locations of their centers. Such conversion was seen between the left two columns in Figure 2.9. The same process is used in Figures 2.32 and 2.33, for laminate [75/15] and [60/0], respectively. In these figures, only the upper and lower bound envelopes in T3/N52 and IM7/977 in the first two columns from the left.

The key steps in the master envelope generation (in the top row for each laminate) are the normalization by radius, and the averaging of 15 CFRP. Both steps are shown with green background in the top row. There are also 2 key steps in the recovery of the original baseline laminates in the bottom row. The first is the recovery of radius normalized envelopes from the master by shifting their center location by Δ° - Δ*, as seen in Figure 2.28. The second step is the de-normalization from the radius-normalized envelopes to the original baseline envelopes. Both steps are shown in boxes with green background in the lower row.

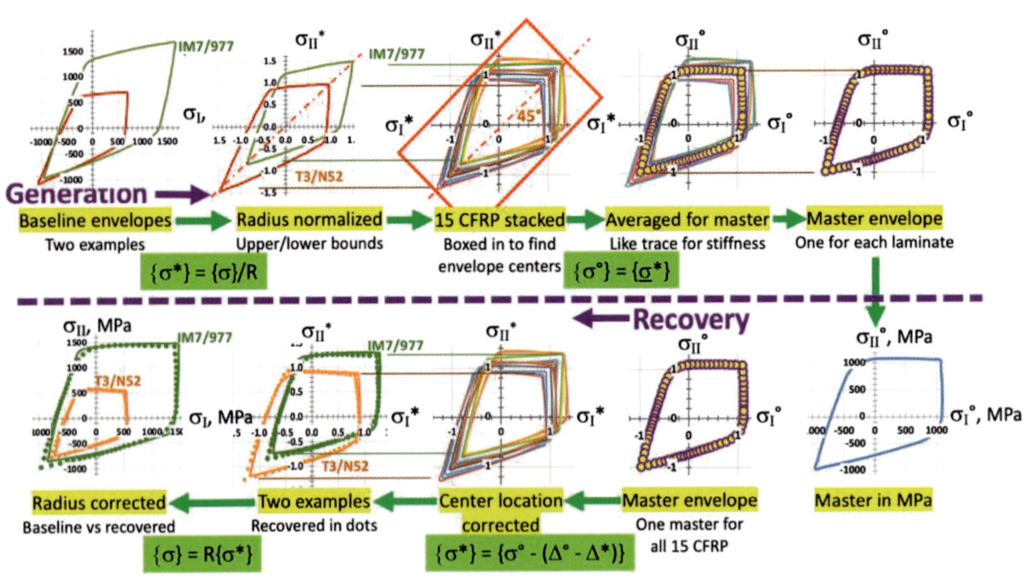

Figure 2.32 Generation and recovery of [75/15]

The slope of the major diameters in the last two figures were 45 and 31 degree, respectively. Values for other DD can be obtained as described in Figure 2.34. They are taken from the master values in the right columns in Figures 2.29 and 2.30. Note the line of complementary about which the sum of two components is equal to 90 degree.

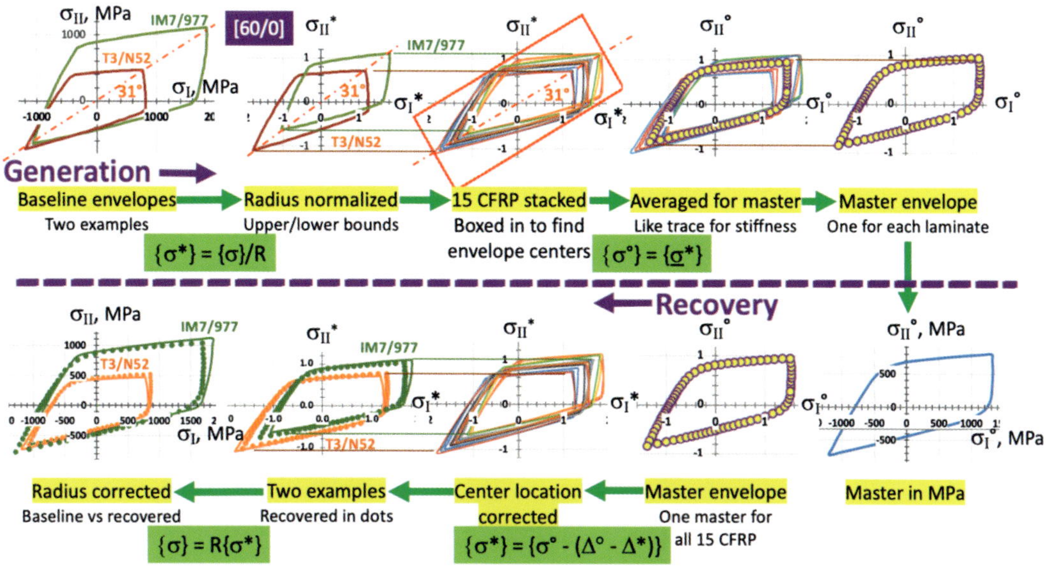

Figure 2.33 Generation and recovery of [60/0]

An example is seen in the lower right table that laminate [60/0] has a slope of 31 degree while its complementary laminate [90/30] with 59 degree. The same complementary relation between Δx^* and Δy^* in the top two tables exists; i.e., the displacement components above and below the "complementary line" are interchangeable; for the same two laminates in [60/0] and [90/30] their values are 0.07 and 0.04 for Δx^* which are just the opposite in 0.04 and 0.07 for Δy^*.

Δx°	0	15	30	45	60	75
15		0.80				
30	0.38	0.36	0.41			
45	0.14	0.14	0.19	0.30		
60	0.07	0.08	0.09	0.12	0.14	
75	0.05	0.05	0.05	0.06	0.07	0.06
90	0.05	0.05	0.04	0.05	0.05	

Δy°	15	30	45	60	75
	0.06				
0.05	0.07	0.14			
0.05	0.06	0.12	0.30		
0.04	0.05	0.09	0.19	0.41	
0.05	0.05	0.08	0.14	0.36	0.80
0.05	0.05	0.07	0.14	0.38	

Δy°/Δx°						
15		0.07				
30	0.14	0.20	0.34			
45	0.34	0.40	0.61	1.00		
60	0.61	0.70	1.00	1.15	2.91	
75	0.88	1.00	1.43	2.48	5.15	14.3
90	1.00	1.15	1.66	2.91	7.12	

Complementary

Slope,deg					
	4				
8	11	19			
19	22	31			
31	35	45	49	71	
41	45	55	68	79	86
45	49	59	71	82	

Mirror symmetry

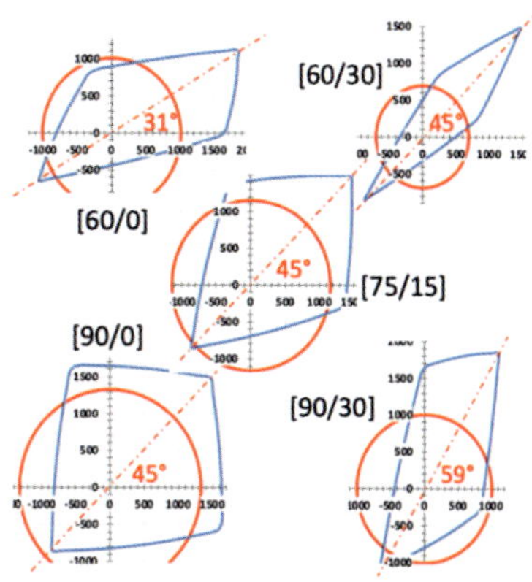

Figure 2.34 Center locations and slope of major diameters of their envelopes

2.8 Interpolation of envelopes

For field-based double-double laminates, it is possible to interpolate or zoom to reach laminates with smaller angle increments than those shown earlier, including that in Figure 2.34 set at 15 degree. Such important feature (not possible with quad that is discrete, not field-based) allows master envelopes to be interpolated in the same way. Interpolation will be demonstrated with different failure criterion and for both FPF and LPF. The actual failure criterion used is the unit circle criterion, instead of the Tsai-Wu used up to this point. With the demonstration of two failure criteria, it is intended to show that the master envelope concept is more general than one specific failure criterion. Note that interpolated envelope is shown in dots and is identical to the theoretical envelope in solid line. The same accuracy in interpolation can be derived from [60/30] and [75/15], as that from [75/30] and [60/15].

Interpolation or zoom can provide opportunities in design, manufacturing and testing. DD being field-based, its properties are continuous functions of the angle increments. One way to find the optimum is to scan all increments until the desired property becomes insensitive to further angle increment change. Another use is to find angles that would be the simplest for manufacturing such as automated layup or fiber placement machine, or filament winding.

DD laminates are much easier for optimization because they are field-based. They have far fewer variables and can easily reach a homogenized laminate for the entire structure. All family members of DD are brothers and sisters. No strangers among them. Their relations are rational, systematic, and fixed. All members of the family can be used. There are no rules that would outlaw certain members of the family such as the 10 percent rule, nesting rule, 45 degree or less between adjacent plies. Quad laminates do not belong to one family. Some are based on 6-, 8-, or 10-ply laminates. They do not fit in tables like members of the DD family in Figure 2.35.

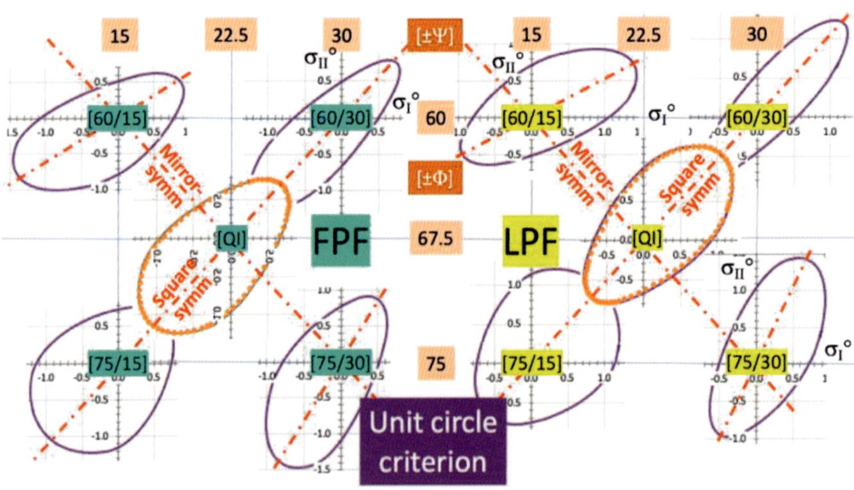

Figure 2.35 Interpolation of [QI] from adjacent laminates for FPF and LPF

The inclinations of envelopes through interpolations in Figure 2.34 based on Tsai-Wu can be changed for the unit circle criterion shown in Figure 2.36.

A more aggressive interpolation of [QI] from [90/0] and [±45] is shown in Figure 2.37. The agreement between the baseline in solid line, and the interpolated one in dots are reasonable, although not as precise as that in Figure 2.35 where the gap between the envelopes is closer.

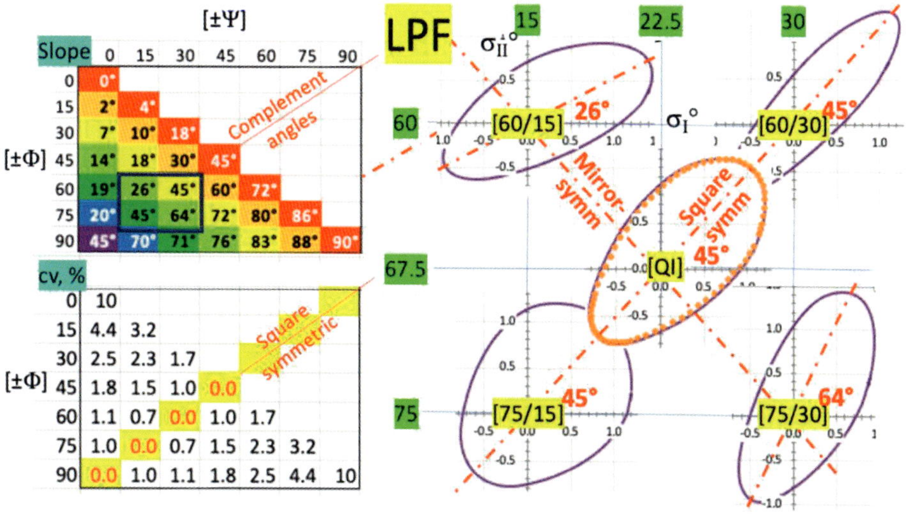

Figure 2.36 Complementary slope of major diameters of envelopes from mirror symmetry

Interpolation or zoom that DD offers is not possible with quad. That is the difference between a field-based and a collection of dispersed quad laminates.

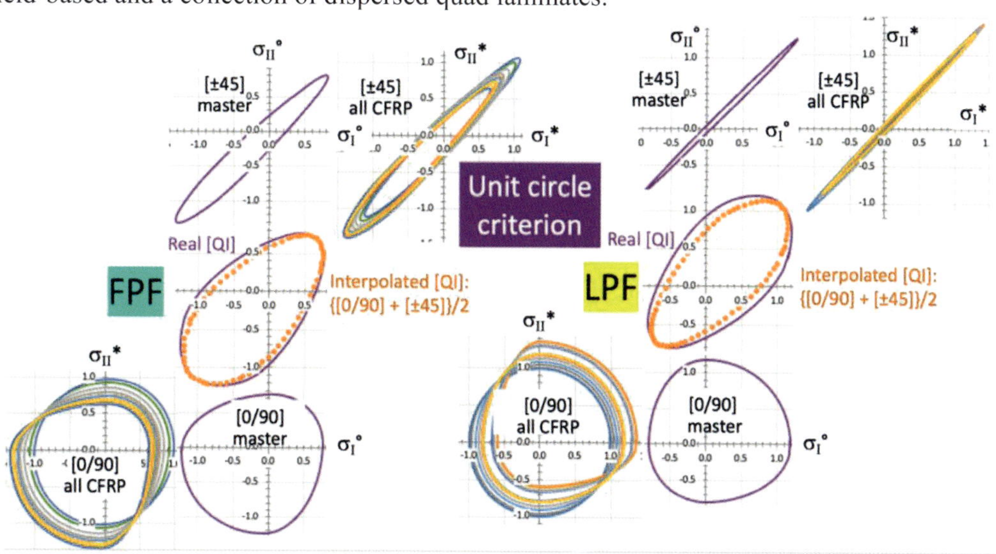

Figure 2.37 Interpolation of [QI] from [90/0] and [±45] laminates for FPF and LPF

Mores example of interpolation are seen in Figure 2.38 where 3 adjacent envelopes show the interpolated middle envelope to be close to the actual value.

Interpolation or zoom is an important feature of field-based DD. It allows simple representation of their properties such as those with 15 degree angle increments used thus far. If wider choices of laminates are desired, the field can be expanded to any degree without limitation. For a composite components that will be produced in large numbers, even a small improvement in properties may be justified.

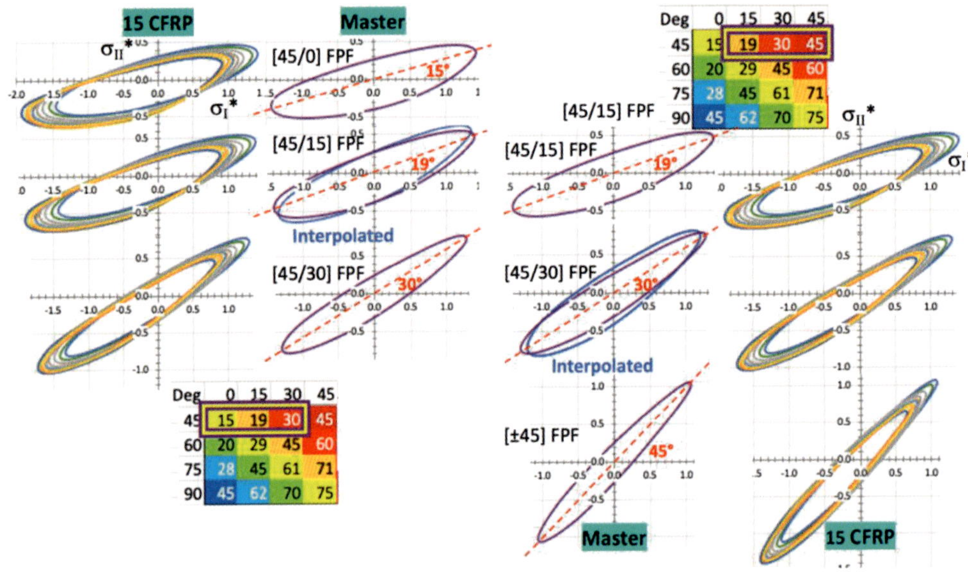

Figure 2.38 Interpolated envelope in the middle in close agreement to theoretical envelope

2.9 Transformation of failure envelopes

Transformation through coordinate axis rotation can establish symmetry of not only materials and laminates but also failure envelopes. For Tsai-Wu and unit circle criteria such transformation can be made through rigid-body rotations. With symmetry, the number of independent envelopes can be deduced and reduced. This is illustrated in Figure 2.39 how 2 laminates in [60/15] and [75/30] are the same and connected by transformation. They are mirror image of one another. There are 2 steps in this transformation. The first for [60/15] is a 180 degree rotation about the y-axis (out of the x-y plane), followed by a 90 degree rotation to the right. The result is laminate [75/30], the complement of [60/15] where 75 + 15 = 60 + 30 = 90. A reverse transformation is also shown in this figure where the first step is a rotation of 180 degree about the x-axis, followed by a rotation of 90 degree to the left. The original starting laminate of [60/0] is thereby recovered. The slopes of the mutually transformed envelopes are complementary in 26 and 64, where their sum is 90 degree. The angles are shown in a box in the table on the lower left. It is important the recognize that with this example of mirror symmetry, the laminates above the complementary angle line (in dashed red line) are mirror image of those below the line. Thus the number of independent laminates reduce from 28 to 16. The fewer independent laminates the simpler optimization and other operations. Every step that can be rational helps us move forward with advances not always expected. Coordinate transformation is fundamental. Failure criteria and their envelopes must be transformable. This means that failure assessment can be done on any reference coordinate system, such as the component or material axes. It is a test of the generality of the failure criteria. Being invariant is best assurance of this generality.

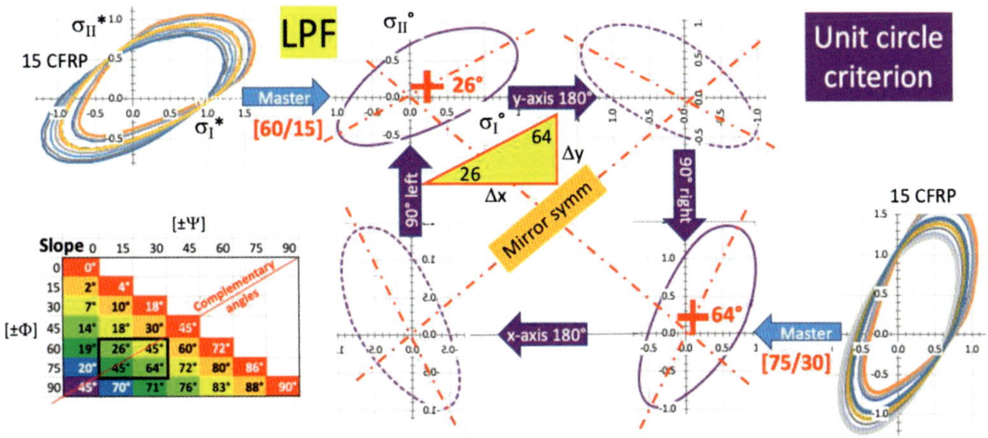

Figure 2.39 Transformation of failure envelopes

Such mirror symmetry can be seen in the master envelopes of 3 materials in T3/N52, C-Ply 64 and IM7/977 in Figure 2.40 based of Tsai-Wu LPF. As seen before, these 3 materials represent the lower bound, medium or master, and upper bound of the 15 CFRP. Those square symmetric envelopes have olive green shades while those with mirror image have blue shades. The remaining 12 CFRP are bounded by the envelopes of the two bounds. The square symmetric laminates are the back bone, and act like mirrors for the symmetries among the remaining laminates.

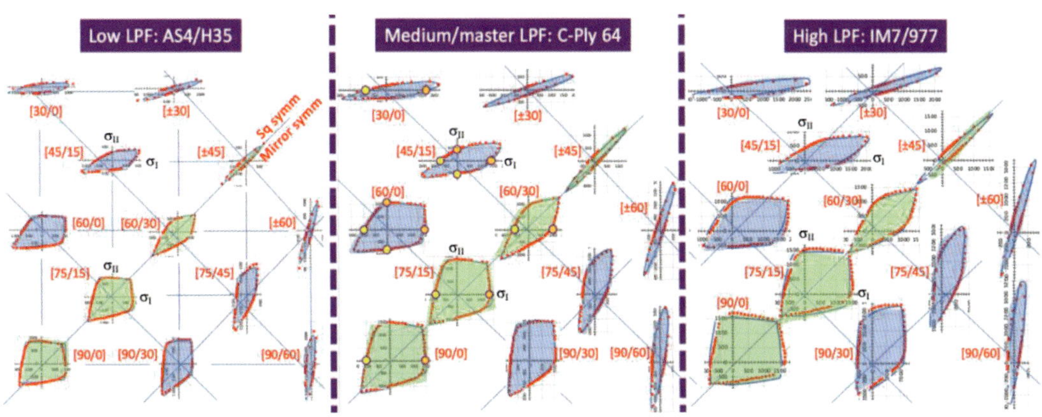

Figure 2.40 Representative sizes and shapes of master LPF failure envelopes

It is possible to show similar transformation for quad laminates. In Figure 2.41, the center is the traditional carpet plot of in-plane stiffness of T800/Cyt. It is divided into two parts by the square symmetric line in brown squares. In the olive green area above this line, the number of plies in [0] is higher than those in [90]. In the blue area below this line, the reverse is true; i.e., more [90] than [0]. Along this line, the number of plies between [0] and [90] are equal.

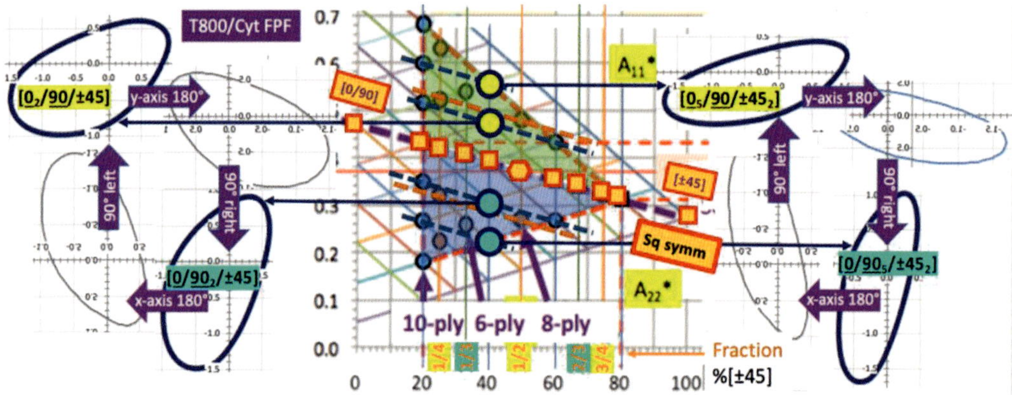

Figure 2.41 Transformation of failure envelope of quad laminates

On the right of the carpet plot, the envelope of the hard laminate is shown. It can be seen that by a 2-step transformation, a mirror symmetric laminate is generated with the [0] and [90] plies interchanged. On the left, a different laminate can be transformed to find its mirror symmetric laminate. Thus envelope transformation can be done with any laminate. For DD, it is useful to reduce the number of independent laminates. For other laminates, like quad and others, they are often unrelated, discrete laminates. Transformation is similarly accomplished. It is therefore possible to generate all envelopes below the square symmetric lines by transformation of the corresponding laminates above it. Such reduction can be the basis of testing fewer laminates for design data generation. The mirror symmetry of failure envelopes is a necessary condition that can verify the validity of two complementary envelopes. If the mirror image cannot be verified, there may be an issue with the failure criterion. This is a good feature to check failure criteria that are not invariant (most criteria are not).

2.10 Nettles failure envelopes

For a long time, Alan Nettles of NASA Marshall Space Flight Center has been advocating the use of notched coupons like one with open hole to generate design allowable than the use of smooth, no-hole coupons. There are several justifications:

1) A notch can represent defects in lamination that would invariably exist in real life
2) Load required to fail coupons is lower than no-hole coupons.
3) Data scatter is smaller. Fewer tests will be needed
4) For DD, the coupon edges can determine their mode of failure. Open hole coupons can force failure away from the edges leading to a more reliable, consistent failure mode.

With master envelopes proposed herein, it is an easy extension to use LPF envelope to cover what we would like to name Nettles envelope. This can be seen in using existing data from Waruna Seneviratne of Wichita State's NIAR in Figures 2.42 and 2.43 where the uniaxial data on smooth, no hole and open-hole (in stress concentration factors) data, are shown in blue and maroon lines, respectively. Next to each of the envelope is the value of radius of same-area circle of the envelope. The ratios of no-hole/open-hole radii are shown in figures in aqua background. Note that they match the average stress concentration factors between the tension and compression strength.

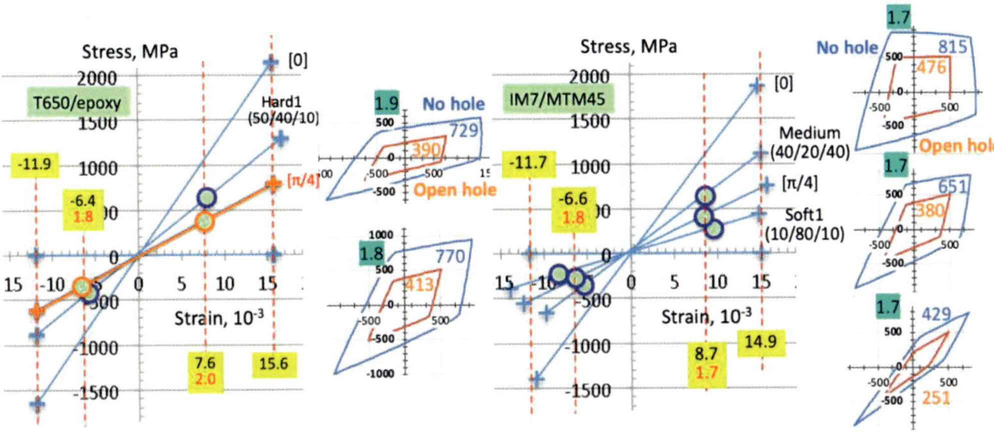

Figure 2.42 Uniaxial data and their implied no-hole and open-hole failure envelopes

Strength data are far more difficult than stiffness data. Master envelopes should help to increase confidence and simplicity. It recognizes that strength, like stiffness, for materials and their laminates are not random, and unconnected. Data generation for each material and laminate need not be a full-service exercise. Master envelope has taught us that every material and every laminate is a member of a family. This is particularly true with DD, and other field-based families. But not so with dispersed laminates like the quad family. As more and more data are acquired, fewer and fewer tests will be required. In the end, the required design allowable will not be the show stopper of innovation that it has been since the beginning of composites technology in the 1960's.

.

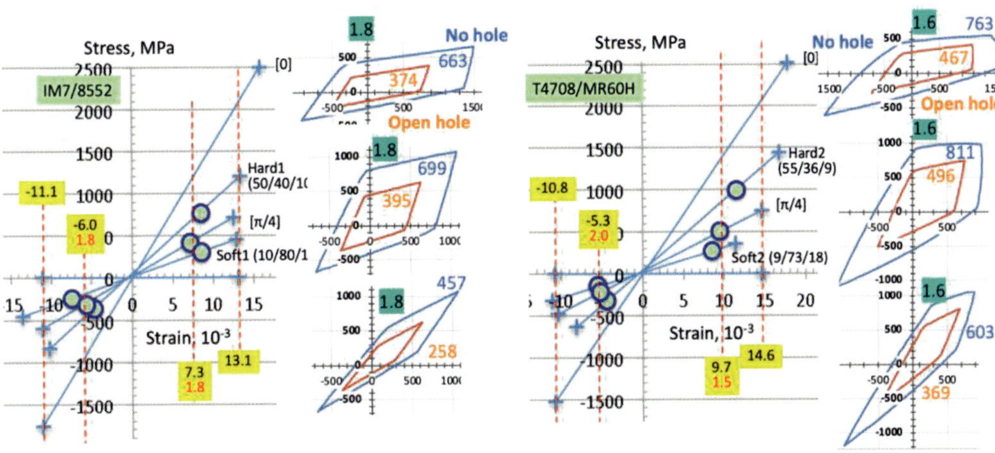

Figure 2.43 Uniaxial data and their implied no-hole and open-hole failure envelopes

The Nettles envelope is an extension of the master envelope from smooth to notched coupons. It is possible that coupons with stress raiser other than open hole can be used. The corresponding envelopes for other notches can be similarly generated without additional assumption. This is another example of simplification that the master envelope can offer.

Figures 2.23 and 2.40 clearly show that material and laminate are equivalent. Just two faces of the same entity. As the volume of data increases, with improved modeling of their failure mechanisms, and test methods, generation of design data can be one of calibration that would be a miniscule effort as compared with a full-service required currently. The process of design and manufacturing of composites can be much more robust and simplified.

There is no long a need to make composites black aluminum. Composites have their own life and methodology. Totally different from metals but are as easy and reliable as metals. DD is the first step toward that utopia. Nettles envelope is another step further our goal for simplicity.

2.11 One parameter

There is a significant advantage of having one parameter to describe the stiffness and strength of composite laminates. It not only simplifies property data collection and use, it makes scaling possible. In case of stiffness, the one parameter that can be used is the trace. In Figure 2.44, four tables with their stiffness components and engineering constants are summarized for all 15 CFRP materials and all DD laminates at 15 degree angle increments. The 4 tables are separated by 15 degree rigid body rotations; i.e., with the 1-axis at 0, 15, 30 and 45 degree. Other CFRP materials are expected to fall within the values presented, and other DD laminates in between 15 degree angle increments may be obtained by interpolation.

Rotation 0 1-axis 0 2-axis 90

	A11*	0	15	30	45	60	75	90	
	0	89							90
	15	83	78						75
	30	71	66	53					60
[±Φ]	45	58	53	40	27				45
	60	50	45	32	19	11			30
	75	47	42	30	17	9	6		15
	90	47	42	29	16	8	6	5	0
		90	75	60	45	30	15	0	A22*
					[±Ψ]				

A66*	0	15	30	45	60	75	90
0	3.1						
15	5.5	8.0					
30	10.4	12.9	17.7				
45	12.9	15.3	20.2	22.6			
60	10.4	12.9	17.7	20.2	17.7		
75	5.5	8.0	12.9	15.3	12.9	8.0	
90	3.1	5.5	10.4	12.9	10.4	5.5	3.1
					A12* = A66* - 1.5		
				[±Ψ]			

A61*	0	15	30	45	60	75	90	
0	0.0							90
15	0.0	0.0						75
30	0.0	0.0	0.0					60
45	0.0	0.0	0.0	0.0				45
60	0.0	0.0	0.0	0.0	0.0			30
75	0.0	0.0	0.0	0.0	0.0	0.0		15
90	0.0	0.0	0.0	0.0	0.0	0.0	0.0	0
	90	75	60	45	30	15	-A62*	
				[±Ψ]				

	E1*	0	15	30	45	60	75	90	
	0	88							90
	15	80	71						75
	30	61	51	30					60
[±Φ]	45	50	41	22	11				45
	60	47	40	24	11	6			30
	75	47	41	27	13	7	5.4		15
	90	47	41	28	14	7	5.4	5.2	0
		90	75	60	45	30	15.0	0.0	E2*
		E1* = 1/a11*			E2* = 1/a22*				

nu/21*	0	15	30	45	60	75	90	
0	0.32							90
15	0.74	1.10						75
30	1.08	1.32	1.43					60
45	0.70	0.83	0.97	0.77				45
60	0.31	0.39	0.51	0.47	0.31			30
75	0.10	0.16	0.26	0.26	0.17	0.08		15
90	0.04	0.09	0.18	0.20	0.13	0.05	0.02	0
	90	75	60	45	30	15	0	nu/12*
	nu/21* = -a21/a11				nu/12* = -a12/a22			

nu/61*	0	15	30	45	60	75	90	
0	0.00							90
15	0.00	0.00						75
30	0.00	0.00	0.00					60
45	0.00	0.00	0.00	0.00				45
60	0.00	0.00	0.00	0.00	0.00			30
75	0.00	0.00	0.00	0.00	0.00	0.00		15
90	0.00	0.00	0.00	0.00	0.00	0.00	0.00	0
	90	75	60	45	30	15	-nu/62*	
	nu/61* = a61/a11				nu/62* = a62/a22			

Rotation 15 | 1-axis 15 | 2-axis 75

A11*	0	15	30	45	60	75	90	
0	78							90
15	74	71						75
30	65	62	53					60
[±Φ] 45	55	52	42	32				45
60	47	44	35	24	17			30
75	43	40	31	20	12	8		15
90	42	38	29	19	11	7	6	0
	90	75	60	45	30	15	0	A22*

[±Ψ]

A66*	0	15	30	45	60	75	90
0	8.0						
15	9.2	10.4					
30	11.6	12.9	15.3				
45	12.9	14.1	16.5	17.7			
60	11.6	12.9	15.3	16.5	15.3		
75	9.2	10.4	12.9	14.1	12.9	10.4	
90	8.0	9.2	11.6	12.9	11.6	9.2	8.0

A12* = A66* - 1.5

[±Ψ]

A61*	0	15	30	45	60	75	90	
0	18.9							90
15	16.0	13.2						75
30	9.9	7.1	1.0					60
45	5.2	2.4	-3.7	-8.4				45
60	4.7	1.9	-4.2	-8.9	-9.4			30
75	7.0	4.2	-1.9	-6.6	-7.1	-4.8		15
90	8.4	5.6	-0.5	-5.2	-5.7	-3.4	-2.0	0
	90	75	60	45	30	15	-A62*	

[±Ψ]

E1*	0	15	30	45	60	75	90	
0	33							90
15	46	53						75
30	54	51	36					60
[±Φ] 45	48	43	27	14				45
60	42	39	27	13	8			30
75	34	35	27	15	8	6.0		15
90	26	31	27	15	8	5.8	5.3	0
	90	75	60	45	30	15.0	0.0	E2*

E1* = 1/a11* E2* = 1/a22*

nu/21*	0	15	30	45	60	75	90	
0	0.35							90
15	0.32	0.47						75
30	0.63	0.88	1.22					60
45	0.55	0.67	0.87	0.72				45
60	0.34	0.37	0.45	0.41	0.25			30
75	0.35	0.28	0.26	0.23	0.13	0.05		15
90	0.47	0.32	0.22	0.18	0.09	0.04	0.06	0
	90	75	60	45	30	15	0	nu/12*

nu/21* = -a21/a11 nu/12* = -a12/a22

nu/61*	0	15	30	45	60	75	90	
0	-2.28							90
15	-1.63	-1.05						75
30	-0.54	-0.07	0.69					60
45	-0.18	0.15	0.70	0.82				45
60	-0.39	-0.09	0.40	0.63	0.60			30
75	-0.98	-0.52	0.11	0.43	0.48	0.39		15
90	-1.56	-0.86	-0.05	0.33	0.41	0.30	0.11	0
	90	75	60	45	30	15	-nu/62*	

nu/61* = a61/a11 nu/62* = a62/a22

Rotation 30 | 1-axis 30 | 2-axis 60

A11*	0	15	30	45	60	75	90	
0	53							90
15	53	53						75
30	52	51	50					60
[±Φ] 45	48	47	46	42				45
60	41	41	40	36	29			30
75	35	35	33	29	23	17		15
90	32	32	31	27	20	14	11	0
	90	75	60	45	30	15	0	A22*

[±Ψ]

A66*	0	15	30	45	60	75	90
0	17.7						
15	16.5	15.3					
30	14.1	12.9	10.4				
45	12.9	11.6	9.2	8.0			
60	14.1	12.9	10.4	9.2	10.4		
75	16.5	15.3	12.9	11.6	12.9	15.3	
90	17.7	16.5	14.1	12.9	14.1	16.5	17.7

A12* = A66* - 1.5

[±Ψ]

A61*	0	15	30	45	60	75	90	
0	26.5							90
15	23.2	19.8						75
30	15.6	12.3	4.8					60
45	9.0	5.7	-1.8	-8.4				45
60	6.6	3.3	-4.2	-10.8	-13.2			30
75	7.5	4.2	-3.3	-9.9	-12.3	-11.4		15
90	8.4	5.1	-2.4	-9.0	-11.4	-10.5	-9.6	0
	90	75	60	45	30	15	-A62*	

[±Ψ]

E1*	0	15	30	45	60	75	90	
0	13							90
15	20	27						75
30	34	40	47					60
[±Φ] 45	40	43	41	26				45
60	34	37	35	21	12			30
75	21	27	30	20	11	8.1		15
90	14	21	27	20	11	7.4	6.0	0
	90	75	60	45	30	15.0	0.0	E2*

E1* = 1/a11* E2* = 1/a22*

[±Ψ]

nu/21*	0	15	30	45	60	75	90	
0	0.32							90
15	0.05	-0.11						75
30	0.00	-0.03	0.23					60
45	0.25	0.26	0.44	0.47				45
60	0.38	0.33	0.28	0.22	0.06			30
75	0.57	0.45	0.26	0.12	-0.01	-0.03		15
90	0.72	0.55	0.30	0.12	0.00	0.02	0.15	0
	90	75	60	45	30	15	0	nu/12*

nu/21* = -a21/a11 nu/12* = -a12/a22

[±Ψ]

nu/61*	0	15	30	45	60	75	90	
0	-1.32							90
15	-1.37	-1.38						75
30	-1.11	-0.99	-0.16					60
45	-0.53	-0.27	0.71	1.55				45
60	-0.41	-0.17	0.52	1.22	1.24			30
75	-0.63	-0.40	0.19	0.79	0.97	0.79		15
90	-0.82	-0.56	0.03	0.61	0.81	0.61	0.32	0
	90	75	60	45	30	15	-nu/62*	

nu/61* = a61/a11 nu/62* = a62/a22

[±Ψ]

Rotation 45 | 1-axis 45 | 2-axis 45

A11*	0	15	30	45	60	75	90	
0	27							90
15	30	32						75
30	35	37	42					60
[±Φ] 45	37	40	44	47				45
60	35	37	42	44	42			30
75	30	32	37	40	37	32		15
90	27	30	35	37	35	30	27	0
	90	75	60	45	30	15	0	A22*

[±Ψ]

A66*	0	15	30	45	60	75	90
0	22.6						
15	20.2	17.7					
30	15.3	12.9	8.0				
45	12.9	10.4	5.5	3.1			
60	15.3	12.9	8.0	5.5	8.0		
75	20.2	17.7	12.9	10.4	12.8	17.7	
90	22.6	20.2	15.3	12.9	15.3	20.2	22.6

A12* = A66* - 1.5

[±Ψ]

A61*	0	15	30	45	60	75	90	
0	20.8							90
15	19.4	18.0						75
30	15.6	14.2	10.4					60
45	10.4	9.0	5.2	0.0				45
60	5.2	3.8	0.0	-5.2	-10.4			30
75	1.4	0.0	-3.8	-9.0	-14.2	-18.0		15
90	0.0	-1.4	-5.2	-10.4	-15.6	-19.4	-20.8	0
	90	75	60	45	30	15	-A62*	

[±Ψ]

E1*	0	15	30	45	60	75	90	
0	8							90
15	11	14						75
30	19	21	27					60
[±Φ] 45	28	32	40	47				45
60	28	33	41	40	27			30
75	18	24	33	32	21	13.6		15
90	11	18	28	28	19	11.1	7.7	0
	90	75	60	45	30	15.0	0.0	E2*

E1* = 1/a11* E2* = 1/a22*

nu/21*	0	15	30	45	60	75	90	
0	0.24							90
15	0.00	-0.15						75
30	-0.11	-0.20	-0.25					60
45	0.10	0.04	-0.02	0.04				45
60	0.37	0.29	0.16	-0.02	-0.25			30
75	0.63	0.51	0.29	0.04	-0.20	-0.15		15
90	0.77	0.63	0.37	0.10	-0.11	0.00	0.24	0
	90	75	60	45	30	15	0	nu/12*

nu/21* = -a21/a11 nu/12* = -a12/a22

nu/61*	0	15	30	45	60	75	90	
0	-0.70							90
15	-0.96	-1.17						75
30	-1.14	-1.33	-1.63					60
45	-0.73	-0.83	-0.96	0.00				45
60	-0.22	-0.21	0.00	0.96	1.63			30
75	-0.03	0.00	0.21	0.83	1.33	1.17		15
90	0.00	0.03	0.22	0.73	1.14	0.96	0.70	0
	90	75	60	45	30	15	-nu/62*	

nu/61* = a61/a11 nu/62* = a62/a22

Figure 2.44 Summary of stiffness components and elastic constants of DD in percent of trace

DD is unique among composite materials is its field-based formulation. It is not a collection or stand-alone laminate that has little or no connection with other laminates. For example it is easy to show the ratio of E_1/E_6 like that can be used in EI/GJ in Figure 2.45 for DD laminates. This is a powerful example where the result, shown with yellow background, is universally valid for all CFRP and all the laminates summarized in one table. Again, values in between those 15 degree angle increments can be obtained by interpolation.

E1, %	0	15	30	45	60	75
15		71			Master	
30	61	51	30			
45	50	41	22	11		
60	47	40	24	11	6.4	
75	47	41	27	13	6.7	5.4
90	47	41	28	14	7.2	

	E6, %	15	30	45	60	75
15		8.0			Master	
30	10	13	18			
45	13	15	20	23		
60	10	13	18	20	18	
75	5.5	8.0	13	15	13	8
90	3.1	5.5	10	13	10	

	E1/E6	15	30	45	60	75
15		8.9			Master	
30	5.9	4.0	1.7			
45	3.9	2.7	1.1	0.5		
60	4.5	3.1	1.4	0.5	0.4	
75	8.5	5.1	2.1	0.9	0.5	0.7
90	15	7.5	2.7	1.1	0.7	

Figure 2.45 Best E_1/E_6 for DD laminates

The extension of stiffness to strength has also been demonstrated. It is summarized in the table in Figure 2.22 that is repeated here in Figure 2.46. These are radius-normalized strength. For absolute strengths for 15 CFRP they are shown in the bottom row of Figure 2.15. For CFRP other than T800/Cyt, the material rating factors can also be found in the bottom row of Figure 2.18. For T3/N52, the factor is 0.78. Materials and laminates are equal as seen in Figure 2.23.

[±Φ] \ [±Ψ]	0	15	30	45	60	75
15		0.42			Master	
30	0.62	0.53	0.36			
45	1.02	0.9	0.59	0.33		
60	1.47	1.34	1.01	0.59	0.36	
75	1.81	1.68	1.34	0.9	0.53	0.42
90	1.93	1.81	1.47	1.02	0.62	

[±Φ] \ [±Ψ]	0	15	30	45	60	75
15		0.22			Norm: [90/0]	
30	0.32	0.27	0.19		Master/1.93	
45	0.53	0.47	0.30	0.17		
60	0.76	0.70	0.52	0.30	0.19	
75	0.94	0.87	0.70	0.47	0.27	0.22
90	1.00	0.94	0.76	0.53	0.32	

[±Φ] \ [±Ψ]	0	15	30	45	60	75
15		0.53			T800/Cyt	
30	0.79	0.66	0.45		1.26*master	
45	1.29	1.13	0.74	0.41		
60	1.85	1.69	1.27	0.74	0.45	
75	2.28	2.12	1.69	1.13	0.66	0.53
90	2.43	2.28	1.85	1.29	0.79	

Figure 2.46 Repeat of Figure 22 of the one-parameter strength of 15 CFRP and 14 DD laminates

The ratio between stiffness and strength of a material is assumed to be fixed. For a given application, a composite material may be limited by either its stiffness or strength. With ability to control both stiffness and strength by their single parameters, it is easy to select a DD laminate that can match precisely the required stiffness and strength so both are used effectively in a specific application.

The strength values of laminates based on radius are taken from the laminate rating in the column on the right in Figure 2.15. Two examples shown in the lower right (with yellow background) in Figure 2.47 are the DD laminates with matched strength and stiffness in E_1. These two materials represent the lower and upper bounds, respectively. All other CFRP are expected to be bounded by them. Interpolation can be used to find the DD angles for desired matching between those listed.

This is another example of DD laminates being continuous and can be easily manipulated.

[±Ψ] (columns), [±Φ] (rows)

[±Φ] Radius	0	15	30	45	60	75
15		243			Master	
30	363	306	208			
45	595	524	341	191		
60	856	783	586	341	208	
75	1053	979	783	524	306	243
90	1125	1053	856	595	363	

[±Φ]	E1, %	15	30	45	60	75
15		71			Master	
30	61	51	30			
45	50	41	22	11		
60	47	40	24	11	6.4	
75	47	41	27	13	6.7	5.4
90	47	41	28	14	7.2	

[±Φ]	R/E1	15	30	45	60	75
15		3.4			Master	
30	5.9	6.0	6.9			
45	12	13	15	17		
60	18	20	24	31	32	
75	22	24	29	40	46	45
90	24	26	31	43	50	

For AS4/H35: R* = 0.75

Trace = 162 GPa

[±Φ] Radius	0	15	30	45	60	75
15		182			AS4/H53	
30	272	229	156		Master*0.75	
45	446	393	256	143		
60	642	587	440	256	156	
75	789	734	587	393	229	182
90	844	789	642	446	272	

[±Φ]	E1, GPa	15	30	45	60	75
15		115			AS4/H53	
30	99	83	49		Master*1.62	
45	81	66	36	18		
60	76	65	39	18	10	
75	76	66	44	21	11	8.7
90	76	66	45	23	12	

[±Φ]	R/E1	15	30	45	60	75
15		1.6			AS4/H53	
30	2.8	2.8	3.2			
45	5.5	5.9	7.2	8.0		
60	8.4	9.1	11	14	15	
75	10	11	13	19	21	21
90	11	12	14	20	23	

For T800/Cyt: R* = 1.26

Trace = 183 GPa

[±Φ] Radius	0	15	30	45	60	75
15		306			T800/Cyt	
30	457	385	262		Master*1.26	
45	750	660	429	240		
60	1079	987	739	429	262	
75	1326	1233	987	660	385	306
90	1418	1326	1079	750	457	

[±Φ]	E1, GPa	15	30	45	60	75
15		130			T800/Cyt	
30	112	93	55		Master*1.83	
45	92	75	40	20		
60	86	73	44	20	12	
75	86	75	49	24	12	10
90	86	75	51	26	13	

[±Φ]	R/E1	15	30	45	60	75
15		2.4			T800/Cyt	
30	4.1	4.1	4.8			
45	8.2	8.8	11	12		
60	13	13	17	21	22	
75	15	16	20	28	31	31
90	16	18	21	29	35	

Figure 2.47 DD laminates with desired match between strength and stiffness

Another example how the uniaxial tensile strength of laminates (shown with X) can be extended to imply their uniaxial compression and shear strength. Selected examples are shown in Figure 2.48 made with 3 quad laminates in QI in [0/±45/90], hard in [0_5/±45_2/90] and soft in [0/±45_4/90] and 3 DD in [±22.5/±67.5], [±0/±50], [±32/±58] laminates with 3 CFRP materials in T4708/MR (master), T800/Cyt (upper), and T300/N52 (lower).

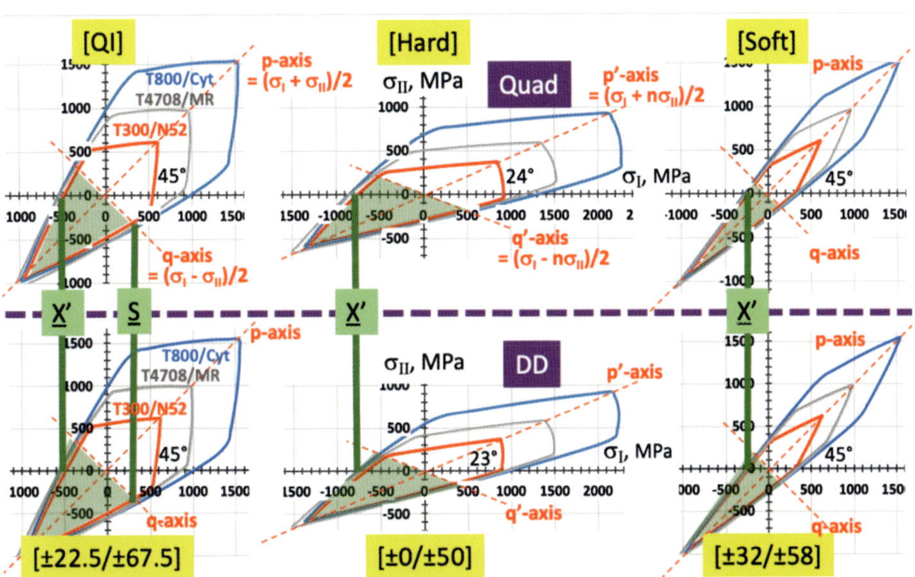

Figure 2.48 Equal strength in negative p-zone of all CFRP in quad and DD

Further simplification in testing for design data can be further achieved. As stated earlier, the A-to-Z testing of selected laminates such as QI, hard and soft would not give confidence in how laminates other than those are in actual use. Master envelopes can not only anticipate in the resulting data (not like blind testing) but also reduction in the number of tests, and to that of calibration and validation.

Figure 2.49 shows the laminate uniaxial strengths of 3 materials in T4708/MR (equal to the master), T800/Cyt (upper bound), and T3/H52 (lower bound). Tensile strengths vary among DD laminates CFRP vary as seen in the coefficients of variations in the last column. Compressive and shear strengths on the right 2 columns however remain essentially close for all three materials with their lower coefficients of variations. They are not inconsistent with those 3 DD shown in Figure 2.48.

#	X	T4708	T8/Cy	T3/H52	Average	cv
2	[75/0]	1300	1920	780	1333	43%
8	[90/0]	1270	1900	780	1317	43%
3	[60/0]	1250	1500	820	1190	29%
1	[75/15]	1130	1600	680	1137	40%
10	[90/15]	1120	1680	780	1193	38%
4	[60/15]	950	1050	700	900	20%
6	[45/0]	750	800	700	750	7%
11	[90/30]	710	1000	420	710	41%
12	[75/30]	700	840	400	647	35%
7	[45/15]	500	500	450	483	6%
5	[60/30]	400	450	350	400	13%
14	[45/0]	300	300	280	293	4%
15	[75/45]	250	400	150	267	47%

#	X'	T4708	T8/C	T3/H52	Average	cv
8	[90/0]	900	850	880	877	3%
2	[75/0]	880	850	780	837	6%
3	[60/0]	810	820	750	793	5%
10	[90/15]	750	750	780	760	2%
1	[75/15]	740	720	680	713	4%
6	[45/0]	700	750	650	700	7%
4	[60/15]	650	650	600	633	5%
7	[45/15]	500	500	500	500	0%
11	[90/30]	450	450	450	450	0%
12	[75/30]	420	420	380	407	6%
5	[60/30]	320	320	350	330	5%
14	[90/45]	200	200	200	200	0%
15	[75/45]	180	150	150	160	11%

	S	T4708	T800/	T3/52	Average	cv
10	[±45]	678	710	619	669	7%
9	[45/30]	596	624	544	588	7%
5	[60/30]	513	537	469	506	7%
13	[±30]	513	537	469	506	7%
7	[45/15]	430	451	394	425	7%
4	[60/15]	348	365	319	344	7%
12	[30/15]	348	365	319	344	7%
6	[45/0]	348	364	319	344	7%
3	[60/0]	265	278	244	262	7%
11	[30/0]	265	278	244	262	7%
1	[75/15]	183	192	170	182	6%
14	[±15]	183	192	170	182	6%
2	[75/0]	100	106	95	100	5%
8	[90/0]	17	20	20	19	9%

Figure 2.49 Uniaxial tensile, compressive, and shear strength of selected DD laminates

Strength data in Figure 2.50 are plotted in Figure 2.49 in columns. Also shown in the upper right the selected compressive and shear strength of all CFRP in DD tables. Their difficult tests may be avoided for most CFRP after extensive data from various CFRP become available. This is like shear modulus that need not be tested (being 3 percent of trace). Tensile tests being the simplest to perform may be adequate for design data generation. Tensile data only is usually reported with metals. With DD, this can be the case also.

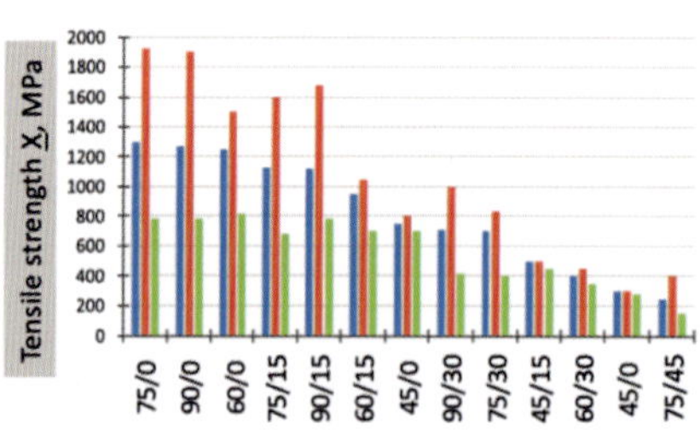

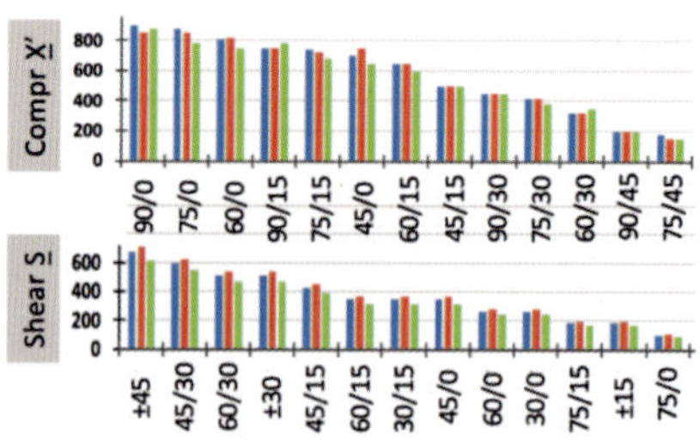

X'	0	15	30	45
45	700	500		
60	793	633	330	
75	837	713	407	160
90	877	760	450	200

S	0	15	30	45
45	344	425	588	669
60	262	344	506	588
75	100	182	344	425
90	19	100	262	344

Figure 2.50 Tensile, compressive, and shear strengths of DD

The sensitivities of the angle increment Δ on the best DD angles and their working stress for 4 multiple loading conditions are shown in Figure 2.51. The load sets are given in unit stress

vectors, as described in Chapter 1. It is a convenient input so the output would be in maximum working stress.
The best DD is highlighted for each loading. The change in Δ is an alternative to interpolation to find the best laminate from a baseline. Among all the examples in Figure 2.51, there is only the lower right case for submarine are there nearly 10 percent variation. It would seem that only if large number of component/part is to be produced angle increment of 15 degree (that is the baseline for this chapter) is good enough as the final choice.

Again, DD being field-based will permit zooming to reach better laminates. It is so different from quad that is a collection of dispersed, unconnected laminates.

Upper wing			Lower wing			Fuselage			Submarine		
-1.00	0.00	0.20	1.00	0.00	-0.20	0.50	1.00	0.00	-0.50	-1.00	0.00
-0.80	0.10	0.00	0.90	0.00	1.00	0.50	0.70	0.10	-0.50	-0.90	0.10
-0.70	0.20	-0.10	0.80	-0.10	0.10	0.50	0.80	0.10	0.00	0.60	0.10
0.20	-0.40	0.00	0.20	-0.40	0.00	0.50	0.90	0.10	0.40	0.00	0.10
0.50	0.00	0.00	-0.50	0.00	0.00	-0.10	0.60	0.10	0.00	0.50	0.00
0.20	0.00	0.20	0.20	0.20	0.20	0.20	0.40	0.30	0.20	0.20	0.10
0.30	-0.50	0.10	0.30	-0.60	0.10	0.10	-0.60	0.20	0.30	-0.30	0.10

Upper wing skin	Unit circle criterion						
Δ	15	13	11	9	7	5	3
[±Φ]	60	117	121	63	56	60	59
[±Ψ]	0	0	0	0	0	0	0
MPa	639	634	638	634	624	639	638

Lower wing skin	IM7/977						
Δ	15	13	11	9	7	5	3
[±Φ]	45	156	132	45	49	45	44
[±Ψ]	15	52	22	18	21	25	18
MPa	245	237	242	247	244	245	247

Fuselage							
Δ	15	13	11	9	7	5	3
[±Φ]	90	87	88	81	77	80	105
[±Ψ]	30	33	33	36	35	35	36
MPa	534	542	542	542	542	543	544

Submarine							
Δ	15	13	11	9	7	5	3
[±Φ]	90	91	88	90	83	90	99
[±Ψ]	30	26	22	27	28	25	36
MPa	708	767	732	758	747	750	762

Figure 2.51 Effect of angle increment Δ on the best DD and their radius for effective strength

2.12 Comparative rating systems

The traditional study of composites usually starts with one or more given materials. The task is usually to find the best laminates for stiffness, strength, toughness, buckling, and natural frequency. This is a point design and will require re-calculation if the material is changed. With one-parameter for stiffness and strength, it may be possible to take advantage of scaling for without re-calculation. DD can be positioned to make this possible because it can be easily homogenized. Most laminates in use currently cannot be homogenized. That makes scaling difficult if not impossible. Homogenization makes manufacturing much simplified also. Instead of hundreds of laminates in a structure, as conventional fuselage and wings will have, one optimum DD for the entire structure. It can be laid with speed and quality totally out of reach by the conventional approach.

Trace is the only constant that is needed for laminate stiffness. Its determination can be obtained from any laminate including [0/90] where $A_{11}^* = E_1^* = 47$ percent; i.e., trace = $E_1^*/0.47$. Shear $E_6^*= 0.03$ x trace should be as confidently accurate as any test data that can be measured. Note that Poisson's ratio of [0/90] is 0.04, or $A_{11}^* = E_1^*/(1 - 0.04^2) = 0.999E_1^*$. There is more significance of having one constant beyond its simplicity. One feature is scaling. The relative values of their trace between two materials, say, 20 percent higher of one over the other, is reflected in all stiffness related properties like deformation, buckling and vibration. Their difference is 20 percent and needs no additional calculation.

Simplifications of the stiffness and strength are made possible with trace, omni envelope, Tsai-Wu and unit circle criteria, and field-based double-double laminates. Lam search, an optimization tool, can select the highest-strength laminate under 4 load sets in Figure 2.52. Their load sets are again in unit stress vectors as used in the last figure so working stress can be obtained directly, as explained in Figure 1.34.

Multiple load sets

Lower wing			Fuselage		
1.00	0.00	-0.20	0.50	1.00	0.00
0.90	0.00	1.00	0.50	0.70	0.10
0.80	-0.10	0.10	0.50	0.80	0.10
0.20	-0.40	0.00	0.50	0.90	0.10
-0.50	0.00	0.00	-0.10	0.60	0.10
0.20	0.20	0.20	0.20	0.40	0.30
0.30	-0.60	0.10	0.10	-0.60	0.20

Upper wing			Wide band		
-1.00	0.00	0.20	1.00	0.00	0.00
-0.80	0.10	0.00	-1.00	0.00	0.00
-0.70	0.20	-0.10	0.00	1.00	1.00
0.20	-0.40	0.00	0.00	-1.00	1.00
0.50	0.00	0.00	0.00	0.00	1.00
0.20	0.00	0.20	1.00	1.00	1.00
0.30	-0.50	0.10	-1.00	-1.00	1.00

		Best DD	[45/15]	[90/30]	[60/0]	[60/30]
Trace	#	CFRP	Lower wing	Fuselage	Upper wing	Wideband
162	1	AS4/H35	212	429	586	167
232	2	IM6/ep	250	537	633	209
168	3	T3/F93	166	353	457	134
206	4	T3/N52	185	371	545	149
139	5	C-Ply 55	252	535	665	207
163	6	C-Ply 64	262	535	752	213
218	7	IM7/977	283	630	673	240
183	8	T800/Cyt	245	534	639	200
180	9	IM7/8552	231	590	516	213
195	10	IM7/MTM	318	754	717	275
144	11	T7/2510	188	393	535	151
160	12	T650/EP	231	482	638	188
181	13	IM7/8552'	274	602	687	227
158	14	T4708/MR	187	385	550	146
168	15	T8S/3900	445	1027	1028	376
		Average	249	544	641	206

Figure 2.52 Highest strength laminates based on 4 multiple load sets

The resulting best laminates are shown in the top row of the table on the right. The working stress in MPa is listed for each load set for each CFRP. For example, for the lower wing skin and AS4/H35 the working stress is 212 MPa, as seen in the first column and top row on the right table. The normalized working stress depends on material, not the laminate.

In Figure 2.53, the columns show the normalized strengths (by their respective average at the bottom of each column in Figure 2.53 in values of 249, 544, 541 and 206) have small coefficients of variations in the last column of each table. It is similar to Figure 2.23 where material contributions to all DD laminates have the same rating.

The normalized strengths can be plotted in Figure 2.53. On the left the stiffness is in ascending order; the right, strength in ascending order. Note their nearly equal strengths when normalized.

What we have done with trace, master failure envelopes, rating and scaling of laminates, particularly those in the DD family, become simple and straightforward because of the discovery of single parameters for stiffness and strength, scalar product in Tsai-Wu failure criterion, and the recognition of the importance of field-based family of laminates. Not only are the theory is fundamental correct, but also achieve simplicity as our primary goal. Quad was conceived in the

1960's by people with limited knowledge and experience in mechanics of composites. With DD, the familiar rules imposed on the quad laminates no longer apply. Our new horizon is based on homogenization, materials/laminates equivalence, rating, scaling, tensile data only, and more.

Stiffness

			[45/15]	[90/30]	[60/0]	[60/30]		
Trace	#	CFRP	Lower	Fuselag	Upper	Wide	Ave	cv
139	5	C-Ply 55	1.01	0.98	1.04	1.00	1.01	2%
144	11	T7/2510	0.76	0.72	0.83	0.73	0.76	7%
158	14	T4708/MR	0.75	0.71	0.86	0.71	0.76	9%
160	12	T650/EP	0.93	0.89	0.99	0.91	0.93	5%
162	1	AS4/H35	0.85	0.79	0.91	0.81	0.84	7%
163	6	C-Ply 64	1.05	0.98	1.17	1.03	1.06	8%
168	3	T3/F93	0.67	0.65	0.71	0.65	0.67	4%
168	15	T8S/3900	1.79	1.89	1.60	1.82	1.78	7%
180	9	IM7/8552	0.93	1.08	0.80	1.03	0.96	13%
181	13	IM7/8552'	1.10	1.11	1.07	1.10	1.10	1%
183	8	T800/Cyt	0.99	0.98	1.00	0.97	0.98	1%
195	10	IM7/MTM	1.28	1.39	1.12	1.33	1.28	9%
206	4	T3/N52	0.74	0.68	0.85	0.72	0.75	10%
218	7	IM7/977	1.14	1.16	1.05	1.16	1.13	5%
232	2	IM6/ep	1.01	0.99	0.99	1.01	1.00	1%

CFRP	Trace	Strength
AS4/H35	162	0.84
IM6/ep	232	1.00
T3/F93	168	0.67
T3/N52	206	0.75
C-Ply 55	139	1.01
C-Ply 64	163	1.06
IM7/977	218	1.13
T800/Cyt	183	0.98
IM7/8552	180	0.96
IM7/MTM	195	1.28
T7/2510	144	0.76
T650/EP	160	0.93
IM7/8552'	181	1.10
T4708/MR	158	0.76
T8S/3900	168	1.78

Strength

			[45/15]	[90/30]	[60/0]	[60/30]		
Trace	#	CFRP	Lower	Fuselag	Upper	Wide	Ave	cv
168	3	T3/F93	0.67	0.65	0.71	0.65	0.67	4%
206	4	T3/N52	0.74	0.68	0.85	0.72	0.75	10%
158	14	T4708/MR	0.75	0.71	0.86	0.71	0.76	9%
144	11	T7/2510	0.76	0.72	0.83	0.73	0.76	7%
162	1	AS4/H35	0.85	0.79	0.91	0.81	0.84	7%
160	12	T650/EP	0.93	0.89	0.99	0.91	0.93	5%
180	9	IM7/8552	0.93	1.08	0.80	1.03	0.96	13%
183	8	T800/Cyt	0.99	0.98	1.00	0.97	0.98	1%
232	2	IM6/ep	1.01	0.99	0.99	1.01	1.00	1%
139	5	C-Ply 55	1.01	0.98	1.04	1.00	1.01	2%
163	6	C-Ply 64	1.05	0.98	1.17	1.03	1.06	8%
181	13	IM7/8552'	1.10	1.11	1.07	1.10	1.10	1%
218	7	IM7/977	1.14	1.16	1.05	1.16	1.13	5%
195	10	IM7/MTM	1.28	1.39	1.12	1.33	1.28	9%
168	15	T8S/3900	1.79	1.89	1.60	1.82	1.78	7%

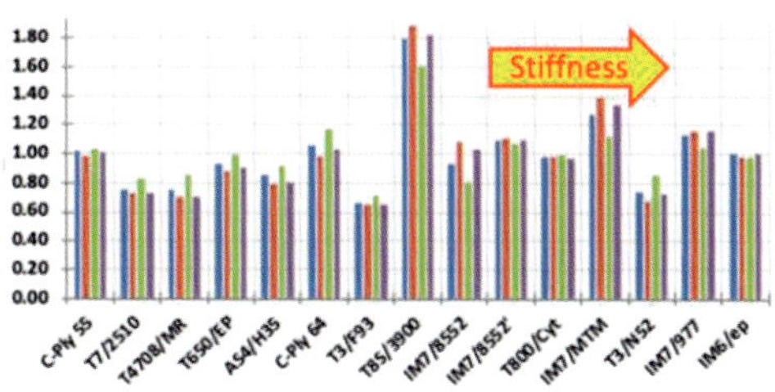

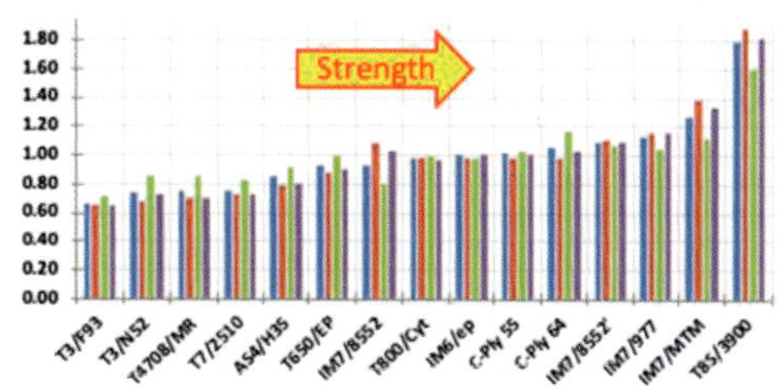

Figure 2.53 Normalized strengths from absolute values in Figure 2.57 in ascending orders

The rating shown in Figure 2.54 can be compared with those listed in lower row of Figure 2.20. These two rating schemes shown in Figure 2.54 are based on different criteria. The first column in blue is the strength of the average controlling load of 4 load cases. The second in maroon is the master strength based on the average of 14 DD laminates. Being close among the two rating scheme gives us confidence in this comparison. Having a single parameter is very important.

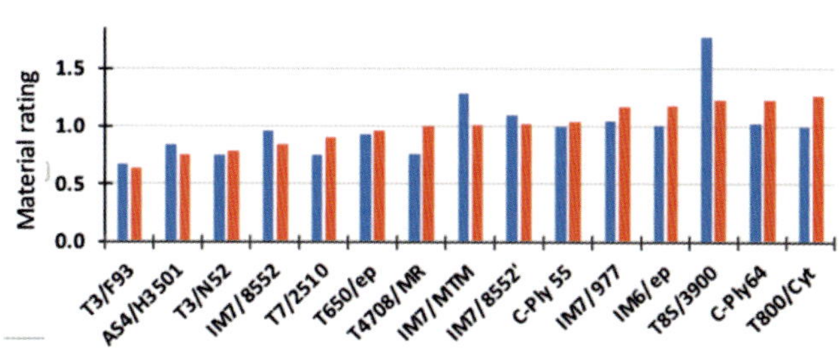

Figure 2.54 Comparative ratings by Lam search (blue) and ascending master envelope (maroon)

Rule of mixtures formula can be applied to the longitudinal stiffness and strength, E_x and X, directly and plotted in Figure 2.55. New with this plot are the recognitions of trace for stiffness and the average of X and X' for strength, from which fiber stiffness and strength are recovered. Also included is the fiber volume fraction that is needed for the rule of mixtures. This plot is more meaningful than the straight plot of fiber stiffness and strength where fiber volume and compressive strength are not included. Fiber stiffness and strength can now be related directly to laminate properties with more credence because compressive and shear strengths be estimated as seen in Figure 2.55. Availability of allowable data as a show stopper may soon be over.

#	3	1	4	11	9	6	12	10	13	14	5	7	2	8	15
CFRP	T3/F9	AS4/H	T3/N5	T7/25	IM7/8	C-Ply6	T650/	IM7/M	IM7/8	T4708	C-Ply	IM7/9	IM6/e	T800/	T8S/3900
Ex	146	138	181	126	171	141	139	175	159	142	121	191	203	162	151
X	1314	1447	1500	2172	2326	2944	2194	2500	2501	2524	2530	3250	3500	3768	3000
X'	1220	1447	1500	1450	1200	1093	1653	1700	1700	1700	1669	1600	1540	1656	2500
v/f	0.60	0.66	0.70	0.59	0.57	0.64	0.59	0.59	0.59	0.59	0.55	0.64	0.66	0.64	0.59
E/f	217	184	228	188	264	194	207	261	261	212	194	263	271	223	229
X/f	2112	2129	2143	3069	3093	3154	3260	3559	3560	3580	3734	3789	3818	4238	4661

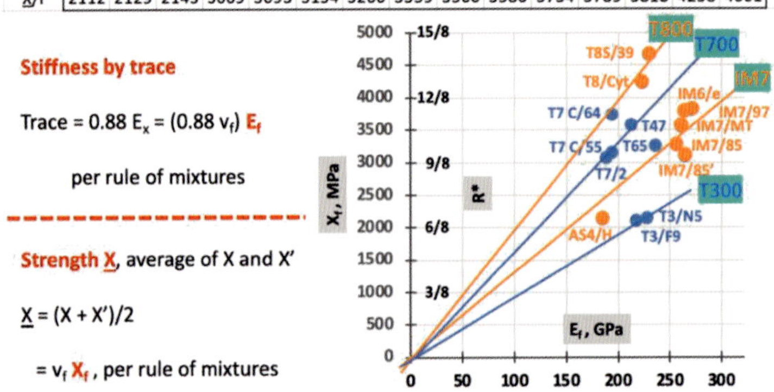

Figure 2.55 Fiber strength versus stiffness for 15 CFRP

A similar plot on DD laminates can also be made in strength versus stiffness. This is shown in Figure 2.56 which has two parts: A_{11} on the right; A_{66} on the left in fractions of trace. The vertical axis is that of the laminate strength measured by its radius in MPa.

	Radius*	Stiffness	
CFRP	Master	A11*	A66*
[90/0]	1125	0.469	0.031
[75/0]	1053	0.472	0.055
[75/15]	979	0.420	0.080
[60/0]	856	0.500	0.104
[60/15]	783	0.448	0.129
[45/0]	595	0.580	0.129
[60/30]	586	0.323	0.177
[45/15]	524	0.527	0.153
[30/0]	363	0.708	0.104
[45/30]	341	0.402	0.202
[30/15]	306	0.656	0.129
[±15]	243	0.781	0.080
[±30]	208	0.531	0.177
[±45]	191	0.274	0.226

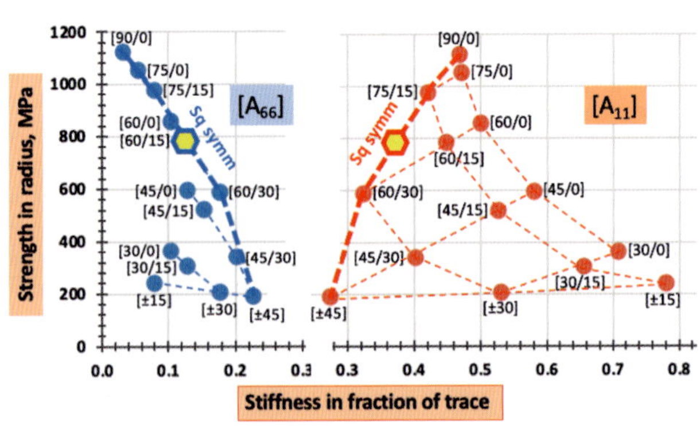

Figure 2.56 Strength in radius and stiffness in fraction of trace for DD laminates

The heavy dashed lines are the square symmetric laminates. The master radius values are taken from those in the upper left table in Figure 2.47, which were taken from the laminate rating in MPa from the right column in Figure 2.15 (normalized by 582, the master value). For specific materials such as AS4/H35 or T800/Cyt, factors from laminate rating must be applied as seen in the first column of Figure 2.47. The factors are 0.75 and 1.26 to be applied to the master envelopes, respectively. The x-axis in fractions of trace remains the same for all materials.

These last two figures show again the equivalence between materials and laminates which is more relevant and rational with DD than those with dispersed laminates.

Trace is the one and only constant needed to define the stiffness of all CFRP laminates. For example, materials in AS4/H35 and T800/Cyt have respective values of 162 and 183 GPa, from the middle tables of Figure 2.47. Their ratio 183/162 = 1.43; T800/Cyt is 43 percent stiffer than AS4/H35 That ratio remains the same for all laminates, i.e., for [45/15], [90/30], [60/0], [60/30],… If a CFRP structure made of many laminates its deformation using the former will be 13 percent stiffer than the latter. No additional calculation is necessary. Thus trace is the one and only one material constant that can be used to rate and scale the stiffness of not only composite laminates but also composite structures made of many laminates of the same material.

The rating and scaling are made easy because the laminates are field-based double-double. All member laminates are evenly spaced such as by a constant angle increment of 7.5 degree used in all the tables in this chapter. They are easily homogenized (in 4 or less repeats). Being discrete, not continuous, quad laminates in 0, ±45, 90 cannot be simplified because they are heterogeneous, i.e., cannot be homogenized. Heterogeneous laminates are very complicated. They have thousands of stacking sequences (as compared with 2 for double-double.). Member laminates in the quad family are not evenly spaced as seen in the carpet plot in Figure 2.41. Laminates can be multiple of 6-, 8- or 10-ply sub-laminates. The highest strength laminates can be from different sub-laminates and are therefore not well-behaving functions as is the case of the field-based double-double. For quad, each design is unique (with ply drops and blending) and cannot be readily rated and scaled. Quad will have size effect while double-double will not. Quad laminates are likely to warp while DD will not. Master envelopes can be used to find the highest strength double-double, the same laminate will be the best for all CFRP, i.e., only need to optimize once. The key is to work with a field-based laminates, of which double-double is probably the most flexible and efficient choice.

2.13 Conclusions

The key to simplification is DD. It will reach the goal because it is field-based. It is single-valued, continuous, and well behaved. It can zoom and optimize. Quad and other dispersed laminates are *ad hoc* and have none of the above quality and utility.

In manufacturing DD can be homogenized, and be produced with one DD, without mid-plane symmetry, without blending, ply drops in singles, no warpage, no internal discontinuities and no wrinkles. Quad laminates have been produced with man-made rules that have inflicted many complications and challenges. They cannot be homogenized because the basic in-family building blocks are too thick, like 16 or 20 plies. Supplemental plies are often needed to round off plies in a laminate. Blending and ply drops make each structure unique, not seen before nor seen again in the future. Every design has to start from scratch. DD is a new world. Elegantly simple, scalable and producible with unmatched quality and speed.

For DD, material and laminate are the two faces of the same entity. With one appropriate parameter each in trace and radius, DD laminates can be rated and scaled. Every material can have all laminates with same rating. If [90/0] is the top laminate, it is the top laminate for all materials with the same numerical rating and superiority. Conversely, every laminate can be made by all materials. The top material is, say, T800/Cyt. It is the top material for all laminates with the same superior numerical rating. So material, laminate, and their stiffness and strength are all orderly defined and connected through rigid rules of transformation. Quad are dispersed, arbitrary and subjective, can have some local connections (same 8- or 10-ply family), but no global connection. Scaling can be a powerful tool that can change material and/or laminate without re-calculation. Such opportunities are inherent with DD, but not so with dispersed laminates like the quad.

For LPF, the only test data needed are X and X'. Their average value is correlated to the radius of the master envelopes. This single parameter for strength provides the criterion for rating of materials and laminates alike. With the single parameter for stiffness in trace, concise and precise description of composites can be consolidated in one table of materials versus laminates, as seen in Figure 2.20. There is no such universality in quad and other dispersed laminates. Design data generation is one of validation or calibration for DD while dispersed laminates are the results of point design and must be tested individually and in full range from A to Z. For DD design data generation can be totally different with tensile tests only. Compression and shear data can be deduced; see Figures 2.48 – 2.50. Tensile tests only may suffice. LPF formulation can be extended to coupons with hole (Nettles envelope) and possibly any stress raisers and damage. Examples can be found in Chapter 10.

DD can provide one laminate for the entire component, while dispersed laminates may take hundreds of laminates. Such complexities are the results of stacking sequence, blending among laminates, ply drops with mid-plane symmetry, and other subjective choices with little or no analytic foundations. DD is disciplined, and with one-parameters in trace and radius, possibly more to come. The resulting structures are systematic, optimizable, scalable, and leading to the simpler design and manufacturing than even metals. Switching from quad to DD could be like the sea changes from particle to wave, from DC to AC, from analog to digital, and so on. In all honesty and modesty, the future is now!

Acknowledgement

The author wishes to thank contributions from many individuals in our global research team on DD, with particular inputs from Antonio Miravete, Daniel Gay, Alan Nettles, Waruna Seneviratne, Dan Wang, and Maj Gen (Ret) Jocelyn Seng, USAF.

Support from Northrop-Grumman Corporation for the publication of this book is gratefully acknowledged.

References

See Chapter 1

MicMac/Duplex can be used to calculate master failure envelopes as described at the end of Chapter 15. Free copies are available from Think Composites.

CHAPTER 3

Unique Manufacturing Opportunity

Dr.-Ing. Erik Kappel

April 2022

Tailorbility is the key advantage of ply-based carbon-fiber-epoxy laminates. More than 50% of the structural weight of today's civil aircrafts, as the Airbus A350 or the Boeing 787, are made from CFRP, while unidirectional (UD) Prepregs are still the baseline. Large-scale components, like fuselage shells, upper and lower wing covers or door surround structures are fabricated using automated processes, such as automated tape laying (ATL) or automated fiber placement (AFP). More automated approaches are also investigated in context of circumferential frames and longitudinal stiffeners, indicating that efficient manufacturing will be even more important for next generation aircrafts. Longitudinal stiffeners, also denoted as stringers, are addressed in the following, to demonstrate unique manufacturing opportunities, enabled by a new family of laminates – so called Double-Double (DD) laminates [0,1].

Zone-based laminate design is common practice. The classical laminate theory (CLT) [2] is used in the design and stress analysis context to define local laminate parameters. One finds extensive use of locally varying stacking sequences in full-scale components, such as wing covers. Hundred and more different stacking sequences in a single component are not uncommon.

However, laminates and all zones are usually composed of a pre-defined set of ply orientations. In most aerospace structures those ply angles are 0°, 45°, -45° and 90°. Thus, tailoring the laminate locally means adapting the relative fractions of those four pre-defined ply orientations. It does not refer to orienting the individual ply angles based on given loads.

Tailoring a single laminate is rather easy. However, when the number of adjacent zones increases, laminate design becomes complex. Compatibility between adjacent zones needs to be assured in order to avoid harmful discontinuities of stiffness and also manufacturability.

In practice, a set of five guidelines, called stacking rules, has been established over years. Those stacking rules are intended to assure robust laminates. They usually refer to the aforementioned 0°, (45°, -45°), 90° ply orientations, while laminates are specified by relative fractions [a%, b%, c%] of those. The most relevant stacking rules are:

- Laminates shall be symmetric to mid plane
- Balance of ply orientations $+\alpha, -\alpha$
- Each ply fraction must be part of the laminate stack, with at least 10% - The 10% rule. (sometimes 8%)
- Ply nesting shall be avoided (no more than four plies of identical orientation in a sub block within the laminate stack)
- ±45° inter-ply angle increments

Figure 1 shows the basic equation of the CLT, with the ABD matrix describing the interaction between strain (ε), curvature (κ), membrane forces (N) and bending moments (M).

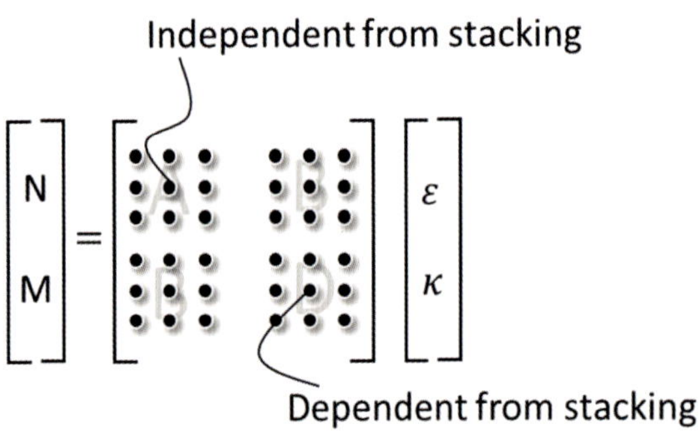

Figure 1: CLT's basic relation

The population of the ABD matrix determines the complexity of the material's behavior. Figure 2 illustrates the effect of the stacking rules on the population, which directly indicates why those rules are used in practice.

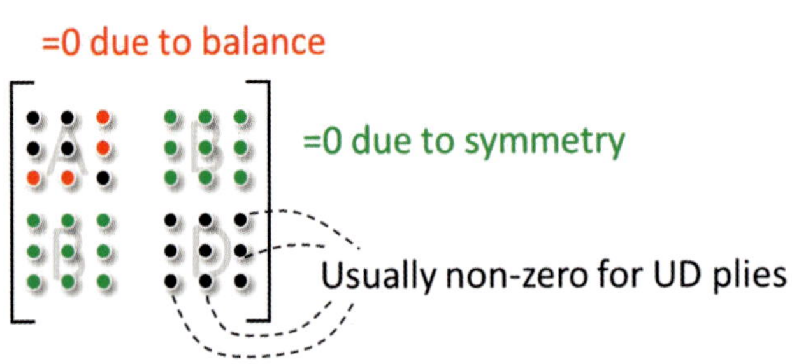

Figure 2: Stacking rule effects on ABD matrix population in CLT

Balance and symmetry assure a rather simple mechanical behavior. It is characterized by no in-plane extension-shear coupling, no extension-bending coupling and no or very little bending-twist coupling. The uncoupled behavior is an essential point for the laminate design!

The ply-nesting rule, in contrast, does not affect the ABD matrix populations directly. It aims to avoid severe laminate inhomogeneity, which is induced by high concentrations of a particular ply orientation within the laminate stack, for example a sub-stack of six or more 0°

plies. The inter-ply-angle increment does not directly affect the ABD matrix population as well. However, the combination of both rules, decreases the number of stacking-rule conform laminates drastically.

The rules assure that CFRP structures act as desired in-service, where membrane forces only lead to in-plane strains and bending moments lead to curvature only. The absence of coupling also assures that parts do not distort after manufacturing, during the cool down at the end of the high-temperature manufacturing process. It is particularly important to highlight that the stacking rules need to be fulfilled for any laminate of a multi-zone structure. Thus, whenever the laminate thickness or stacking is adapted locally, laminate transitions need to be developed and compatibility needs to be in mind. Late-entry laminate changes in part design, for example, can be difficult to handle, as in some cases, multiple zone-specific stackings need to be updated, which in turn also affects tool shapes. As will be shown, DD offers interesting design and manufacturing opportunities, which promote easy local adaptations.

Figure 3 shows an example of a prepreg-made, autoclave-cured CFRP C-profile, which allows for demonstrating the selected challenges of conventional laminates. The structure consists of a thin section (16 plies) a thick section (32 plies) and a transition zone in between.

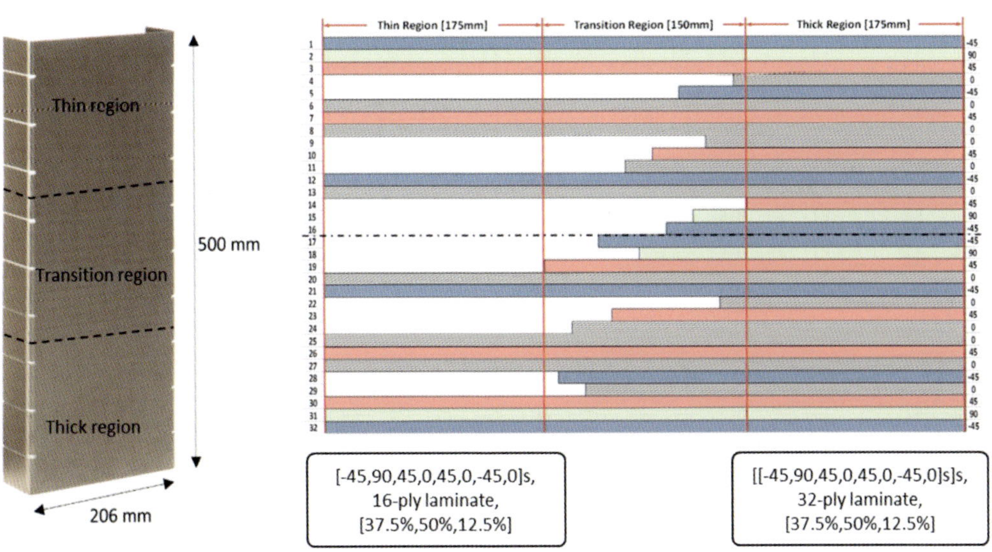

Figure 3: CFRP C-channel, prepreg made with a conventional 0°,45°, -45°, 90° laminate

The ply-drop-offs are distributed within the transition zone. What already looks quite complex for the simple, one-dimensional transition, becomes significantly more challenging when transition regions need to be designed in two dimensions, or when different transition zones intersect each other. It should also be noted that ply drop offs in conventional laminates are usually located within the laminate stack. This induces a considerable number of little resin pockets and also local fiber waviness within the inner laminate architecture, which for sure do not improve performance.

3.1 Double-Double family of laminates

The Double-Double family of laminates is characterized by a building block (BB) and the repetition parameter r. A DD laminate is usually asymmetric but balanced. The DD building block is composed of four plies, as for example

$$[\varphi, -\varphi, \psi, -\psi]_{rT}$$

4-UD-Ply building block ‚repeating index'

At first sight, the DD facts sound weird for composite experts. It is well known that structures with asymmetric laminates distort already during the cool down at the end of the autoclave process. Reaching tight dimensional tolerances appears impossible with asymmetric stacking and complex mechanical interactions are to be expected, as coupling terms in the [A] and [D] matrix as well as the [B] matrix are populated. Recall that all those issues are eliminated when laminates are designed conform to the stacking rules.

Homogenization is the key of the DD concept.

When DD laminates are examined in detail, one will find the important key-role of the repetition parameter r, as increasing r leads to homogenization.

Table 1 visualizes the process of homogenization as a function of the repetition parameter, while the decreasing size of the red dots symbolizes the approaching of zero.

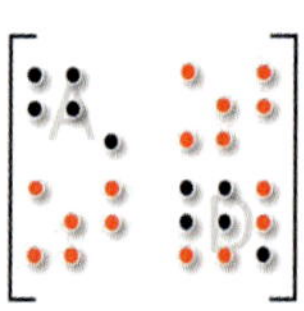

DD, UD-ply based, r=1

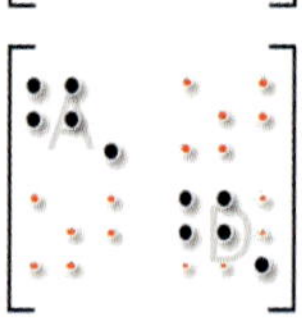

DD, UD-ply based, r>>1

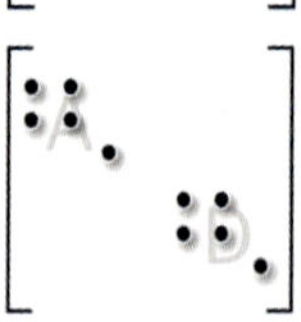

Isotropic material

Table 1: ABD matrices for DD configurations and an isotropic material

When a DD laminate, with a single BB, is examined, which is equivalent to r=1, one will find well populated [A], [B], [D] matrices and all the implied difficulties. Even though, the coupling terms in the [A] matrix are zero, due to DD's balanced character, complex laminate behavior must be anticipated, likely leading to a technically useless laminate. However, when the number of BB repeats is increased one finds clear trends for the [B] entries and the coupling terms of the [D] matrix. It can be shown that the [B] entries scale with 1/r, while the coupling terms in [D] even scale with $1/r^2$.

One can conclude that the negative aspects of the asymmetry in the laminate approach zero for increasing number of repeats. This process is denoted as homogenization, as the mechanical behavior of a DD laminate with multiple repeats approach the characteristic of an isotropic media in terms of coupling. Laminate stiffness and bending resistance can still differ between the laminate's principal directions.

One can further conclude that, when coupling diminishes, as in isotropic materials, manufacturing of DD composite structures with a high level of dimensional fidelity must be possible. But how many repeats to do we need in practice?

As will be shown later in this section, $r \geq 4$ already leads to parts without critical, unwanted distortions after manufacturing. Nonetheless, more repeats are recommended.

But even for the lower threshold of $r=4$, the minimum DD laminate consists of 16 plies. Most CFRP aerospace structures are made from low-, medium-, or high-grade prepregs, which refer to cured ply thicknesses of around 0.125 mm, 0.188 mm or 0.250 mm, respectively. Thus, a 16-ply laminate show a thickness of two up to four mm. In case thinner laminates are in focus for a structure or the minimum number of repeats shall be higher, DD laminates are often discussed in context of thin-ply technologies. When super thin prepregs are examined [9], with cured ply thicknesses of only 0.02 mm for example, the 16-ply laminate would be 0.32 mm thick only, which offers additional opportunities and hardly any limits, referring to the minimum DD laminate thickness.

To demonstrate, that no special materials are mandatory for DD, all parts presented hereafter are made from conventional, state of the art, unidirectional low-grade prepreg, indicating that the opportunities related to DD can be used already with the materials available today.

In conventional multi-zone laminates the laminate properties are adapted by locally varying the ply-orientation fractions within the laminate stack. This allows for changing the laminate characteristic from a quasi-isotropic to a more orthotropic one, for example, by dropping 90° plies or by adding 0° plies. However, changing the characteristics is always accompanied by mismatches between adjacent zones in terms of thermal expansion (CTEs) and discrete changes of in-plane stiffness and out-of-plane bending resistance.

DD laminate design pursues a novel strategy, utilizing the described mechanism of the homogenization. In a DD part, the ply orientations in the building block remain constant throughout the whole part. The exact angles in the BB are defined based on the load scenario at hand [3] or they are determined by a direct conversion of a conventional laminate stacking. In a DD structure, only the number of repeats is adapted in the zones locally to adjust stiffness or bending properties in a specific region. This is a remarkable parallel to sizing processes in metal structures, where material thickness is adapted locally. At the same time, it is a remarkable difference to the design of conventional laminates (CLs), in which a local change always affects adjacent areas. Figure 4 visualizes the stacking scheme of the C-profile shown above and a corresponding DD equivalent, which mimics the CL properties of the slightly orthotropic C-profile laminate.

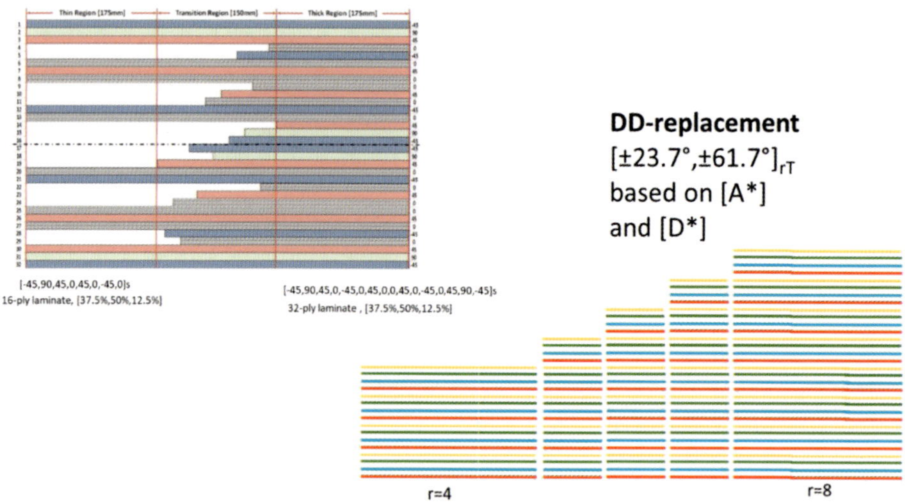

Figure 4: Simplified stacking scheme due to DD concept

In Figure 4 BB drop offs are located on the top surface. Drop-offs can also be located on the bottom surface or within the laminate stack, as indicated in Figure 5.

Figure 5: Layer drop-off scenarios. bottom surface (left), within laminate stack (right)

However, the latter solution is questionable, as the four-ply BB architecture of DD, will lead to resin pockets of increased size and stronger fiber waviness, as outlined for CL.

A quasi-isotropic [25%,50%,25%] laminate is a special conventional laminate, which is still often used. In fact, a QI laminate can also be considered a symmetric DD laminate, when the whole stacking is rotated by ±22.5°. Thus, [0°,45°, -45°,90°] turns into [22.5°,67.5°, -22.5°, -67.5°] or [-22.5°,22.5°, -67.5°, 67.5°] as shown in Figure 6. When more orthotropic laminates are in focus, the difference between φ and ψ within the DD BB deviates from 45°, as shown in Figure 6 as well.

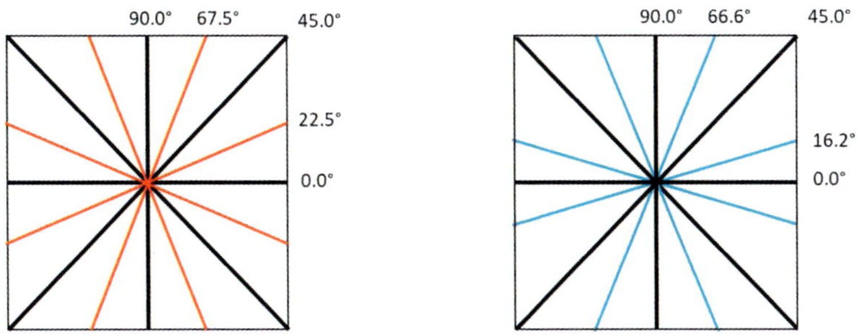

Figure 6: DD replacements for QI laminate (left) and for more-orthotropic one (right), blue in Table 2

To understand the described advantageous aspects and differences to CLs, the DD concept is applied on part level hereafter.

3.2. Double-Double omega stringer

Longitudinal stiffeners are common components in shell-like lightweight structures in aerospace applications. They increase global stiffness and the resistance against buckling. While T-stringers are often used for wing covers, hat-shaped or omega stringers are used to reinforce fuselage shells in longitudinal direction. Such an omega stringer serves hereafter to demonstrate the combination of a modern prepreg manufacturing technique with the unique design features of DD laminates.

Figure 7 shows an omega stringer section made from conventional low-grade prepreg. The stringer shape has been in focus of dedicated studies on single-sided diaphragm forming [4] in the past, which focused on combining flat prepreg deposition with AFP technology and the subsequent forming. The omega stringer is made conform to stacking rules, featuring [0°, 45°, -45°, 90°] ply orientations. The red-marked region (in Figure 7) is additionally reinforced with two UD layers aligned along the part's longitudinal (0°) direction.

Figure 7: Omega stringer section with constant laminate thickness in feet regions and the red-marked area for local reinforcement

Figure 8 outlines selected specifications of the stringer reference and shows the quality of the uncured formed full-prepreg-laminate stack prior curing. The stringer in Figure 7 has been autoclave cured while the colored plot in Figure 8 indicates that the laminate-thickness tolerance was met.

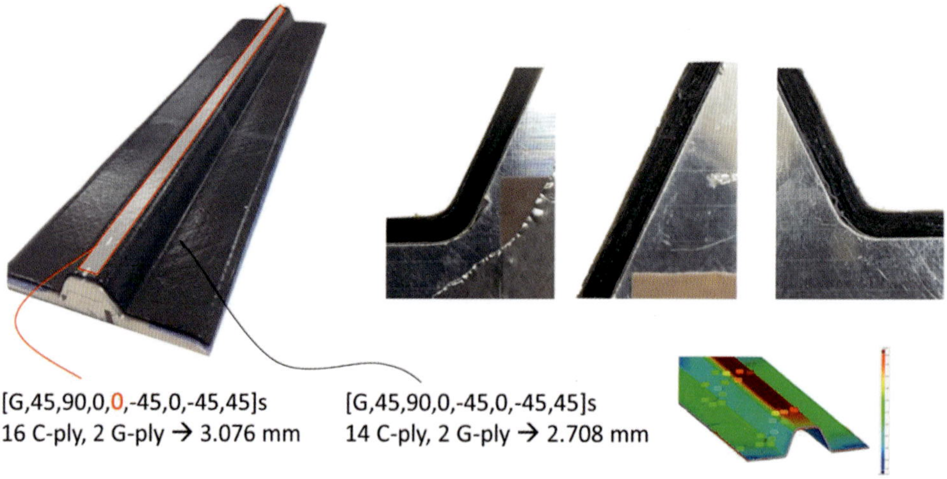

Figure 8: Specification of conventional-laminate omega stringer and formed prepreg prior curing

Stringer feet usually have a constant laminate thickness today. Those feet are attached to the fuselage's skin laminate in a co-bonding process. Even though it is well known that peel stress issues are promoted by this assembly [5] it is common practice today. Reducing the feet thickness towards their outer edges is a recommended design to reduce the feet stiffness (see Figure 9), which alleviates the peel issue. In small scale, this is executed today by creating little chamfers at the edges during the EEOP trimming procedure.

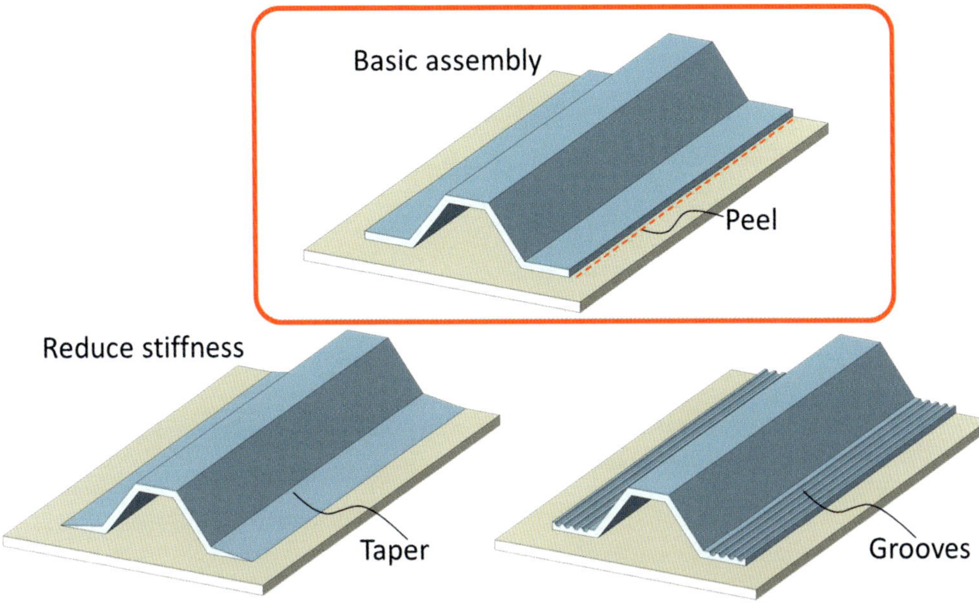

Figure 9: Peel-stress issue and proposed design recommendations. Pictures inspired from [5.]

Strong tapering along the whole stringer feet, with conventional, stacking-rule conform laminates is impossible without violating those rules. In fact, the combination of the symmetry-, balance- and 10% rule impede tapering. The unique laminate architecture concept of DD allows for tapering. BB repeats can be reduced towards the outer feet edges by dropping BBs on the outer or inner surface. Of course, the recommended minimum number of repeats needs to be kept in mind, but the following experiments show that small areas with two or only a single BB do not produce critical deformations on stringer level.

3.3 Tapering by "card sliding"

The DD omega stringer presented hereafter pursues a fourfold ambition. It combines:

- DD layup – homogenizing a quasi-isotropic and a more orthotropic layup zone
- State of the art low-grade carbon-fiber-epoxy prepreg + autoclave curing
- Flat material deposition and subsequent forming using single-sided diaphragm forming
- Laminate tapering for minimum peel at stringer feet – by so called "card sliding"

It is the best choice to determine φ and ψ in the DD BB based on the load scenario at hand [3]. However, DD replacements for conventional laminates can also be determined directly based on coefficients of [A], [D] or [A] and [D] simultaneously. Excellent matching DD equivalents can be identified until the 0° ply-orientation fraction for example in the CL exceeds 44%.

Table 2 provides a list of six conventional laminates and their corresponding DD replacements. The laminates refer to a fuselage shell example presented in [6]. The DD replacements are determined based on thickness-normalized stiffness matrix entries [A*] =

[A]/t_{lam} here. A DD replacement is determined individually for each conventional laminate, by adjusting the BB angles in order to minimize $|A_{11}-A_{11}^{DD}| + |A_{22}-A_{22}^{DD}| + |A_{66}-A_{66}^{DD}|$. It shall be noted, that it is also possible to use a multi-objective formulation to determine DD replacements.

Plies			percentages			Stiffn. normal.			DD angles		stiff. normal.		
n [0]	m [±45]	l [90]	% [0]	% [±45]	% [90]	A_{11}^{*}	A_{66}^{*}	A_{22}^{*}	±Φ	±Ψ	$A_{11}^{*,DD}$	$A_{66}^{*,DD}$	$A_{22}^{*,DD}$
3	8	4	20	53	27	55	22	64	25.9	69.5	55	22	64
9	12	6	33	44	22	69	19	56	16.2	64.6	69	19	55
8	10	8	31	38	31	64	17	64	18.9	70.7	64	18	64
13	16	10	33	41	26	68	18	59	16.2	66.6	68	18	58
4	8	4	25	50	25	60	21	60	22.4	67.4	60	21	60
4	10	3	24	59	18	62	24	53	23.0	62.7	61	24	54
Aver.:			27.7	47.5	24.8	63	20.2	59.3	20.4	66.9	62.8	20.3	59.2
Std. Dev.:			5.4	7.9	4.4	5.2	2.6	4.3	4.0	3.0	5.3	2.4	4.3

Table 2: Selected conventional laminates and corresponding DD replacements. QI laminate (yellow) and more orthotropic laminate (blue)

It can be seen that the A_{ii}*match between the individual CLs and their DD replacements is excellent, while the φ and ψ angles in the specific DD BB differ drastically. Homogenization is the key of DD. The simplest approach to homogenize the group of laminates is to determine the average DD angles from the set of individual laminates. Table 2 provides the data for the average configuration and the corresponding standard deviations for the angles in the BB, as well as the stiffness matrix entries. For the case at hand, the average DD laminate BB would be [±20.4°, ±66.9°] for example.

Alternative approaches for determining the best DD replacement, for the six provided conventional laminates, can be set up as a multi-objective problem or a single objective problem, wherein the six individual objectives are combined into a single one, leading to the aim to minimize $(\sum Obj\_i)$.

When the latter solution is executed, the BB angles are determined [±20.9°, ±66.4°]. Those are remarkably close to the BB angles of the simple averaging procedure in Table 2. This observation has been verified in context of other structures, which is in line with the idea of homogenization. This finding, in context of averaging, is utilized for the DD omega stringer. A more orthotropic laminate is considered for the middle chord [33%,41%,26%] (blue marked in Table 2), while a quasi-isotropic laminate is assumed for the rest of the stringer [25%,50%,25%] (yellow marked in Table 2). BB angles of [±16.2°, ±66.6°] and [±22.4°, ±67.4°] are determined for the individual zones, respectively. Those two sets of angles are averaged, leading to the [±19.3°, ±67.0°] BB angles for the DD-omega stringer.

As indicated above, laminate-thickness changes, in DD part, are generated by locally varying the BB repeats. Thus, ramps can be easily created. The so-called "card sliding" technique has been proposed [0,1] to realize DD parts with tapered regions.

The core of "card sliding" is the fact that all plies and sub laminates (BB) in the full laminate stack have the same dimensions, as it is the case for a stack of cards. Figure 10 shows the simple idea of the card sliding technique.

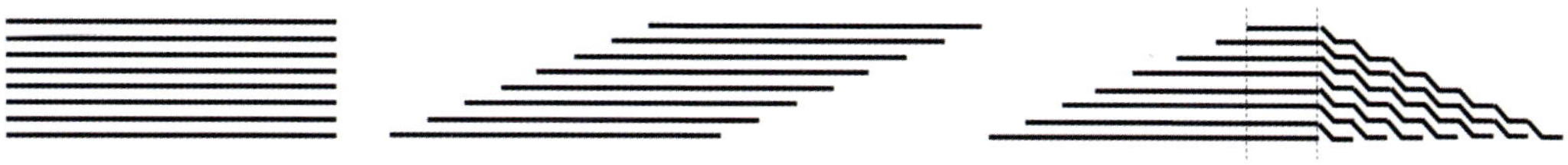

Figure 10: Idea of card sliding and its application to taper laminates

Each layer in this stack is relatively moved by a fixed or an individual shift (see Figure 12). The constant-shift approach, creates a continuous ramp, while the individual shifts allows for adapting ramp characteristics. Card sliding is utilized for the DD omega, while Figure 11 shows how the concept is applied, while each colored line indicates a single four-ply sub laminate.

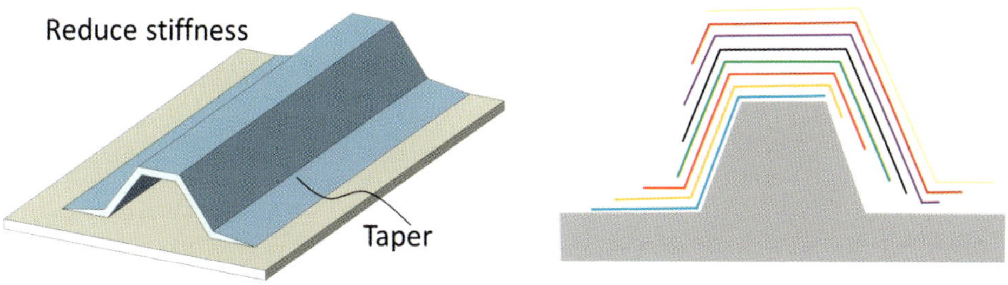

Figure 11: Card sliding applied for omega stiffener

The fixed shift between the sub laminates and the resulting ramping is a relevant design parameter for the omega. In the present case, the fixed shift has been selected, in order to assure the maximum laminate thickness (32 plies) along the middle chord width, while the tapering develops towards the outer feet edges.

Figure 12 summarizes the procedure for the DD omega stringer. The DD stringer laminate thickness varies between four and 32 plies (0.5 – 4.0 mm). It consists of one to eight sub laminates. The card-sliding stack is composed flat and formed subsequently using the single-sided diaphragm technique [4].

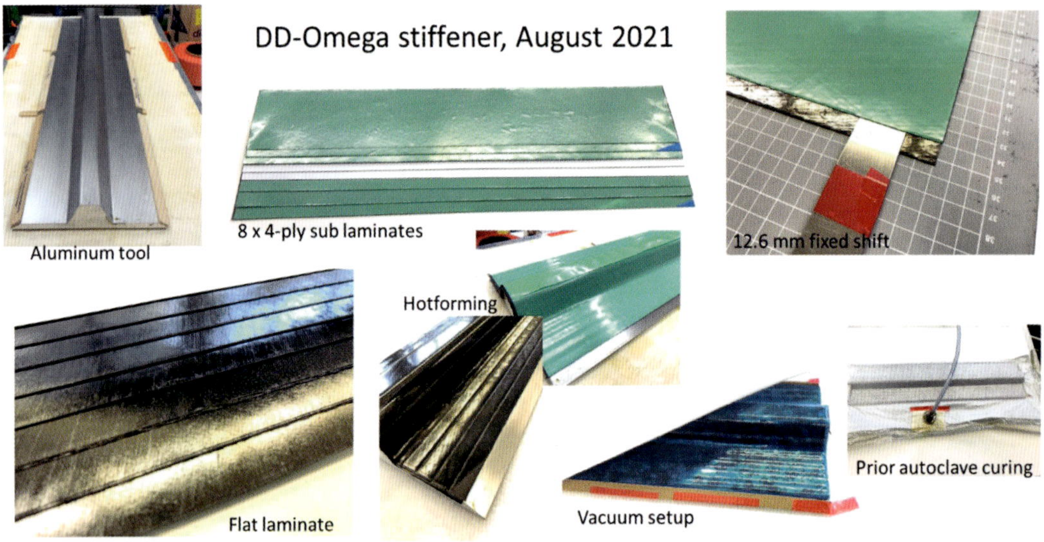

Figure 12: Manufacturing procedure for DD omega stringer

The stringer has been autoclave cured on a single-sided aluminum tool after prepreg preforming, while Figure 13 shows a fabricated stringer after trimming.

Figure 13: DD omega stiffener after trimming

The DD concept is based on homogenization, which directly depends on the number of BB repeats. The exact stacking sequence within the building block remains widely disregarded, as it has no effect on the stiffness matrix [A] entries, as those are stacking independent. However, the process of homogenization, is affected by the BB stacking sequence. This can be shown analytically, but on part level it is more impressive.

Figure 14 shows the first manufactured omega stringer, immediately after the demolding process. The first stringer has been manufactured using the stacking sequence [19.3°, -19.3°, 67°, -67°]$_{rT}$. The manufactured part showed relevant twist, which was surprising at first, as the number of repeats was considered sufficiently high.

Figure 14: First manufactured DD stringer with [19.3°, -19.3°, 67°, -67°]$_{rT}$ layup

A simple thermo-elastic FE model was set up in ABAQUS CAE to investigate the observation. The surface-based model, uses shell elements and simulates the cool down from the curing temperature of 180°C to room temperature of 20°C. Thus, asymmetry effects, in particular the fully populated [B] matrix, affect the part shape after the cool down. Figure 15 shows FEA results of two different configurations, with the first manufactured configuration on the right [19.3°, -19.3°, 67°, -67°]$_{rT}$ and [19.3°,-67°,-19.3°,67°]$_{rT}$ layup on the left.

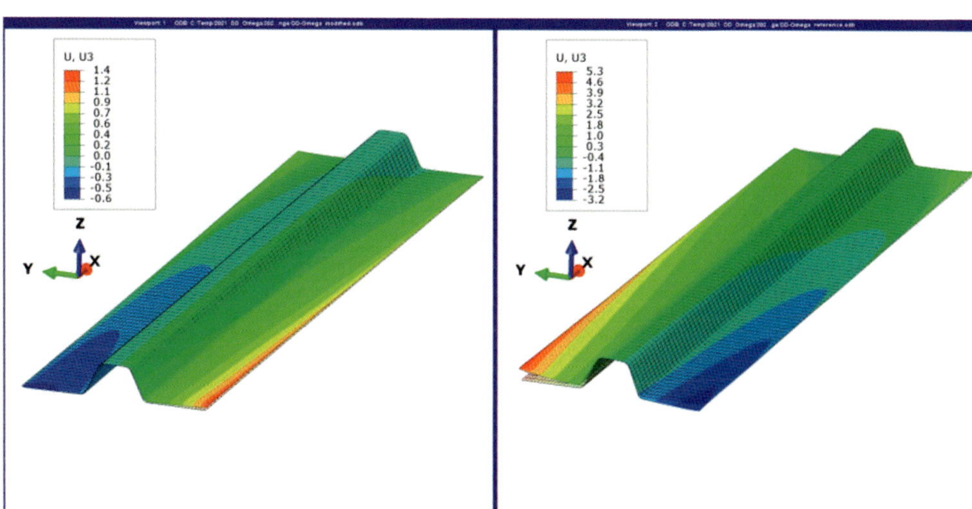

Figure 15: Deformations due to cool down. Right: [19.3°,-19.3°,67°,-67°]$_{rT}$ layup, left: [19.3°,-67°,-19.3°,67°]$_{rT}$ layup

The direct comparison visualizes the effects of stacking sequence within the DD building block. It can be seen that the [19.3°, -67°, -19.3°, 67°] produces significantly less twist/deformation than the initial layup, which experimentally substantiated the analytics presented in [0], wherein the fastest homogenization is predicted for the $[\varphi, -\psi, -\varphi, \psi]_{rT}$ BB.

3.4 Saving weight by tapering

Saving weight is a typical aim for structural optimization. Carbon fiber epoxy composites are well known for their excellent density-specific properties. Saving weight, by locally minimizing the laminate thickness of a part, without impeding its structural integrity is an omnipresent task, many companies and researchers deal with. DD laminates offer new potentials, which can contribute to even lighter structures. However, we cannot expect that all structures will be lighter with DD. But tapering, for example, as it was demonstrated for the DD omega, allows for saving weight, as laminate areas, which experience little material efforts can be saved or minimized. This is true for the omega and the effect is expected to be more relevant for multi-zone-shell structures presented hereafter.

For the DD omega, 41.5% weight could be saved due to consequent feet tapering. Structural tests need to be executed to substantiate the remarkable weight reduction from a strength perspective. However, the peel stress issue, as outlined in [5], could be addressed by pursuing the DD concept.

Figure 16 shows the cross section of the manufactured DD stringer, the corresponding layer stack, as well as areas (green triangles) where material was saved.

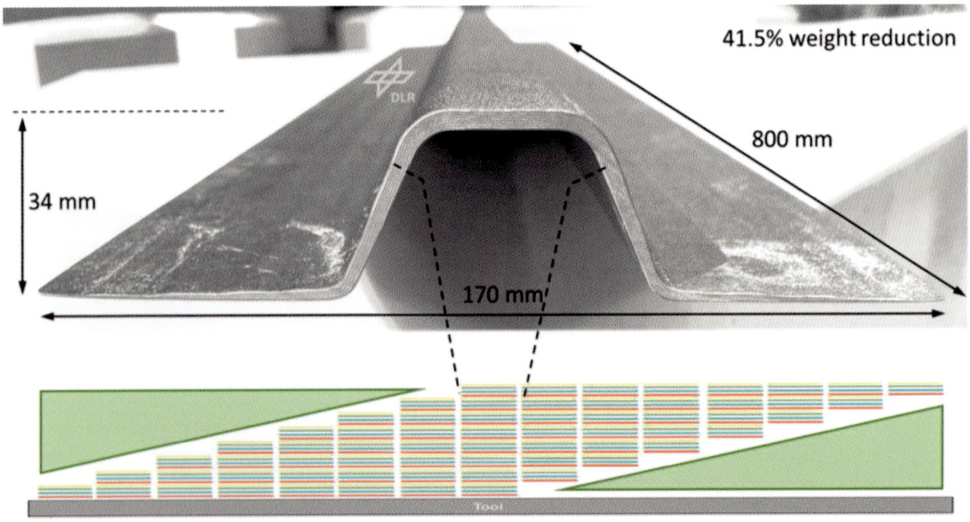

Figure 16: Feet tapering allows for saving weight in peel critical areas

3.5 Laminate quality and the effect of ply thickness

Adequate laminate quality is essential for aerospace components, independent whether we look at DD or conventional laminates. Figure 17 shows section cuts of the manufactured DD stringer. Reflections on the cut surface allow for following the ply paths. Red stars mark BB drop-offs on the outer and inner part surface. The observed laminate quality looks excellent, even though a BB drop off refers to a local 0.5 mm laminate thickness reduction.

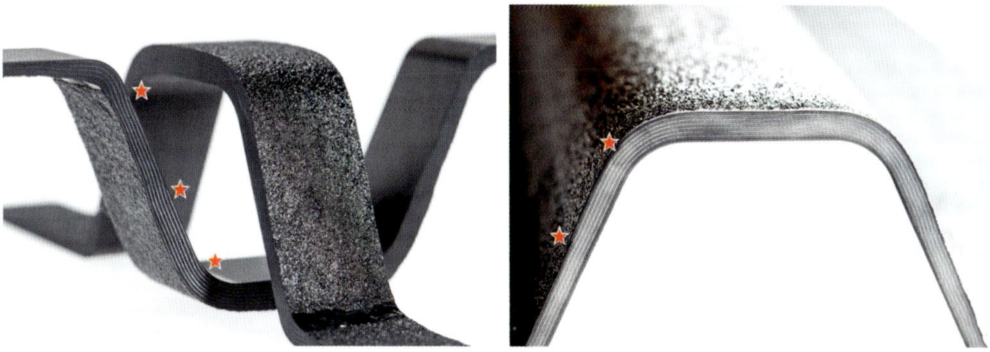

Figure 17: BB drop-offs (red stars) along the stringers cross section. Tool-sided surface (left), bag side (right). Reducing the ply thickness is a conceivable measure to even improve the inner laminate quality. Simultaneously, the number of repeats can be increased which should minimize deformations.

The created FE model, described above, serves to demonstrate the effect of repeats and prepreg ply thickness on part level. The initial reference FE model, with the [19.3°, -19.3°, 67°, -67°]$_{rT}$ has been used to generate two additional virtual stringer configurations. Those feature the same laminate thickness as the reference. In the first case, the assumed ply thickness is reduced by 50%, while the number of repeats is doubled. The second case assumes a 25% prepreg ply thickness (0.03 mm), while the repeats are increased to 4r. Figure 18 shows the results of the analyses.

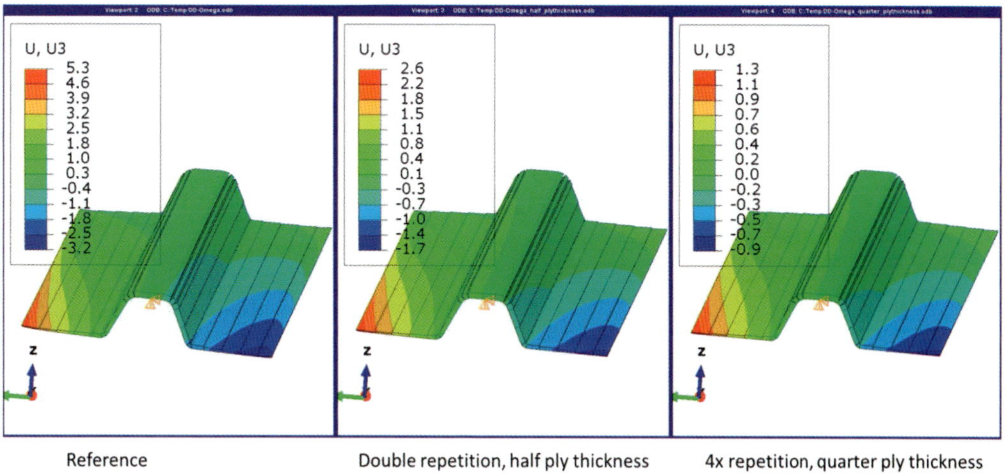

Figure 18: FE model result – The effect of ply thickness reduction or increase of BB repeats

It is found that the 1/r character, outlined for the [B] matrix entries, transfers to the part deformations for the omega in focus. At the same time, the results underline the important role of the appropriate selection of the BB stacking sequence. The [19.3°, -67°, -19.3°,67°]$_{rT}$ stacking produces only 1/3 (see Figure 15) of process-induced deformations compared to the manufactured reference with [19.3°, -19.3°, 67°, -67°]$_{rT}$.

DD in combination with card-sliding allows for elegant tapering the stringer feet in transverse direction. The DD concept also allows for tapering stiffeners or other components in longitudinal direction, as well as tapering in both directions in combination with additional local patches. Figure 19 shows corresponding DD stringer samples, which have been manufactured in accordance with the aforementioned process.

Figure 19: Longitudinal tapering using DD (left) and bi-directional tapering with patches stringer (right)

3.6 Shell components for aerospace application

Fuselage shells and door surround structures (DSS) in particular are ideal examples for zone-based laminate design. Figure 20 visualizes the complex laminate architecture, the corresponding AFP manufacturing procedure and the laminate characteristics and thicknesses of the structure.

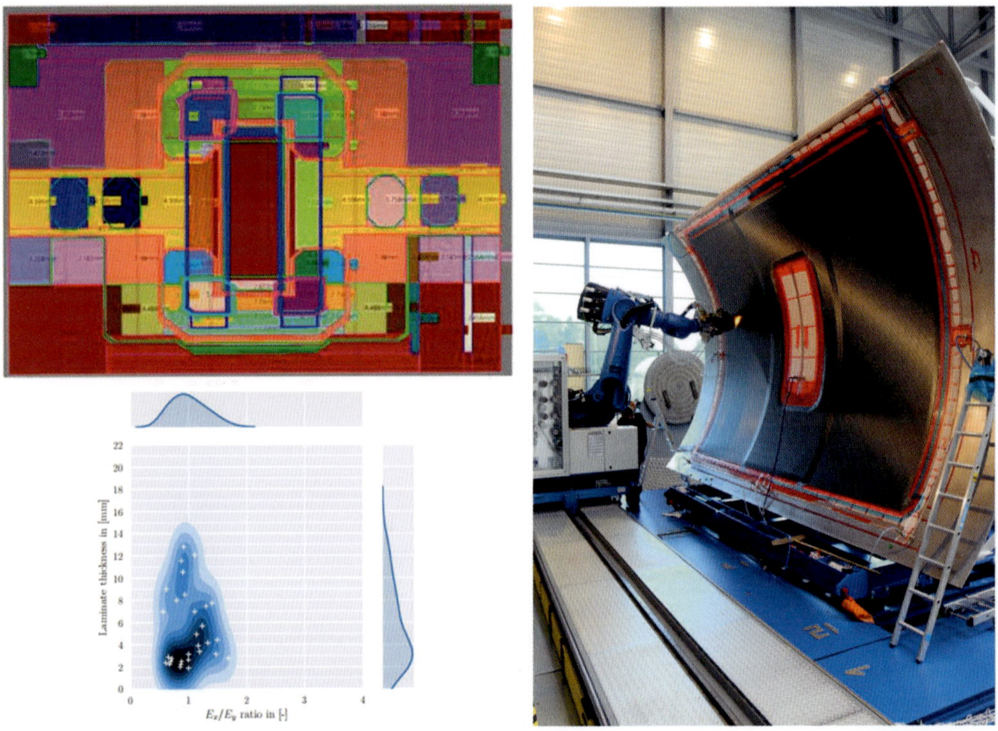

Figure 20: Door-surround structure (Image from [7]), manufactured and analyzed at DLR

The illustrated fuselage section refers to a conventional laminate design, featuring the usual [0°,45°, -45°,90°] plies only. The inhomogeneity due to stacking differences, with zone-specific varying ply-orientation fractions and the laminate thickness ranges are also visualized. Due to the strong laminate-thickness variations throughout the whole part, ramp and transition zone design is a challenging task for this kind of structure. A DD configuration of the illustrated DSS would feature a single BB stacking sequence and locally varying repeats. As DD allows for straight-forward local laminate-thickness adaptation, by simply increasing repeats locally, the transition-zone-design challenges are relaxed drastically, as compatibility between adjacent zone is fulfilled per definition due to the DD laminate architecture.

A full DSS has not yet been realized with a DD laminate. As a first step, a flat DD skin section is realized, which aims to demonstrate the unique DD features for multi-zone shell structures. The specifications of the 400 mm x 300 mm demonstrator are shown in Figure 21.

The flat DD structure features three up to ten repeats, which refers to realistic laminate thicknesses between 1.5 and 5.0 mm. Two different ramp characteristics, in particular 1:10, 1:20 are realized while others are easy possible. The DD stringer layup has been adopted for the flat part, which has been manufactured with a single-sided tool concept in an autoclave process. All DD layer run outs are located on the vacuum-bag side of the part.

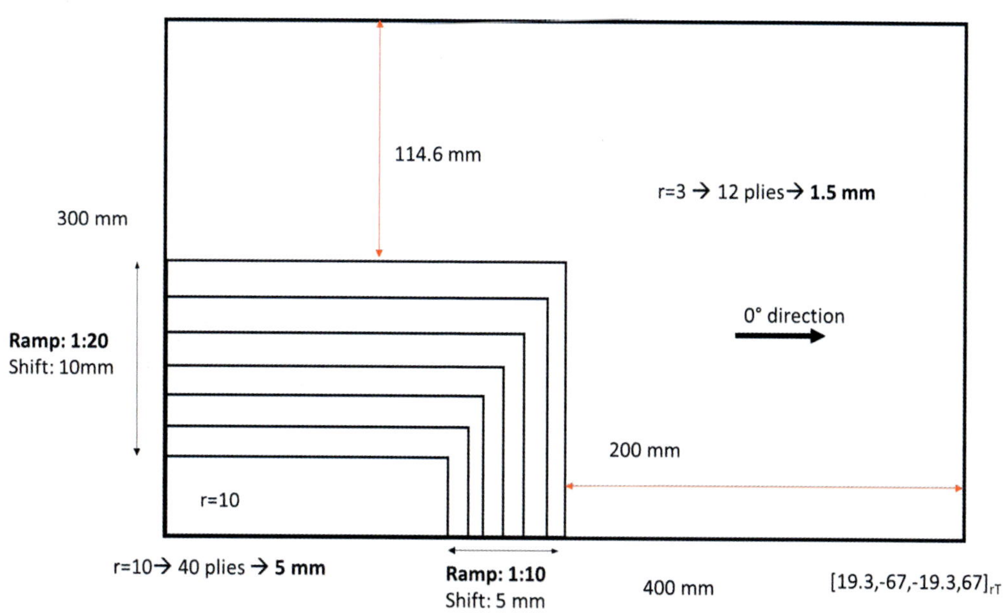

Figure 21: Flat DD sample with three up to ten repeats and 1:20 and 1:10 ramp designs

Figure 22 shows the manufactured flat DD part. Critical, asymmetry-induced distortions were not observed after manufacturing. As drop offs are located on the outer surfaces, the DD laminate is free of internal defects as resin pockets or ply waviness (see detail view in Figure 22), which is considered a positive aspect from a damage initiation and crack propagation perspective.

Figure 22: Flat DD demonstrator, 1.5-5.0 mm thick, constant BB

It should be noted that the DD concept also allows for droppingplies within the laminate stack. However, as four plies dropped at once, the effect of internal drop offs need further investigation and testing efforts. Locating drop offs on the outer surfaces, without impeding the laminate properties, is a unique DD feature. For the part at hand it is considered the preferred solution to locate drop-offs on the outer surface, which can (if necessary) be covered by a single glass-fabric-prepreg layer, as it is common for aerospace CFRP components today. Figure 22 summarizes the DD-specific design opportunities in context of laminate optimization [3], which in fact show direct parallels to material-thickness optimization, as it is executed in sizing processes for metal structures.

3.7 Special-purpose structures and applications

DD laminates can be thought in context of non-aerospace specific structures as well. Crash elements, for example, are a conceivable application, whose creation process can benefit from the thickness-tailoring opportunities of DD. Figure 23 illustrates different configurations, while the three on the right side are realizable using DD.

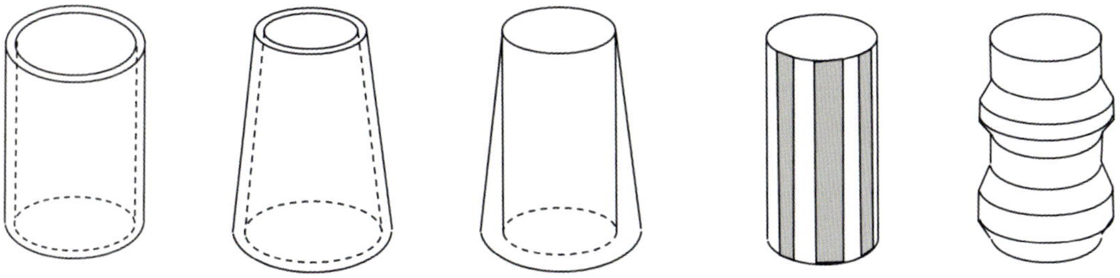

Figure 23: Conventional crash structures (1,2) and novel configurations enabled by DD (3,4,5)

Crash elements have often constant material or laminate thickness [8]. The crash characteristic, as for example a progressive force-over-crushing-length relation is realized by increasing the element's diameter [8], leading to cone-like shape. At the same time, the necessity for cone-shaped mandrels is generated. The easy thickness tailoring of DD allows for creating a progressive laminate thickness increase along the part's longitudinal axis, without the necessity for a conical mandrel.

Combining a cylindrical base laminate with additional patches in longitudinal direction offers new design features to affect buckling and crash behavior. Even more complex laminate-thickness variations are conceivable (see cases 4, 5 in Figure 23), which can be used to locally increase the part's stiffness in hoop direction or to trigger a certain collapse shape when the element is loaded axially.

From the manufacturing perspective, the card sliding technique can also be used as a preliminary step to create DD cylinders for example. In this case, the fixed shift of the layer can be designed in a way that after wrapping the laminate stack around a cylindrical tool, a DD cylinder with a constant laminate thickness is produced. This procedure has been executed and Figure 24 shows a DD cylinder, which has been made from a flat DD laminate stack, created with the card sliding technique. The cylinder has a diameter of 132 mm and an axial length of 200 mm.

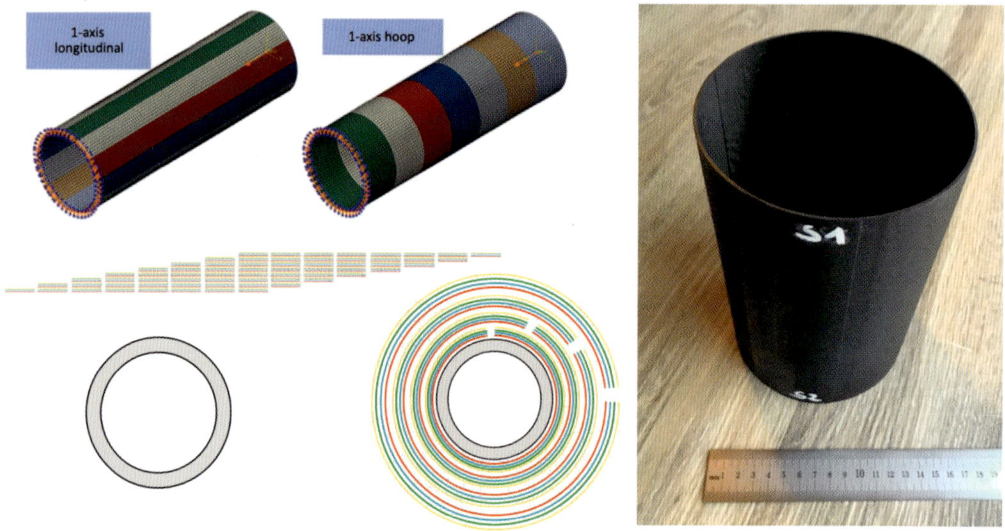

Figure 24: Card sliding, also suited for different wrapping techniques

The manufactured cylinder represents a first DD manufacturing trial, which does not focus on crash. Instead, it is focused on another remarkable property of the DD family of laminates. Figure 25 shows how laminate CTEs in x- and y-direction develop, when the angle φ is varied in the considered laminate stack $[\varphi, -\varphi, -\varphi, \varphi]_{rT}$. The CTE development is symmetric with respect to the 45° as it can be expected.

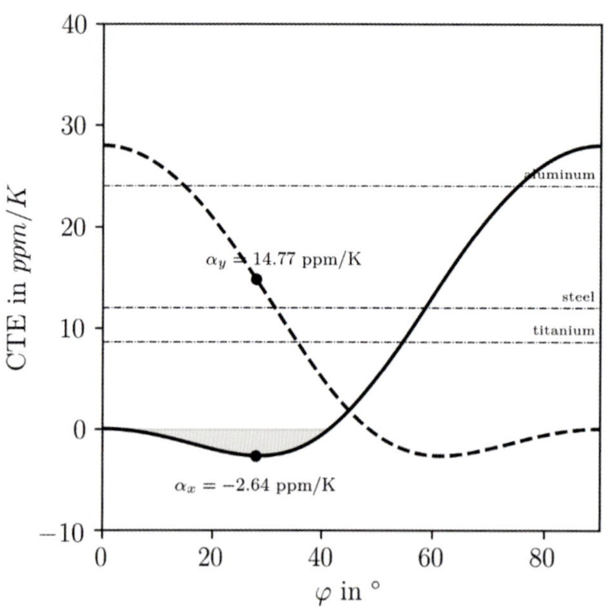

Figure 25: Negative laminate CTE domain for $[\varphi, -\varphi, -\varphi, \varphi]_{rT}$ *laminate*

The gray marked area in Figure 25 is of particular interest for the application at hand.

It can be seen that the laminate CTE in x-direction reaches a minimum at $\varphi = 30°$. The corresponding CTE is negative, with a remarkable magnitude of -2.64 ppm/K. This shall be highlighted, as the related physical behavior is hardly achievable with other materials in engineering.

A flat [30, -30]8T laminate sample has been created from Hexcel's M21E/IMA UD prepreg to verify the negative CTE. Figure 26 shows the created raw data from the tests and the determined linear coefficients of thermal expansion in all three principal laminate directions. The analysis substantiates the negative CTE in the laminate's x-direction and a comparably high CTE in the in-plane transverse direction. The CTE in the laminate's through-thickness direction is found comparable to regular quad laminates.

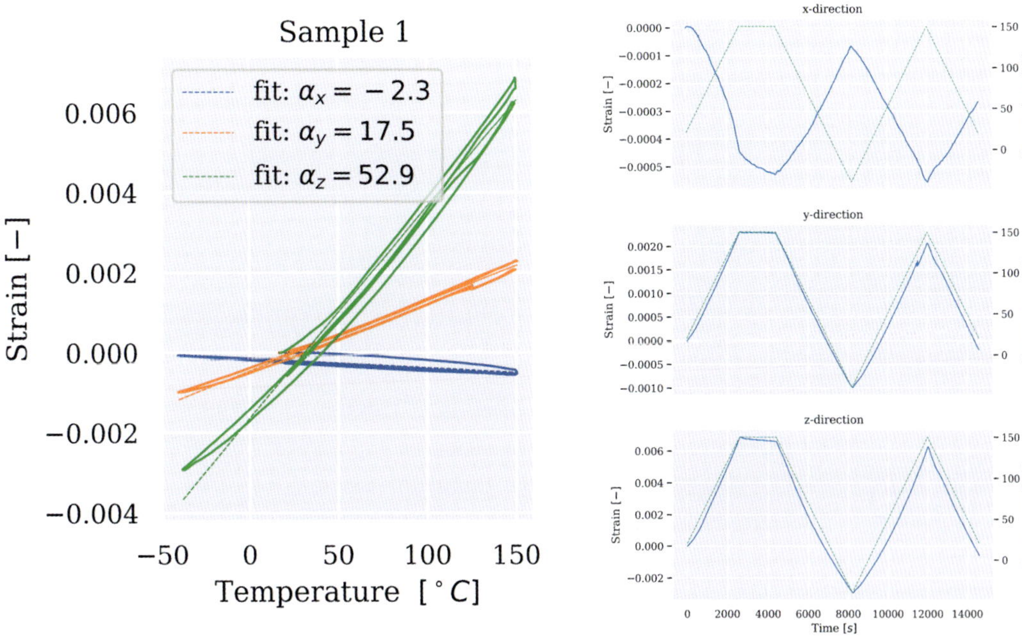

Figure 26: [30,-30]8T M21E/IMA laminate CTEs, determined with thermo-mechanical analysis

A part with negative CTE acts counterintuitive. It contracts during heating and expands when cooled down. This particular effect has been in focus of the making the DD cylinder shown in Figure 24. Aligning the laminate's global x-direction with the cylinder's hoop direction, as shown in Figure 26, leads to the fact, that the cylinder diameter increases when the part is cooled down and wise versa.

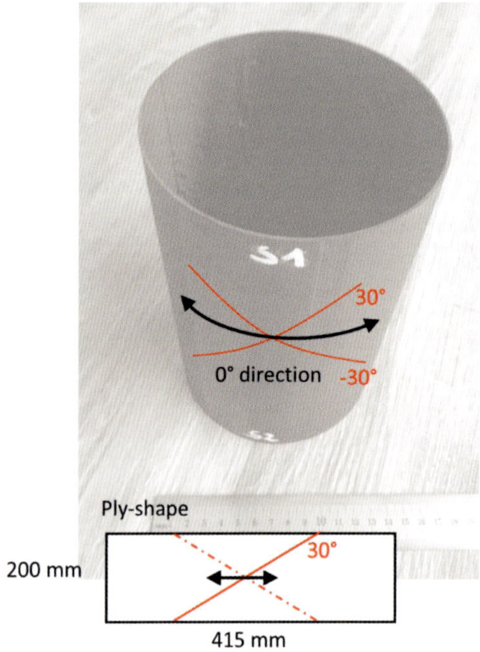

Figure 26: Global laminate alignment to utilize negative CTE to produce a counterintuitive behavior under thermal loads

The manufactured cylinder has an inner diameter of 132.88 mm at RT, which is 0.28 mm higher, than the tool diameter at the maximum process temperature of 180°C. At room temperature, a diameter deviation between the aluminum tool and the part is 0.78 mm, which is equivalent to 2.45 mm increase in hoop direction. High CTEs of the CFRP laminate in laminate thickness direction, which are usually in the range of 35-75 ppm/K, cannot explain the observed effect. Thus, the laminate's negative CTE must be the driver. In conceptual design, a negative CTE can be used in different contexts. NCTE cylinders, can be used to generate glue-less joints, while the parts can act as sleeves or plugs. First assembly test has already been executed as shown in Chapter 10.

3.8 Conclusion & Outlook

Conventional laminates, composed of 0°, 45°, -45°,90° plies, are still the baseline for composite structures in aerospace. Stacking rules are widely accepted and adopted, as laminates created in accordance, do not show complex coupling mechanisms while a satisfying high degree of dimensional fidelity is achieved.

Double-Double laminates represent a new family of laminates, which promises advantageous aspects in terms of optimization and manufacturing in particular, which are outlined in the present section.

DD laminates won't be better in all facets. For some structures, for example highly orthotropic ones, DD won't be the best solution. As long it is tried to mimic conventional laminates, the mechanical or weight benefits are expected rather small. Nonetheless, with those ambitions one can still benefit from DD's unique manufacturing opportunities.

When the ply angles in the DD laminate are defined based on load scenarios, disregarding CLs conventions and restrictions, the full potential can be entrapped and the combination of simplified optimization and novel manufacturing techniques promise considerable advantages.

With DD, we'll likely lose some percent compared to the structural optimum with arbitrary layup, but with DD, we simplify the creation of multi-zone laminates. Complex permutation discussions can be avoided and the tricky challenge to assure compatibility and manufacturability between adjacent zones is eased. Drop-off staggering and transition zone design diminish or can be drastically simplified in addition.

Laminate thickness changes are described by a single integer parameter in the DD concept, which is easy to optimize and which is a strong similarity to established sizing processes for metal designs. It has been shown that equivalent DD laminates can be determined for most technically relevant laminates in aerospace applications. Replacements can be identified with focus on membrane or bending properties or based on a combination of both.

However, DD is still a quite new family of laminates. Many procedures in aerospace established today are focused on or created for traditional laminates. Joint-sizing methods for mechanical fasteners as rivet or bolts, for example, are defined and are only valid for stacking-rule conform laminates. In order to extend the application of DD, those sizing methods need to be extended to DD or novel approaches must be developed. Locating BB drop offs on the outer Part surfaces minimizes laminate-internals defects as waviness and resin pockets for example. However, CAI studies are mandatory to prove this advantageous aspect and to assure that no new challenges are generated in terms of damage initiation and propagation.

In total, DD laminates offer new interesting design features, simplified optimization and unique manufacturing opportunities. Those can help to reduce costs, save weight, shorten part development times, while the concept is still open for well-known late-entry laminate changes.

References

[0]	Tsai SW. Double–Double: New Family of Composite Laminates. AIAA JOURNAL Vol. 59, No. 11, November 2021
[1]	Kappel E. DD-laminates and manufacturing. Stanford's Composite Design Workshop XXII, January 2022
[2]	Nettles AT. Basic Mechanics of Laminated Composite Plates. NASA Reference Publication 135. 1994
[3]	Kappel E. Double–Double laminates for aerospace applications — Finding best laminates for given load sets. Composites Part C: Open Access 8 – 100244 – 2022
[4]	Kappel E, Albrecht M. A controlled recipient evacuation process to form composite profiles from flat multi-angle prepreg stacks – infrastructure and C-profile verification, Advanced Manufacturing: Polymer & Composites Science 2019
[5]	Da Silva LFM et al. Design Rules and Methods to Improve Joint Strength – Handbook of Adhesion Technology. Pp. 689-723, 2011
[6]	Tsai et al. Composites Part C: Open Access 8 (2022) 100244 Low Weight/Low Cost Design and Manufacturing, 2019
[7]	Stüve J et al. Automated Production of an CFRP-Aircraft Side Shell using AFP and Inductive Co-bonding technology. 11th Airtec Congress, Munich, 2016
[8]	Zhu et al. On energy-absorbing mechanisms of metal/WF-CFRP hybrid composite columns. Polymer Composites, 2020
[9]	North Thin Ply Technology NTPT THINPREG™ 402 DATA SHEET, Version 1.6 – May 11th 2017

Abbreviations

AFP	Automated fiber placement
ATL	Automated tape laying
BB	Building Block
CAI	Compression after impact
CFRP	Carbon-Fiber-Reinforced-Plastics
CL	Conventional laminate, composed of 0°, 45°, -45°, 90° plies
CLT	Classical Laminate Theory
CTE	Coefficient of thermal expansion
DD	Double-Double
DLR	German Aerospace Center
DSS	Door-surround-structure
EEOP	Engineering edge of part
NCTE	Negative CTE
QI	Quasi-isotropic
RT	Room temperature
UD	Unidirectional

CHAPTER 4
Preferred Stacking Sequences for Homogenization

Antonio Miravete
miravete@mit.edu

This chapter deals with the analysis of the different stacking sequences of the double – double building blocks. A novel layup method discovered by Stephen W. Tsai, called double – double, uses 4-ply [±Φ/±Ψ] building blocks, which are stacked upon each other [1 -12]. For the last 60 years, most conventional layups that are widely used in the aerospace industry comprise plies oriented at 0°, 90°, +45° and −45° enforcing symmetry. These laminates are called “quads”.

There is a basic difference between double – double and quads. The double – double laminates are composed of several repeated building blocks formed by 4 angles with no symmetry, while the quad building blocks have a variable number of angles, 4 or higher than 4, and symmetry is enforced.

The double – double method can play a fundamental role in making the most of the possibilities of current manufacturing processes [13].

The automated placement of tape by AFP/ATL machines allows orienting the layers in any direction. However, most laminates used in the aerospace industry have the fibers oriented in just four directions. By double – double, it is possible to take full advantage of placing fibers in the most efficient orientations, making more lightweight and lower cost parts.

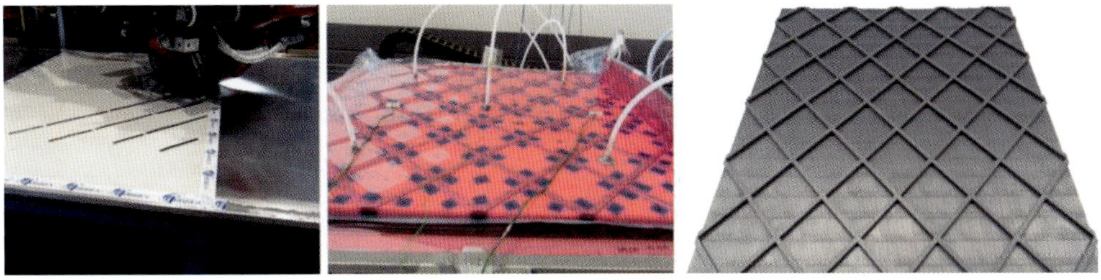

Figure 4.1 Automated placement of fiber and binder (left), resin infusion (center) and aerospace grade final skin/grid part (right) [14].

A promising alternative to autoclave processing is the dry fiber process, which involves the automated lay-up of a dry fiber preform in a mold, infusion of liquid resin through the preform and resin cure, combining the advantages of the AFP and the infusion process. The double – double method has been successfully applied to a grid/skin part in collaboration with MTorres (Figure 4.1) [14].

Although still in the early stage, composite 3D printing is gaining traction within the manufacturing industry. It provides a quick and automated approach to manufacturing composite parts, which used to be labor intensive and require highly skilled operators to make.

Here again, by double – double, it is possible to take full advantage of printing resin/fiber systems in any orientation, making more lightweight and lower cost parts.

The optimal winding angle for cylindrical structures under internal pressure is near 55 degrees. By double – double, this example can be generalized to any geometry and any loading condition because of its possibility of winding the fibers, not only in four directions, but in anyone.

That double – double laminates are formed by small, repeated building blocks presents an enormous advantage to manufacture compared to quads, formed by larger non-repeated sublaminates including enforced symmetry, their manufacturing being complicated and prone to error.

As double-double, nature also "manufactures" structures composed of repeated building blocks. We find repeated building blocks in the DNA of all living things (Figure 4.2) [15-20].

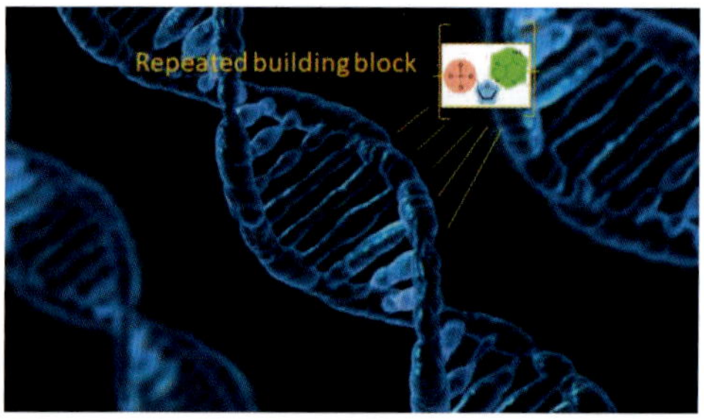

Figure 4.2 Repeated building blocks in the DNA of living things.

We find repeated building blocks in two different levels in trees and plants. Both DNA (cell level) and polymer backbone (molecular level) are formed by repeated building blocks (Figure 4.3) [21-27].

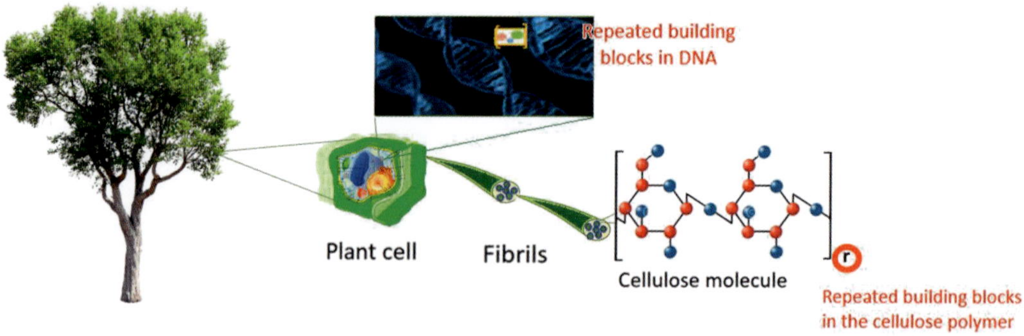

Figure 4.3 Repeated building blocks at two different levels in trees and plants. Both DNA (cell level) and polymer backbone (molecular level) are formed by repeated building blocks.

Figure 4.4 shows the sequencing of a) DNA, b) cellulose molecule, c) double – double and d) quads.

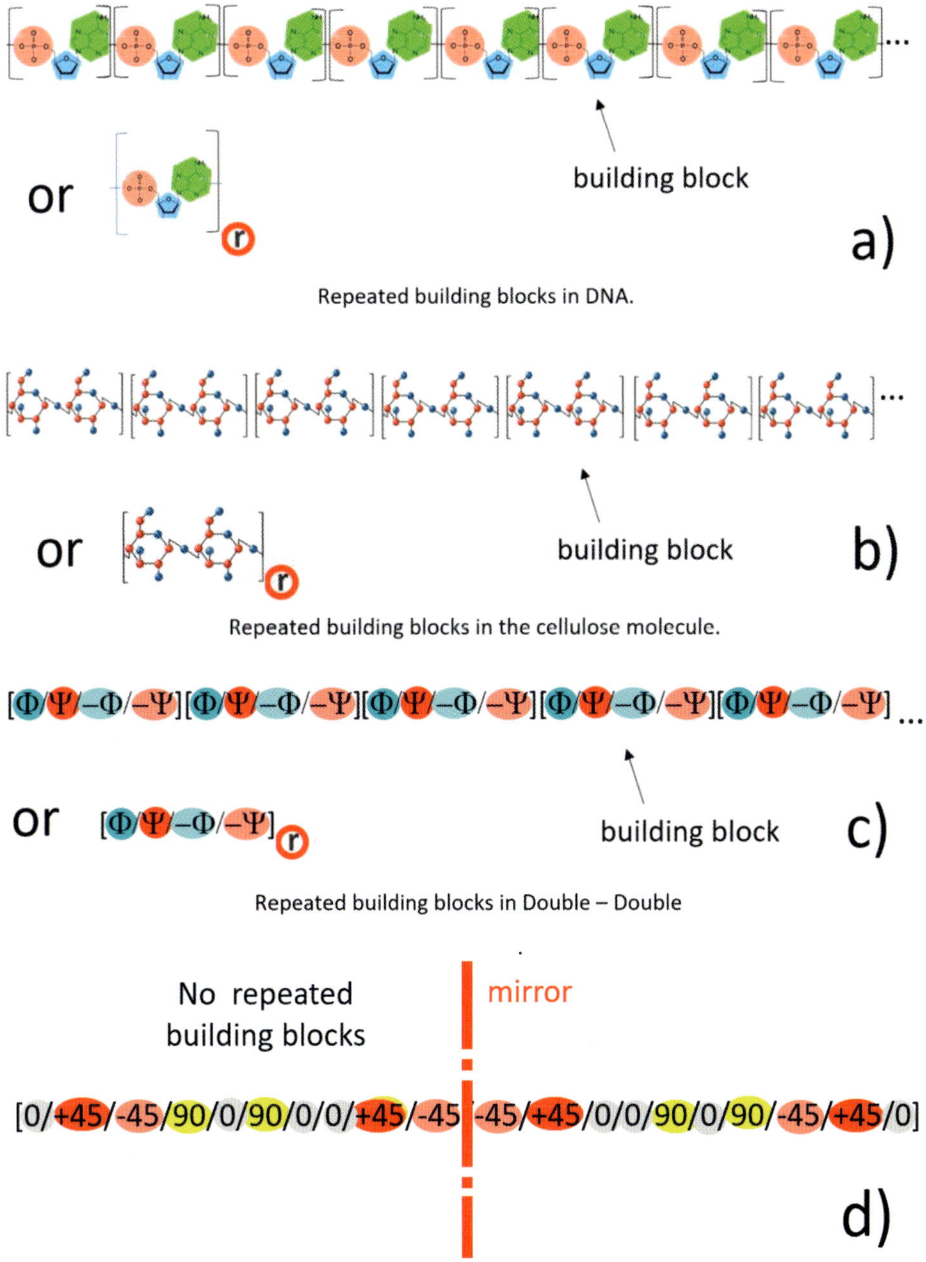

Figure 4.4 Repeated building blocks in a) DNA, b) cellulose molecule and c) double – double. No repeated building blocks in quads. Symmetry is enforced.

4.1 Homogenization, that great unknown

In most textbooks on composite materials, stiffness, strength and damage tolerance are discussed in depth. However, very few textbooks talk about that great unknown that is homogenization.

Homogenization of a laminate is defined by 2 conditions:

$$[A^*] = [D^*];\ [B^*] = 0 \qquad \text{Eq. 4.1}$$

Where $[A^*] = [A]/h$; $[D^*] = 12[D]/h^3$; $[B^*] = 2[B]/h^2$; h = laminate thickness

Homogenization is an important property because a homogeneous laminate lacks the weak properties of orthotropic laminates while maintaining all their advantages.

As we will see in this chapter, repeated building block methods, like double – double reach homogenization quickly. The resulting homogenized double – double laminates exhibit the following advantages:

- ✓ Simple to design and manufacture as an orthotropic metal. Building block formed for just 4 plies. Only six stacking sequences. The resulting component would be lightweight, with a taper formed by single ply drops.
- ✓ Simple to layup along the 1-axis or use card sliding, with no internal defects since ply drops are on the exterior surfaces, easy inspection for having each ply visible, easy local reinforcement and repair with in-family laminate. Aggressive taper that can be as high as 10:1 to save weight.
- ✓ Intrinsically tough with finely dispersed, highly repeated building blocks
- ✓ No thermal warpage
- ✓ No tension – bending coupling
- ✓ No tension – shear coupling
- ✓ No bending – twisting coupling
- ✓ Free edge effect negligible
- ✓ Maximum interlaminar shear stress = 1.5 shear stress applied

Quads present the following disadvantages:

- ✓ Inflicted with complexities that include the 10% rule to deal with unknown secondary loads
- ✓ Challenge of blending of quad laminates among adjacent elements. Impossible selection of best laminates among thousands of stacking sequence permutations done with no guarantee of uniqueness. The number of permutations follows an exponential function with the number of plies in the quad (Figure 4.5).

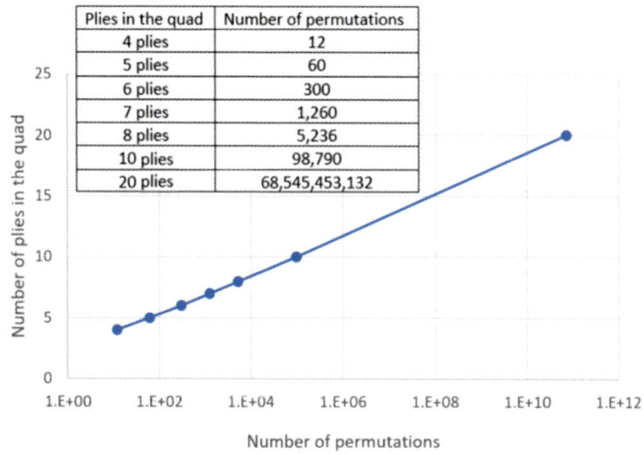

Plies in the quad	Number of permutations
4 plies	12
5 plies	60
6 plies	300
7 plies	1,260
8 plies	5,236
10 plies	98,790
20 plies	68,545,453,132

Figure 4.5 Number of permutations versus number of plies in the quad.

- ✓ layup pattern that must maintain mid-plane symmetry and conflicting mismatches among elements.
- ✓ Quads enforce symmetry and symmetry delays homogenization significantly.
- ✓ Bending – twisting coupling is always present.
- ✓ Interlaminar stress analysis is very complex.
- ✓ Free edge effect is not negligible, difficult to analyze because of the mathematical singularity.
- ✓ Maximum interlaminar shear stress varies, its through thickness distribution is non-parabolic.

4.2 Lamination Parameter Plots

Tsai and Pagano (1968) and Tsai and Hahn (1980) demonstrated that [A*], [B*] and [D*] can be expressed as linear functions of the invariants (U_1, U_2, U_3, U_4, U_5) and the so-called lamination parameters (in red).

$$\begin{bmatrix} A_{11}^* \\ A_{22}^* \\ A_{12}^* \\ A_{66}^* \\ A_{16}^* \\ A_{26}^* \end{bmatrix} = \begin{bmatrix} 1 & \xi_1^A & \xi_2^A & 0 & 0 \\ 1 & -\xi_1^A & \xi_2^A & 0 & 0 \\ 0 & 0 & -\xi_2^A & 1 & 0 \\ 0 & 0 & -\xi_2^A & 0 & 1 \\ 0 & \xi_3^A/2 & \xi_4^A/2 & 0 & 0 \\ 0 & \xi_3^A/2 & -\xi_4^A/2 & 0 & 0 \end{bmatrix} \begin{bmatrix} U_1 \\ U_2 \\ U_3 \\ U_4 \\ U_5 \end{bmatrix}$$

$$\begin{bmatrix} B_{11}^* \\ B_{22}^* \\ B_{12}^* \\ B_{66}^* \\ B_{16}^* \\ B_{26}^* \end{bmatrix} = \begin{bmatrix} 1 & \xi_1^B & \xi_2^B & 0 & 0 \\ 1 & -\xi_1^B & \xi_2^B & 0 & 0 \\ 0 & 0 & -\xi_2^B & 1 & 0 \\ 0 & 0 & -\xi_2^B & 0 & 1 \\ 0 & \xi_3^B/2 & \xi_4^B/2 & 0 & 0 \\ 0 & \xi_3^B/2 & -\xi_4^B/2 & 0 & 0 \end{bmatrix} \begin{bmatrix} U_1 \\ U_2 \\ U_3 \\ U_4 \\ U_5 \end{bmatrix}$$

$$\begin{bmatrix} D_{11}^* \\ D_{22}^* \\ D_{12}^* \\ D_{66}^* \\ D_{16}^* \\ D_{26}^* \end{bmatrix} = \begin{bmatrix} 1 & \xi_1^D & \xi_2^D & 0 & 0 \\ 1 & -\xi_1^D & \xi_2^D & 0 & 0 \\ 0 & 0 & -\xi_2^D & 1 & 0 \\ 0 & 0 & -\xi_2^D & 0 & 1 \\ 0 & \xi_3^D/2 & \xi_4^D/2 & 0 & 0 \\ 0 & \xi_3^D/2 & -\xi_4^D/2 & 0 & 0 \end{bmatrix} \begin{bmatrix} U_1 \\ U_2 \\ U_3 \\ U_4 \\ U_5 \end{bmatrix} \qquad \text{Eq. 4.2}$$

Being A*, B*, D*, U_1, U_2, U_3, U_4 and U_5 (Figure 4.6):

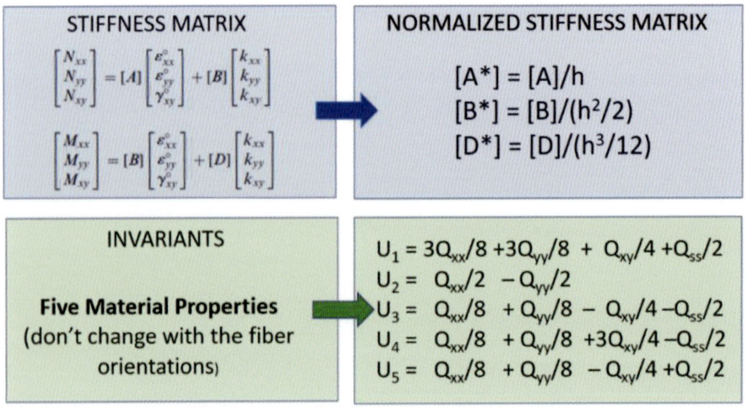

Figure 4.6 A*, B*, D*, U_1, U_2, U_3, U_4 and U_5.

The on-axis plane stress stiffness [Q] of a unidirectional or fabric ply can be computed from the engineering constants as follows:

$$[Q] = \begin{bmatrix} \frac{E_x}{1-\nu_x\nu_y} & \frac{\nu_y E_x}{1-\nu_x\nu_y} & 0 \\ \frac{\nu_x E_y}{1-\nu_x\nu_y} & \frac{E_y}{1-\nu_x\nu_y} & 0 \\ 0 & 0 & E_s \end{bmatrix} \qquad \text{Eq. 4.3}$$

For all laminates:

$$\xi_1^A = (A_{11}^* - A_{22}^*) / 2U_2$$
$$\xi_2^A = (A_{11}^* + A_{22}^* - 2U_1) / 2U_3$$
$$\xi_1^D = (D_{11}^* - D_{22}^*) / 2U_2 \qquad \text{Eq. 4.4}$$
$$\xi_2^D = (D_{11}^* + D_{22}^* - 2U_1) / 2U_3$$

The Feasible domain of all laminates in a ξ_1^A and ξ_2^A plot can be represented graphically (Figure 4.7).

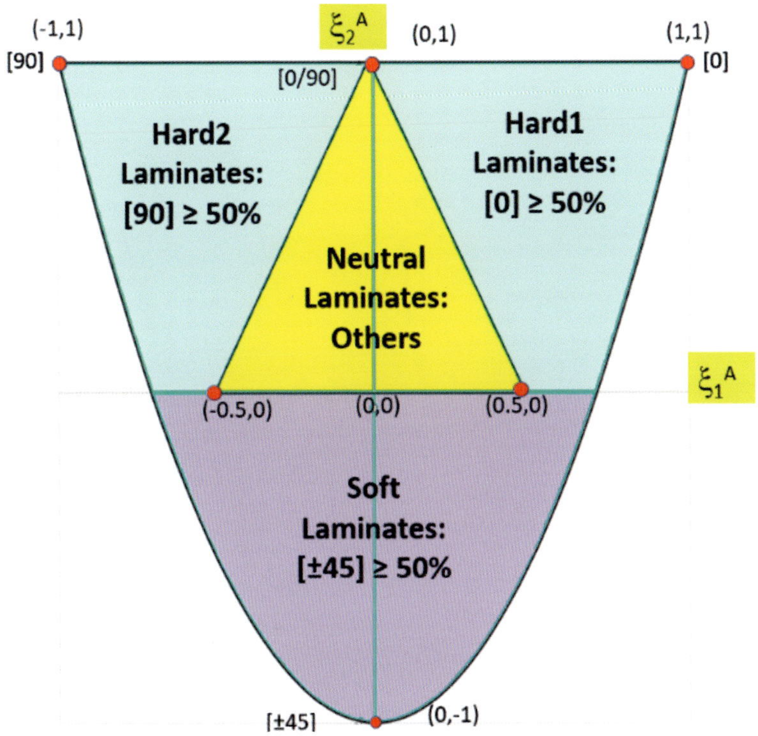

Figure 4.7 Feasible domain of all laminates in a ξ_1^A and ξ_2^A plot.

Carbon fiber, aero and non-aero composite applications are plotted in Figures 4.8 and 4.9, respectively.

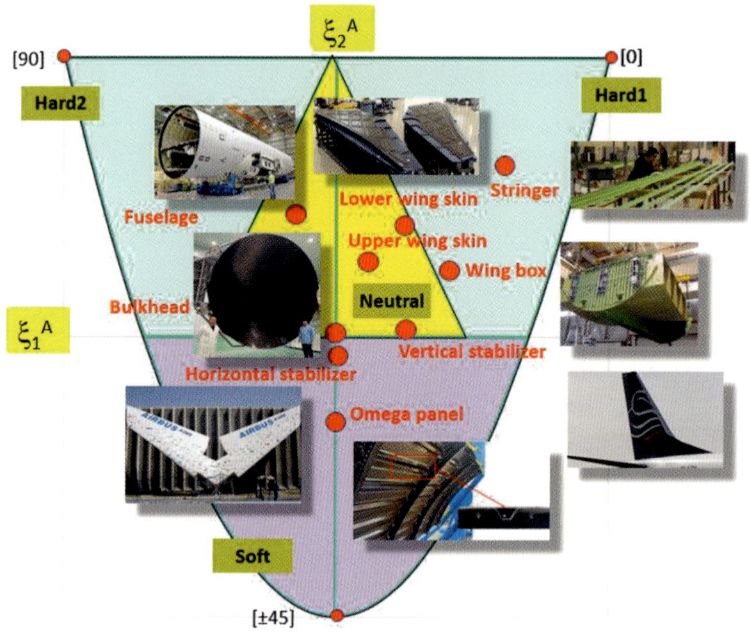

Figure 4.8. LPP of carbon fiber, aero composite applications.

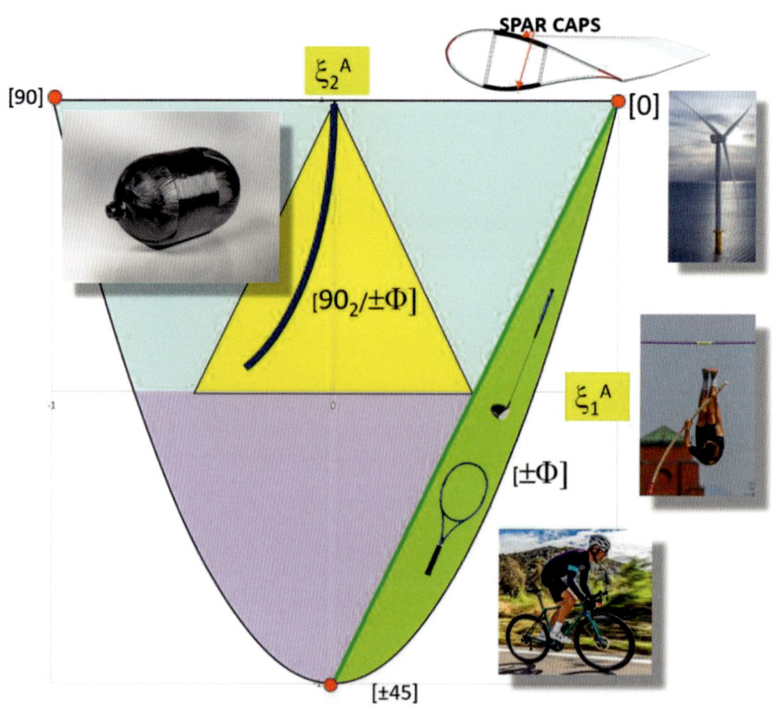

Figure 4.9 LPP of carbon fiber, non-aero composite applications.

Glass fiber composite applications are plotted in Figure 4.10.

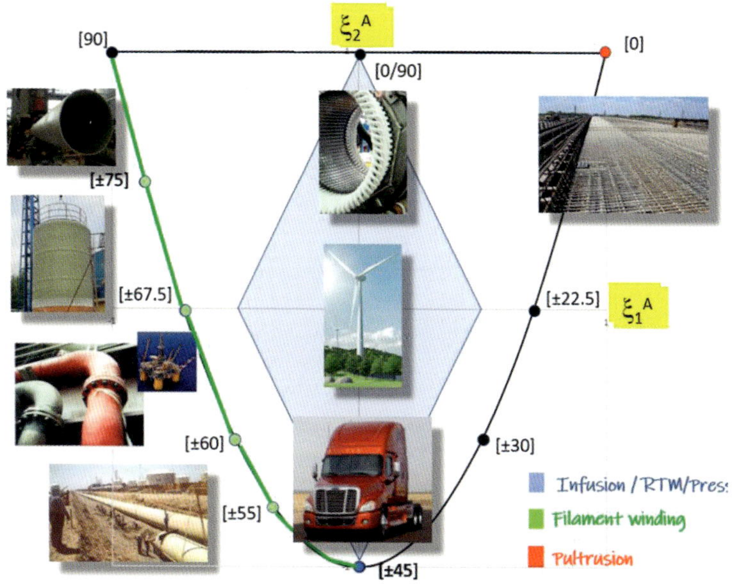

Figure 4.10 LPP of glass fiber composite applications.

A summary of composite application is plotted in Figure 4.11.

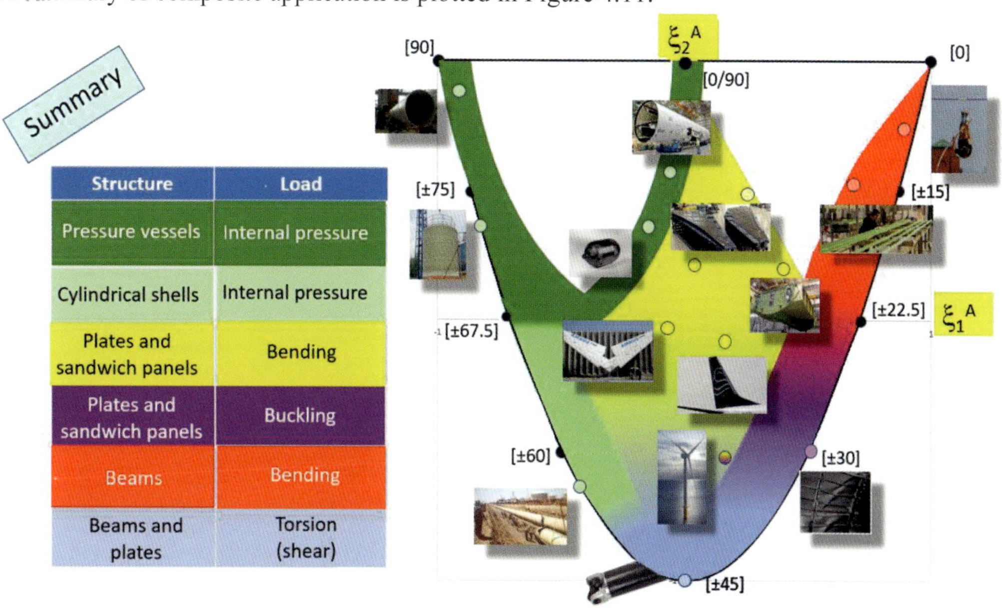

Figure 4.11 LPP of summary of composite applications.

4.3 Lamination properties: repetitive building blocks, feasible domain and dispersion.

Let's differentiate the properties of materials and lamination. It's well known that materials have key properties for composites design, such as:

- Stiffness
- Strength

However, there are other properties, not so well known, which are essential for laminated composites. They are called lamination properties and they are related to homogenization:

- Building Blocks
- Feasible domain
- Dispersion

Composites laminates formed by building blocks enhance a fast homogenization. The smaller the building block, the faster homogenization is achieved. The concept of feasible domain is critical to determine the minimum size of the building block. Finally, dispersion is important on two scales. Inside the laminate, finely dispersion repeated building blocks facilitate a fast homogenization. Inside a building block, that more disperse the stacking sequence, the faster to reach homogenization. Figure 4.12 shows the differences between quads and building block-type laminates in terms of homogenization.

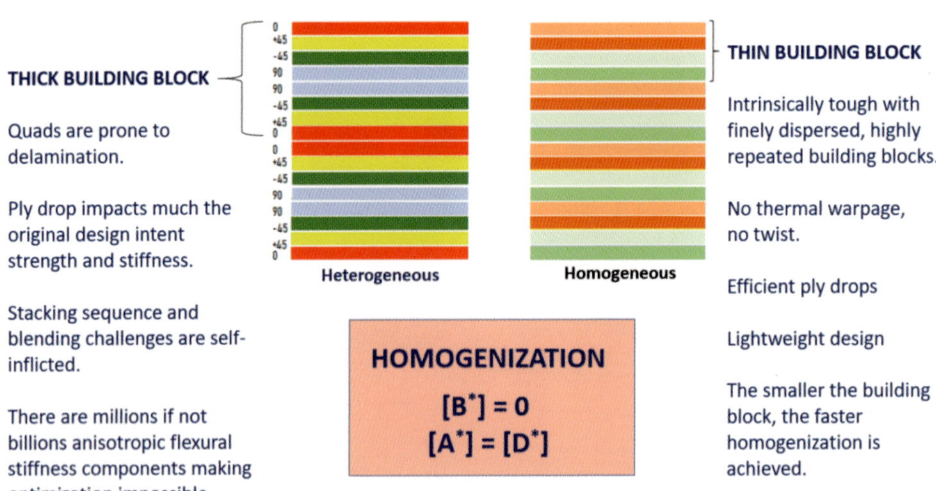

Figure 4.12 Differences between quads and building block-type laminates in terms of homogenization.

Current wind blades are formed by 1-, 2- and 3-ply sublaminates (Figure 4.13).

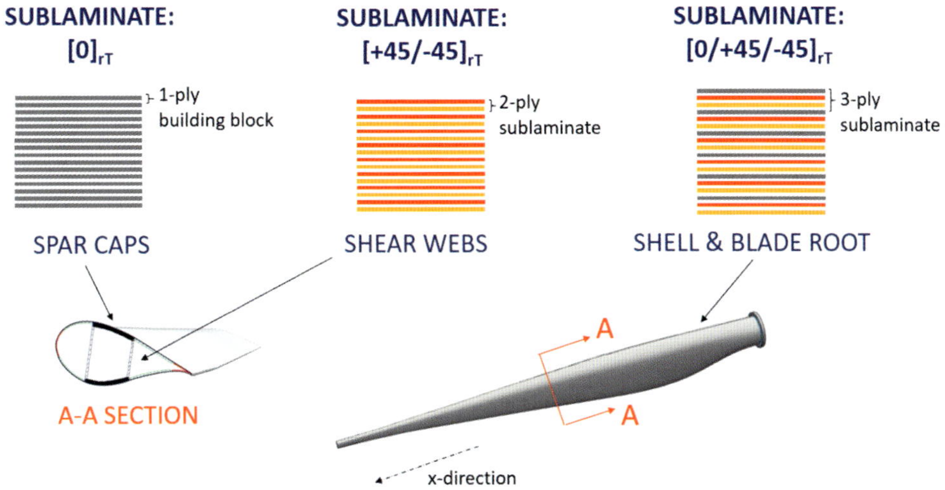

Figure 4.13 Current wind blades are formed by 1-, 2- and 3-ply sublaminates.

However, the feasible domains of 1- , 2- and 3-ply sublaminates are just dots and parabolic curves (Figure 4.14)

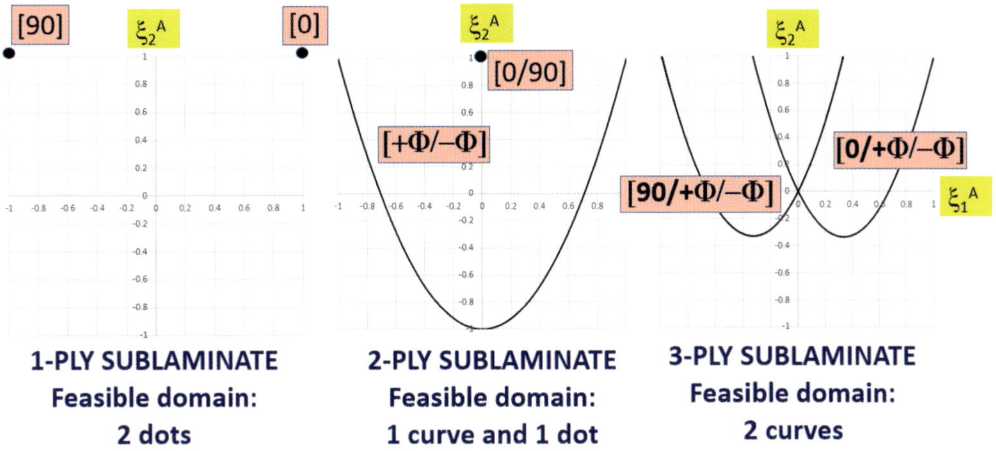

Figure 4.14 Feasible domains of 1- , 2- and 3-ply sublaminates.

Figure 4.15 shows nine families of laminates with various sizes of building blocks. DD is the only configuration using a 4- ply sublaminate. The rest of configurations use 5-, 6-, 8- or 10-ply sublaminates.

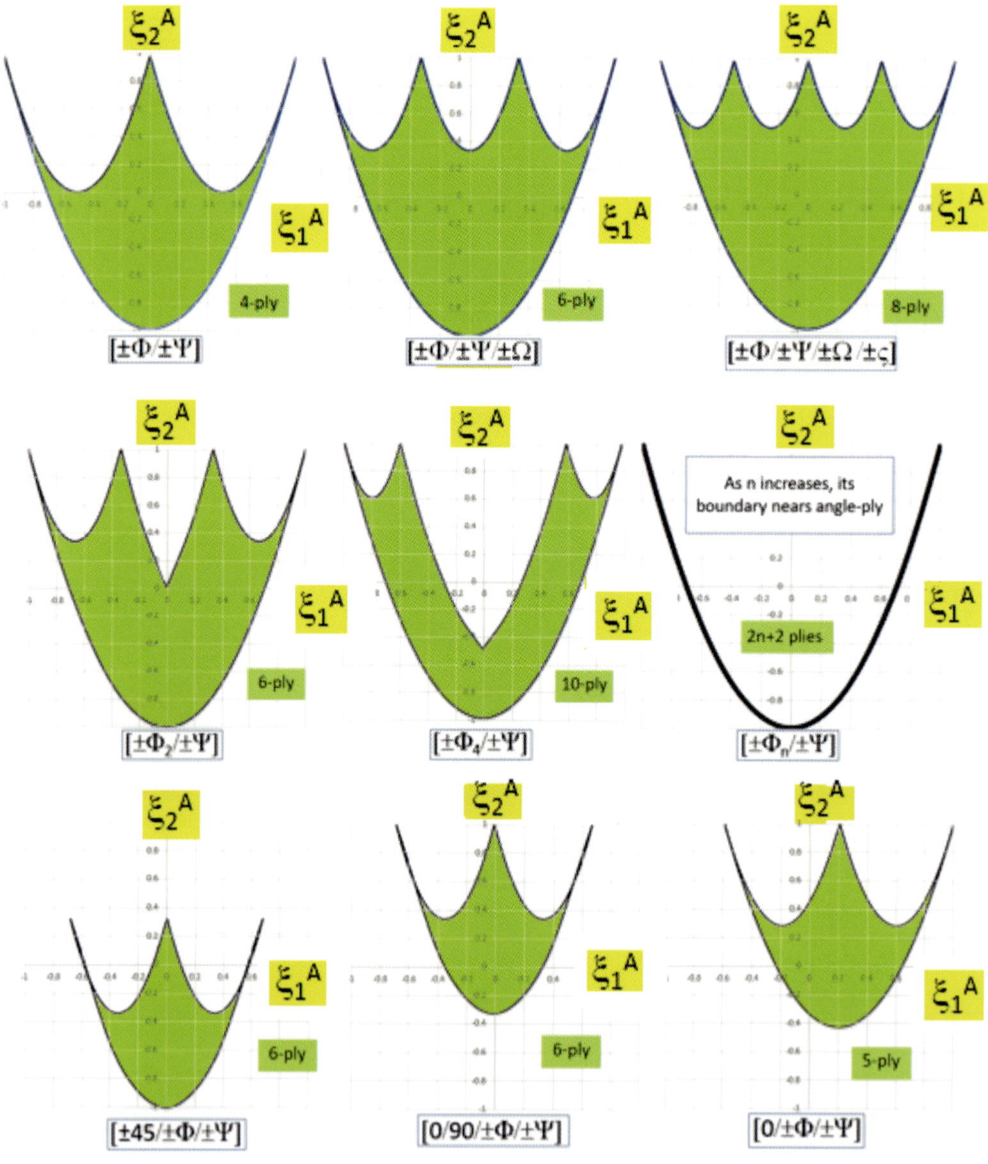

Figure 4.15 Nine families of laminates with 4-, 5-, 6- 8- and 10-ply building blocks.

Figure 4.16 shows a LPP of double – double and their boundaries.

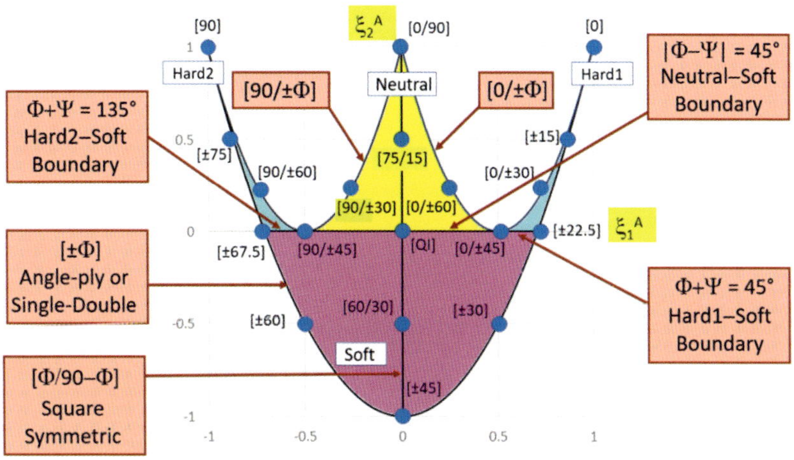

Figure 4.16 LPP of double – double and their boundaries.

For double – double:

$\xi_1^A = (\cos 2\Phi + \cos 2\Psi) / 2, \quad \xi_1^D = \xi_1^A$ if homogenized Eq. 4.5

$\xi_2^A = (\cos 4\Phi + \cos 4\Psi) / 2, \quad \xi_2^D = \xi_2^A$ if homogenized

Figure 4.17 shows the comparison of the LPP of double-double and the summary of composite applications.

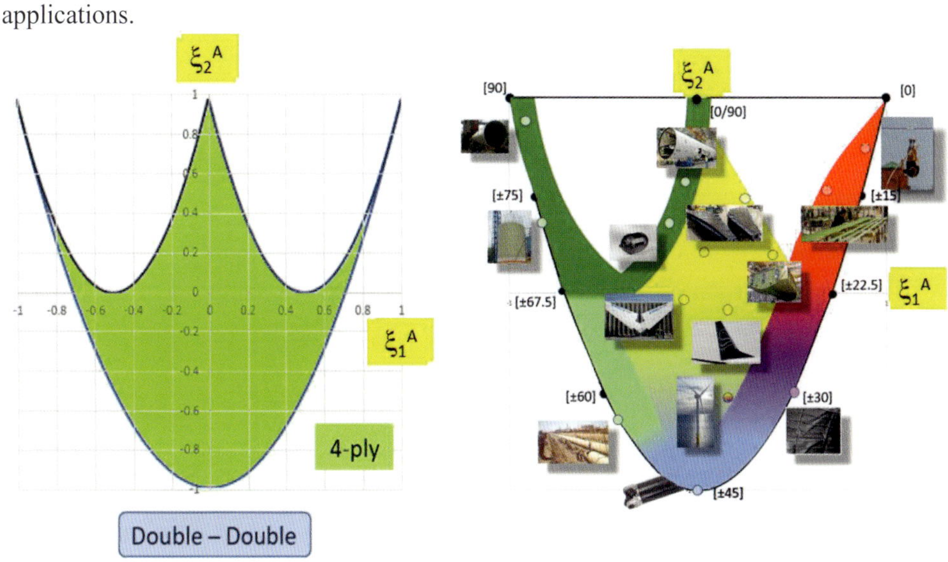

Figure 4.17 Comparison of the LPP of double-double and the summary of composite applications.

4.4 Quad feasible domains

Figure 4.18 shows a LPP of 20-ply quads.

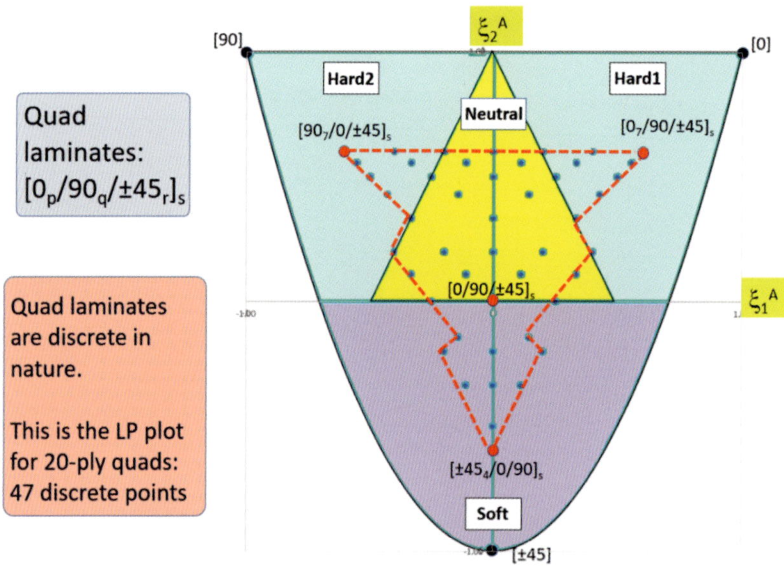

Figure 4.18 LPP of 20-ply quads.

Figure 4.19 represents the four areas of a quad LPP. Only the aqua area is the feasible domain for quads applying the 10% rule.

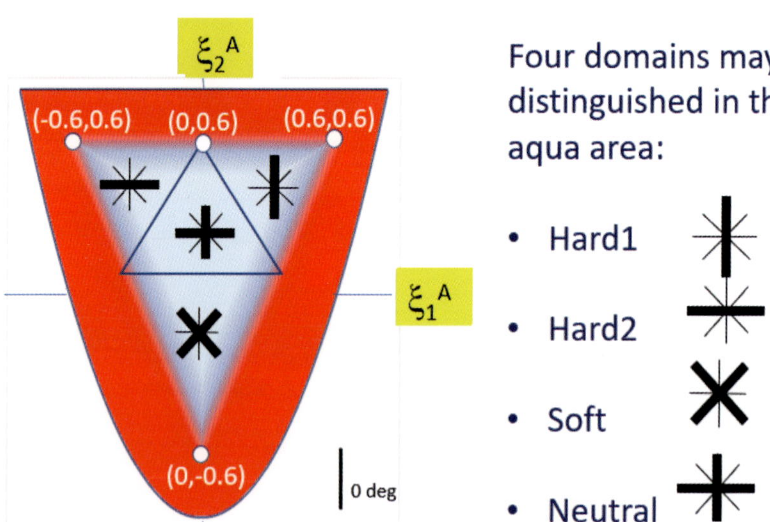

Figure 4.19 Representation of the four areas of a quad LPP. Only the aqua area is the feasible domain for quads applying the 10% rule. Quad feasible domain is very limited.

Figure 4.20 shows the laminate parameters of 16-, 20- and 24-ply quads. The only way to increase the dot density is to thick the building blocks.

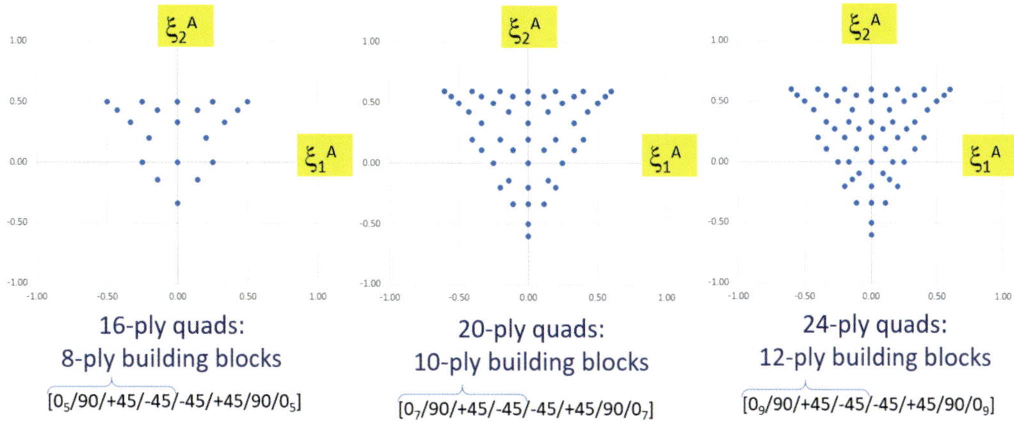

Figure 4.20 Laminate parameters of 16-, 20- and 24-ply quads.

Figure 4.21 shows the evolution of quads from a 3D pyramid LPP.

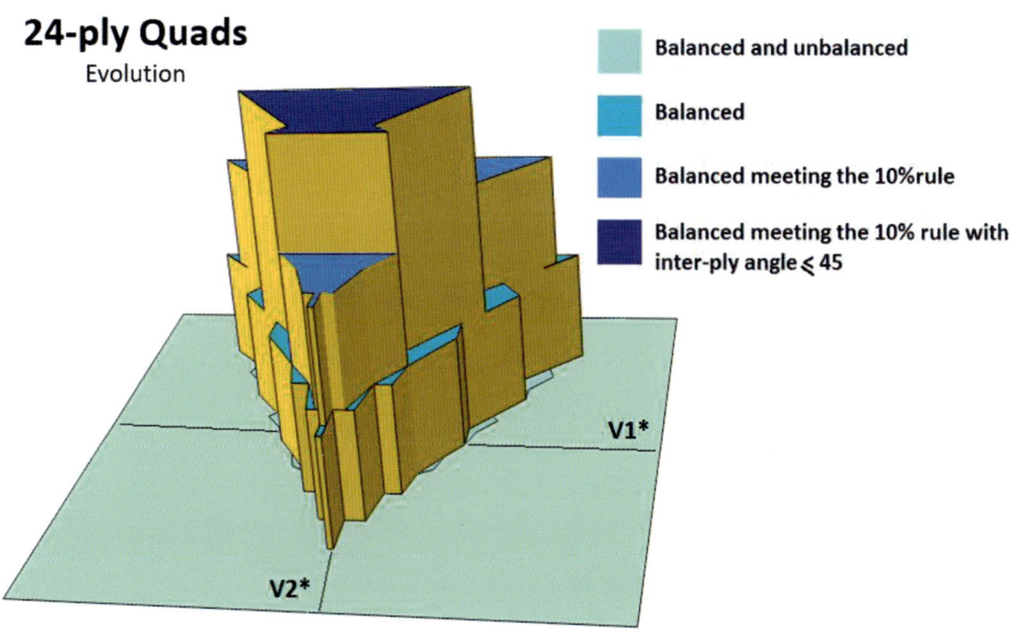

Figure 4.21 Evolution of quads from a 3D pyramid de laminate parameter plot.

4.5 Homogenization criterion. Validation

There are thousands of stacking sequences in quads. This is an enormous advantage of DD for having only four stacking sequence permutations, instead of thousands:

1) Staggered 1 in [Φ/–Ψ/–Φ/Ψ]
2) Staggered 2 in [Φ/Ψ/–Φ/–Ψ]
3) Staggered 3 in [Φ/–Ψ/Ψ/–Φ]
4) Paired in [Φ/–Φ/Ψ/–Ψ]

It is, however, to know with these six stacking sequences how many repeats will be needed to reach homogenization. We can set the definition of homogenization by the value of the largest absolute component of [B*] and [A]* - [D]*.

Homogenization criterion validation

The criterion used is the following:

$$|[A_{ij}^*] - [D_{ij}^*]| < 2\% \text{ Tsai}$$
$$|[B_{ij}^*]| < 2\% \text{ Tsai} \qquad \text{Eq. 4.6}$$
$$i = 1,2,6 \text{ and } j = 1,2,6$$
$$\text{with Tsai} = A_{11}^* + A_{22}^* + 2A_{66}^*$$

An experimental warping study was carried out in [28]. Two samples of an omega part were made out using two double – double stacking sequences and tested to determine the warping after cure. The part was tapered to get weight savings (Figure 4.22).

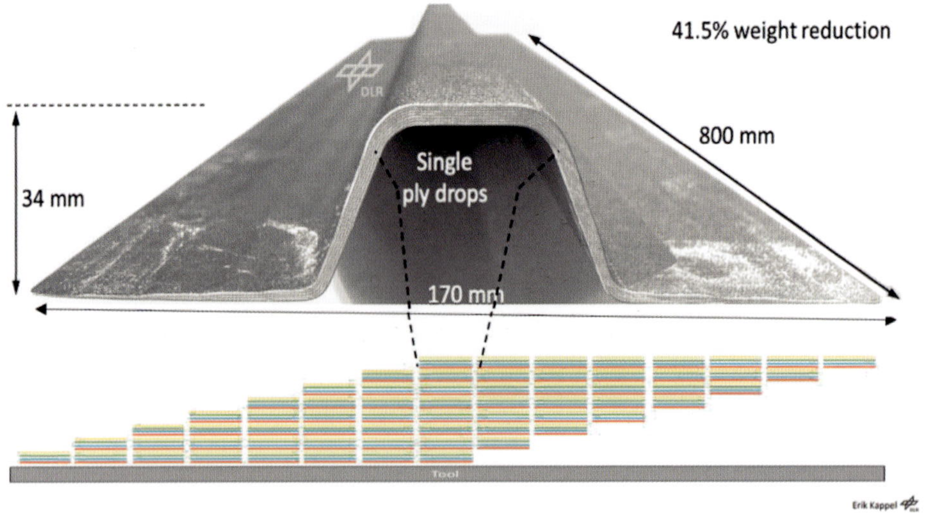

Figure 4.22 Tapered omega made of double – double laminates [28].

The cooldown was from 180°C to 20°C. The number of repeats was 8 at the center and 1 at the edges, so the average number of repeats was: (8+1)/2 = 4.5.

The two sequences used were:

- ✓ Staggered 1: [19.3/-67/-19.3/67]
- ✓ Paired : [19.3/-19.3/67/-67]

Figure 4.23 shows the following results:

- ✓ Staggered 1 [19.3/-67/-19.3/67]: the part showed hardly any noticeable twist/warpage, the FE model gave U3=1.4mm (Figure 4.23, left)
- ✓ Paired [19.3/-19.3/67/-67]: the part showed twist deformations, the FE model gave U3=5.3mm (Figure 4.23, right)

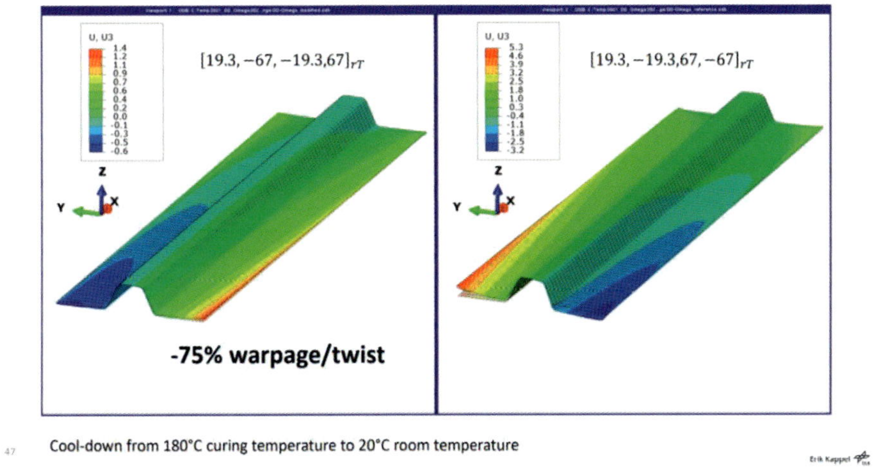

Figure 4.23 FE model results [28].

When the Staggered 1 sequence was used (left), the part showed hardly any noticeable twist/warpage (4.5 average number of repeats were used and the minimum number needed was 4, Figure 4.23, left) and when the paired sequence was used (right), the part showed twist deformations (4.5 average number of repeats were used and the minimum number needed to reach homogenization was 8, Figure 4.23, right). Table 4.1 shows the minimum repeats for homogenization using the homogenization criterion.

Table 4.1 Minimum repeats for homogenization using the homogenization criterion.

Sequence	Test results	Criterion results
[19.3/-67/-19.3/67]$_{rT}$	4.5 repeats : No warpage	Minimum 4 repeats
[19.3/-19.3/67/-67]$_{rT}$	4.5 repeats: Warpage	Minimum 8 repeats

4.6 Minimum number of plies to reach homogenization

Figures 4.24 and 4.25 show the amination parameter expressions for Staggered 1 and Staggered 2, and Staggered 3 and Paired, respectively.

Homogenization	Staggered 1	Staggered 2
	$V_1^* = (\cos 2\Phi + \cos 2\Psi)/2$	$V_1^* = (\cos 2\Phi + \cos 2\Psi)/2$
[A] In-plane	$V_2^* = (\cos 4\Phi + \cos 4\Psi)/2$	$V_2^* = (\cos 4\Phi + \cos 4\Psi)/2$
	$V_3^* = 0$	$V_3^* = 0$
	$V_4^* = 0$	$V_4^* = 0$
[D] Bending	$W_1^* = (\cos 2\Phi + \cos 2\Psi)/2$	$W_1^* = (\cos 2\Phi + \cos 2\Psi)/2$
$\lvert W_i^* - V_i^* \rvert < 2\%$ Tsai to	$W_2^* = (\cos 4\Phi + \cos 4\Psi)/2$	$W_2^* = (\cos 4\Phi + \cos 4\Psi)/2$
reach homogenization and	$W_3^* = 3(\sin 2\Phi + \sin 2\Psi)/8r^2$	$W_3^* = 3(\sin 2\Phi - \sin 2\Psi)/8r^2$
make it possible to taper	$W_4^* = 3(\sin 4\Phi + \sin 4\Psi)/8r^2$	$W_4^* = 3(\sin 4\Phi - \sin 4\Psi)/8r^2$
[B] stretching/bending coupling	$Y_1^* = (-\cos 2\Phi + \cos 2\Psi)/8r$	$Y_1^* = (-\cos 2\Phi + \cos 2\Psi)/8r$
$\lvert Y_1^* \rvert$ and $\lvert Y_2^* \rvert < 2\%$ Tsai	$Y_2^* = (-\cos 4\Phi + \cos 4\Psi)/8r$	$Y_2^* = (-\cos 4\Phi + \cos 4\Psi)/8r$
[B] stretching/ twisting coupling	$Y_3^* = (-\sin 2\Phi + \sin 2\Psi)/4r$	$Y_3^* = (-\sin 2\Phi - \sin 2\Psi)/4r$
$\lvert Y_3^* \rvert$ and $\lvert Y_4^* \rvert < 2\%$ Tsai	$Y_4^* = (-\sin 4\Phi + \sin 4\Psi)/4r$	$Y_4^* = (-\sin 4\Phi - \sin 4\Psi)/4r$

Figure 24. Lamination parameter expressions for Staggered 1 and Staggered 2.

Homogenization	Staggered 3	Paired
	$V_1^* = (\cos 2\Phi + \cos 2\Psi)/2$	$V_1^* = (\cos 2\Phi + \cos 2\Psi)/2$
[A] In-plane	$V_2^* = (\cos 4\Phi + \cos 4\Psi)/2$	$V_2^* = (\cos 4\Phi + \cos 4\Psi)/2$
	$V_3^* = 0$	$V_3^* = 0$
	$V_4^* = 0$	$V_4^* = 0$
[D] Bending	$W_1^* = (\cos 2\Phi + \cos 2\Psi)/2 + 3(\cos 2\Phi - \cos 2\Psi)/8r^2$	$W_1^* = (\cos 2\Phi + \cos 2\Psi)/2$
$\lvert W_i^* - V_i^* \rvert < 2\%$ Tsai to	$W_2^* = (\cos 4\Phi + \cos 4\Psi)/2 + 3(\cos 4\Phi - \cos 4\Psi)/8r^2$	$W_2^* = (\cos 4\Phi + \cos 4\Psi)/2$
reach homogenization and	$W_3^* = 0$	$W_3^* = 3(\sin 2\Phi + \sin 2\Psi)/8r^2$
make it possible to taper	$W_4^* = 0$	$W_4^* = 3(\sin 4\Phi + \sin 4\Psi)/8r^2$
[B] stretching/bending coupling	$Y_1^* = 0$	$Y_1^* = (-\cos 2\Phi + \cos 2\Psi)/4r$
$\lvert Y_1^* \rvert$ and $\lvert Y_2^* \rvert < 2\%$ Tsai	$Y_2^* = 0$	$Y_2^* = (-\cos 4\Phi + \cos 4\Psi)/4r$
[B] stretching/ twisting coupling	$Y_3^* = (-3\sin 2\Phi + \sin 2\Psi)/8r$	$Y_3^* = (-\sin 2\Phi + \sin 2\Psi)/8r$
$\lvert Y_3^* \rvert$ and $\lvert Y_4^* \rvert < 2\%$ Tsai	$Y_4^* = (-3\sin 4\Phi + \sin 4\Psi)/8r$	$Y_4^* = (-\sin 4\Phi + \sin 4\Psi)/8r$

Figure 4.25 Lamination parameter expressions for Staggered 3 and Paired.

Figure 4.26 shows the regions no meeting the homogenization criterion (in red) for the five sequences.

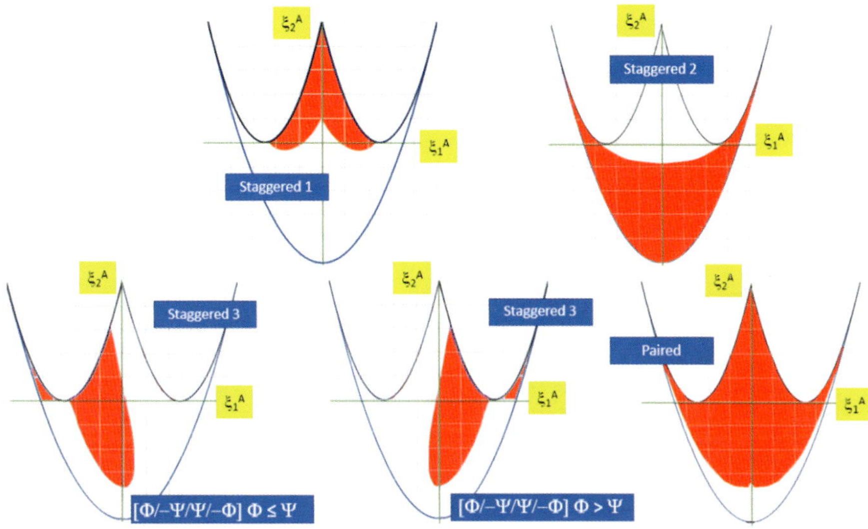

Figure 4.26 Regions no meeting the homogenization criterion (in red).

The only solution to meet the homogenization criterion for 4 repeats in the whole LP plot is a combination of Staggered 1 and Staggered 3 sequences (Figure 4.27).

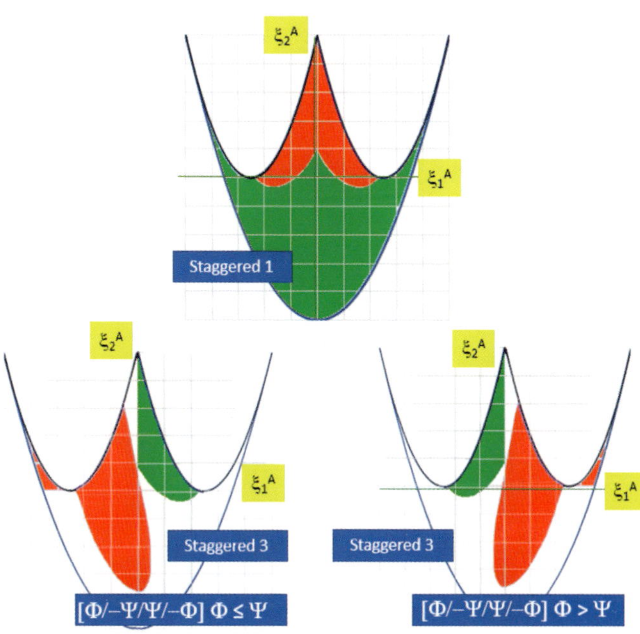

Figure 4.27 Regions meeting the homogenization criterion (in green).

Figure 4.28 represents the application of the homogenization criterion to Staggered 1 for 4 repeats. (ξ_1^B/ξ_2^B) lamination parameters.

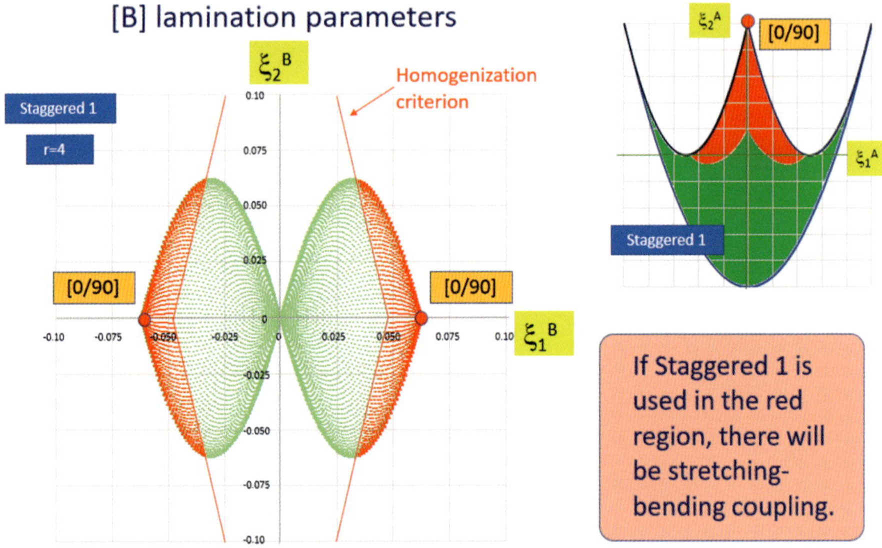

Figure 4.28 Application of the homogenization criterion to Staggered 1 for 4 repeats. (ξ_1^B/ξ_2^B) lamination parameters.

Figure 4.29 represents the application of the homogenization criterion to Staggered 1 for 3 repeats. (ξ_1^B/ξ_2^B) lamination parameters.

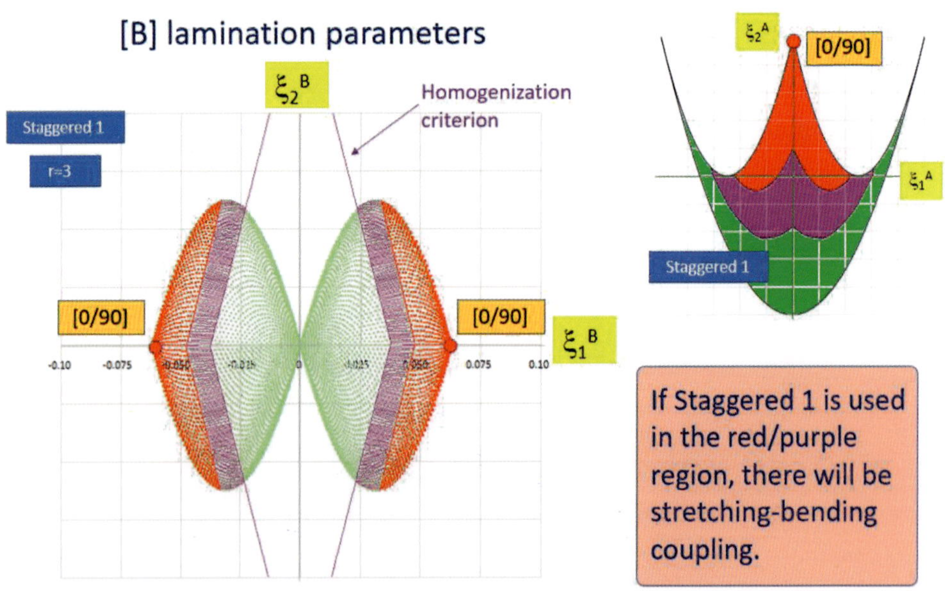

Figure 4.29 Application of the homogenization criterion to Staggered 1 for 3 repeats. (ξ_1^B/ξ_2^B) lamination parameters.

Figure 4.30 represents the application of the homogenization criterion to Staggered 1 for 4 repeats.(ξ_3^B/ξ_4^B) lamination parameters.

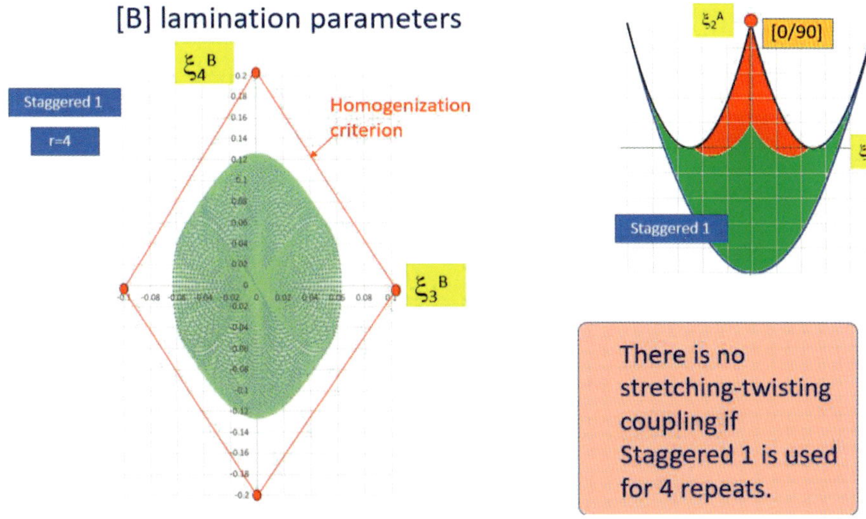

Figure 4.30 Application of the homogenization criterion to Staggered 1 for 4 repeats. (ξ_3^B/ξ_4^B) lamination parameters.

Figure 4.31 shows the application of the homogenization criterion to Staggered 1 for 4 repeats. (ξ_3^B/ξ_4^B) lamination parameters.

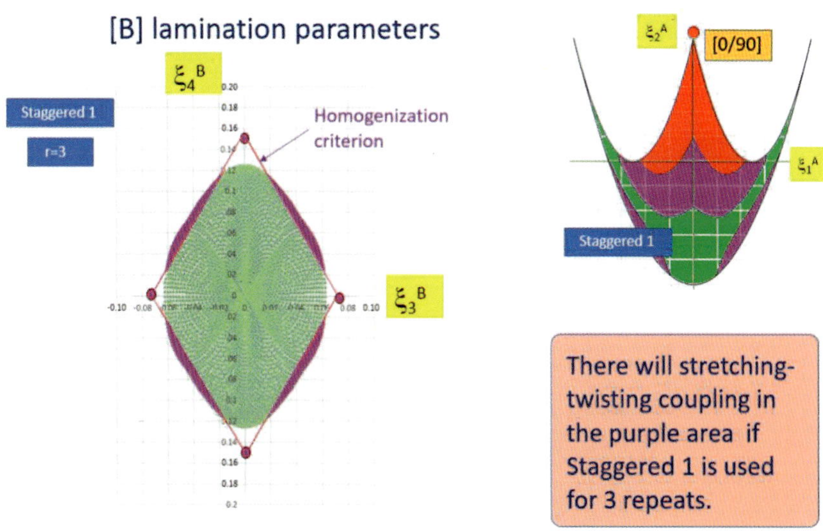

Figure 4.31 Application of the homogenization criterion to Staggered 1 for 3 repeats. (ξ_3^B/ξ_4^B) lamination parameters.

Figure 4.32 represents the application of the homogenization criterion to Staggered 1 for 4 repeats. (ξ_3^A - ξ_3^D/ ξ_4^A - ξ_4^D) lamination parameters.

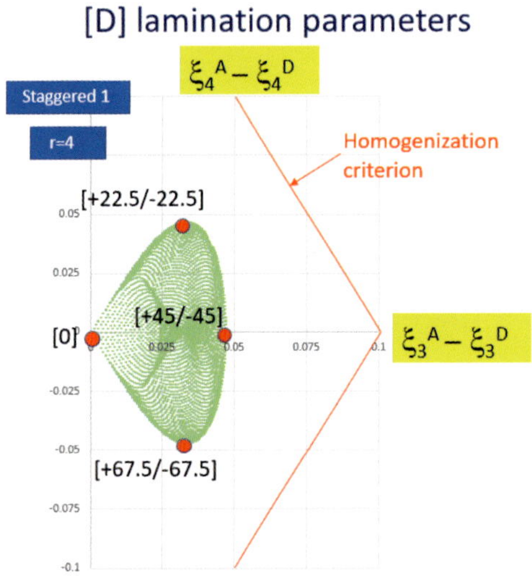

Figure 4.32 Application of the homogenization criterion to Staggered 1 for 4 repeats. (ξ_3^A -ξ_3^D/ ξ_4^A -ξ_4^D) lamination parameters.

Figure 4.33 shows the application of the homogenization criterion to Staggered 1 for 3 repeats. (ξ_3^A - ξ_3^D/ ξ_4^A - ξ_4^D) lamination parameters.

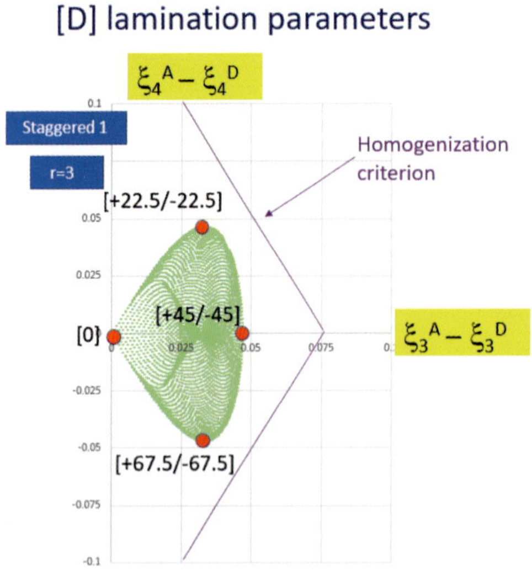

Figure 4.33 Application of the homogenization criterion to Staggered 1 for 3 repeats. (ξ_3^A -ξ_3^D/ ξ_4^A -ξ_4^D) lamination parameters.

Figure 4.34 shows the application of the homogenization criterion to Staggered 3 for 4 repeats. (ξ_3^B/ξ_4^B) lamination parameters and $\Phi \leq \Psi$.

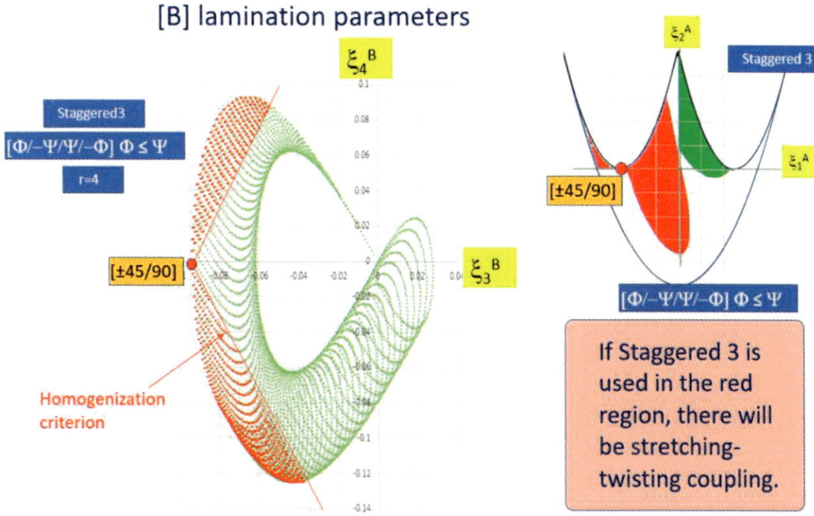

Figure 4.34 Application of the homogenization criterion to Staggered 3 for 4 repeats. (ξ_3^B/ξ_4^B) lamination parameters and $\Phi \leq \Psi$.

Figure 4.35 shows the application of the homogenization criterion to Staggered 3 for 4 repeats. (ξ_3^B/ξ_4^B) lamination parameters and $\Phi > \Psi$.

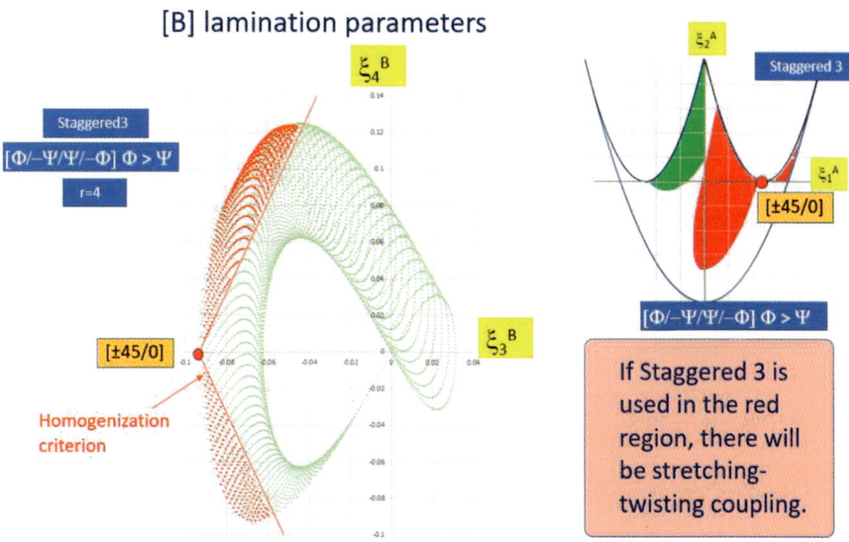

Figure 4.35 Application of the homogenization criterion to Staggered 3 for 4 repeats. (ξ_3^B/ξ_4^B) lamination parameters and $\Phi > \Psi$.

Figure 4.36 represents the application of the homogenization criterion to Staggered 3 for 4 repeats. ($\xi_1^A - \xi_1^D / \xi_2^A - \xi_2^D$) lamination parameters.

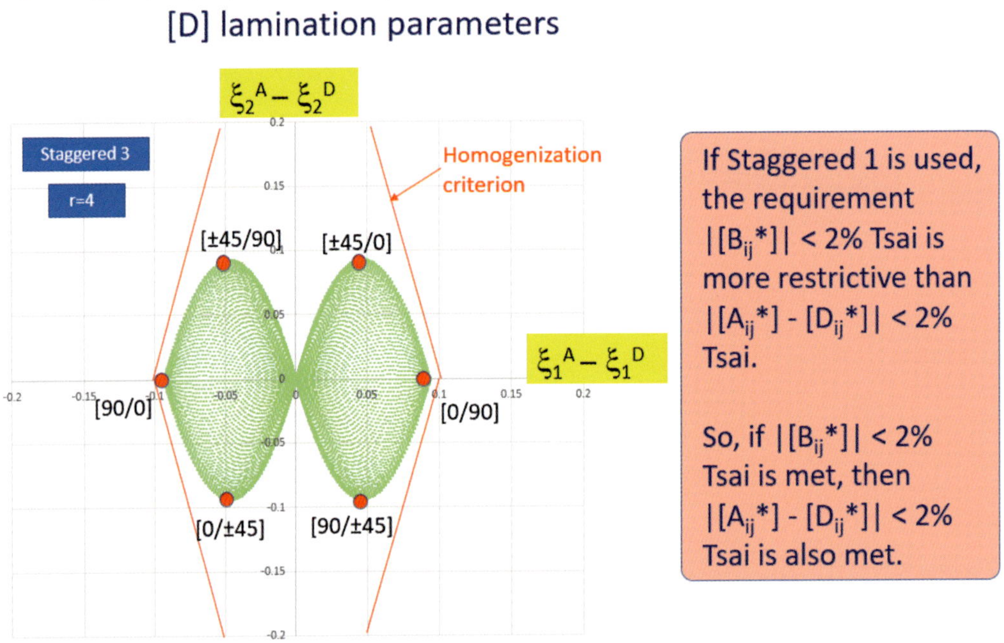

Figure 4.36 Application of the homogenization criterion to Staggered 3 for 4 repeats. ($\xi_1^A - \xi_1^D / \xi_2^A - \xi_2^D$) lamination parameters.

Figure 4.37 shows the minimum number of double-double plies to reach homogenization.

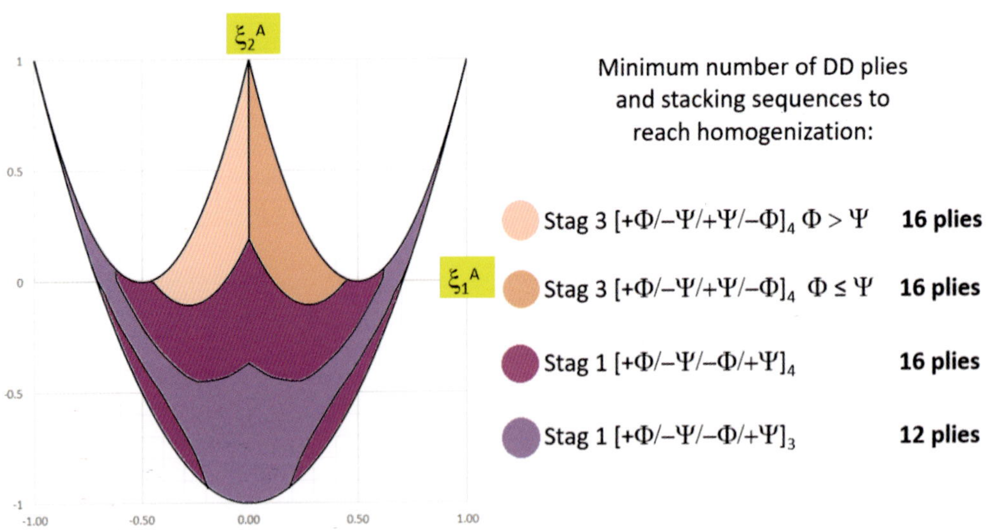

Figure 4.37 Minimum number of double-double plies to reach homogenization.

4.7 Convergence rates to homogenization

Stacking sequences having the same stacking except in the outer 2 plies:

- Staggered 1 [Φ/−Ψ/−Φ/Ψ] and Staggered 2 [Φ/Ψ/−Φ/−Ψ] have the same stacking sequence except in the outer 2 plies
- Staggered 3 [Φ/−Ψ/Ψ/−Φ] and Paired [Φ/−Φ/Ψ/−Ψ] have the same stacking sequence except in the outer 2 plies
- The convergence rate of the four sequences varies. Staggered 1 and Staggered 3 are the fastest sequences to converge because they are the most disperse ones.

There is symmetry regarding the axes Φ=Ψ and Φ+Ψ = 90. The four sequences converge to the same [ABBD]* with [B]*=0 and [A]*=[D]*, the only difference is the convergence rate. A key factor is the absolute difference between |Φ − Ψ|.

In order to determine the converge rate for the five sequences, five angle differences will be analyzed:

Φ= 80 and Ψ= 0

Φ= 60 and Ψ= 20

Φ= 40 and Ψ= 40

In order to determine the converge rate, we are going to define the "variation regarding A*=D* and B*= 0" as:

Maximum value of
- Maximum value of $|B_{ij}^*| / A_{11}^*$ for i=1,2,3, j=1,2,3
- Maximum value of $|B_{ij}^*| / D_{11}^*$ for i=1,2,3, j=1,2,3
- Maximum value of $|A_{ij}^* - D_{ij}^*| / A_{11}^*$ for i=1,2,3, j=1,2,3
- Maximum value of $|A_{ij}^* - D_{ij}^*| / D_{11}^*$ for i=1,2,3, j=1,2,3

It makes sense that Staggered 3 and Staggered 4 show the fastest convergence rates since Φ=80, Ψ=0 is very close to Φ=90, Ψ=0 where Staggered 3 and Staggered 4 are naturally symmetric: $[0/90/90/0]_{rT}$. Staggered 1 and Staggered 2 are not naturally symmetric at Φ=90, Ψ=0, therefore they are slower. Paired is not a disperse sequence, it's the slowest (Figure 4.38).

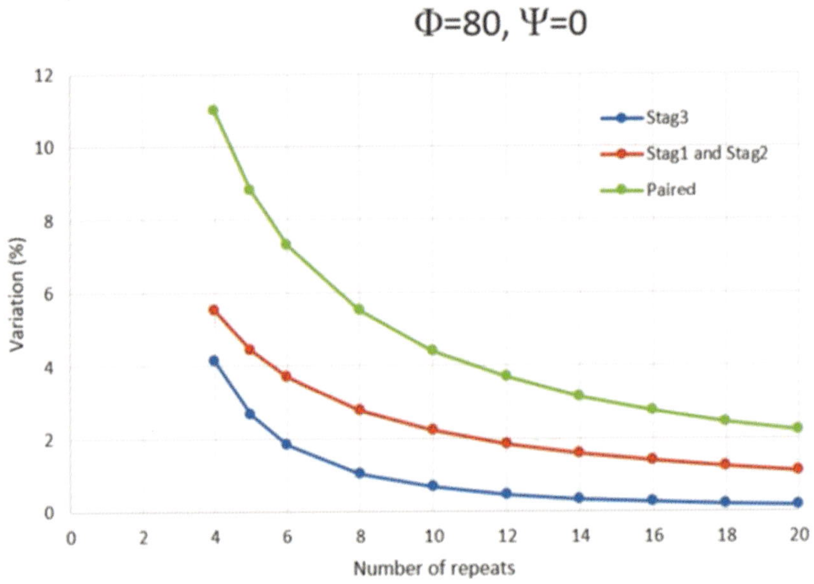

Figure 4.38 Convergence rate of the five sequences for Φ = 80 and Ψ= 0.

Φ=60, Ψ=20 is on the border between Staggered 1 and Staggered 3, therefore, Staggered 1, Staggered 2, Staggered 3 and Staggered 4 are close. Staggered 1 presents the fastest converge rate. We are already in the yellow area (Staggered 1 domain), (Figure 4.39).

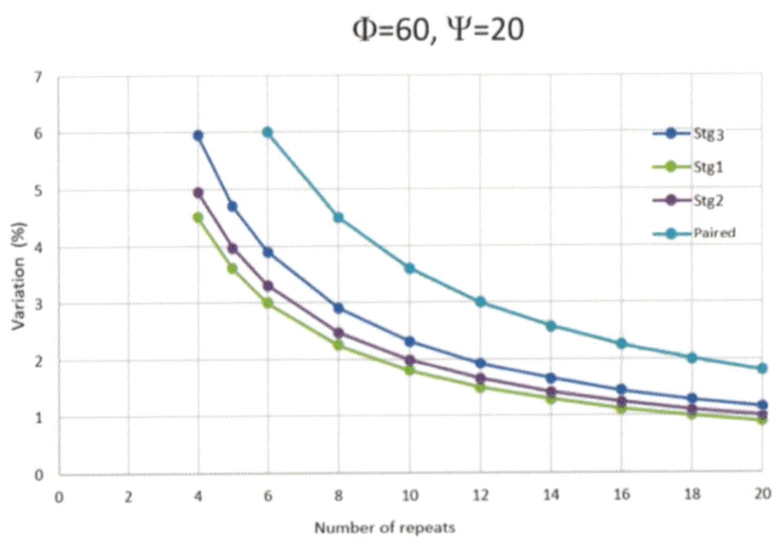

Figure 4.39 Convergence rate of the five sequences for Φ = 60 and Ψ= 20.

As expected, Staggered 1 shows the fastest convergence rate since Φ=40, Ψ=40 is very close to Φ=45, Ψ=45 where Staggered 1 is naturally symmetric $[45/\text{-}45/\text{-}45/45]_{rT}$., (Figure 4.40).

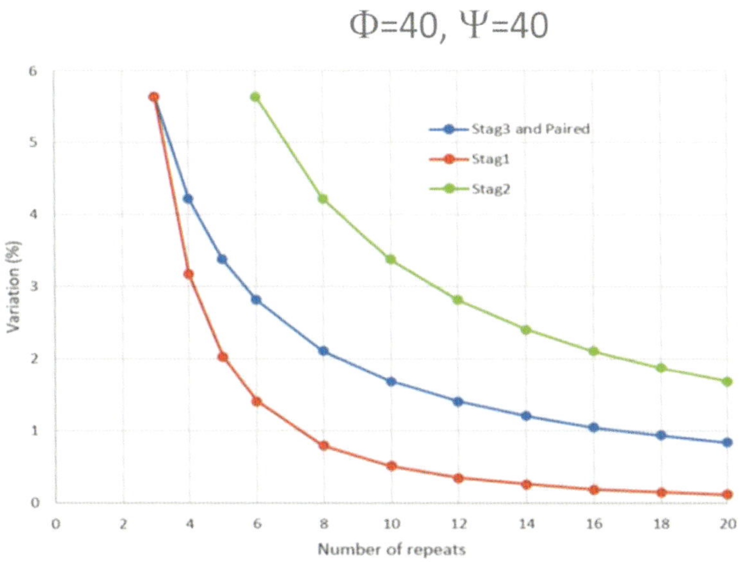

Figure 4.40 Convergence rate of the five sequences for Φ= 40 and Ψ= 40.

4.8 Well-dispersed Double-Double angles enhances homogenization

The sequences that need fewer repeats to reach homogenization are those whose couples are composed of different angles and different signs.

Only Staggered 1 [Φ/–Ψ/–Φ/Ψ] and Staggered 3 [Φ/–Ψ/Ψ/–Φ] meet this requirement:

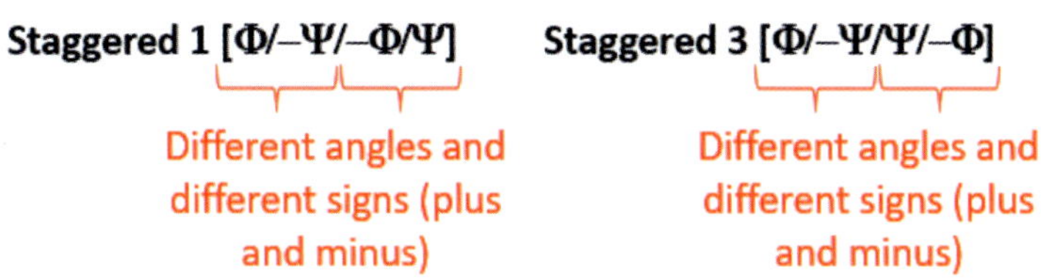

It makes sense then that these sequences are the fastest to reach homogenization because they are the most dispersed, and therefore the most favorable to reach homogenization using the fewest number of repeats.

4.9 Homogenization enhances weight savings through tapering

Let's analyze the ply drop of a hard laminate from 24 to 16 plies (Figure 4.41).

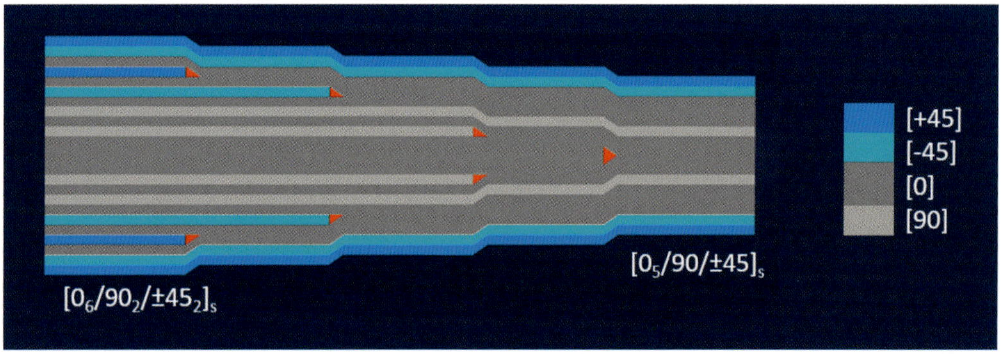

Figure 4.41 Scheme of a quad taper.

Since quads are a collection of discrete laminates, there is no continuity, stiffness and strength change along the ply drop (Figure 4.42).

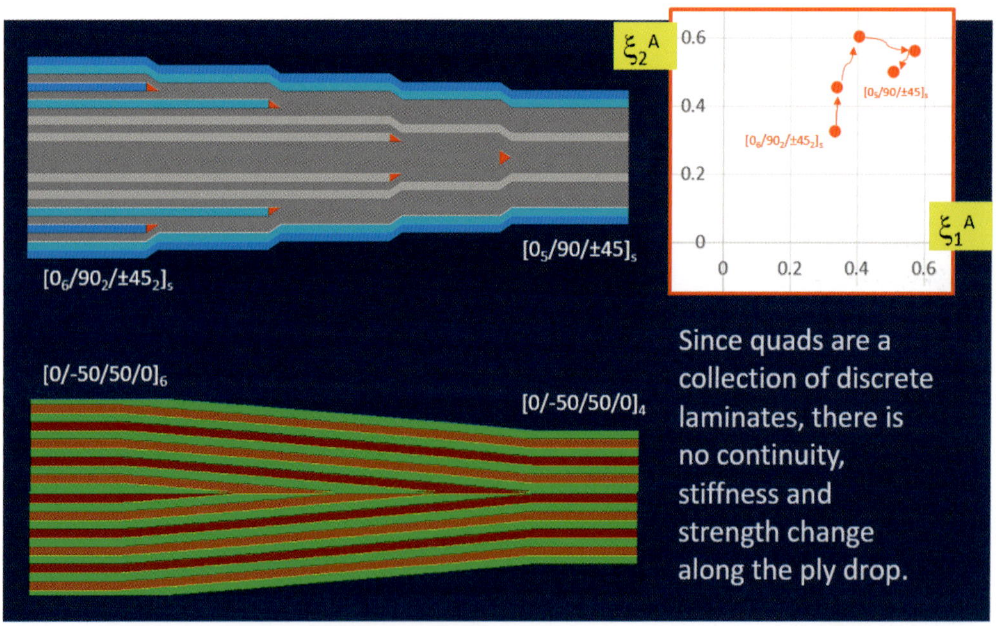

Figure 4.42 Stiffness and strength change along the ply drop in a quad taper.

Double-Double has no such a problem (Figure 4.43).

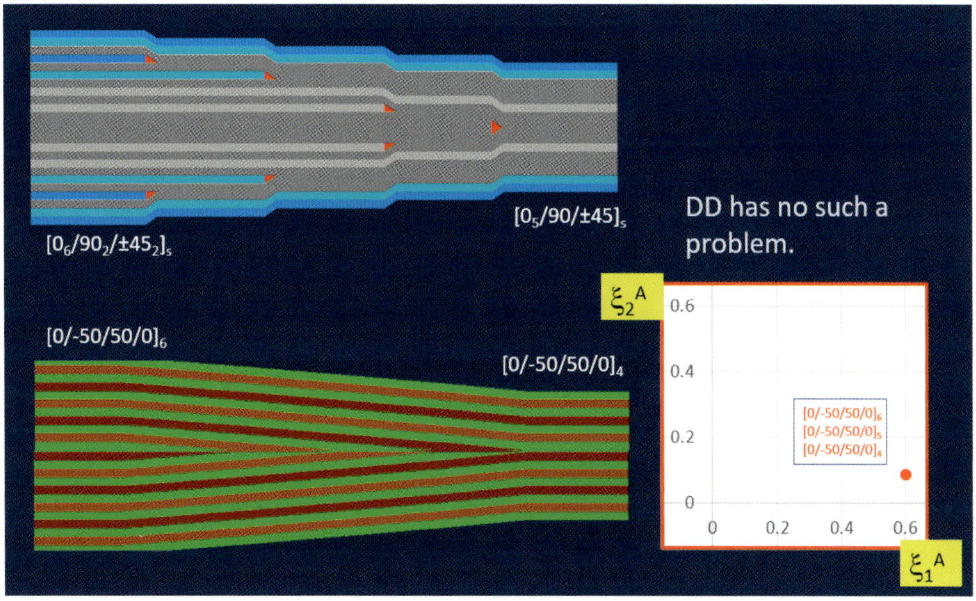

Figure 4.43 Stiffness and strength don't change along the ply drop in a DD taper.

It's not possible to do single ply drops when using quads. The ply drop must be symmetric. Double-double has no such a problem (Figure 4.44).

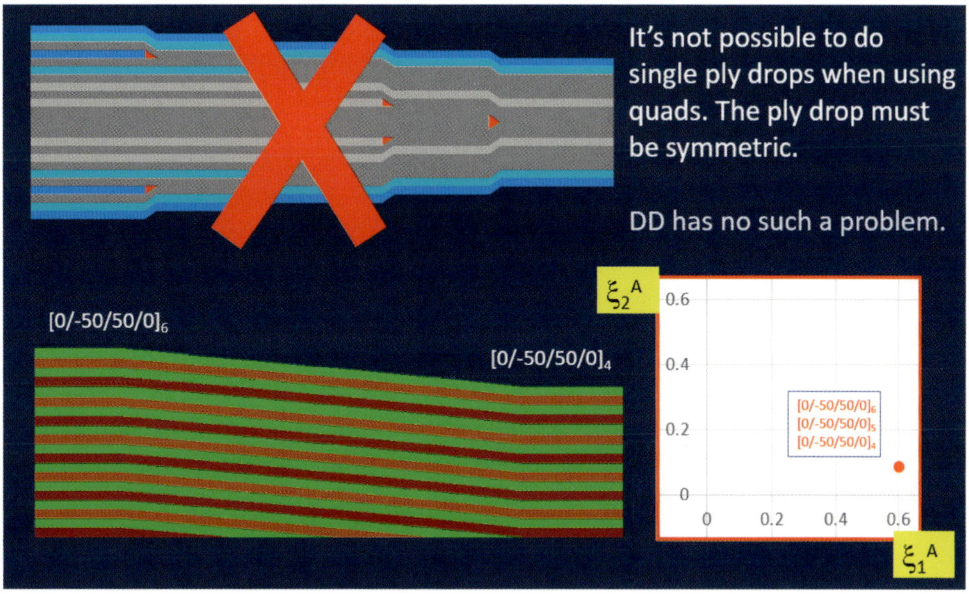

Figure 4.44 It's not possible to do single ply drops when using quads. The ply drop must be symmetric. Double-double has no such a problem

Quad ply drops create out-of-plane load effects and resin pockets (in red) are prone to develop delamination paths. Using double-double, fewer failure modes are encountered and no pockets are created (Figure 4.45).

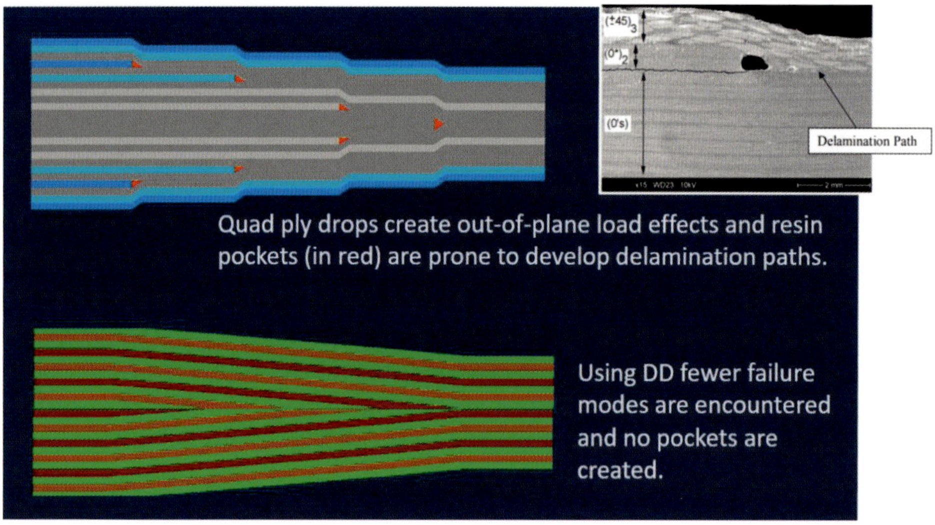

Figure 4.45 Quad ply drops create out-of-plane load effects and resin pockets (in red) are prone to develop delamination paths.

Quad ply drops have blocks of plies with the same fiber orientation, which create stress concentrations. Inter-ply angle = 90° are common, decreasing interlaminar strength and toughness. Double-double has no such a problem (Figure 4.46).

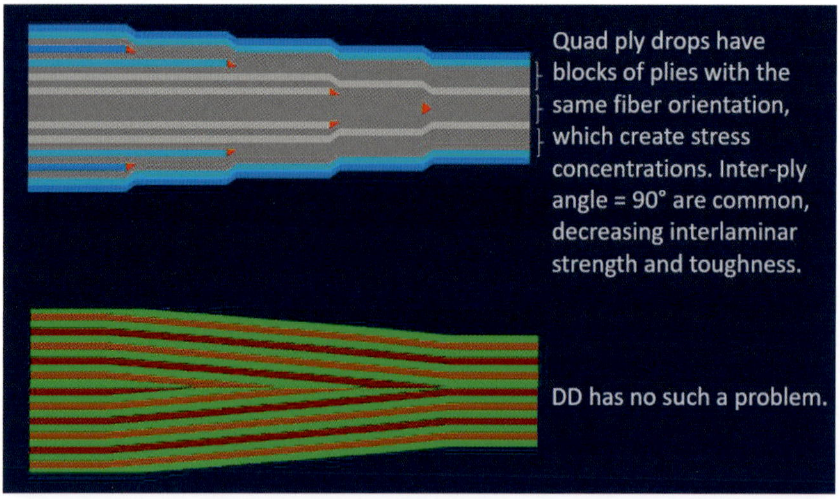

Figure 4.46 Quad ply drops have blocks of plies with the same fiber orientation, which create stress concentrations.

4.10 Quad replacement

Figure 4.47 shows the quad replacement for neutral laminates.

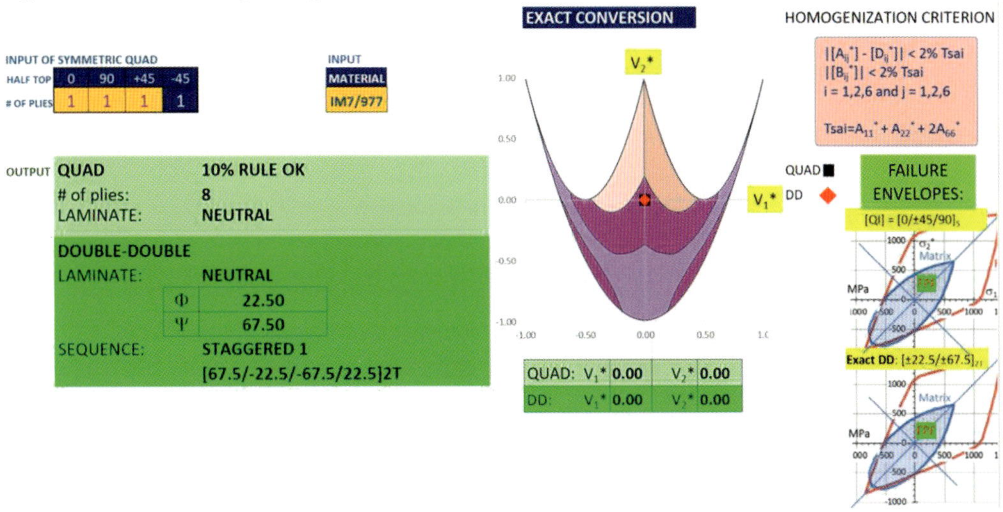

Figure 4.47 Quad replacement for neutral laminates.

Figure 4.48 represents the quad replacement for soft laminates.

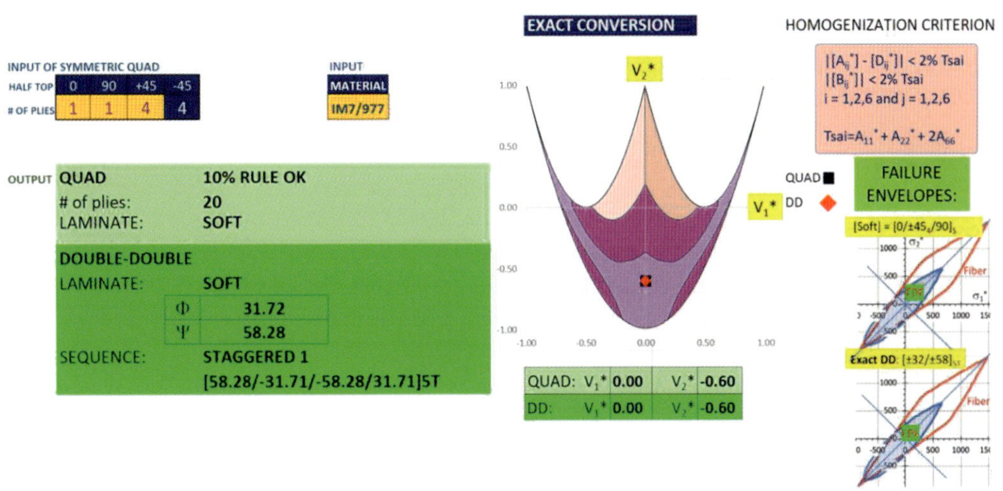

Figure 4.48 Quad replacement for soft laminates.

Figure 4.49 shows the quad replacement for hard laminates.

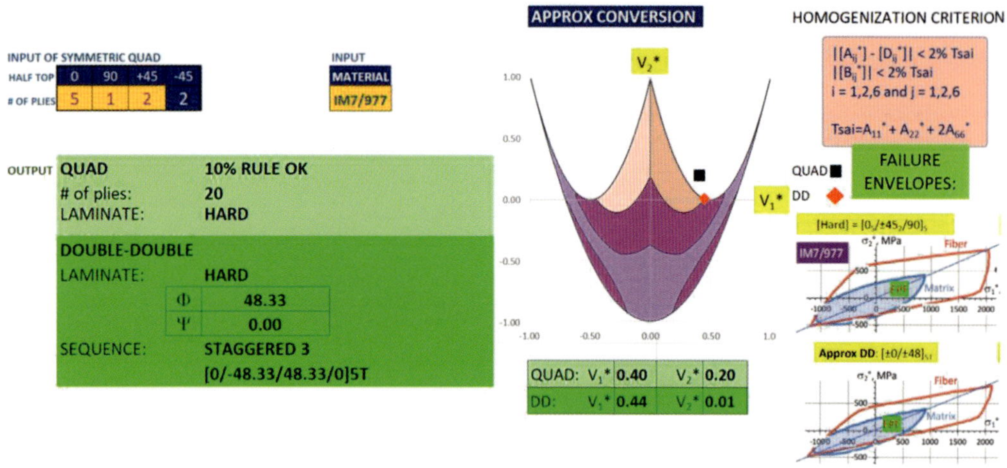

Fig 4.49 Quad replacement for hard laminates.

4.11 The Single-Double case

A particular case is the called single – double (Φ = Ψ). Figure 4.50 shows the normalized stiffness matrix for paired [+25/-25]$_6$ and Staggered 1 [+25/-25/-25/25]$_3$. For the paired sequence, [A] = [D] and for the Staggered 1, [B] = 0. Paired stacking sequence maintains single-thickness plies while Staggered 1 has plies in double thickness, which is an advantage for paired stacking sequence.

However, Staggered 1 is naturally symmetric and therefore after cure there will be no warpage. Homogenization is another practical way where warpage can also be limited by high repeats. This sequence exhibits a bending-twisting coupling. Paired sequence may generate some warpage after cure because of the in plane-bending coupling and no bending-twisting coupling will be done. It's recommended to use the Staggered 1 sequence in tapered structures in order to suppress any warpage at the edges. It's also recommended in parts subjected to in-plane loads, since no bending will be generated.

Paired stacking is recommended in closed parts with no free edges like cylinders or conical structures, subjected to bending, since no twisting will be generated. That paired stacking sequence maintains single-thickness plies, which will allow to refine the part thickness.

Homogenization is very important to composite laminates in general, and DD in particular. In general, when a laminate is homogenized by having many repeats, tapering can reduce weight, say, from 10, to 9, to 8, to 7, … If it is not homogenized, tapering can still be done to reduce weight, but the laminate properties will change as plies are dropped. It is no longer a simple way to achieve an optimum laminate with constantly varying thickness and properties.

ABBD* for $[\pm25]_6$

21.21	4.47	0	0	0	0.73
4.47	2.95	0	0	0	0.18
0	0	4.96	0.73	0.18	0
0	0	2.19	21.21	4.47	0
0	0	0.54	4.47	2.95	0
2.19	0.54	0	0	0	4.96

ABBD* for $[+25/-25/-25/25]_3$

21.21	4.47	0	0	0	0
4.47	2.95	0	0	0	0
0	0	4.96	0	0	0
0	0	0	21.21	4.47	0.73
0	0	0	4.47	2.95	0.18
0	0	0	0.73	0.18	4.96

Figure 4.50 Normalized stiffness matrices for paired $[+25/-25]_6$ (left) and Staggered 1 $[+25/-25/-25/25]_3$ (right) sequences.

A homogenized laminate is naturally symmetric. There is no need to enforce mid-plane symmetry that is needed for quad to prevent warpage. Ply layup can be continuous, nonstop (not keeping track of where mid-plane is and then reverse the stacking sequence). Speed can be 6X, less prone to error, and much simplified layup.

Ply drops can now be one at a time, not in pairs with mid-plane symmetry. Ply drop can then be placed strategically at the neutral axis, or on the exterior surface of laminates. When ply drops are removed from the interior, resin rich pockets, fiber breaks, ply wrinkles, voids can all be reduced if not eliminated totally. Ply compaction and resin infiltration of plies that we do continuously, without interruptions/terminations, will be easier to achieve and with better quality. With internal uniformity (no disruptions by ply drops), there is no tendency for warpage. Inspection is easier because all plies are visible, not buried inside laminates.

With DD, 1-axis layup can be easily determined and implemented. There is no tendency to delaminate when there is no cross-plying. This approach may be more effective than use of toughened resin. Lower material and processing cost, and better cured product can be expected. Less warpage also because there is no cross-plying.

The smaller the number of repeats to reach homogenization, the lower the minimum gauge of the material. When the minimum gauge of quad is limited to 10 times the comparable DD, the number of applications can be directly related to it. Lower gauge will open up applications for small, hand-held devices like computers, phones, acoustic and medical equipment, and sporting good that quad is simply too thick to be competitive.

DD concept can be equally effective for other manufacturing processes, such as braiding and 3D printing. Mastering single-double or double-double would be the key. Quad is simply too complicated to follow. Unidirectional is too weak to carry off-axis or shear load. The art of tapering is a critical key for DD. It should equally apply to braiding and 3D printing.

With our discovery that minimum repeat to reach homogenization will set a new standard for use of DD effectively. A laminate that is not necessarily the strongest can be selected because its minimum repeats suit the application better. It is another design variable that will make DD more powerful and higher precision as a material system with flexibilities never offered by any previous systems.

4.12 Symmetric Double-Double: enforced symmetry delays homogenization

Figures 4.51 – 4.53 represent the number of repeats for two double – double sequences with and without enforced symmetry.

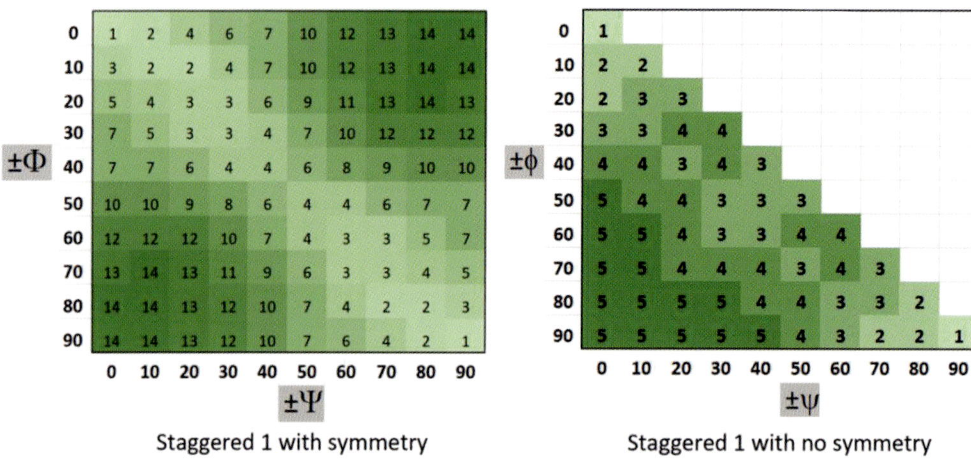

Figure 4.51 Number of repeats for Staggered 1 sequence with and without enforced symmetry. The symmetric case needs up to 14 repeats against 5 repeats for the double – double with no symmetry.

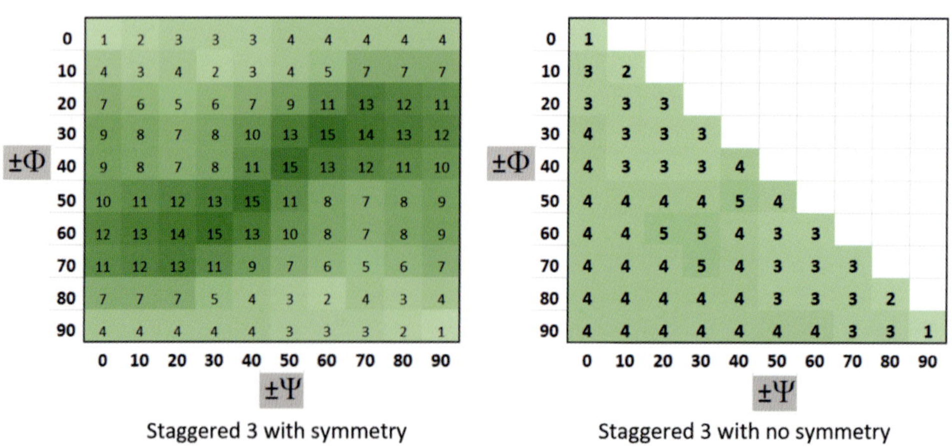

Figure 4.52 Number of repeats for Staggered 3 sequence with and without enforced symmetry. The symmetric case needs up to 15 repeats against 4 repeats for the double – double with no symmetry.

Figure 4.53 shows the best combination of sequences for double – double with and without enforced symmetry. In both cases, the best sequences are Staggered 1 and Staggered 3. The symmetric case needs up to 13 repeats against 4 repeats for the double – double with no symmetry.

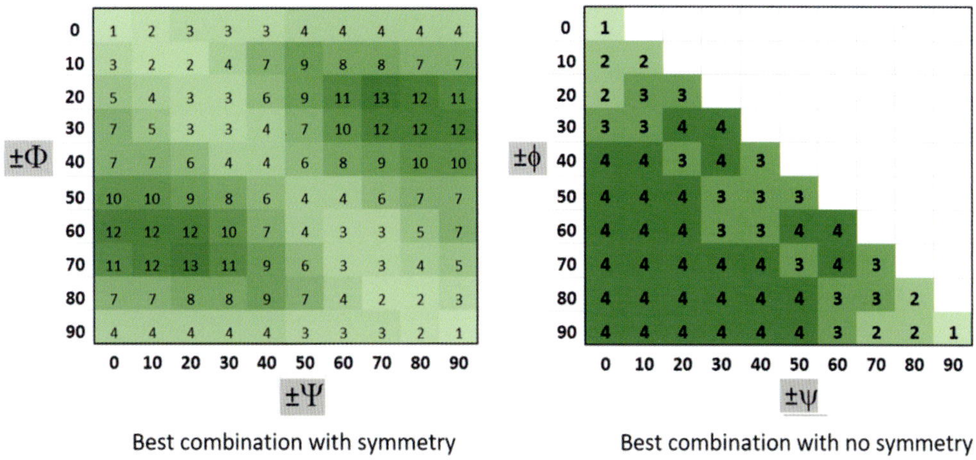

Figure 4.53 Number of repeats for the best combination of sequences with and without enforced symmetry. The symmetric case needs up to 13 repeats against 4 repeats for the double – double with no symmetry.

4.13 Bending – twisting coupling

The strains of a symmetric balanced laminate subjected to bending will be given by $[D]^{-1}$. Equations 4.6 show matrix [D] and $[D]^{-1}$.

$$[D] = \begin{bmatrix} D_{11} & D_{12} & D_{16} \\ D_{12} & D_{22} & D_{26} \\ D_{16} & D_{26} & D_{66} \end{bmatrix}$$

$$[D]^{-1} = \frac{1}{D_0} \begin{bmatrix} D^{\circ}_{11} & D^{\circ}_{12} & D^{\circ}_{16} \\ D^{\circ}_{12} & D^{\circ}_{22} & D^{\circ}_{26} \\ D^{\circ}_{16} & D^{\circ}_{26} & D^{\circ}_{66} \end{bmatrix} \qquad \text{Eq. 4.7}$$

Where:

$$
\begin{aligned}
D_0 &= D_{11}(D_{22}D_{66} - D_{26}^2) - D_{12}(D_{12}D_{66} - D_{16}D_{26}) + D_{16}(D_{12}D_{26} - D_{22}D_{16}) \\
D_{11}^{\circ} &= (D_{22}D_{66} - D_{26}^2) \\
D_{12}^{\circ} &= -(D_{12}D_{66} - D_{16}D_{26}) \\
D_{16}^{\circ} &= (D_{12}D_{26} - D_{22}D_{16}) \\
D_{22}^{\circ} &= (D_{11}D_{66} - D_{16}^2) \\
D_{26}^{\circ} &= -(D_{11}D_{26} - D_{12}D_{16}) \\
D_{66}^{\circ} &= (D_{11}D_{12} - D_{12}^2)
\end{aligned}
\qquad \text{Eq. 4.8}
$$

If a bending moment M_{xx} is applied:

$$
k_{xx} = \frac{D_{11}^{\circ}}{D_0} M_{xx} \qquad k_{yy} = \frac{D_{12}^{\circ}}{D_0} M_{xx} \qquad k_{xy} = \frac{D_{16}^{\circ}}{D_0} M_{xx} \qquad \text{Eq. 4.9}
$$

Thus, even though no twisting moment is applied, the plate would twist unless:

$$
D_{16}^{\circ} = (D_{12}D_{26} - D_{22}D_{16}) = 0 \qquad \text{Eq. 4.10}
$$

This is possible only if the balanced symmetric laminate contains fibers in the 0 and 90 directions only, which is not the case for the quads. Therefore, quads exhibit bending- twisting coupling (Figure 4.54).

In the double – double case, the stacking sequence Staggered 3 is naturally uncoupled in bending- twisting. One repeat is enough to get uncoupled (Figure 4.54).

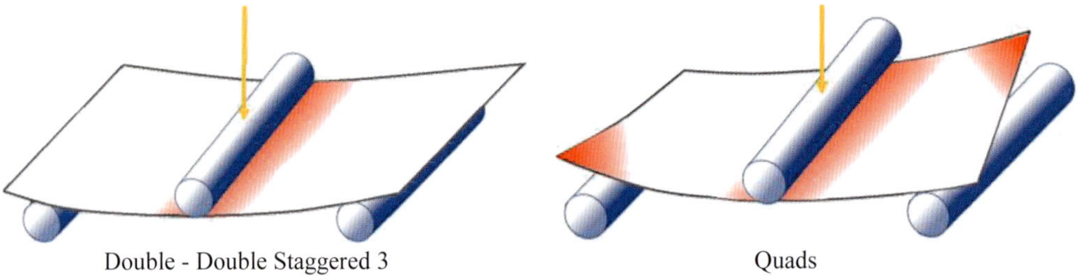

Figure 4.54 No twisting in the Staggered 3 sequence (left) and bending – twisting coupling for quads (right).

The max $(D_{16}, D_{26}) / D_{11}$ values are represented in Figure 4.55 for the 12 stacking sequences of 4-ply quads and 4 to 10 repeats.

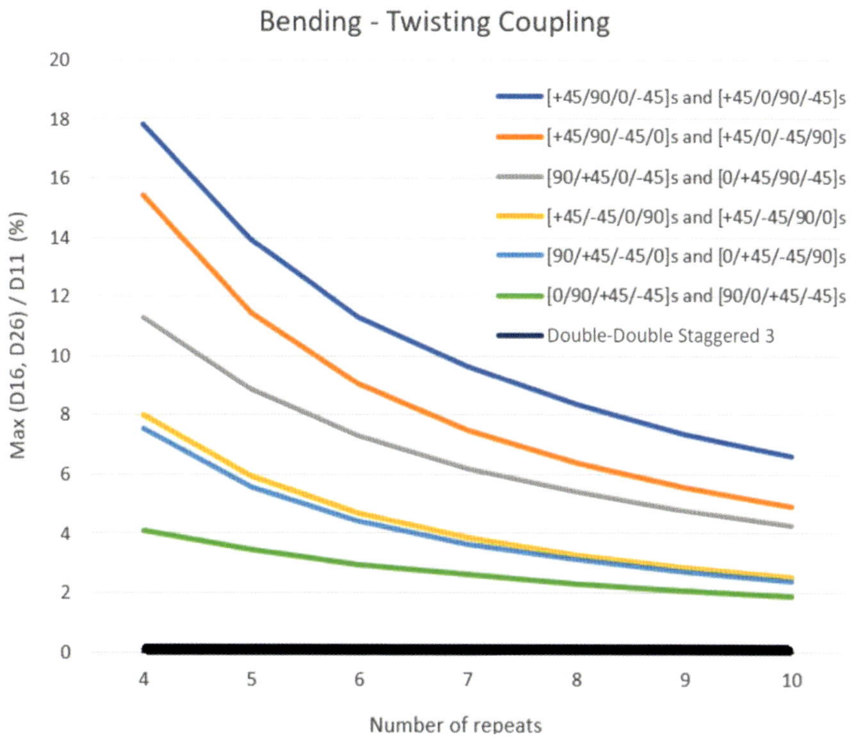

Figure 4.55 Max (D_{16}, D_{26}) / D_{11} values for the 12 sequences of 4-ply quads and 4 to 10 repeats.

4.14 Minimum gauge

DD comprises stacking 4-ply sub-laminates as building blocks. They are made even thinner when thin plies of one half thickness are used. Instead of being 0.50 mm thick with regular plies, the building block would be 0.25 mm with thin plies. This is only 1/10 of the thickness of a hard laminate as $[0_5/\pm45_2/90]_s$ that is 2.5 mm thick. The DD replacement of this hard quad would have the same laminate thickness in $[\pm0/\pm50]_{10T}$; i.e., with thin-plies, DD will have 10 repeats [29].

4.15 DD2

So far, the double – double laminate analyzed is called DD1. There is another DD type, called DD2, which is represented in Figure 4.56.

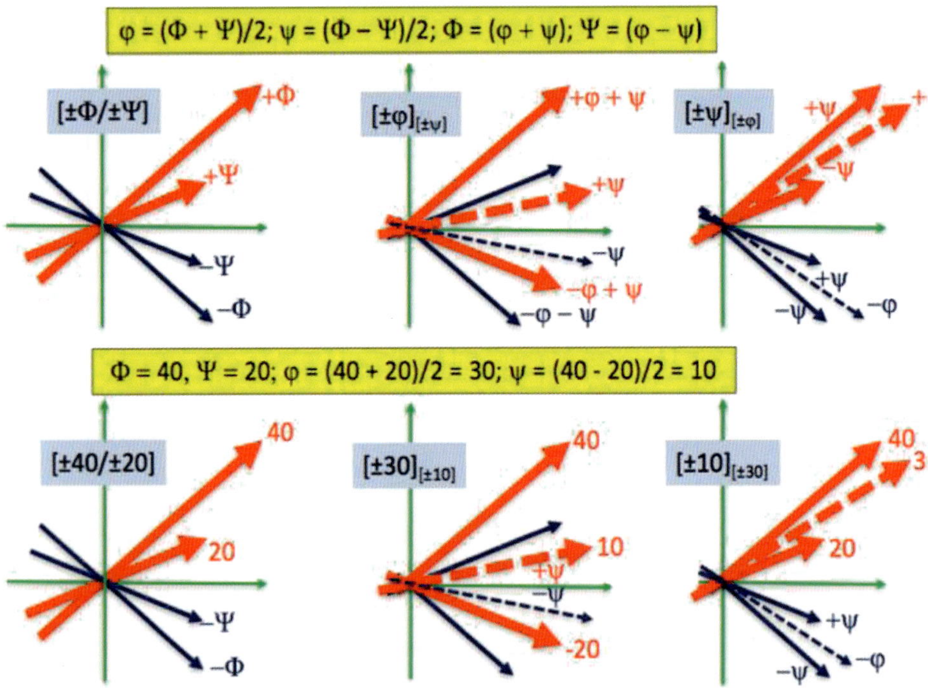

Figure 4.56 Three representations of double – double helix.

Two tacking sequences can be distinguished in DD2:

- ✓ Direct $[+\phi/-\phi]_{[+\psi/-\psi]}$
- ✓ Inverse $[+\phi/-\phi]_{[-\psi/+\psi]}$

Figure 4.57 represents the number of repeats for the direct sequence $[+\phi/-\phi]_{[+\psi/-\psi]}$.

±ϕ \ ±ψ	0	10	20	30	40	50	60	70	80	90
0	1	1	2	3	3	3	3	2	1	1
10	2	2	3	3	4	4	3	2	2	2
20	3	3	3	4	5	4	3	3	3	3
30	5	4	4	4	4	4	4	3	4	5
40	6	5	5	4	4	4	4	4	6	6
50	6	6	4	4	4	4	4	5	5	6
60	5	4	3	4	4	4	4	4	4	5
70	3	3	3	3	4	5	4	3	3	3
80	2	2	2	3	4	4	3	3	2	2
90	1	1	2	3	3	3	3	2	1	1

Figure 4.57 Number of repeats for the direct sequence $[+\phi/-\phi]_{[+\psi/-\psi]}$.

The maximum number of repeats is 6. The only symmetry found is:

Repeats for $[\phi]_{[\psi]}$ = repeats for $[90-\phi]_{[90-\psi]}$.

Figure 4.58 represents the number of repeats for the inverse sequence Inverse $[+\phi/-\phi]_{[-\psi/+\psi]}$.

±ϕ \ ±ψ	0	10	20	30	40	50	60	70	80	90
0	1	2	3	3	3	3	3	3	2	1
10	2	2	2	3	3	3	3	2	2	2
20	3	3	2	3	4	4	3	2	3	3
30	5	4	3	4	4	4	4	3	4	5
40	6	6	4	4	4	4	4	4	6	6
50	6	6	4	4	4	4	4	4	6	6
60	5	4	3	4	4	4	4	3	4	5
70	3	3	2	3	4	4	3	2	3	3
80	2	2	2	3	3	3	3	2	2	2
90	1	2	3	3	3	3	3	3	2	1

Figure 4.58 Number of repeats for the direct sequence $[+\phi/-\phi]_{[+\psi/-\psi]}$.

The maximum number of repeats is 6. There is symmetry regarding the axes $\phi = 45°$ and $\psi = 45°$.

Figure 4.59 represents the ratio of the number of repeats of DD2 Inverse/Direct.

±ϕ \ ±ψ	0	10	20	30	40	50	60	70	80	90
0	1	2	1.5	1	1	1	1	1.5	2	1
10	1	1	0.7	1	0.8	0.8	1	1	1	1
20	1	1	0.7	0.8	0.8	1	1	0.7	1	1
30	1	1	0.8	1	1	1	1	1	1	1
40	1	1.2	0.8	1	1	1	1	1	1	1
50	1	1	1	1	1	1	1	0.8	1.2	1
60	1	1	1	1	1	1	1	0.8	1	1
70	1	1	0.7	1	1	0.8	0.8	0.7	1	1
80	1	1	1	1	0.8	0.8	1	0.7	1	1
90	1	2	1.5	1	1	1	1	1.5	2	1

Figure 4.59 Ratio of number of repeats of DD2 Inverse/Direct.

Sometimes, DD2 Inverse needs fewer repeats that DD2 direct and in some other cases needs more.

The number of repeats needed for DD2 – direct and inverse sequences – are higher than DD1. Sometimes, 6 repeats are needed.

4.16 References

1) Tsai S. W., Double–Double: New Family of Composite Laminate, AIAA Journal Vol. 59, No. 11, November 2021, pp.4293,4306.
2) Tsai, S. W., Melo, J. D. D., Sihn, S., Arteiro, A., and Rainsberger, R., Composite Laminates: Theory and Practice of Analysis, Design and Automated Layup, Stanford Univ. Composites Design Group, Stanford Univ., Stanford, CA, 2017.
3) York, C., Laminate Stiffness Tailoring for Improved Buckling Performance, Mendeley Data, V1, 2020. https://doi.org/10.176632/v2h5h5vdcc.1
4) Nielson, M. W. D., Butler, R., and Rhead, A. T., Minimum Mass Laminate Design for Uncertain In-Plane Loading, Composites Part A, Vol. 115, Dec. 2018, pp. 348–359.
5) Guin, W. E., and Nettles, A. T., A Straightforward Approach to Thickness Tailoring in Composite Structures Using Non-Traditional Layups, NASA T&M.
6) Swanson, S. R., and Trask, B. C., Strength of Quasi-Isotropic Laminates Under Off-Axis Loading, Composites Science and Technology, Vol. 34, 1989, pp. 19–34.
7) Tsai, S. W., and Melo, J. D. D., Composite Materials Design and Testing, Stanford Univ. Composites Design Group, Stanford, CA, 2015.
8) Shrivastava, S., Sharma, N., Tsai, S. W., and Mohite, P. M., D and DD Drop Layup Optimization of Aircraft Wing Panels Under Multi-Load Case Design Environment, Composite Structures, Vol. 248, Sept. 2020, Paper 112518.
9) Vermes, B., Tsai, S. W., Riccio, A., Di Caprio, F., and Roy, S., Application of the Tsai's Modulus and Double-Double Concepts to the Definition of a New Affordable Design Approach for Composite Laminates, Composite Structures, Vol. 251, Oct. 2020, Paper 113246.
10) Tsai, S. W., and Melo, J. D. D., A Unit Circle Failure Criterion for Carbon Fiber Reinforced Polymer Composites, Composite Science Technology, Vol. 123, 2016, pp. 71–78.
11) Tsai, S. W., Sharma, N., Arteiro, A., Roy, S., and Rainsberger, R., Composite Double-Double and Grid/Skin Structures, Stanford Univ. Composites Design Group, Stanford, CA, 2019.
12) Arteiro, A., Sharma, N., Melo, D., Ha, S. K., Miravete, A., Miyano, Y., Massard, T., Shah, P. D., Roy, S., and Hahn, H. T., "A Case for Tsai's Modulus, an Invariant-Based Approach to Stiffness," Composite Structures, Vol. 252, Nov. 2020, Paper 112683.
13) Miravete A., Materials and Processes, Stanford Composites Design Workshop XXII, Jan 24-28, 2022.
14) Idarreta I., Dry Fiber and Potential Applications, Carbon Fiber Conference, Knoxville, Tennessee, November 19, 2019.
15) Panni, S., Rombo, S.E.: Searching for repetitions in biological networks: methods, resources and tools. Brief. Bioinform. 16(1), 118–136 (2015) 10.
16) Palopoli L, Rombo SE, Terracina G. Flexible pattern discovery with (extended) disjunctive logic programming. In: 15th International Symposium on Foundations of Intelligent Systems (ISMIS'05). Springer-Verlag Berlin Heidelberg, 2005, 504–13.

17) Carvalho AM, Freitas AT, Oliveira AL, et al. An efficient algorithm for the identification of structured motifs in DNA promoter sequences. IEEE/ACM Trans Comput Biol Bioinform 2006;3(2):126–40.
18) Fassetti F, Leone O, Palopoli L, et al. Ip6k gene identification in plant genomes by tag searching. BMC Proc 2011; 5(Suppl 2):S1.
19) Grossi R, Pietracaprina A, Pisanti N, et al. MADMX: a strategy for maximal dense motif extraction. J Comp Biol 2011;18(4):535–45.
20) Rombo SE. Extracting string motif bases for quorum higher than two. Theor Comput Sci 2012;460:94–103.
21) Jawaid M., Asim M., Phenolic Polymers Based Composite Materials, SpingerLink Editors, 2021.
22) Vasconcelos A., Pomin V. H., Marine Carbohydrate-Based Compounds with Medicinal Properties, Mar Drugs. 2018 Jul; 16(7): 233.
23) Pomin V.H. Antimicrobial sulfated glycans: Structure and function. Curr. Top. Med. Chem. 2017;**17**:319–330.
24) Pomin V.H., Bezerra F.F., Soares P.A.G. Sulfated glycans in HIV infection and therapy. Curr. Pharm. Des. 2017;**23**:3405–3414.
25) Vacquier V.D. Evolution of gamete recognition proteins. Science. 1998;**281**:1995–1998. doi: 10.1126/science.281.5385.1995.
26) Vilela-Silva A.C.E.S., Alves A.P., Valente A.P., Vacquier V.D., Mourão P.A.S. Structure of the sulfated α-L-fucan from the egg jelly coat of the sea urchin Strongylocentrotus franciscanus: Patterns of preferential 2-O and 4-O-sulfation determine sperm cell recognition. Glycobiology.
27) Pomin V.H. Sulfated glycans in sea urchin fertilization. Glycoconj. J. 2015;**32**:9–15.
28) Kappel E., DD-laminates and manufacturing, Stanford Composites Design Workshop XXII, Jan 24-28, 2022.
29) Papila M., Design of and with thin-ply non-crimp fabric as building blocks for composites, Sci Eng Compos Mater 2018; 25(3): 501–516.

Brian Falzon with partner Tara Walsh, and friend, the late Adrian Mouritz, at a Peruvian Restaurants in Melbourne, December 19, 2022

CHAPTER 5

Optimization of a Regional Aircraft Fuselage Barrel by Re-Designing with Double-Double Laminates

A. Riccio[1], F. Di Caprio[2] and S.W. Tsai[3]

[1] Department of Engineering, University of Campania "Luigi Vanvitelli", Aversa 81031, Italy

[2] C.I.R.A. (Italian Aerospace Research Centre), Capua 81043, Italy

[3] Department of Aeronautics and Astronautics, Stanford University, Stanford, CA 94305, USA

Abstract

The performance constraints and the rising of increasing need of environmental sustainability, encourage the weight reduction in aeronautical structural design. For the last twenty years, fibre-reinforced composite materials have been considered the main opportunity to meet the demands for lightweight high-performance structures. However, many limitations in applying the real potential of composite laminates have been found which are, for the most, related to the real chances to optimize laminates stacking sequences. Indeed, the typical stacking sequences, normally adopted with classic quad laminates, e.g. quasi-isotropic [0°,90°, ±45°] layups, require symmetry and balancing. These constrains, usually, reduce optimization chances leading to an over-dimensioning of the aeronautical structural components with respect to the real service loading requirements. The limitation in design and optimization introduced by classical quasi isotropic laminates can be overcome by adopting a new class of laminates: the Double-Double. Double-double, first introduced by prof. S.W. Tsai, can be seen as a new concept of laminates, with plies not limited to specific angle orientations, able to guarantee small building blocks, easy homogenization and tapering. In this paper, a newly developed FEM tool, namely the "DD automated Design Tool" is presented, able to provide optimised layups of doble-double laminates, by interacting with the optimization engine Lamsearch, according to material properties, realistic applied loads and boundary conditions. Actually, a strong interaction with iterations between the FEM software and the Lamsearch engine allows to rapidly find the best composite Double-Double laminates angles to meet specific requirements in terms of strength. To demonstrate the double-double weight saving potential, the optimization of a real composite fuselage barrel, subjected to realistic loading conditions has been performed. A strong weight reduction has been achieved, with respect to the initial standard quad composite design, by introducing Double-Double laminates for the skin.

Keywords: Double-Double, Tsai modulus, fuselage barrel, weight optimization, FEM

5.1 Introduction

Weight saving has always been a must for the design of aeronautical structures. Improved performance, low fuel consumption and reduced emissions can offer great advantages in an increasingly competitive market [1-3] and the introduction of composite materials has made possible amazing progresses to fulfil these requests [4-6]. Initially, composites were used only for design of secondary components, due to the difficulties in predicting and controlling their damage behaviour; while, in the last two decades, due to an increased and consolidated knowledge on composites failure mechanisms and thanks to the availability of effective computational tools able to correctly simulate their mechanical behaviour [7-11], a more and more increasing number of load-bearing aircraft composite components have been designed. This step forward in designing with Fibre Reinforced Polymers allowed to achieve remarkable advantages, in terms of structural weight saving (20%-30% reduction), if compared to metallic structures, with a consequent decrease in fuel consumption [12-14]. Furthermore, these lighter components have proven to ensure enhanced stiffness and strength, if compared with metallic structures, and lower assembly costs, as they drastically reduce the need for joints, such as fasteners and bolts. Indeed, in most applications, different metallic components can be replaced by a single composite part. Nevertheless, the inherent strength and stiffness properties of composite materials suggest they can guarantee, from material properties perspectives, more than 20%-30% in weight saving. Hence the full potential of composite materials has not been exploited.

Recently, stronger requirements, in terms of environmental sustainability, have been issued by regulations' authorities, which demand not only the reduction of exhaust emissions but also for the decrease of manufacturing waste, less power consumption and, hence, further weight saving [15].

In this context, the need for optimization of products and processes to satisfy the severe aeronautical legislation conditions assumes a relevant importance. With composite structures, optimization studies have been focused on weight and manufacturing costs reduction [16-22], as well as on stacking sequence as design variables [23-25]. The use of permutation operators has revealed to be the most common layup optimization methodologies [26-28] and genetic algorithms have been frequently adopted in stacking sequence optimization problems; however, they have been proven to be extremely time consuming and to require huge data storage capabilities. As a matter of fact, most practical laminate stacking sequences are typically limited to a small set of discrete angles, such as 0°, ±45° and 90°, with symmetry and balancing constraints, which makes the optimization problem and the use of permutation operators very complex and limited. Indeed, to meet the desired requirements, the use of restricted plies orientations and constraints, inevitably leads to an increase in the laminate thickness, contradicting the weight reduction advantage associated with the use of composites.

These limitations can be faced by introducing the newly developed Double-Double (DD) laminates. DD laminates have been developed by Prof. S.W. Tsai and can be identified as the most powerful technology capable of producing really optimised composite structures, in terms of weight and strength [29-32]. These newly proposed laminates completely removes the limits on ply orientation angles and laminate symmetry, making homogenisation very easy to be achieved in design. The combination of DD laminates with newly developed theories on laminates stiffness, such as the Tsai's modulus proposed in Refs. [33-34], opens an incredibly large number of opportunities in composite structural design and manufacturing. Indeed, while standard laminates are constituted by 0°,90° and ±45° oriented plies organised in symmetric and balanced building block with minimum 8 plies, Double-Double are laminates made by very thin building blocks with [±Φ, ±Ψ] stacking sequence, where Φ and Ψ are two arbitrarily and continuously chosen angles between 0° and 90° without restrictions on ply orientation. These thin building blocks can be easily characterised by adopting the Tsai's modulus instead of the typical four stiffness properties required for the definition of quasi-isotropic laminates.

In this chapter, an automated design tool for DD laminates, is proposed. The tools uses iterates between a FE solver and an optimization analytical tool, namely "Lamsearch" to determine the optimal DD angles depending on specific material properties, geometry, boundary conditions and applied loads [30-31]. The DD automated design tool considers as input data the thicknesses of the analysed structure at specific locations and the stress states computed through FE analyses. The tool finds out the best combination of [±Φ, ±Ψ] angles and, therefore, the updated thicknesses at any location which minimises the margin on the chosen failure criteria. The DD automated design tool effectiveness has been tested by performing the redesign the skin of a regional aircraft composite fuselage barrel. The reduction of the skin weight has been evaluated with respect to the original standard laminated composites design.

In section 2, the background on Double-Double laminates and related theoretical concepts, such as the Tsai's modulus, are introduced. In section 3, the proposed DD- automated design tool is briefly described. Finally, in section 4 the application to the composite fuselage barrel is detailed and the main results are discussed.

5.2 Background on Double-Double (DD) Laminates

Double-Double (DD) laminates introduce a totally new approach to design with composite materials. DD laminates, proposed by Prof. S.W. Tsai in [32], have been demonstrated to be effective in replacing the standard quasi-isotropic stacking sequences in laminates design by allowing considerable structural weight saving and easy of production. The DD laminates key characteristic is the easy of homogenization of stiffness properties. In order to understand the advantages offered by homogenization of the DD laminates, the Tsai's modulus, which is a scalar entity allowing to fully characterize the stiffness of a composite lamina, needs to be introduced. The Tsai's modulus allows to introduce the master ply concept which represents a significant step towards the simplification of the design with composites. In this section, the advantages of the DD laminates over standard quasi isotropic laminates are highlighted, the Tsai's modulus and the master ply concepts are briefly described and the advantages of laminates homogenization are commented.

5.2.1. *Advantages of Double-Double laminates over Quasi-isotropic quad laminates*

The quasi-isotropic laminates, also known as quads, are based on fixed four plies orientation angles (0,90,±45). For these laminates, mid-plane symmetry and balancing of angles are mandatory to avoid coupling of out of plane and in plane effects which may be undesirable in the manufacturing phase. These constraints lead to a minimum thickness of the quad building block of 8 plies (*1 mm* if *0.125 mm* thick plies are considered). Hence, quad laminates are typically 20 or more plies thick, at least *2.5 mm,* which is more than the reasonable thickness of a fuselage skin, but often unsatisfactory for applications in other industrial fields, such as transportation, electronics and medical. Indeed, with quad laminates, to attain high properties in a specific direction, the number of plies in that direction needs to be increased, but this increase must be operated by respecting the minimum set (0,90, ±45) and the symmetry. Thus, an incredibly high number of plies is introduced which are not directly correlated to the desired increase of properties. It is easy to understand that quad laminates lead to over dimensioning of structural components and make the design and manufacturing of composite components very complex. Furthermore, stacking sequence optimization with quads is nearly impossible as the number of permutations starts with 24 when considering only the 4-ply sub-laminate and growths with the increase of plies numbers, reaching the impractical value of 166,080 when considering 10 laminae in each building block.

Double-Double laminates are introduced to overcome these limitations. The sub-laminate block is made of two sets of angles [±Φ, ±Ψ] and is considerably thinner that the quads building block. Furthermore, mid-plane symmetry is not required, but is naturally satisfied by homogenization, as later discussed. The number of stacking sequence permutations remains constant at 4, as DD contains only two each ±Φ and ±Ψ, this allows to easily optimize the layup to gain desirable requirements. Ply drop is another key aspect that characterize the DD laminates. Indeed, the mid-plane symmetry requirement of the quasi-isotropic laminates leads to non-optimal ply drop profile, due to manufacturing difficulties. This leads to compromise in terms of strength and toughness of the resulting structures. The homogenization characteristic of the DD laminates, which make them naturally symmetric, permits to place the drop in any position, leading to a simplified production of tapered components and, consequently, promoting weight saving.

5.2.2. *Tsai's modulus and master ply concept*

Tsai's modulus is a material property which can be used to define the stiffness of composite plies. It is given by the sum of the diagonal components of the stiffness matrix [Q] of a lamina, the so called *trace*, according to Equation (5.1). This definition has been proposed by Tsai and Melo in [32], which also demonstrated the invariance according to Equation (5.2), where A^* and D^* are the in-plane and flexural laminate stiffness matrices normalized by the trace.

$$tr(Q) = Q_{11} + Q_{22} + 2Q_{66} \tag{5.1}$$

$$tr(A^*) = tr(D^*) = tr\ (Q) \tag{5.2}$$

They also found that the components of the stiffness matrix for several Carbon Fiber Reinforced Polymers systems, normalized by the trace, turn to be approximately the same, introducing thereby the Master Ply concept. The relationship between the orthotropic in-plane stiffness matrix components and the Tsai 's modulus of all the CFRP is expressed in Equation (5.3)

$$\begin{cases} Q_{11} = Q_{11}^* \cdot tr(Q) \\ Q_{22} = Q_{22}^* \cdot tr(Q) \\ Q_{12} = Q_{12}^* \cdot tr(Q) \\ Q_{66} = Q_{66}^* \cdot tr(Q) \end{cases} \tag{5.3}$$

The plane stress stiffness components, normalized by the Tsai's modulus, of the Master Ply can be used to compute the stiffness matrix of a particular laminate A^*, according to the relationships in Equation (5.4).

$$E_x^* = \frac{E_x}{tr[A^*]};\ E_y^* = \frac{E_y}{tr[A^*]};\ G_{xy}^* = \frac{G_{xy}}{tr[A^*]} \tag{5.4}$$

The coefficient of variation of the trace-normalized stiffness parameters, which refers to the fibre direction, is up to 2% for different groups of materials, such high modulus/standard modulus carbon/epoxy and aramid/epoxy, as reported in [35], while glass/epoxy, which is more sensitive to small variation of the longitudinal Young's modulus, cannot be represented using the master ply concept. The impressive accomplishment is represented by the possibility to evaluate the mechanical behaviour of a composite component by using the master ply properties, then the values of resulting stiffness can be easily scaled, adopting the Tsai's modulus as scaling parameter, for each material system belonging to the master ply group.

5.2.3. Laminate homogenization

Laminate homogenization is the powerful feature which characterize DD laminates. Homogenization means uniform distribution of the plies in a laminate. In quads, homogenization is usually obtained with a very large number of repeated building blocks, because quad building blocks contains plies (0,90,±45) with large differences in orientation. The homogenization condition is mathematically represented by Equation (5.5), where matrix [A*] is the in-plane stiffness matrix and matrix [D*] is the flexural stiffness matrix, both normalized by the Tsai's modulus, while matrix [B] is the flexural-inplane coupling matrix.

$$[A^*] = [D^*] \quad and \, [B] = 0 \tag{5.5}$$

From equation (5.5) , it can be easily deduced that homogenization can mitigate the warping that characterize the non-symmetric layups as demonstrated in [31-32]. Since DD laminates are characterised by very thin building blocks with ply orientation chosen according to the "useful" directions homogenization can be easily achieved without the need for a large number of building block and with laminates with very small thickness.

5.3 DD - Design optimization tool

As already remarked, Double-Double laminates have been demonstrated to provide stronger and lighter design solutions than the standard quasi-isotropic quad laminates. In this section, an efficient DD-Design optimization tool is proposed, capable to find optimal design solutions for a composite structure with Double-Double laminates, by considering user defined loading and boundary conditions. Actually, the proposed tool is capable of replacing a Quad composite structure with a more efficient, lighter and feasible design with [±Φ; ±Ψ] stacking sequences fulfilling all the stiffness and strength requirements. The proposed design tool enforces iterations between a FE model of the structure and the analytical Lamsearch optimizer to find the optimal distribution of the [±Φ; ±Ψ] angles and the optimal distribution of the thicknesses. In this section, the Lamsearch analytical tool, detailed explained in [31], is briefly described and the workflow of the proposed DD-Design optimization tool is presented.

5.3.1. Lam search tool

Lam search is an analytical tool capable of finding the best combination of [±Φ; ±Ψ] angles to satisfy specific strength requirements. It is based on a set of input data, such as material properties, thickness and stress distributions. The Classic Laminate Theory (CLT) is the base of the calculations with the addition of several well-known failure criteria (for example, Max-Stress, Tsai-Wu, etc.). Lam Search development started years ago and it has been continuously updated with additional options and features, for example new failure criteria, buckling capabilities, etc. The Lamsearch optimizer makes use of the Visual Basic editor for Excel with a user-friendly interface where different parameters can be given as input, such as material, stress condition in up to 49 cells, ply thickness, failure criteria. The optimizer provides as output the most efficient DD laminate configuration with the best [±Φ; ±Ψ] angles and the thickness in each cell capable of minimising the safety margins on the chosen failure criteria. In order to correctly scale the thickness in each cell the strength ratio *R*, defined in equation (5.6), is introduced.

$$R = \frac{\sigma^{allowable}}{\sigma^{applied}} \tag{5.6}$$

Figure 5.1 shows the workflow with input and outputs quantities for the Lamsearch optimizer. The stress components and the initial thickness in each of the considered 49 cells are considered as input for the optimizer; while the best DD angles minimizing the thickness in cells and the thicknesses in each cell, found according to a selected failure criterion, are considered outputs for the procedure. First, the cell with the minimum R value (above unity) is selected; this cell is defined as the controlling cell, its value of R is defined as the controlling strength ratio and the angles of this cell as taken as the best DD angles output. Then the thicknesses in each cell are scaled according to the R value found for that cell to obtain the new output thicknesses.

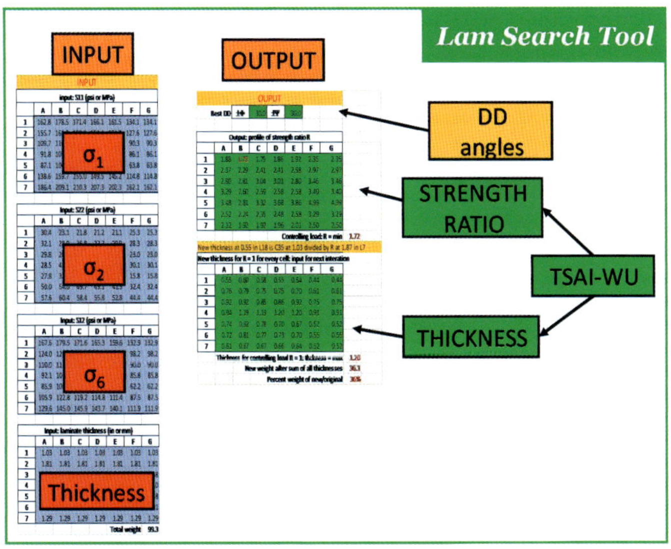

Figure 5.1. Lamsearch schematic sheet.

5.3.2. DD-Design workflow

The DD-Design tool is a fully automated easy-to-use iterative procedure for Finite Elements (FE) models, which create iterations between the FE analyses and the Lam Search analytical sheet, with the aim to search for the best DD layups, and the best thicknesses distribution for any composite structure. The tool updates the stress distributions, by means Finite Element analyses, by updating the laminate stacking sequence [±Φ; ±Ψ] and the thickness in cells provided by the Lamsearch. The proposed design tool provides a verification of the assumption made by Lamsearch to extend the DD angles of the controlling cell to all the other cells.

Figure 5.2 shows the workflow of the DD-automated design tool, highlighting the main steps needed to run the optimization process. First, the numerical Finite Element model of the structure needs to be defined and discretized according to the finite elements' methodology rules. Then the elements are grouped in cells which will have the same DD-layup after the optimization (this choice extremely impacts the optimization results). At the end of a preliminary FE analysis, the mean stress components in each cell are evaluate and passed to the Lamsearch optimizer together with the thickness distribution. Lamsearch finds the best DD angles and the new thickness in cells which can be used as an input for the FE solver and the procedure is iterated until no changes are observed in best [±Φ; ±Ψ] angles and until the strength ratio within each cell is close to unity but above unity everywhere.

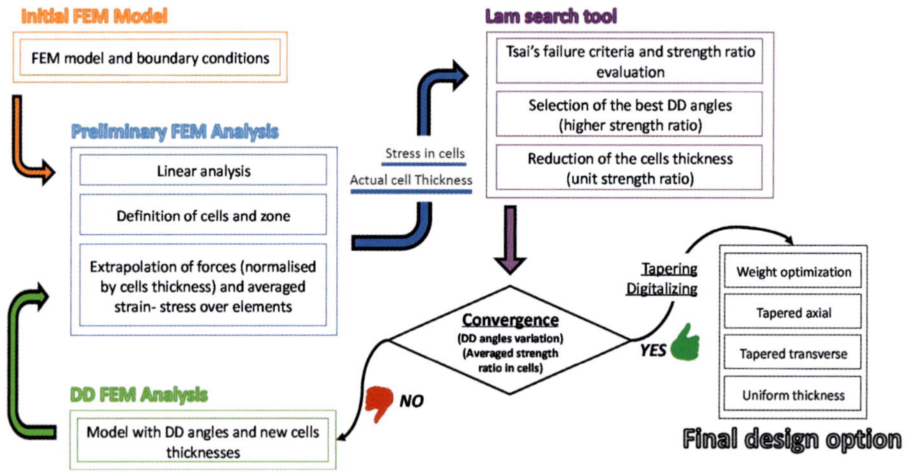

Figure 5.2. Workflow of the DD-Design tool.

When convergence is reached, the output solution takes into account the specific boundary and loading conditions of the Finite Element model and the DD laminates characteristics. The solution, also, provides a measure of the mass reduction obtained. However, the optimised version of the FE model is still characterised by continuously scaled thicknesses in cells (operated by Lamsearch), which are not directly applicable to composite materials, usually, characterised by discrete thicknesses variations related to the presence of a discrete number of plies and building blocks. Hence a "digitizing" preliminary operation is needed, on the optimised design solution, to discretize conservatively, the thicknesses, according to building block thickness. In performing the digitizing transformation of the FE model, part of the mass reduction benefits gets lost. Alternative DD design configurations, considering specific manufacturing procedures, such as axial and transvers tapering, are provided as a result, together with the reduction in terms of weight saving associated to the manufacturing oriented designs.

The tool can work entirely in background mode, useful for design optimization with possible controlling external code and with computational effort saving. Any Finite Element platform can be adopted for the implementation of the tool. For the application presented in this chapter, Ansys and Abaqus implementations have been performed. Indeed, the DD-automated tool is intended to optimize the DD angles and the thickness at specific locations according to strength constraints verified by using failure criteria. As a further development, secondary tools for the optimization of DD laminates, under specific buckling and stiffness requirements, have been established. The buckling optimization tool allows to select the DD layup according to desired critical buckling load, though eigenvalues analyses; while the stiffness optimization tool permits to define a select as optimization criterions a desired stiffness distribution (useful in aeroelastic design) or a maximum allowed displacement at specific locations. Figure 5.3 shows the flowchart of the advanced DD automated design tool obtained by including the additional optimization tools previously introduced.

Actually, a first FE-Lamsearch iteration is carried out to find the DD angles and the thickness distribution satisfying the strength requirements, then the thickness at specific locations are modified by the additional optimization tools in order to fulfil stiffness and buckling requirements.

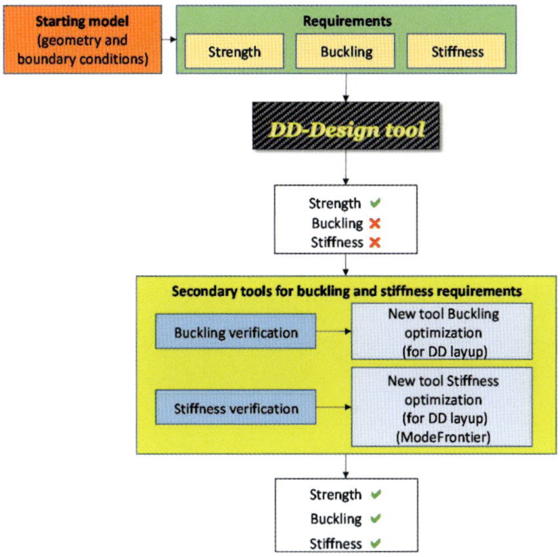

Figure 5.3. Advanced DD automated design tool - flowchart.

5.4 Application to a Regional Aircraft Composite Fuselage Barrel

To preliminary demonstrate the effectiveness of the proposed DD Automated Design Tool, with respect to strength requirements, a numerical application, on the redesign of a regional aircraft composite fuselage barrel, is introduced in this section. The fuselage barrel, considered for this application, is a real fuselage barrel of a regional aircraft characterised by quasi-isotropic QUAD stacking sequences, with different thicknesses. With the aim to lighten the skin fuselage weight, the DD-Design tool has been adopted to replace the standard QUADS laminates with the more efficient Double-Double layups in the skin, according to strength requirements (buckling and stiffness requirements have been neglected in this work).

5.4.1. Geometrical and FE model description

The numerical model of the fuselage barrel is shown in Figure 5.4.

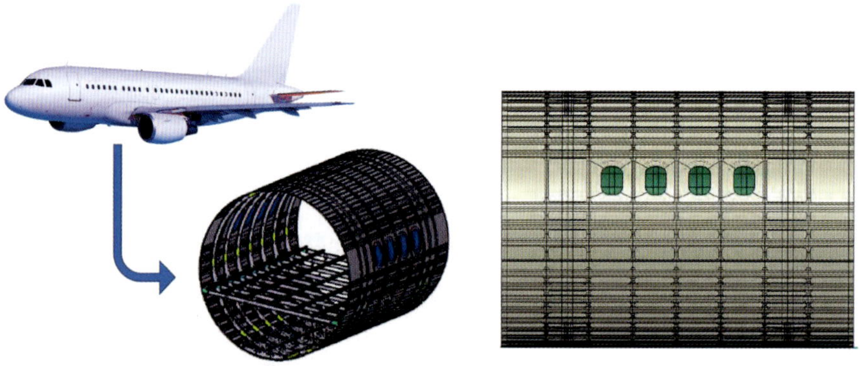

Figure 5.4. Fuselage barrel numerical model.

Internal pressure and mass loads have been considered as loading conditions. Indeed, the structure, the passengers, the baggage, the forward section and cockpit mass have been considered, with a normal load factor greater than 1, in computations. Loading and boundary conditions applied to the FEM model are schematically represented in Figure 5.5. Mass and pressure loads have been applied in reference points, linked with the fuselage barrel parts through rigid coupling constraints.

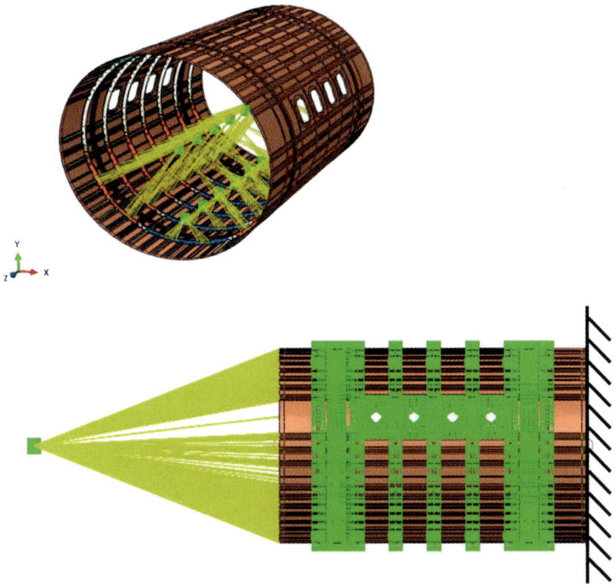

Figure 5.5. Fuselage barrel loading and boundary conditions.

The barrel is characterised by a radius of 1811 mm and a length of 4926 mm. due to the structural and loading symmetry, half the model has been considered in computations. The half Finite Element model of the skin has been divided in 42 cells, each including elements with the same stacking sequence, as shown in Figure 5.6. The skin model has been divided in 7 regions with the same layup.

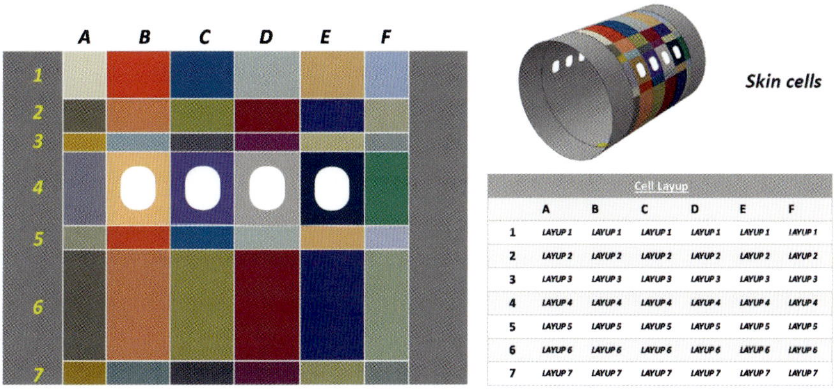

Cell Layup

	A	B	C	D	E	F
1	LAYUP 1	LAYUP 1	LAYUP 1	LAYUP 1	LAYUP 1	LAYUP 1
2	LAYUP 2	LAYUP 2	LAYUP 2	LAYUP 2	LAYUP 2	LAYUP 2
3	LAYUP 3	LAYUP 3	LAYUP 3	LAYUP 3	LAYUP 3	LAYUP 3
4	LAYUP 4	LAYUP 4	LAYUP 4	LAYUP 4	LAYUP 4	LAYUP 4
5	LAYUP 5	LAYUP 5	LAYUP 5	LAYUP 5	LAYUP 5	LAYUP 5
6	LAYUP 6	LAYUP 6	LAYUP 6	LAYUP 6	LAYUP 6	LAYUP 6
7	LAYUP 7	LAYUP 7	LAYUP 7	LAYUP 7	LAYUP 7	LAYUP 7

Figure 5.6. Definition of the cells.

The T700 C-Ply64 Composite Fibre Reinforced Plastic (CFRP) material system has been considered for the skin. The mechanical properties of This material system are listed in Table 5.1.

Material Properties – T700 CPLY64					
E_{11} = 140800 [MPa]	E_{22} = 9300 [MPa]	G_{12}=G_{13}= 5800 [MPa]	G_{23} = 4060 [MPa]	ν_{12}=ν_{13}= 0.3 [-]	F_{1t}=2944 [MPa]
F_{2t}=60 [MPa]	F_{1c}=1983 [MPa]	F_{2c}=220 [MPa]	S_{12}=93 [MPa]	S_{22}=93 [MPa]	Thickness = 0.125 [mm]

Table 5.1. Mechanical properties of T700 C-Ply 64 (CFRP).

The numerical model has been discretized by means of general-purpose shells with reduced integration that allow finite membrane deformation and large rotations and include the effects of transverse shear deformation and thickness change. Contacts elements have been used to tie the different components of the model. Linear analyses have been performed.

5.4.2. *Optimization process*

The results of the numerical simulation, representing the deformed shape with the contour plot of the midplane strain components, in the laminate reference system, are shown in Figure 5.7 for the starting configuration with quasi-isotropic QUAD stacking sequences.

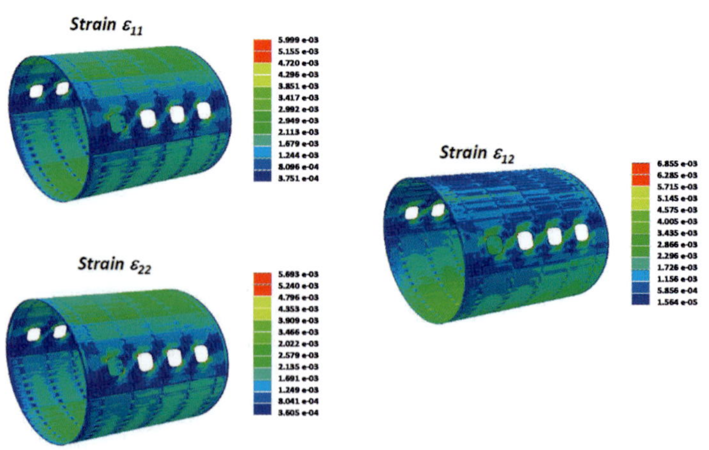

Figure 5.7. Starting Quad configuration - Contour plot of strain distribution .

The strain in the jth cell has been evaluated by averaging the values among all its elements, as defined by Equation (5.7), where n represents the number of elements in the cell. The corresponding stresses have been evaluated by multiplying the strains by the thickness normalized $[A]^*$ laminate matrix, according to Equation (5.8).

$$\varepsilon_{11_{cell_j}} = \frac{1}{n} \cdot \Sigma_{i=1}^{n} abs\left(\varepsilon_{11_i}\right) \tag{5.7}$$

$$\begin{Bmatrix} \sigma_{11} \\ \sigma_{22} \\ \sigma_{12} \end{Bmatrix} = [A]^* \cdot \begin{Bmatrix} \varepsilon_{11} \\ \varepsilon_{22} \\ \varepsilon_{12} \end{Bmatrix} \tag{5.8}$$

The computed average stress values in each cell, obtained by the preliminary FE analysis on the QUAD configuration, are given as input to the lamsearch optimizer, which, at the end of the first iteration (Iteration 0), provides, as output, an optimal single-D building block of [Φ±30;Ψ±30], as shown in Figure 5.8. The starting fuselage skin mass has been assumed as the reference value (corresponding to the 100% of the weight) while the mass of the resulting configurations has been evaluated as a fraction of the reference value. A mass reduction of 59.3% has been found after this first optimization step.

Lam Search Tool

INPUT

Stress - S11 [MPa]

	A	B	C	D	E	F
1	162.82	178.46	171.43	166.14	161.49	134.09
2	155.68	161.23	153.57	154.39	154.73	127.59
3	109.68	110.70	102.48	103.61	111.15	90.29
4	91.78	107.68	105.20	105.16	106.34	86.05
5	87.14	106.57	92.09	83.80	80.92	63.78
6	138.59	159.66	155.05	149.52	145.17	114.76
7	186.36	209.09	210.34	207.32	202.28	162.07

Stress - S22 [MPa]

	A	B	C	D	E	F
1	167.59	179.49	171.61	165.27	159.65	132.86
2	123.99	126.76	120.18	119.84	119.72	98.24
3	110.94	110.97	102.54	103.35	110.54	90.05
4	92.07	107.43	104.23	104.17	105.61	85.77
5	85.95	105.57	90.73	82.39	79.45	67.74
6	105.88	122.83	119.18	114.78	111.43	87.52
7	129.60	144.96	145.88	143.71	140.13	111.87

Stress - S12 [MPa]

	A	B	C	D	E	F
1	30.40	23.09	21.82	21.24	21.08	25.27
2	32.12	28.94	26.85	27.70	29.93	28.31
3	29.80	28.09	25.76	26.55	29.63	23.05
4	28.52	41.96	43.33	43.79	43.23	30.06
5	27.82	35.28	28.58	25.39	22.80	15.78
6	50.03	54.03	49.73	45.11	41.53	32.40
7	57.63	60.36	58.44	55.79	52.78	44.45

LAMINATE THICKNESS [mm]

	A	B	C	D	E	F
1	1.03	1.03	1.03	1.03	1.03	1.03
2	1.81	1.81	1.81	1.81	1.81	1.81
3	2.58	2.58	2.58	2.58	2.58	2.58
4	3.10	3.10	3.10	3.10	3.10	3.10
5	2.58	2.58	2.58	2.58	2.58	2.58
6	1.81	1.81	1.81	1.81	1.81	1.81
7	1.29	1.29	1.29	1.29	1.29	1.29

OUTPUT

Best DD	±Φ	30	±Ψ	30

OUTPUT: Profile of strength ratio R

	A	B	C	D	E	F
1	1.88	1.72	1.79	1.86	1.92	2.35
2	2.37	2.29	2.41	2.41	2.58	2.97
3	2.80	2.81	3.04	3.01	2.80	3.46
4	3.29	2.60	2.59	2.58	2.58	3.40
5	3.48	2.81	3.32	3.68	3.86	4.99
6	2.52	2.24	2.35	2.48	2.58	3.29
7	2.12	1.92	1.92	1.96	2.01	2.50

OUTPUT: New cells thickness for R=1 – INPUT OF NEXT ITERATION

	A	B	C	D	E	F
1	0.55	0.60	0.58	0.55	0.54	0.44
2	0.76	0.79	0.75	0.75	0.70	0.61
3	0.92	0.92	0.85	0.86	0.92	0.75
4	0.94	1.19	1.19	1.20	1.20	0.91
5	0.74	0.92	0.78	0.70	0.67	0.52
6	0.72	0.81	0.77	0.73	0.70	0.55
7	0.61	0.67	0.67	0.66	0.64	0.52

Thickness for controlling load R=1: thickness=max	1.20
New weight after sum of all thicknesses	40.7
Percent weight of new/original	59.3

40.7% of initial mass

Figure 5.8. DD Automated design tool - Iteration 0.

The DD angles and the thicknesses of cells, resulting from the first step, have been given in input to the FE solver to obtain the updated stress distributions in cells. These values have been given in input to Lamsearch to start a second iteration. The new DD configuration [Φ±30;Ψ±37.5] has been found with an overall fuselage skin mass reduction to 36%. Further iterations have been performed until the same DD angles between two iterations have been found and no strength ratio below 1 in any cell has been recorded, according to Table 5.2.

Configuration	Layup	Strength ratio	Weight
Initial	Quasi-isotropic	-	100%
Iteration 0	[Φ±30;Ψ±30]	R>1	40.7%
Iteration 1	[Φ±30;Ψ±37.5]	R<1	36%
Iteration 2	[Φ±37.5;Ψ±37.5]	R<1	41.9%
Iteration 3	[Φ±37.5;Ψ±45]	R<1	39.4%
Iteration 4	**[Φ±37.5;Ψ±45]**	**R>1**	**35.4%**

Table 5.2. Evolution of the controlling parameters during iterations.

Indeed, at iteration 4, all cells are characterized by strength ratios over unity and the optimum DD angles of [Φ±37.5;Ψ±45], which are the same as the ones obtained in the previous iteration. Hence, the optimization process has been stopped by achieving a reduction of 64.6% for the overall fuselage skin weight.

To better understand the evolution of strength ratio values in cells during iteration, in Figure 5.9 the values of strength ratios below unity have been highlighted. As it can be appreciated, for the last iteration, all the R values are above unity and are very close to unity providing an effective design solution in terms of mass saving.

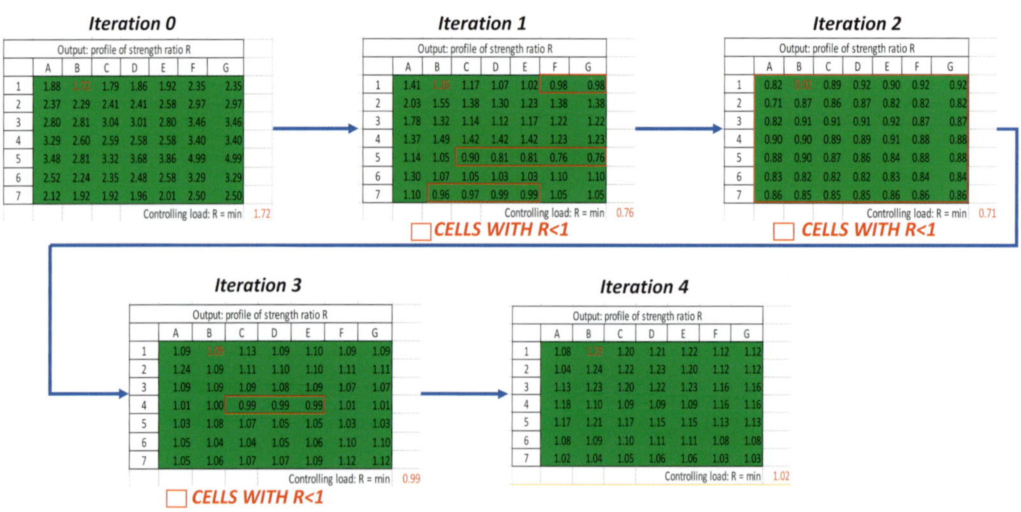

Figure 5.9. Evolution of strength ratio value during iterations

As previously remarked, the DD automated design tool, starting from the weight optimized design configuration, provides as output, several design solutions which are characterised by different levels of manufacturability and weight saving. In Figure 5.10, the Design Options available as output of the DD automated design tool, starting from the weight optimized configuration, are introduced. As it can be noted, passing trough digitizing and tapering operation the weight saving provided by DD laminates decreases and the easy of manufacturing increases; however, for the uniform thickness configuration, which is very easy to be manufactured, the weight saving is still very close to 50% with respect to the initial Quad configuration. This trend can be also appreciated in Figure 5.11, where the different design options are compared in terms of final skin mass.

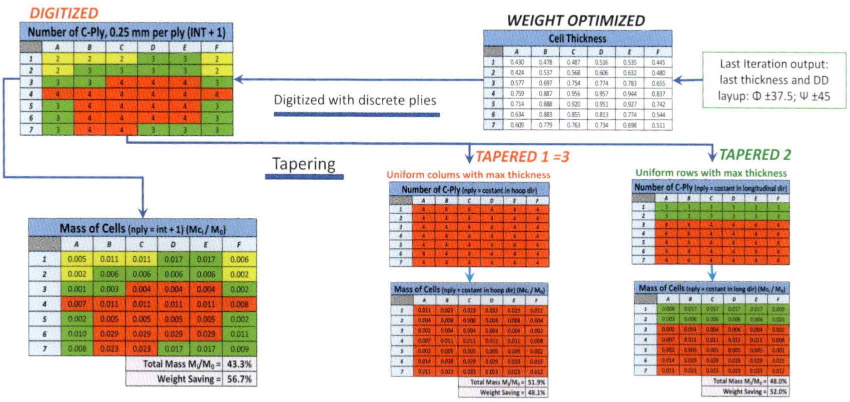

Figure 5.10. Available Design Options starting from the weight optimized configuration

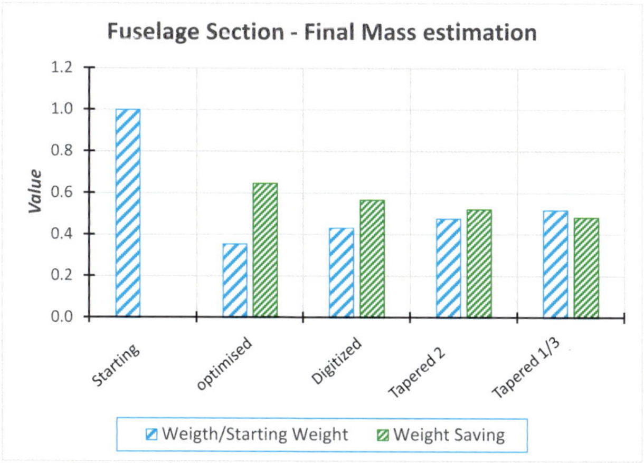

Figure 5.11. Available Design Options starting from the weight optimized configuration

5.5 Application of the master envelopes for design scaling

5.5.1 One parameter material scaling for composite laminates

Characterizing the stiffness and strength of a composite laminate typically requires nine parameters: four for stiffness and five for strength. This complexity makes it difficult to easily and directly compare composite laminates. Progresses have been made by defining the trace, which reduces the number of stiffness parameters to one. Using a single stiffness parameter is crucial because it allows for the direct comparison and evaluation of stiffness between materials and laminates. The concept of a single parameter to characterize composites has also been extended to strength, with the radius-normalized strength parameter.

Comparing the failure envelopes and consequently the radii of equal-area circles for different materials and layups allows a correlation between material performances to be determined. This provides rating factors for materials that enable comparison of strength performance independently of the layup. Using a single-parameter approach, different materials can be easily rated according to their strength performance, as shown with material rating factor in Figure 5.12.

Radius	T3/F93	AS4/H3501	T3/N52	IM7/8552	T7/2510	T650/ep	T4708/MR	IM7/MTM	IM7/8552'	C-Ply 55	IM7/977	IM6/ep	T8S/3900	C-Ply64	T800/Cyt
[75/15]	0.64	0.72	0.77	0.87	0.88	0.95	1.01	1.04	1.04	1.03	1.18	1.18	1.25	1.19	1.25
[75/0]	0.64	0.72	0.77	0.87	0.88	0.95	1.01	1.04	1.04	1.03	1.18	1.18	1.25	1.19	1.25
[60/0]	0.64	0.72	0.77	0.87	0.88	0.95	1.01	1.04	1.04	1.03	1.18	1.19	1.24	1.19	1.26
[60/15]	0.63	0.72	0.77	0.87	0.88	0.95	1.01	1.04	1.03	1.03	1.18	1.19	1.24	1.20	1.26
[60/30]	0.64	0.73	0.77	0.86	0.88	0.96	1.00	1.04	1.03	1.03	1.18	1.18	1.24	1.20	1.25
[45/0]	0.63	0.73	0.77	0.86	0.88	0.95	1.01	1.03	1.03	1.03	1.18	1.19	1.24	1.20	1.26
[45/15]	0.63	0.73	0.77	0.86	0.88	0.95	1.01	1.03	1.03	1.03	1.18	1.19	1.24	1.21	1.27
[90/0]	0.64	0.72	0.77	0.87	0.88	0.95	1.01	1.04	1.04	1.03	1.18	1.18	1.25	1.19	1.25
[45/30]	0.63	0.75	0.78	0.85	0.89	0.96	1.00	1.01	1.02	1.04	1.17	1.18	1.22	1.23	1.26
[±45]	0.64	0.84	0.84	0.78	0.92	0.98	0.98	0.96	1.00	1.06	1.14	1.15	1.18	1.30	1.25
[30/0]	0.63	0.75	0.78	0.84	0.90	0.96	1.00	1.01	1.02	1.04	1.16	1.18	1.22	1.23	1.27
[30/15]	0.63	0.77	0.79	0.83	0.90	0.96	1.00	0.99	1.01	1.04	1.16	1.17	1.21	1.25	1.28
[±30]	0.63	0.82	0.82	0.78	0.93	0.97	0.99	0.95	0.99	1.06	1.14	1.16	1.18	1.30	1.28
[±15]	0.64	0.81	0.81	0.79	0.93	0.98	1.00	0.95	0.99	1.06	1.12	1.15	1.19	1.30	1.28
Material Rating	0.63	0.75	0.78	0.84	0.89	0.96	1.00	1.01	1.02	1.04	1.17	1.18	1.23	1.23	1.26

Figure 5.12. Material rating factors

There are significant advantages in using one parameter to describe the stiffness and strength of composite laminates. This simplifies the material evaluation process and facilitates the selection of appropriate materials for specific applications, improving the performance and reliability of the final product. Moreover, it enables the concept of scaling. The ability to characterize composites by using a single parameter allows manufacturers to scale laminates with properties finely tuned for their intended use.

Indeed, the strength and failure behaviour of composite laminates is traditionally characterized using failure criteria requiring multiple parameters. However, defining a "master envelope", grouping all the strength characteristics in a single parameter, allows the derivation of a rating factor describing the strength performance of a laminate. Actually, the master envelope can be used to derive the radii of equal-area circles and consequently the rating factors that characterize the relative strength performance of different laminate configurations independently of the specific shape of their failure envelopes. Using a single rating factor value simplifies comparisons between layups and materials, while still capturing important performance differences. The master envelope thus enables a single-parameter approach for evaluating composite strength.

This study proposes a single-parameter approach for characterizing and scaling the strength of composite laminates using master failure envelope and rating factors. The procedure flowchart is depicted in Figure 5.13 and herein described:

- The strength ratios *R* of the starting configuration are determined in each cell using the Tsai-Wu failure criterion. These strength ratios represent the stress levels in each cell relative to the failure loads of the laminate.
- The strength ratios *R* for a new material are computed by multiplying the initial strength ratios by the rating factor of the new material. The rating factor captures how efficiently the new material uses its full strength potential.

- The cell thicknesses of the scaled configuration are computed according to the equation (5.9). Scaling the thickness of each cell by the revised strength ratio for the new material results in a scaled laminate configuration with strengths matched to the new material.

$$Tk_{new} = \frac{Tk_{initial}}{R} \tag{5.9}$$

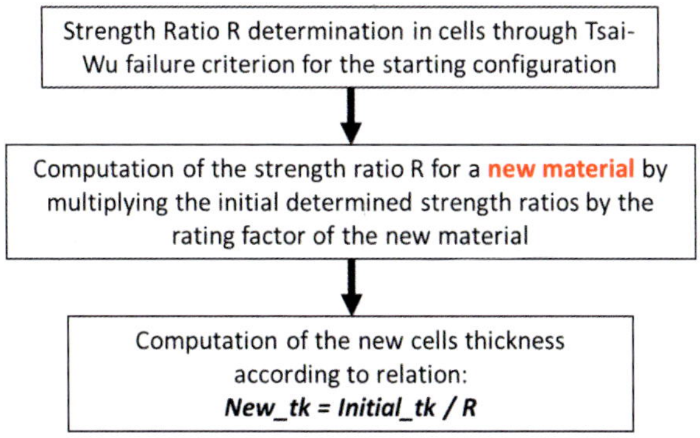

Figure 5.13. Proposed procedure flowchart

In this way, the procedure uses a single rating factor to scale an existing laminate design for a new composite material, without requiring new calculations. In the next subparagraphs, the proposed scaling procedure is applied to a flat composite plate under shear loading conditions to evaluate the effects of material system change on the mass of the plate. The effectiveness of the proposed scaling approach has been evaluated by comparing the thickness distribution obtained from the scaling with the thickness distribution of a new design obtained from recalculations with the new material system. The scaling approach have been also applied to the investigated fuselage barrel by evaluating the effects of material system change on the mass variation of the barrel.

5.5.2 Application of material scaling to a composite flat plate

To preliminarily demonstrate the effectiveness of the proposed approach for material scaling of composite laminates, a numerical application on the redesign of a composite flat plate is introduced in this sub-section. The plate is characterised by a width of 400 mm and a length of 800 mm, as showed in Figure 5.14. The stacking sequence [±60/±30], based on the Double-Doubles family, is used for the present component. The T4708MR Composite Fibre Reinforced Plastic (CFRP) material system with degraded properties has been considered for the plate (see table 5.3 for material properties). The DD-Design tool has been used to optimize the thicknesses associated to each cell as a function of the strength ratio assessed by means of the Tsai-Wu criterion.

Degraded Material Properties – T4708MR					
E_{11} = 142000 [MPa]	E_{22} = 1748 [MPa]	G_{12}=G_{13}= 924 [MPa]	G_{23} = 924 [MPa]	ν_{12}=ν_{13}= 0.051 [-]	F_{1t}=2524 [MPa]
F_{2t}=39.6 [MPa]	F_{1c}=1700 [MPa]	F_{2c}=165 [MPa]	S_{12}=48 [MPa]	S_{22}=48 [MPa]	Thickness = 0.125 [mm]

Table 5.3. Degraded Mechanical properties of T4708MR (CFRP).

The numerical model has been discretized by means of general-purpose shell elements with reduced integration scheme that allow for finite membrane deformation and large rotations and include the effects of transverse shear deformation and thickness change. Loading and boundary conditions applied to the FEM model are schematically represented in Figure 5.14a. Two sides of the plate have been fixed, while loads have been distributed along the opposite edges. The plate has been subdivided in cells, as shown in Figure 5.14b, in order to compute average stresses and related strength ratios R in different locations.

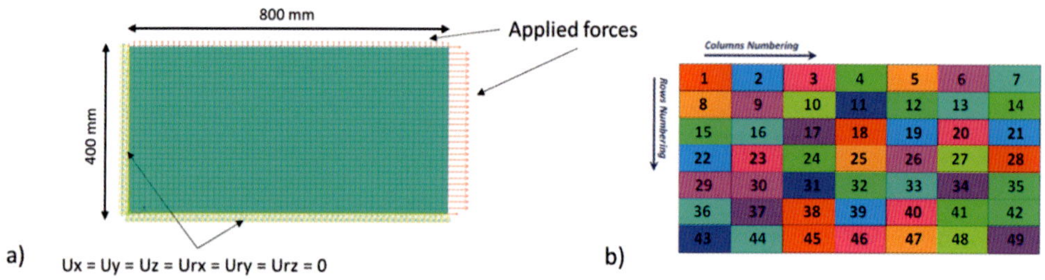

Figure 5.14. FE model: a) flat plate geometry and boundary conditions; b) cells subdivision

The considered flat plate has been involved in an optimisation process of the thickness, fixing the material and stacking sequence. The final thickness of the cells obtained by the DD design tool is related to the strength ratio distribution in the cells. The variable thickness map, in agreement with the applied loading condition and in compliance with the strength requirements related to the chosen material system, provided a plate weight of 52% if compared to the ideal plate with uniform 1 mm Thickness. The order of magnitude of this result is in agreement with the DD design tool optimization result presented for the fuselage barrel.

Starting from this configuration, the proposed scaling procedure has been applied to roughly evaluate the difference in weight of the plate related to a material system change without the need of recalculation. Actually, the starting T4708MR Composite Fibre Reinforced Plastic (CFRP) material system, characterised by a rating factor 1 (in Figure 5.12), has been replaced by the CPLY64 Composite Fibre Reinforced Plastic (CFRP) material system, characterised by a rating factor 1.23 (in Figure 5.12). To perform the material change, the determined strength ratios have been multiplied by the rating factor of the new material, extending the margin of safety in relation to its strength capabilities. The updated strength ratios are used to modify the cell thicknesses without recalculations. The strength ratios determined for the starting material at the end of the optimization process are multiplied by the CPLY64 rating factor 1.23 and, consequently, the thicknesses are updated, as showed in Figure 5.15. The breakthrough of this approach lies in the possibility to scale the component with respect to a material change without any structural recalculation.

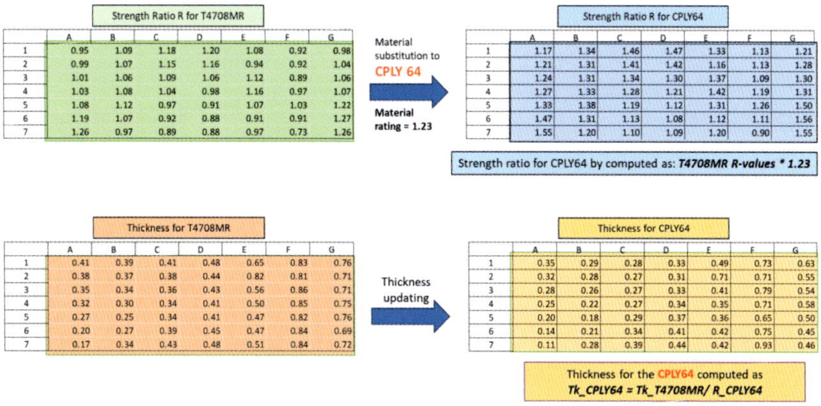

Strength Ratio R for T4708MR

	A	B	C	D	E	F	G
1	0.95	1.09	1.18	1.20	1.08	0.92	0.98
2	0.99	1.07	1.15	1.16	0.94	0.92	1.04
3	1.01	1.06	1.09	1.06	1.12	0.89	1.06
4	1.03	1.08	1.04	0.98	1.16	0.97	1.07
5	1.08	1.12	0.97	0.91	1.07	1.03	1.22
6	1.19	1.07	0.92	0.88	0.91	0.91	1.27
7	1.26	0.97	0.89	0.88	0.97	0.73	1.26

Strength Ratio R for CPLY64

	A	B	C	D	E	F	G
1	1.17	1.34	1.46	1.47	1.33	1.13	1.21
2	1.21	1.31	1.41	1.42	1.16	1.13	1.28
3	1.24	1.31	1.34	1.30	1.37	1.09	1.30
4	1.27	1.33	1.28	1.21	1.42	1.19	1.31
5	1.33	1.38	1.19	1.12	1.31	1.26	1.50
6	1.47	1.31	1.13	1.08	1.12	1.11	1.56
7	1.55	1.20	1.10	1.09	1.20	0.90	1.55

Thickness for T4708MR

	A	B	C	D	E	F	G
1	0.41	0.39	0.41	0.48	0.65	0.83	0.76
2	0.38	0.37	0.38	0.44	0.82	0.81	0.71
3	0.35	0.34	0.36	0.43	0.56	0.86	0.71
4	0.32	0.30	0.34	0.41	0.50	0.85	0.75
5	0.27	0.25	0.34	0.41	0.47	0.82	0.76
6	0.20	0.27	0.39	0.45	0.47	0.84	0.69
7	0.17	0.34	0.43	0.48	0.51	0.84	0.72

Thickness for CPLY64

	A	B	C	D	E	F	G
1	0.35	0.29	0.28	0.33	0.49	0.73	0.63
2	0.32	0.28	0.27	0.31	0.71	0.71	0.55
3	0.28	0.26	0.27	0.33	0.41	0.79	0.54
4	0.25	0.22	0.27	0.34	0.35	0.71	0.58
5	0.20	0.18	0.29	0.37	0.36	0.65	0.50
6	0.14	0.21	0.34	0.41	0.42	0.75	0.45
7	0.11	0.28	0.39	0.44	0.42	0.93	0.46

Figure 5.15. Strength ratio and thickness updating through material scaling methodology

The new proposed design, with modified material and thicknesses resulted in a further 17% reduction in component weight as compared to the optimised design with the starting material leading to a plate weight of 41% if compared to the ideal plate with uniform 1 mm Thickness. The introduction of a stronger material system, with enhanced strength features, allows to increase the safety margin with respect to the starting material and to further reduce the thickness and weight of the component, without any recalculation or new optimization process. In a preliminary design context, this methodology provides the opportunity to easily change the material employed in the design of the component and quickly establish its strength performance without the need for further analysis or recalculations.

The obtained scaled design has been finally verified by performing a new DD design tool optimization with the new material system CPLY64 (The [±60/±30] stacking sequence has been retained). The results in terms of cells thickness have been compared to the scaled configuration.

As Figure 5.16 demonstrates, the new optimization process produced a non-uniform thickness distribution across the cells, with thicker areas where higher stresses necessitated additional material to maintain adequate margin of safety. Conversely, cells with lower stresses are characterized by reductions in thickness.

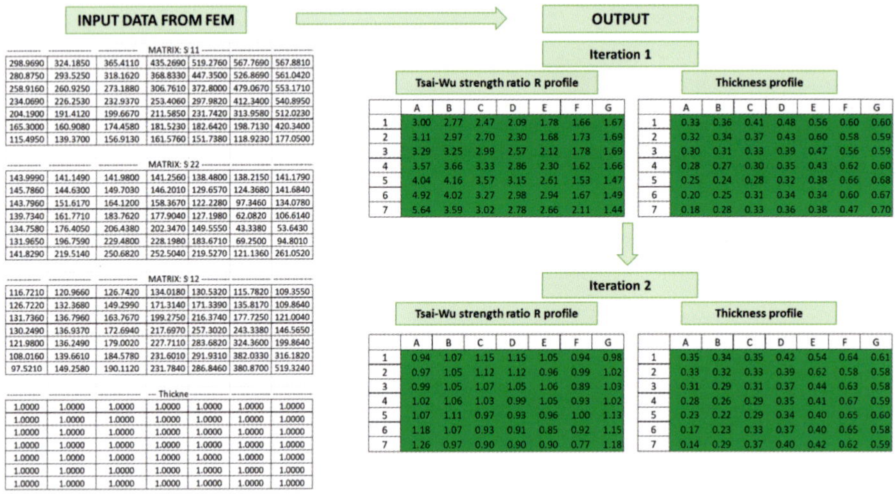

Figure 5.16. Thickness optimization process on the flat plate in CPLY64

A rigorous comparison has been performed between the thickness distribution in cells achieved by the full optimization process and the thickness distribution proposed in the previous section using the material scaling methodology. Remarkably, an average difference of only 10% has been found in the cell thicknesses among all cells when comparing the scaled design based on material rating factors to the fully optimized design. This result is noticeable considering that the thickness distribution obtained using the material scaling methodology has been achieved without any additional recalculation, analysis, or optimization beyond the initial optimized design for the starting material. Yet the scaled design for the new material exhibited a moderate error of only 10% on average in the computed thicknesses across all cells.

5.5.3 Application of the material scaling procedure to the composite fuselage barrel section

In this sub-section, the potential of the proposed material scaling methodology for material selection in composite design is explored through a practical application to the optimized composite fuselage skin presented in Section 5.4. The fuselage skin application provides a real-world test case for validating the proposed material scaling approach and illustrating its potential for aiding composite material selection and development within a simplified, single rating factor-based design framework. In Section 5.4, the DD-Design tool was used to perform a thorough thickness optimization of the fuselage skin, determining the optimal layup and non-uniform thickness distribution for the cells that best aligned with the strength requirements under the applied loading conditions. This optimized design utilized the CPLY64 material system. The resulting optimized design for the CPLY64 material serves as a baseline for evaluating the performance of the scaled design produced using the material scaling methodology. By applying the rating factor of a new material system to the strength ratios determined for the CPLY64 material, a scaled thickness distribution for that new material can be obtained without additional analyses.

For the scaled design, a less performing material system was chosen: the T4708MR material, characterized by a lower rating factor of 0.83 compared to the CPLY64 material, supposed to have a rating factor of 1.

Starting from the strength ratio distribution determined for the cells of the fuselage skin in Iteration 4 of subsection 5.4.2 - representing the optimal CPLY64 design - the strength ratios for the T4708MR material were calculated by multiplying the original CPLY64 strength ratios by the 0.83 rating factor of the T4708MR material.

The scaled strength ratios for the T4708MR material have been then used to recalculate the thickness of each cell, as shown in Figure 5.17, producing a non-uniform thickness distribution tailored for the intended T4708MR material.

Results of the optimization process performed in Section 5.4 with CPLY64 material system.

Strength Ratio Distribution

	A	B	C	D	E	F
1	1.08	1.23	1.20	1.21	1.22	1.12
2	1.04	1.24	1.22	1.23	1.20	1.12
3	1.13	1.23	1.20	1.22	1.23	1.16
4	1.18	1.10	1.09	1.09	1.09	1.16
5	1.17	1.21	1.17	1.15	1.15	1.13
6	1.08	1.09	1.10	1.11	1.11	1.08
7	1.02	1.04	1.05	1.06	1.06	1.03

Thickness Distribution

	A	B	C	D	E	F
1	0.43	0.48	0.49	0.52	0.53	0.45
2	0.42	0.54	0.57	0.61	0.63	0.48
3	0.58	0.70	0.75	0.77	0.78	0.65
4	0.76	0.89	0.96	0.96	0.94	0.84
5	0.71	0.89	0.92	0.95	0.93	0.74
6	0.63	0.88	0.85	0.81	0.77	0.54
7	0.61	0.78	0.76	0.73	0.70	0.51

Material scaling to T4708MR

Material rating = 0.83

Strength Ratio Distribution

	A	B	C	D	E	F
1	0.90	1.02	0.99	1.00	1.01	0.93
2	0.86	1.03	1.01	1.02	0.99	0.93
3	0.94	1.02	1.00	1.01	1.02	0.96
4	0.98	0.91	0.90	0.90	0.90	0.97
5	0.97	1.00	0.97	0.96	0.95	0.94
6	0.89	0.90	0.91	0.92	0.92	0.89
7	0.85	0.86	0.87	0.88	0.88	0.85

Thickness Distribution

	A	B	C	D	E	F
1	0.48	0.47	0.49	0.51	0.53	0.48
2	0.49	0.52	0.56	0.59	0.64	0.51
3	0.62	0.68	0.76	0.76	0.77	0.68
4	0.78	0.98	1.06	1.06	1.05	0.87
5	0.74	0.89	0.95	0.99	0.97	0.79
6	0.71	0.98	0.94	0.88	0.84	0.61
7	0.72	0.90	0.88	0.83	0.79	0.60

Figure 5.17. Strength ratio and thickness updating through material scaling methodology

As the T4708MR material exhibits lower strength features than the starting CPLY64 material, it is characterized by a rating factor less than unity when compared with CPLY64. As a result, the newly computed cell thicknesses using the T4708MR rating factor will be higher if compared to the starting CPLY64 thicknesses. In this case, some of the original weight reduction achieved through optimization for the CPLY64 material is lost.

However, this change from a higher performing material to a less performing one should be viewed from the perspective of considering not only mechanical performance aspects of a material but also economic and availability aspects. The proposed procedure can be used to evaluate the design of a component with a less performing but cheaper material, without performing recalculations or new analyses on the structure.

The new proposed design, with modified material and thicknesses resulted in 8% increase in barrel skin weight if compared to the optimised design with the starting material obtained by the DD-design tool. The introduction of a weaker material system, with lower strength features, bring to a decrease of the safety margins with respect to the starting material and to an increase in the thickness and weight of the component, without any recalculation or new optimization process.

Analysing the decrease in weight obtained for the plate scaling and the increase of weight obtained for the fuselage barrel, it can be deduced that the variation of mass obtained with the scaling procedure is, somehow, connected to the starting design and to starting strength ratios distribution other than the scaling factor associated to the new material system considered for the scaling. It is worth to notice that this scaling procedure, with scaling factors based on table 5.12, can be applied also for stacking sequence variations.

As a Final consideration, the proposed scaling procedure can be considered well-suited for preliminary design studies, material selection and feasibility evaluations, representing an important first step toward modern composite laminate design.

5.6 Conclusions

Double-Double laminates, introduced by prof. S.W. Tsai, represent a considerable step forward in designing with composite materials. DD laminates allow to obtain remarkable weight savings as they can achieve desired requirements in terms of strength and stiffness with stacking sequences constituted by a very limited number of thin building blocks. Indeed, differently from standard Quad laminates, DD laminates are obtained by means of 4 plies [±Φ; ±Ψ] building blocks without requirements on symmetry and balancing since they can be easily homogenized. Double-double laminates together with Tsai's Modulus, failure Criteria, Strength Ratio and Master ply concept makes it possible to design with composites as with metals (scaling one factor) and can be easily optimized and tapered. In this chapter an innovative DD automated optimization tool is presented, capable of finding the optimal [±Φ; ±Ψ] angles combination to fulfil desired constraints and loading conditions by introducing iterations between a FE solver (whatever commercial FE platform can be used) and the Lamsearch optimizer. In order to test the effectiveness of the DD automated design tool and to demonstrate the remarkable potential in terms of weight saving, offered by Double-double laminates, the redesign with DD laminates of a real regional aircraft fuselage barrel has been carried out. The obtained results are very encouraging since a [Φ±37.5;Ψ±45] Double-Double layup with 64.6% weight reduction with respect to the quasi isotropic Quad laminates configuration, has been obtained. This weight reduction decreases to 48.1 % for the very easily manufacturable uniform thickness skin configuration. This result demonstrates the effectiveness of DD automated design tool and the potential of DD laminates in terms of weight saving in composite design.
In addition to the introduction of the DD-design tool, a new single parameter based scaling approach has been introduced for interchangeability of material systems in design. The proposed scaling procedure, which takes its theoretical bases from the definition of the Failure Master Envelopes, has been demonstrated to be effective for scaling design of composite components, with respect to material systems, in a preliminary design stage without the need of further recalculation and providing accurate results (10% error) with respect to refined finite elements analyses. Future works will be addressed to the definition of a scaling procedure with respect to laminate stacking sequences.

5.7 References

[1] Kennedy, G. J., & Martinsy, J. R. R. A. (2012). A comparison of metallic and composite aircraft wings using aerostructural design optimization. Paper presented at the 12th AIAA Aviation Technology, Integration and Operations (ATIO) Conference and 14th AIAA/ISSMO Multidisciplinary Analysis and Optimization Conference.
[2] Lapeña-Rey, N., Mosquera, J., Bataller, E., Ortí, F., Dudfield, C., & Orsillo, A. (2008). Environmentally friendly power sources for aerospace applications. Journal of Power Sources, 181(2), 353-362.
[3] Leifsson, L., Ko, A., Mason, W. H., Schetz, J. A., Grossman, B., & Haftka, R. T. (2013). Multidisciplinary design optimization of blended-wing-body transport aircraft with distributed propulsion. Aerospace Science and Technology, 25(1), 16-28.

[4] Mangalgiri, P.D. Composite materials for aerospace applications. Bull Mater Sci 22, 657–664 (1999).
[5] R.N. Yancey, 2 - Challenges, opportunities, and perspectives on lightweight composite structures: Aerospace versus automotive, Editor(s): J. Njuguna, Lightweight Composite Structures in Transport, Woodhead Publishing, 2016, Pages 35-52.
[6] Denault, J., Dumouchel, M. Consolidation Process of PEEK/Carbon Composite for Aerospace Applications. Advanced Performance Materials 5, 83–96 (1998).
[7] Camanho, P. P., Dávila, C. G., & De Moura, M. F. (2003). Numerical simulation of mixed-mode progressive delamination in composite materials. Journal of Composite Materials, 37(16), 1415-1438.
[8] Turon, A., Camanho, P. P., Costa, J., & Dávila, C. G. (2006). A damage model for the simulation of delamination in advanced composites under variable-mode loading. Mechanics of Materials, 38(11), 1072-1089.
[9] Hou, J. P., Petrinic, N., Ruiz, C., & Hallett, S. R. (2000). Prediction of impact damage in composite plates. Composites Science and Technology, 60(2), 273-281.
[10] Russo A., Riccio A., Sellitto A. A robust cumulative damage approach for the simulation of delamination under cyclic loading conditions (2022) Composite Structures, 281, art. no. 114998.
[11] Riccio A., Palumbo C., Acanfora V., Sellitto A., Russo A. Influence of failure criteria and intralaminar damage progression numerical models on the prediction of the mechanical behavior of composite laminates (2021) Journal of Composites Science, 5 (12), art. no. 310.
[12] J.R. Duflou, J. De Moor, I. Verpoest, W. Dewulf, Environmental impact analysis of composite use in car manufacturing, CIRP Annals, Volume 58, Issue 1, 2009, Pages 9-12.
[13] F W J van Hattum, F Regel & M Labordus (2011) Cost reduction in manufacturing of aerospace composites, Plastics, Rubber and Composites, 40:2, 93-99.
[14] Dinesh Kumar Koli, Geeta Agnihotri, Rajesh Purohit, Advanced Aluminium Matrix Composites: The Critical Need of Automotive and Aerospace Engineering Fields, Materials Today: Proceedings, Volume 2, Issues 4–5, 2015, Pages 3032-3041.
[15] Nöding M, Bertsch L. Application of Noise Certification Regulations within Conceptual Aircraft Design. Aerospace. 2021; 8(8):210.
[16] C.A.C. Antonio, A.T. Marques, A.V. Soeiro. Optimization of laminated composite structures using a bilevel strategy. Composite Structures, 33 (1995), pp. 193-200.
[17] P. Conti, S. Luparello, A. Pasta Layer thickness optimisation in a laminated composite. Composite Part B – Engineering, 28 (3) (1997), pp. 309-317.
[18] Wiggenraad, I.F.M., Arendsen, P., da Silva Pereira, J.M, "Design Optimization of Stiffened Composite Panels With Buckling And Damage Tolerance Constraints," AIAA Paper 98-1750, AIAA, Reston, VA. 1998.
[19] Edwards, D. A., Williams, F. W., Kennedy, D., "Cost optimization of Stiffened Panels using VICONOPT," AIAA J., Vol. 36, No. 2, Feb. 1998, 267-272.
[20] Crossley, W. A., Laananen, D. H., "Genetic Algorithm-Based Optimal Design of Stiffened Composite Panels for Energy Absorption," In Proc. of the American Helicopter Society 52nd Annual Forum, Washington, D. C., June 4-6, 1996, 1367-1376.
[21] J.P. Blasques, M. Stolpe. Maximum stiffness and minimum weight optimization of laminated composite beams using continuous fiber angles. Struct Multidiscip Optim, 43 (4) (2011), pp. 573-588.
[22] H. Ghasemi, R. Brighenti, X. Zhuang, J. Muthu, T. Rabczuk. Optimization of fiber distribution in fiber reinforced composite by using NURBS functions. Comput Mater Sci, 83 (2014), pp. 463-473.
[23] Akira Todoroki, Raphael T. Haftka, Stacking sequence optimization by a genetic algorithm with a new recessive gene like repair strategy, Composites Part B: Engineering, Volume 29, Issue 3, 1998, Pages 277-285.
[24] Ching-Chieh Lin, Ya-Jung Lee, Stacking sequence optimization of laminated composite structures using genetic algorithm with local improvement, Composite Structures,
[25] Volume 63, Issues 3–4, 2004, Pages 339-345.
[26] Boyang Liu, Raphael T. Haftka, Mehmet A. Akgün, Akira Todoroki, Permutation genetic algorithm for stacking sequence design of composite laminates, Computer Methods in Applied Mechanics and Engineering, Volume 186, Issues 2–4, 2000, Pages 357-372.
[27] Zhao Jing, Qin Sun, Vadim V. Silberschmidt, Sequential permutation table method for optimization of stacking sequence in composite laminates, Composite Structures, Volume 141, 2016, Pages 240-252.
[28] Chang, N., Wang, W., Yang, W. et al. Ply stacking sequence optimization of composite laminate by permutation discrete particle swarm optimization. Struct Multidisc Optim 41, 179–187 (2010).

[29] Sachin Shrivastava, Naresh Sharma, Stephen W. Tsai, P.M. Mohite, D and DD-drop layup optimization of aircraft wing panels under multi-load case design environment (2020) Composite Structures, 248, art. no. 112518.
[30] Vermes, B., Tsai, S.W., Riccio, A., Di Caprio, F., Roy, S. Application of the Tsai's modulus and double-double concepts to the definition of a new affordable design approach for composite laminates (2021) Composite Structures, 259, art. no. 113246.
[31] Vermes, B., Tsai, S.W., Massard, T., Springer, G.S., Czigany, T. Design of laminates by a novel "double–double" layup (2021) Thin-Walled Structures, 165, art. no. 107954.
[32] Tsai, S.W. Double–double: New family of composite laminates (2021) AIAA Journal, 59 (11), pp. 4293-4305.
[33] Tsai SW, Melo JDD. An invariant-based theory of composites. Compos Sci Technol 2014;100:237–43.
[34] Albertino Arteiro, Naresh Sharma, Jose Daniel D. Melo, Sung Kyu Ha, Antonio Miravete, Yasushi Miyano, Thierry Massard, Pranav D. Shah, Surajit Roy, Robert Rainsberger, Klemens Rother, Carlos Cimini Jr., Jocelyn M. Seng, Francisco K. Arakaki, Tong-Earn Tay, Woo Il Lee, Sangwook Sihn, George S. Springer, Ajit Roy, Aniello Riccio, Francesco Di Caprio, Sachin Shrivastava, Alan T. Nettles, Giuseppe Catalanotti, Pedro P. Camanho, Waruna Seneviratne, António T. Marques, Henry T. Yang, H. Thomas Hahn, A case for Tsai's Modulus, an invariant-based approach to stiffness, Composite Structures, Volume 252, 2020, 112683.
[35] Ha SK, Cimini CA. Theory and validation of the master ply concept for invariant-based stiffness of composites. Journal of Composite Materials. 2018;52(12):1699-1708.

With George Springer and Jocelyn Patterson Seng, both life long collaborators

CHAPTER 6

Studies on the Impact and Compression-After-Impact Response of 'Double-Double' Carbon- Fibre Reinforced Composite Laminates

Scott L.J. Millen[a], M. Ali Aravand[a], Zahur Ullah[a], Brian G. Falzon[a,b]
[a] *Advanced Composites Research Group (ACRG), School of Mechanical and Aerospace Engineering, Queen's University Belfast, Ashby Building, Belfast, Northern Ireland, BT9 5AH, U.K.*
[b] *School of Engineering, STEM College, RMIT University, Melbourne, VIC, 3001, Australia*
scott.millen@qub.ac.uk, m.aravand@qub.ac.uk, z.ullah@qub.ac.uk, brian.falzon@rmit.edu.au

This chapter presents comparative high fidelity computational finite element (FE) modelling of low-velocity impact (LVI) damage and compression-after-impact (CAI) strength of both 'Legacy Quad Laminates' (LQL) and 'Double-Double' DD laminates, using a validated FE modelling approach. An in-house intralaminar damage model, capturing both fibre-dominated and matrix-dominated damage, along with an available interlaminar cohesive model are used within an explicit dynamic formulation. LQL results are validated using available experimental data before modelling the corresponding DD laminates. Results show that DD laminates are effective replacements for UD or hybrid woven/UD LQL under both LVI and CAI loading conditions. It is shown that matrix and delamination damage may be reduced with DDs. Potential weight savings can be achieved with the use of DD laminates as well as simplifying manufacturing processes. In all the cases presented herein, CAI residual strength increased with the use of a DD laminate over a traditional LQL.

In this study we compare results for both 'Legacy Quad Laminates' (LQL) and equivalent 'Double-Double' DD laminates for low velocity impact (LVI) and compression after impact (CAI) loading conditions. To-date, high fidelity computational modelling of DD laminates has been limited. Vermes et al. [1] used a combination of *Lam search* [2], a search engine tool to find the best laminate based on the strength of un-notched coupons, and finite element method (FEM) tools. The authors demonstrated that the use of *Tsai's modulus* (the trace of the plane stress stiffness matrix [3]) and DD laminates led to significant weight savings compared to LQLs in standard and open-hole tension and compression specimens. Melo et al. [4] evaluated traditional and invariant-based design approaches for smooth and open-hole specimens. To-date only a single reference has compared the the LVI response of a DD laminate and a LQL of equivalent stiffness and thickness [5]. However, this study compared only the standard cross-ply or paired DD sub-laminate [±φ/±Ψ] with a LQL.

Recently, Tsai [6], following the work of Nielsen et al. [7], noted that the standard cross-ply or paired DD sub-laminate [±φ/±Ψ] can make homogenisation more difficult in some cases. The criteria for homogenisation was defined as a ≤2% variation in $D^*_{11} - A^*_{11}$, $D^*_{66} - A^*_{66}$, or D^*_{16}, A^*_{11}or $[B^*]$ e.g. $|B^*_{11}| \leq 0.02$. Therefore, Tsai proposed that the [±φ/±Ψ] sub-laminate building block can be modified to three staggered sub-laminates [φ/-Ψ/-φ/Ψ], [φ/Ψ/-φ/-Ψ] or [φ/-Ψ/Ψ/-φ]. The particular scenario under investigation by Nielsen et al. [7] had a recommended stacking sequence of $[\varphi/-\Psi/-\varphi/\Psi]_{rT}$, where *r* is the number of repeats and *T* stands for the total number of plies in the laminate. *Tsai's Modulus* and invariants, can facilitate the selection of DD equivalents through comparison of *Tsai's Modulus* normalised stiffness matrices for LQL and DD layups. No direct comparison has been presented in literature for LQLs and DDs for the same material properties. However, the use of high-fidelity modelling and CAI analysis can provide the specimen strength without the need to plot failure envelopes.

DD laminates are of growing interest to the research community and industry alike. Thorough analysis and comparison between DD laminates and LQLs is still nascent. Initial comparisons have been made in terms of strength and in-plane stiffness, including open-hole tension and compression specimens. However, one area which has not been thoroughly investigated is LVI and subsequent CAI. Therefore, this chapter presents DD laminate replacements for a variety of LQLs. LVI/CAI simulations are subsequently conducted to compare the performance of DD laminates and LQLs in terms of LVI damage predictions and CAI residual strength.

6.1 Double-Double (DD) laminate generation

In order to generate the equivalent DD laminates, two analytical tools were used. These were the *Lam search* tool developed by Tsai et al. [2,8], based on classical laminate theory (CLT) and the maximum strain failure criterion, and the simplified design tool generated in refs. [9,10]. These tools were used to find the DD laminate which would match the required in-plane stiffness of an existing LQL, based on a set of material properties (*Tsai's Modulus*, longitudinal tensile and compressive strengths and longitudinal modulus).

These methods have focussed on in-plane stiffness-based design using the extensional stiffness matrix, $[A]^*$, to select an equivalent DD sub-laminate, specifically matching the normalised longitudinal stiffness, A^*_{11}, for both the LQL and equivalent DD. However, given that this work focusses on impact, which is heavily influenced by bending, the bending stiffness matrix $[D]^*$has greater influence on the resulting laminate behaviour. Therefore, for each test case both the $[A]^*$ and $[D]^*$matrices were compared between the LQL and DD. If the difference in $D^*_{11}$was greater than 5%, the DD layup was optimised to reduce this difference. This optimisation had the potential to introduce some membrane-coupling effects. Therefore, the paired and staggered sub-laminate arrangements proposed by Tsai [6], to achieve homogenisation, were also compared.

6.2 LVI/CAI simulations and specimens

The finite element plate models used to simulate LVI/CAI measured 150 x 100 mm and were placed on a picture frame support leaving an unsupported region of 125 mm x 75 mm, following the Airbus Industries Test Method (AITM 1-0010 [11]). The impactor was modelled as a spherically shaped, analytical rigid solid. The general simulation arrangement for both LVI and CAI is shown in Figure 6.1. The specimen mesh was refined in a region around the impact zone, matching the mesh used for each test specimen in the previously published literature [12,13]. Specimens were meshed using C3D8R reduced integration elements with enhanced hourglass and distortion control. Fibre-aligned meshing was not pursued as a result of a recent study [14] exploring the merits of different meshing strategies concluded that this was not required. All simulations were executed using ABAQUS/Explicit.

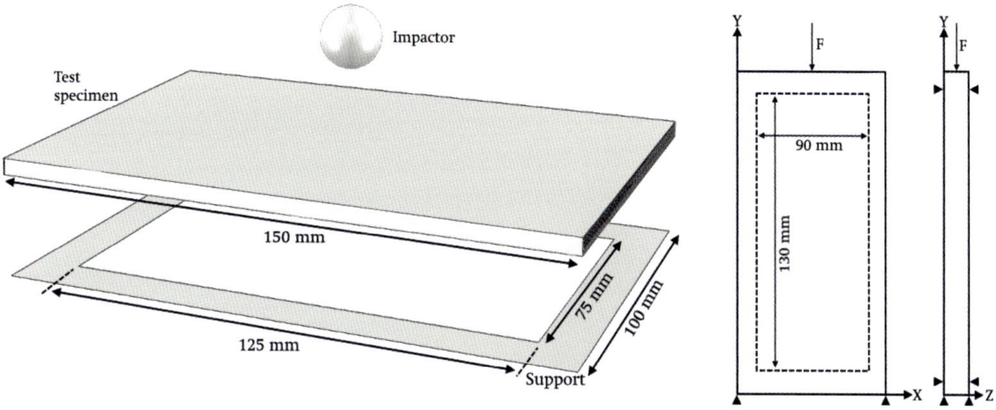

Figure 6.1 – Impact (left) and CAI (right) simulation geometry and boundary conditions

An in-house intralaminar damage model was used in this work. This damage model can capture both fibre-dominated damage (through a quadratic strain-based failure criterion) and matrix-dominated damage (through a failure criterion modified and adapted by Catalanotti et al. [15]), and permits load reversal. It has an advanced characterstic length calculation ($l_{fib} = V_0^e/A_e$, where V_0^e is the undeformed volume of the element, and A_e is the fracture surface area) calculated using an approach proposed by Chiu et al. [16] based on a search algorithm which maximises damage inititation functions. A non-linear shear model with kinematic hardening [17], and robust element deletion control (if $d_{11} > 0.99, or\ det\boldsymbol{F} < 0.1 \lor det\boldsymbol{F} > 3.0$ where $det\boldsymbol{F}$ yields the ratio of the deformed volume, V^e, to the undeformed volume, V_0^e). This intralaminar damage model has been used extensively for LVI/CAI modelling and composite crushing and more details about the theoretical foundations of the model and validation can be found in refs. [12,18–22].

The general contact algorithm was used to simulate contact between the impactor and composite laminate. Cohesive surfaces were used to model ply-to-ply contact and capture delamination using a bilinear traction-separation law. All LVI/CAI simulations were completed on a Linux cluster with 20 CPUs with double precision used to minimise round-off errors.

6.3 Specimens

The two specimens modelled were taken from the literature [12,13]. These were chosen to provide a variety of materials, shown in **Error! Reference source not found.**, layups and ply thicknesses, discussed below.

Table 6.1 – Material properties for numerical simulations

Property	T700GC/M21	IMS 60 Carbon-fibre/epoxy	AS4 woven carbon-fibre/epoxy
Elastic moduli (Gpa)			
E_{11}	130	152	65.4
E_{22}	7.70	8.71	65.4
E_{33}	7.70	8.71	8.71
Poisson's Ratios			
$\nu_{12}=\nu_{13}$	0.33	0.30	0.04
ν_{23}	0.35	0.35	0.33
Shear moduli (Mpa)			
$G_{12}=G_{13}$	4800	4140	3590
G_{23}	3800	3230	3270
Failure strengths (Mpa)			
X_T	2080	1930	862
X_C	1250	962	689
Y_T	60	41.4	862
Y_C	290	276	689
S_{12}	110	82.1	110
Critical energy release rates (mJ/mm^2)			
G_{fib}^T	133	775	201
G_{fib}^C	40	87	92
G_{mat}^T	0.6	0.46	0.32
$G_{mat}^C = G_{mat}^{12} = G_{mat}^{13} = G_{mat}^{23}$	2.1	1.51	2.01
Non-linear shear properties (Mpa) $\tau(\gamma_{ij}) = c_1\gamma_{ij}^3 - sgn(\gamma_{ij})c_2\gamma_{ij}^2 + c_3\gamma_{ij}$			
C_1	37,833	66.5	77.2
C_2	16,512	3.2	3.6
C_3	2334.3	62.4	65.2
Interface fracture energies (mJ/mm^2)			
G_{IC}	0.6	0.46	0.32
G_{IIC}	2.1	1.51	2.01
η	1.45	1.89	2.09
Interface strengths (Mpa)			
σ_I	20	17	20
$\sigma_{II}=\sigma_{III}$	36	30	34

Specimen 1 - The first specimen reported by Tan et al. [12], had a stacking sequence of $[0_2/45_2/90_2/\text{-}45_2]_s$, i.e. the percentage of (0°/±45°/90°) being (25/50/25), and was manufactured from T700GC/M21 with a ply thickness of 0.26 mm. The impact energy was 25J and the impactor had a diameter of 16 mm and mass of 2 kg. Following the completion of the impact step in Specimen 1, the impact boundary conditions were replaced with new boundary conditions for CAI simulation, shown in Figure 6.1.

Specimen 2 - The second specimen was reported by Liu et al. [13]. This specimen had a stacking sequence of [$0/90/0_2/\pm45/0_3/\mp45/0_3/\pm45/0_2/90/0$], (60/30/10), was manufactured from IMS60 Carbon-fibre/epoxy and had a ply thickness of 0.267 mm. However, the upper and lower cross-plies (0/90) had a total thickness of 0.35 mm (0.175 mm each) . These outer plies were used as an alternative to woven plies in a hybrid laminate with layup [$5HS/0_2/\pm45/0_3/\mp45/0_3/\pm45/0_2/5HS$], where 5HS refers to a 5 Harness Satin weave, to compare predictions of a hybrid specimen with equivalent UD and DD laminates. The impact energy was 25J and the impactor measured 12.7 mm in diameter with a mass of 6.4 kg. As before, following the completion of the impact step, new boundary conditions for the CAI simulation were imposed.

In the case of DD laminates, simulations started with a DD laminate equal to, or approximately equal to, the thickness of the equivalent LQL. These simulations were then repeated, removing one DD sub-laminate at a time and comparing the damage predictions until perforation. The best alternative DD laminate was then identified for each LQL, assumed to be the thinnest specimen with the largest residual CAI strength, indicating the structural integrity of the specimen. Results were also compared using intralaminar and interlaminar damage area measurements.

6.4 Initial model validation

Prior to the comparison of DD impact simulation results with those for LQLs, the LQL model for each specimen was validated against the available experimental data. This exercise would validate the material data and intralaminar damage model used in this work with existing experimental data. Validation of Specimen 1 was completed using the experimental data from refs. [12,23]. Validation of Specimen 2 was completed using the experimental data from refs. [13,24]. Results of this validation are shown in Figure 6.2 where impact and CAI predictions from the simulations are plotted against the experimental data for both specimens at different energies. Excellent agreement was obtained between the experiments and simulations for impact force-time curves while the predicted CAI strength for specimen 1 was within 5% of the relevant experiment and within 3% for specimen 2. Given this high level of predictive capability of the underlying intralaminar damage model and simulation boundary conditions, modifications to the layup and thickness of plies could be made with confidence.

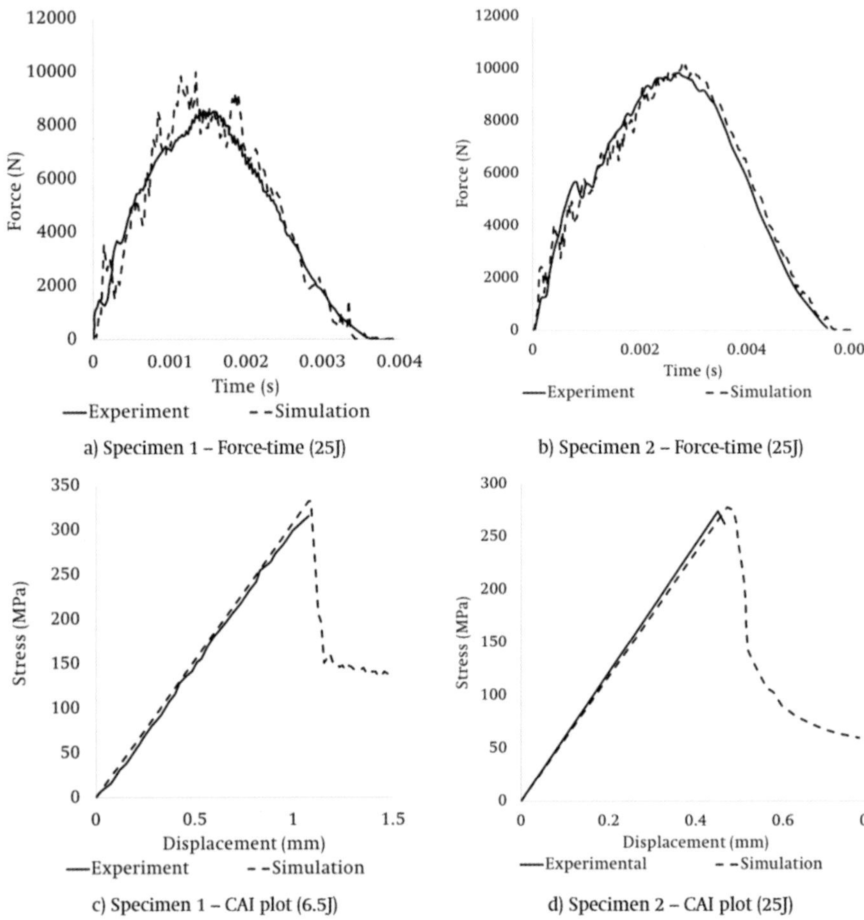

Figure 6.2 – Validation of simulations using LQLs and experimental data [12,13]

6.5 Double-Double (DD) laminate generation

Using a combination of *Lam search* [2,8] and the simplified design tool generated in refs. [9,10] the DD laminate for Specimen 1 LQL $[0_2/45_2/90_2/-45_2]_s$, based on the extensional stiffness matrix was $[\pm22.5/\pm67.5]_{4T}$. A perfect match was achieved with the $[A]^*$ matrix of the LQL. Larger variation occurred for the $[D]^*$ matrix, for example D_{11}^* was 47% lower for the DD laminate, Table 6.2. Therefore, this DD laminate was optimised to match the bending stiffness of the LQL and the resulting DD sub-laminate building block was $[\pm9.5/\pm39.5]$. The difference in D_{11}^*between the LQL and this $[\pm9.5/\pm39.5]_{4T}$ DD was reduced to less than 1% but $A_{11}^*$was 47% higher for the optimised $[\pm9.5/\pm39.5]_{4T}$ DD. In both $[\pm22/\pm67]_{4T}$ and $[\pm9.5/\pm39.5]_{4T}$ DDs, membrane coupling occurred, Table 6.2, and therefore, these laminates could not be considered fully homogenised. Therefore, the staggered DD laminate layups proposed by Tsai [6] were also compared. Table 6.2 shows that the "staggered 3" DD laminate, $[9.5/-39.5/39.5/-9.5]_{4T}$ in this case, was able to achieve a value of D_{11}^*which was 1.6% larger than the LQL but eliminated membrane coupling. Therefore, this DD laminate could be considered homogenised and a suitable replacement for the LQL.

Table 6.2 – Specimen 1 layups and $[ABD]^*$ values

		Paired	Paired	Staggered 1	Staggered 2	Staggered 3
	LQL	DD	DD	DD	DD	DD
		$[\pm\varphi/\pm\Psi]_{rT}$	$[\pm\varphi/\pm\Psi]_{rT}$	$[\varphi/\text{-}\Psi/\text{-}\varphi/\Psi]_{rT}$	$[\varphi/\Psi/\text{-}\varphi/\text{-}\Psi]_{rT}$	$[\varphi/\text{-}\Psi/\Psi/\text{-}\varphi]_{rT}$
(Gpa)	(25/50/25)	$[\pm22/\pm67]_{4T}$	$[\pm9.5/\pm39.5]_{4T}$	$[9.5/\text{-}39.5/\text{-}9.5/39.5]_{4T}$	$[9.5/39.5/\text{-}9.5/\text{-}39.5]_{4T}$	$[9.5/\text{-}39.5/39.5/\text{-}9.5]_{4T}$
A^*_{11}	56.5	55.8	89.0	89.0	89.0	89.0
A^*_{22}	56.5	54.3	19.0	19.0	19.0	19.0
A^*_{66}	20.5	19.1	20.1	20.1	20.1	20.1
A^*_{12}	17.0	16.8	17.8	17.8	17.8	17.8
$\|B^*_{11}\|$	0.0	5.5	4.4	2.2	2.2	0.0
$\|B^*_{22}\|$	0.0	5.4	1.4	0.7	0.7	0.0
$\|B^*_{66}\|$	0.0	0.1	1.5	0.8	0.8	0.0
$\|B^*_{12}\|$	0.0	0.1	1.5	0.8	0.8	0.0
D^*_{11}	90.1	55.8	89.0	89.0	89.0	90.6
D^*_{22}	32.4	54.3	19.0	19.0	19.0	18.5
D^*_{66}	15.7	19.1	20.1	20.1	20.1	19.5
D^*_{12}	10.6	16.8	17.8	17.8	17.8	17.3

Ch.6

In order to assess the effect of different DD layups and homogenisation on impact and CAI strength, the following laminates would be compared for Specimen 1: LQL: $[0_2/45_2/90_2/\text{-}45_2]_s$, DD: $[\pm22/\pm67]_{4T}$, $[\pm9.5/\pm39.5]_{4T}$ and $[9.5/\text{-}39.5/39.5/\text{-}9.5]_{4T}$.

Comparing the results for Specimen 2, the LQL layup was $[0/90/0_2/\pm45/0_3/\mp45/0_3/\pm45/0_2/90/0]$ and the DD sub-laminate building block based on A^*_{11} was $[0_2/\pm39]$. In this case A^*_{11} was 0.65% lower and D^*_{11} was 0.32% lower for the DD sub-laminate. This DD sub-laminate could be considered an appropriate substitute without further optimisation focussed on bending. The optimum arrangement to achieve homogenisation was $[0/\text{-}39/39/0]_{rT}$. Therefore, the following laminates would be compared for Specimen 2: LQL: $[0/90/0_2/\pm45/0_3/\mp45/0_3/\pm45/0_2/90/0]$, DD: $[0/\text{-}39/39/0]_{rT}$.

6.6 Results - DD laminate LVI/CAI results

The assessment of impact resistance of DD laminates is based on plots of force-time and force-displacement, and predictions of matrix and delamination damage areas after impact. Residual strength is shown using stress-displacement curves.

6.6.1 Specimen 1 - Comparison of DD results based on A^*_{11} and D^*_{11}

An initial comparison was completed on Specimen 1 to show the difference in results when using the paired DD laminate found using A^*_{11}, $[\pm22/\pm67]_{4T}$, and the paired DD laminate found optimising for D^*_{11}, $[\pm9.5/\pm39.5]_{4T}$. Results for impact force-time and CAI residual strength plots are shown in Figure 6.3. Results show that the peak force, Figure 6.3a, for the $[\pm9.5/\pm39.5]_{4T}$ specimen optimised to match the bending stiffness of the LQL and extensional stiffness based specimen $[\pm22/\pm67]_{4T}$ were comparable and were 10% and 14% higher than the LQL, respectively. Using the $[\pm9.5/\pm39.5]_{4T}$ DD laminate optimised to match the bending stiffness of the LQL, the peak displacement was 8% larger than the LQL while the extensional stiffness based sub-laminate [±22/±67]4T reduced the peak displacement by 12%. CAI residual strength was 16% higher for the extensional stiffness based DD [±22/±67]4T when compared with the LQL. However, the CAI residual strength was 40% higher for the [±9.5/±39.5]4T specimen optimised to match the bending stiffness of the LQL. This is likely due to the increase of A^*_{11} by 47% over the LQL.

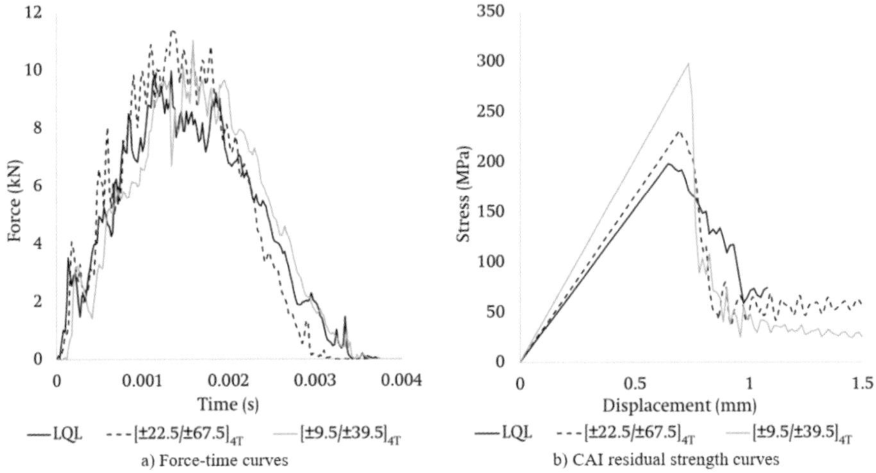

Figure 6.3 – Impact and CAI results for Specimen 1 LQL, paired DD based A^*_{11} and paired DD based on D^*_{11}

6.6.2 Specimen 1 - Comparison of DD results minimising coupling

Figure 6.4 compares the graphical results for the LQL, paired DD [±9.5/±39.5]4T and homogenised DD [9.5/-39.5/39.5/-9.5]4T. It can be seen that the use of the homogenised DD [9.5/-39.5/39.5/-9.5]4T produces minor variations in all measures from the paired DD [±9.5/±39.5]4T. For example, the peak force was 3% lower, peak displacement was equal and CAI strength was 3% higher for the homogenised DD [9.5/-39.5/39.5/-9.5]4T. Figure 6.5 shows the superimposed interlaminar delamination plots after both impact and CAI for the LQL, paired DD [±9.5/±39.5]4T and the homogenised, staggered DD [9.5/-39.5/39.5/-9.5]4T. It can be seen that in all three specimens, CAI damage propagated through the impact damaged region at the centre of the panel. It can also be seen that the delamination damage predictions after impact for both DD laminates were more focussed around the centre of the specimen. A similar pattern is observed for CAI delamination which was caused by more localised delamination damage around the centre of the specimen at the end of the impact step.

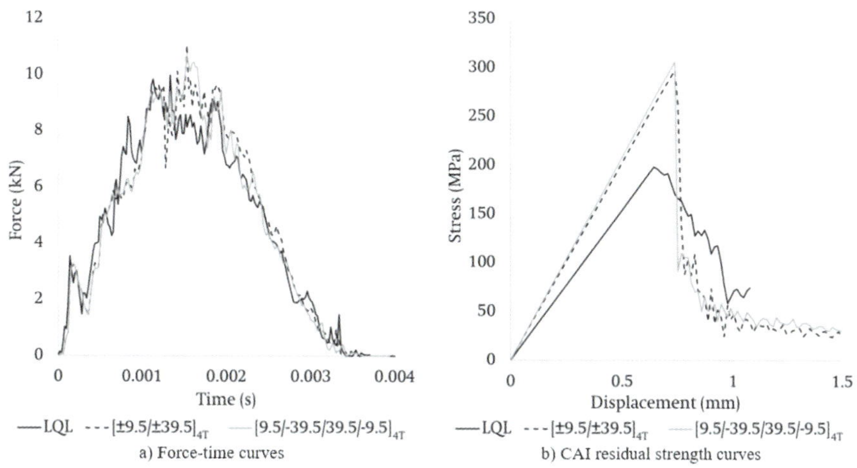

Figure 6.4 – Impact and CAI results for Specimen 1 LQL, paired DD optimised for bending and homogenised DD

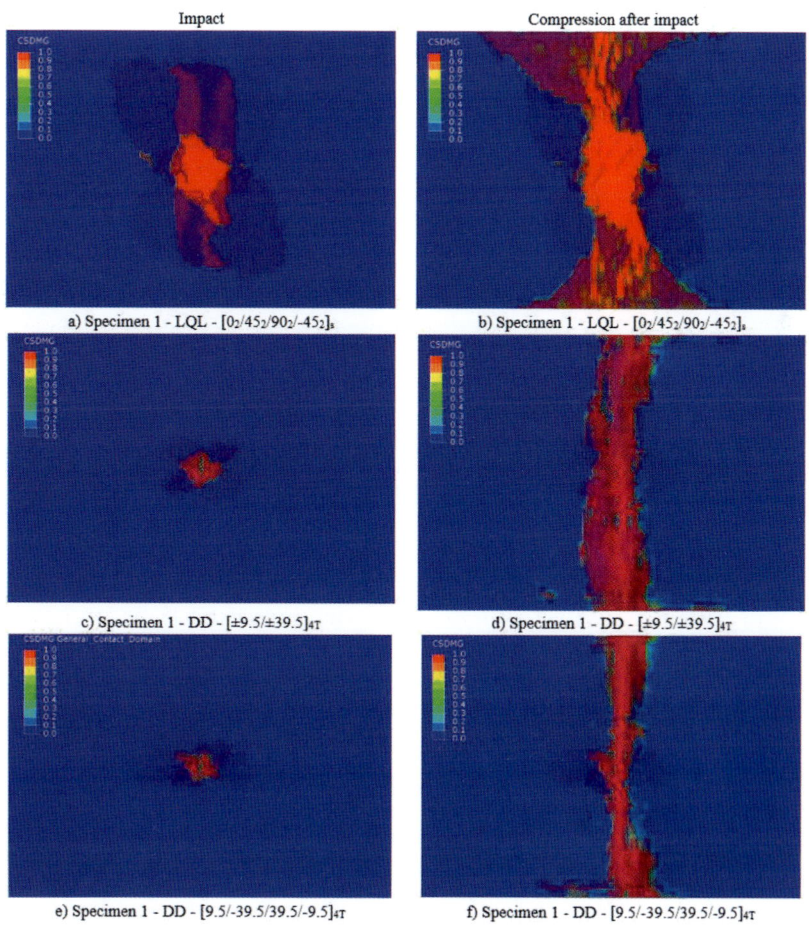

Figure 6.5 – Comparison of predicted impact and CAI delamination for Specimen 1 LQL, optimised and homogenised DDs

In summary, design based on the bending stiffness matrix is more appropriate for both LVI and CAI loading. Figure 6.6 and Figure 6.7 compare the post impact delamination and resulting CAI residual strength for DD specimens designed using $[A]^*$ and $[D]^*$) matrices, respectively ([±22/±67]$_{4T}$ vs [9.5/-39.5/39.5/-9.5]$_{4T}$). Both results show the improvements that can be achieved with a specimen chosen based on the $[D]^*$matrix.

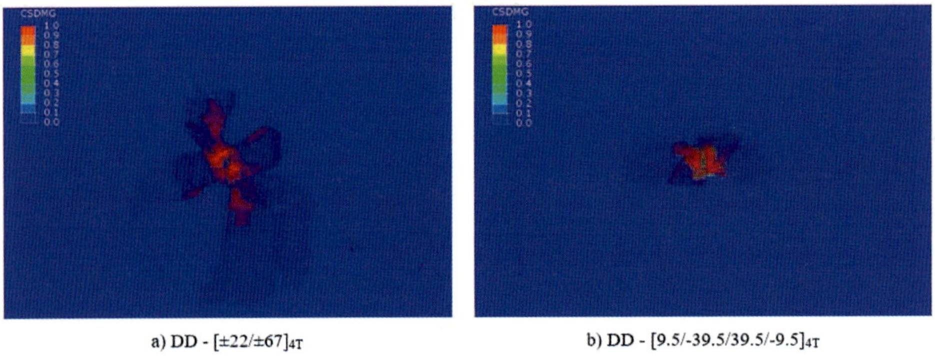

Figure 6.6 - Delamination comparison post impact for DD laminates based on $[A]^*$ and $[D]^*$, matrices

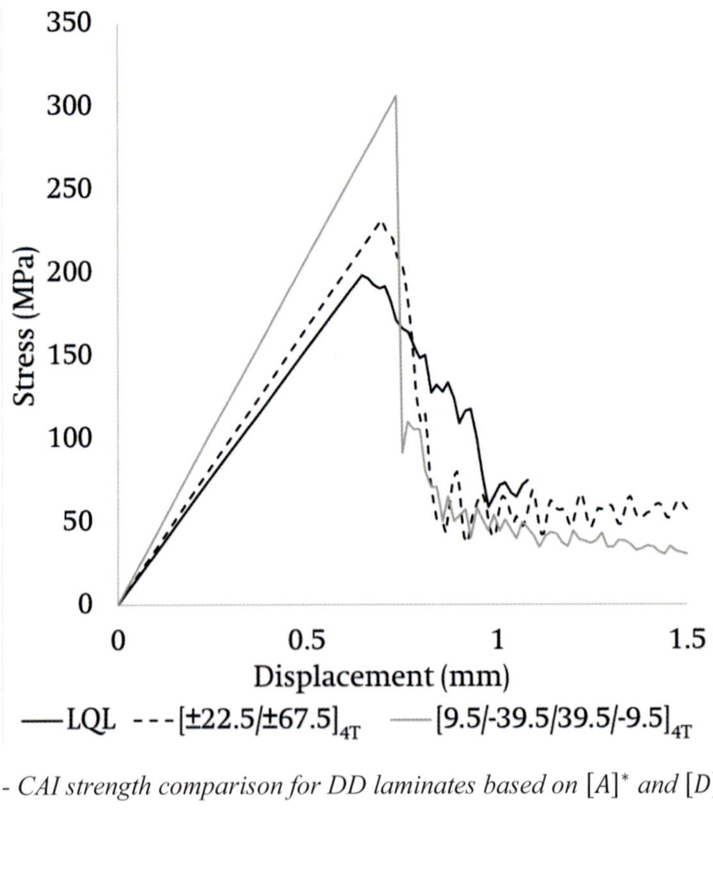

Figure 6.7 - CAI strength comparison for DD laminates based on $[A]^*$ and $[D]^*$, matrices

6.6.3 Specimen 1 - Comparison of DD results when removing sub-laminates

Focusing on the DD laminate results for the homogenised specimen optimised to match the bending stiffness of the LQL, ($[9.5/-39.5/39.5/-9.5]_{rT}$), these illustrate the effect of starting with a DD laminate equal to the thickness of the relevant LQL, consisting of four sub-laminates, before removing sub-laminates one by one. Initially, a DD laminate of the same thickness produced a higher peak force, +7.3%, Figure 6.8, when compared with the LQL. Removing one [9.5/-39.5/39.5/-9.5] sub-laminate produced a lower peak force than the first DD, but with a 32% increase in peak displacement over the experiment. Removing a second sub-laminate resulted in complete failure of the specimen under an impact load of 25J. Figure 6.9 shows the predicted delamination damage. Figure 6.8a and Figure 6.9c show that even with the removal of one sub-laminate, the DD specimen could still provide adequate resistance to LVI.

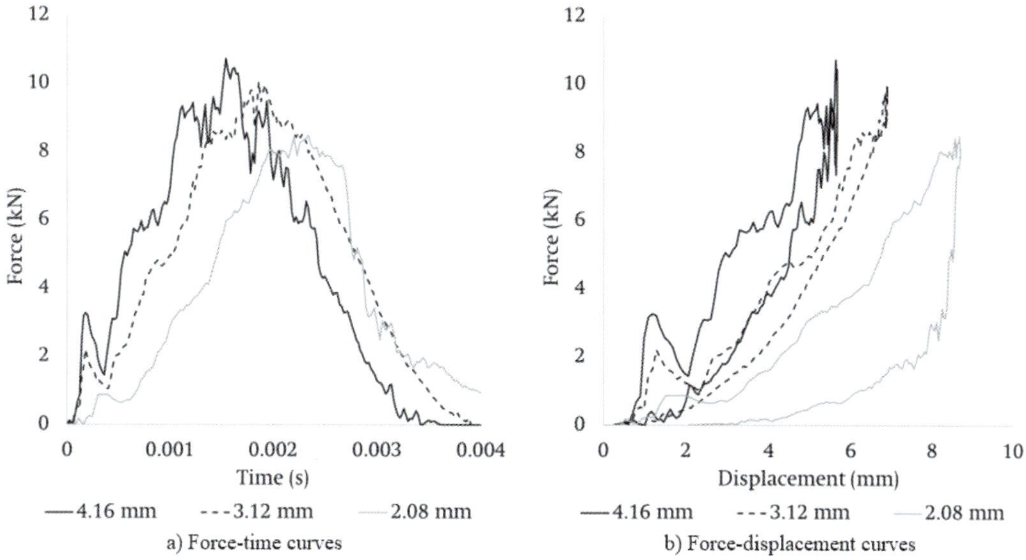

Figure 6.8 - Force-time/force-displacement plots showing the effect of the removal of one sub-laminate at a time for Specimen 1 ($[9.5/-39.5/39.5/-9.5]_{rT}$)

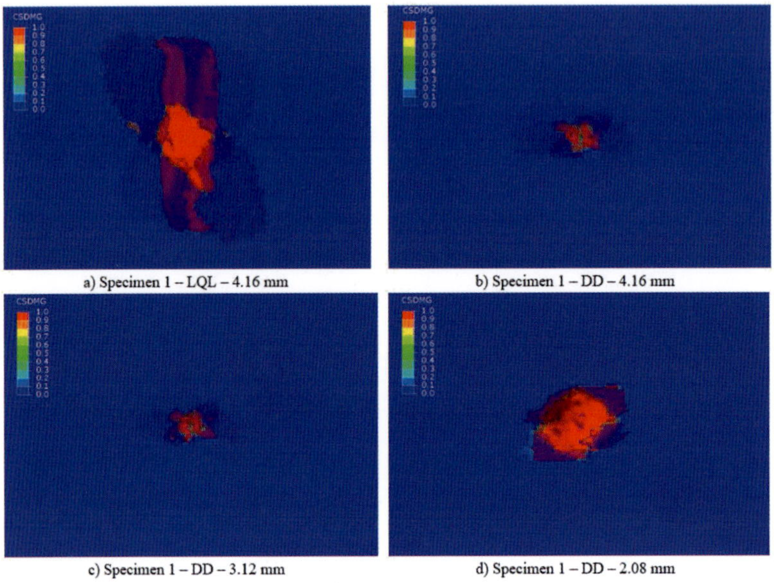

Figure 6.9 - Effect of the removal of one sub-laminate at a time for Specimen 1 ([9.5/-39.5/39.5/-9.5]$_{rT}$) - delamination predictions

When removing sub-laminates it is again important to consider the effect on membrane-bending coupling. Therefore, the change in $[B]^*$matrices was compared for the paired DD [±9.5/±39.5]$_{rT}$ and staggered DD [9.5/-39.5/39.5/-9.5]$_{rT}$ layups of Specimen 1, with progressive sub-laminate removal and results compared. Table 6.3 shows the values for the $[ABD]^*$matrices for each version of Specimen 1 i.e. $[0_2/45_2/90_2/\text{-}45_2]_s$, manufactured from T700GC/M21 with a ply thickness of 0.26 mm, while Figure 6.10 shows a graphical comparison of normalised $[B]^*$ matrix values for different DD layups of Specimen 1. It can be seen that coupling effects are larger than the LQL for the paired DD laminate [±9.5/±39.5]$_{rT}$ and with each sub-laminate removal the coupling effects progressively increase. However, results show that for the staggered DD laminate [9.5/-39.5/39.5/-9.5]$_{rT}$, coupling effects are effectively eliminated, even with the removal of two sub-laminates, since the laminate is approaching symmetry.

Table 6.3 - Specimen layups and $[ABD]^*$ values

	Specimen 1						
	LQL	DD	DD	DD	DD	DD	DD
(GPa)	(25/50/25)	$[\pm9.5/\pm39.5]_{4T}$	$[\pm9.5/\pm39.5]_{3T}$	$[\pm9.5/\pm39.5]_{2T}$	$[9.5/-39.5/39.5/-9.5]_{4T}$	$[9.5/-39.5/39.5/-9.5]_{3T}$	$[9.5/-39.5/39.5/-9.5]_{2T}$
A^*_{11}	55.0	89.0	89.0	89.0	89.0	89.0	89.0
A^*_{22}	55.0	19.0	19.0	19.0	19.0	19.0	19.0
A^*_{66}	19.1	20.1	20.1	20.1	20.1	20.1	20.1
A^*_{12}	16.8	17.8	17.8	17.8	17.8	17.8	17.8
$\lvert B^*_{11}\rvert$	0.0	4.4	5.9	8.9	0.0	0.0	0.0
$\lvert B^*_{22}\rvert$	0.0	1.4	1.8	2.7	0.0	0.0	0.0
$\lvert B^*_{66}\rvert$	0.0	1.5	2.0	3.1	0.0	0.0	0.0
$\lvert B^*_{12}\rvert$	0.0	1.5	2.0	3.1	0.0	0.0	0.0
D^*_{11}	89.2	89.0	89.0	89.0	90.6	91.9	95.6
D^*_{22}	31.5	19.0	19.0	19.0	18.5	18.1	17.0
D^*_{66}	13.7	20.1	20.1	20.1	19.5	19.1	17.8
D^*_{12}	11.5	17.8	17.8	17.8	17.3	16.8	15.5

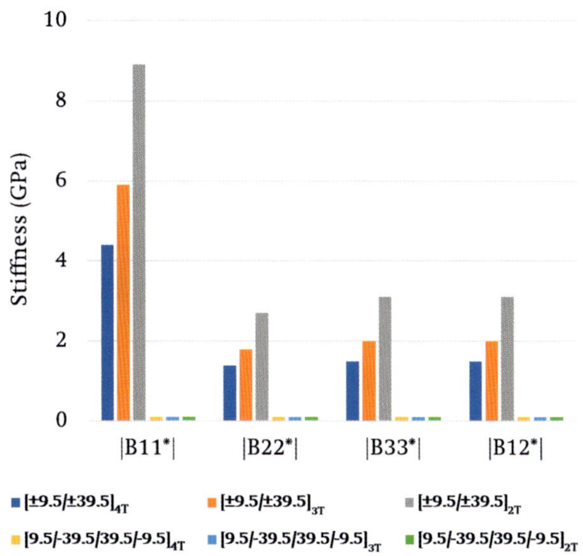

Figure 6.10 - Graphical comparison of normalised $[B]^*$ matrix values for different DD layups of Specimen 1

6.6.4 Specimen 1 - CAI results

Comparing CAI results for Specimen 1, Figure X.11 shows the superimposed intralaminar matrix and interlaminar delamination damage contours for two specimen variations (4.16 mm thick LQL and 3.12 mm thick DD $[9.5/-39.5/39.5/-9.5]_{3T}$ both with 0.26 mm thick plies), at ultimate failure. Compressive fibre-dominated damage contours, following CAI, are shown in Figure X.11e and f. Complete fibre failure was observed through the width of the LQL specimen, starting from the vicinity of the impact damage. However, there was significantly less fibre damage in the DD specimen.

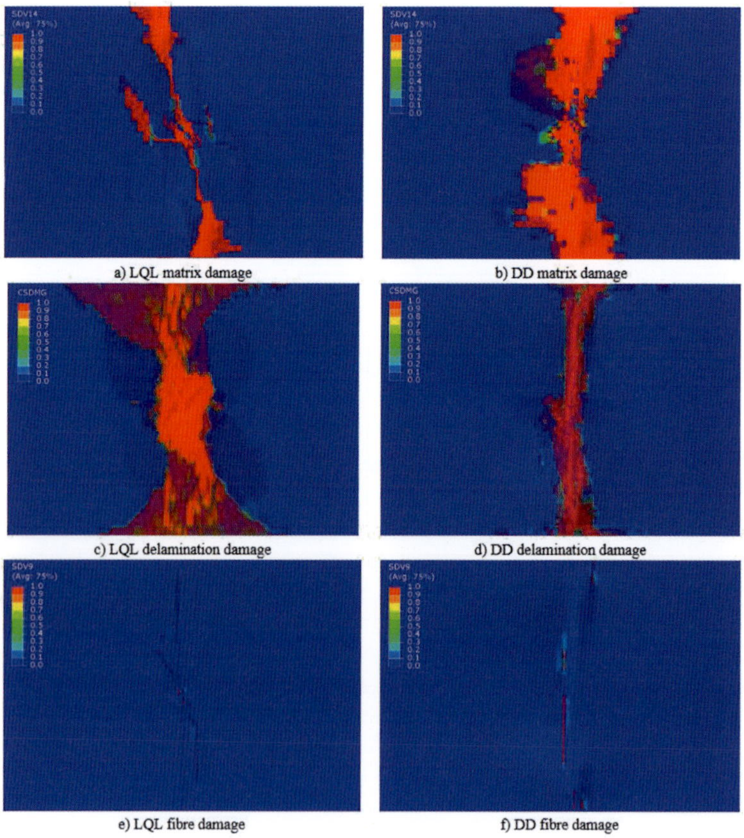

Figure X.11 - CAI results for Specimen 1

Figure 6.12 shows the stress-displacement curves for variations of Specimen 1. Figure 6.12a compares the residual strength predictions for LQL and DD $[9.5/-39.5/39.5/-9.5]_{4T}$ variations The presence of impact damage reduced the residual strength of both laminates. Therefore, the damaged LQL residual strength was 198 MPa while the damaged DD $[9.5/-39.5/39.5/-9.5]_{4T}$ residual strength was 55% higher at 306 MPa.

Figure 6.12b compares the $[9.5/-39.5/39.5/-9.5]_{4T}$ 4.16 mm thick DD with the thinner 3.12 mm thick DD $[9.5/-39.5/39.5/-9.5]_{3T}$. The residual strength for the damaged 3.12 mm thick DD was 321 MPa, 62% higher than the damaged LQL and 5% higher than the 4.16 mm thick DD laminate.

Therefore, the most impact resistant DD laminate for this case was $[9.5/-39.5/39.5/-9.5]_{3T}$ given that it had the lowest thickness, small impact damage predictions but had the highest specific residual strength. The use of a $[9.5/-39.5/39.5/-9.5]_{3T}$, 3.12 mm thick layup versus the LQL $[0_2/45_2/90_2/-45_2]_s$, 4.16 mm thick, reduced the 150 x 100 mm specimen mass by approximately 25%.

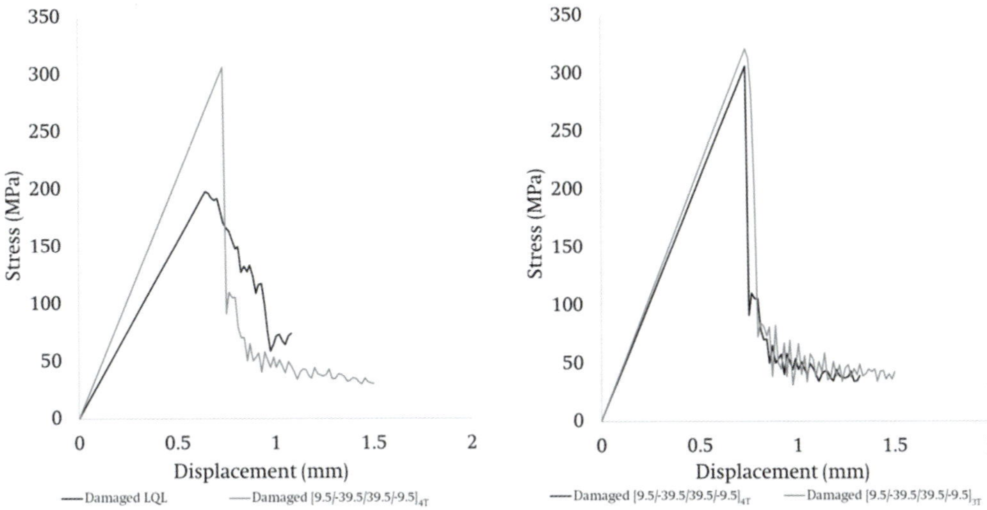

a) DD laminate thickness = LQL thickness (4.16 mm) b) DD laminate thickness < LQL thickness

Figure 6.12 - CAI stress-displacement curves for variations of Specimen 1

6.6.5 Specimen 1 – use of thinner plies

The results presented previously used the same ply thickness, 0.26 mm, for both the LQL and DD laminate. However, DD also presents the opportunity to use thinner plies. Therefore, further simulations were undertaken, reducing the DD ply thickness by half, to an intermediate thickness of 0.13 mm and again to 0.065 mm for thin plies, to investigate any change in behaviour arising from these significant changes in ply thickness. Using a ply thickness of 0.13 mm, a laminate thickness of 2.60 mm ($[9.5/-39.5/39.5/-9.5]_{5T}$) was required to prevent total failure under a 25J impact while using 0.065 mm plies required a laminate thickness of 1.82 mm $[9.5/-39.5/39.5/-9.5]_{7T}$. Figure 6.13 shows the CAI residual strength predictions for the LQL and all three DD laminates with ply thicknesses of 0.26 mm, 0.13 mm and 0.065 mm, respectively. These results show that use of thinner plies increased the CAI residual strength by 6% from 312 MPa to 332 MPa for 0.13 mm and 8% from 312 MPa to 337 MPa for 0.065 mm plies.

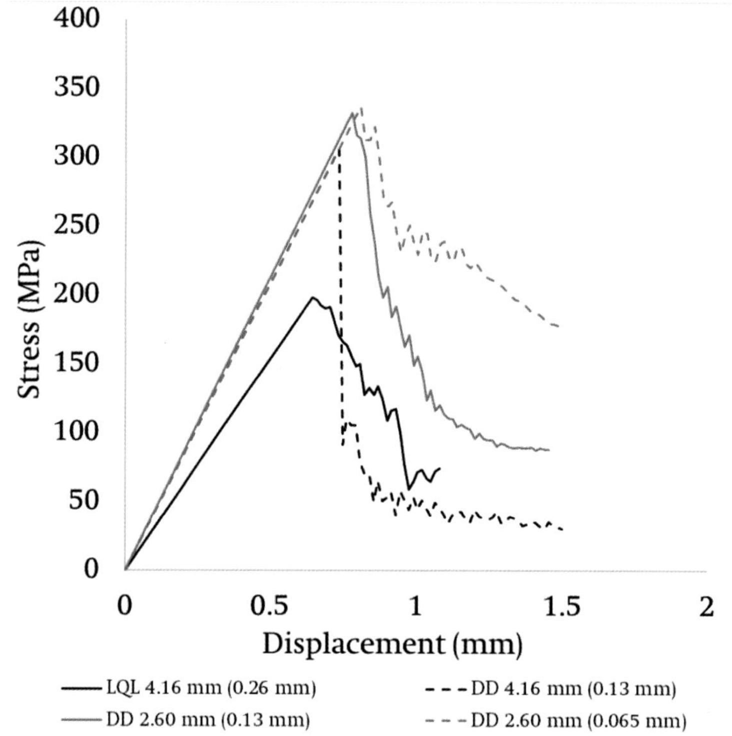

Figure 6.13 - Comparison of CAI plots for Specimen 1 with thick, intermediate and thin plies (numbers outside brackets represent laminate thickness, numbers in brackets represent ply thickness)

6.7 Results - Specimen 2

Specimen 2 had a stacking sequence of $[0/90/0_2/\pm45/0_3/\mp45/0_3/\pm45/0_2/90/0]$, was manufactured from IMS60 Carbon-fibre/epoxy and had a ply thickness of 0.267 mm. The upper and lower cross-plies (0/90) had a total thickness of 0.35 mm (0.175 mm each).

Results for Specimen 2 are shown in Figure 6.14 and Figure 6.15. Figure 6.14a shows how the original hybrid UD-woven specimen (experiment), equivalent UD LQL and DD laminate ($[0/-39/39/0]_{3T}$) force-time plots compare. These plots show that although the UD LQL predicted higher peak force, the DD laminate could closely match the hybrid UD-woven specimen but with lower laminate thickness (3.204 mm vs 5.34 mm thick). The peak force for this DD was within 1% of the hybrid UD-woven specimen. Figure 6.14b compares the CAI stress-displacement plots for the same three specimens. It can be seen that in this case, the DD laminate (318 MPa, +14%) and UD LQL (348 MPa, +25%) specimens had higher residual strengths than the hybrid UD-woven specimen, 279 MPa. This was due to the smaller damage after impact which can be seen in Figure 6.15. However, in this case the CAI strength of the DD laminate was 8% lower than the pure UD specimen. This is in contrast to Specimen 1 where the CAI strength of the DD laminate was significantly higher than the LQL. This may be explained by studying the $[A^*]$ matrices for this specimen. For example, for Specimen 1 A^*_{11} of the DD laminate was 45% larger than the LQL, however, for Specimen 2 A^*_{11} of the DD laminate was 0.5% lower than the LQL.

Comparing the predicted matrix and delamination damage plots for the hybrid UD/woven, equivalent UD LQL and DD laminates, Figure 6.15, delamination was smallest for the DD laminate, closely followed by the hybrid UD-woven specimen. While the matrix damage area for the DD laminate was larger than the UD LQL, both were much smaller than the hybrid laminate.

The use of a $[0/-39/39/0]_{3T}$ layup, 3.204 mm thick, versus the hybrid UD-woven laminate of $[5HS/0_2/\pm45/0_3/\mp45/0_3/\pm45/0_2/5HS]$, 5.34 mm thick, has the potential to reduce the specimen mass by approximately 40%, assuming a 150 x 100 mm specimen.

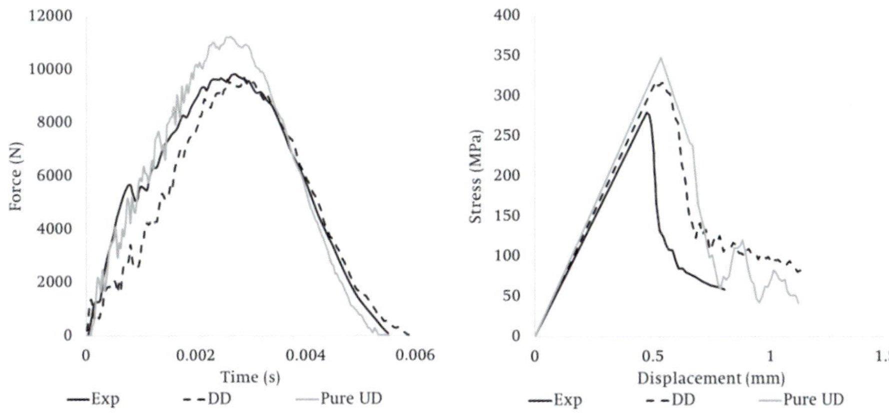

a) Force-time curves for LQL/DD/Pure UD

b) Stress-displacement CAI plots

Figure 6.14 - Impact and compression-after-impact plots for Specimen 2

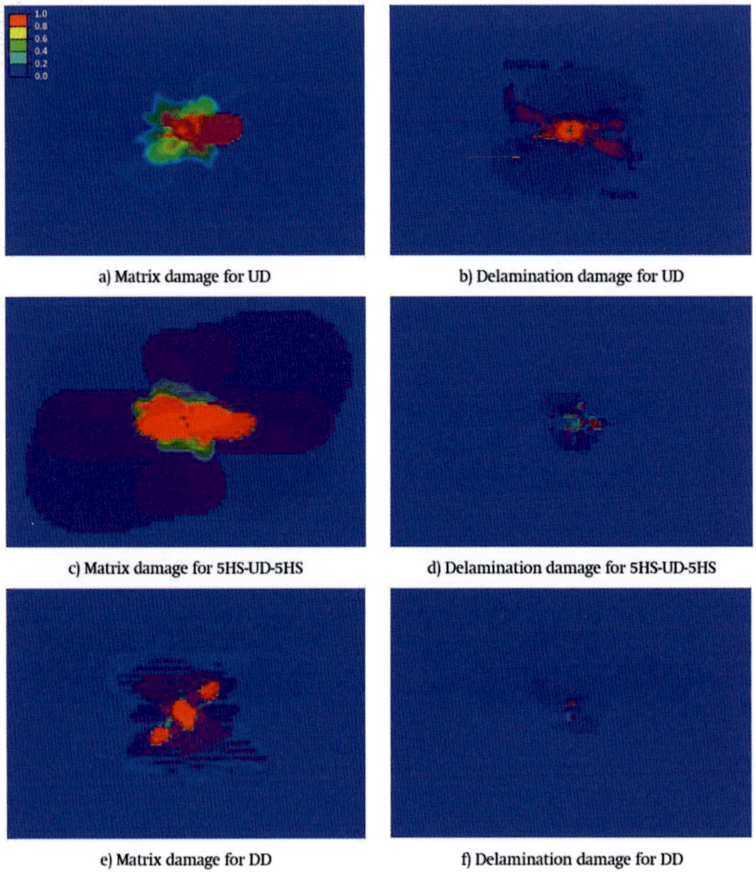

Figure 6.15 – Matrix and delamination predictions after impact for variants of Specimen 2

6.8 Conclusions

Double-double (DD) laminates have been shown to be capable of replacing legacy quad laminates (LQLs) effectively for either UD or hybrid woven/UD arrangements under both LVI and CAI loading conditions. Choosing a DD sub-laminate to closely match the normalised bending stiffness matrix, $[D]^*$ of the LQL produced significant improvements in laminate performance. However, careful consideration of the effects of membrane-coupling are required. It has been shown that matrix and delamination damage can be maintained or reduced when comparing DD and LQLs under LVI. Under CAI, the residual strength of the DD was close to or larger than the LQL for the cases studied herein, however, the magnitude of the increase varied depending on the material and angles used.

It was found that in all cases reported in this chapter, a thinner laminate made of a DD layup could achieve comparable or improved results over the LQL. For example, using the typical approach to select a DD sub-laminate using extensional stiffness matrix matching, CAI residual strength increased by 16% with a 25% reduction in specimen mass. However, for a design focussed on bending stiffness matching and homogenisation, A_{11}^* increased by 62% over the LQL and CAI residual strength increased by up to 62% with a 25% reduction in specimen mass. While this chapter has focussed on computational modelling, future work aims to further support these findings through an LVI/CAI experimental programme to compare legacy and DD layups and further validate the simulations in this work.

6.9 Acknowledgements

This study was conducted within the Belfast Maritime Consortium UKRI Strength in Places project, 'Decarbonisation of Maritime Transportation: A return to Commercial Sailing' led by Artemis Technologies, Project no. 107138. Significant knowledge transfer from Professor Stephen W. Tsai and the team at Stanford University is also gratefully acknowledged. Support and assistance by PhD student Aidan Hawkins is also acknowledged.

6.10 References

[1] Vermes B, Tsai SW, Riccio A, Di Caprio F, Roy S. Application of the Tsai's modulus and double-double concepts to the definition of a new affordable design approach for composite laminates. Compos Struct 2020:113246. https://doi.org/10.1016/j.compstruct.2020.113246.

[2] Tsai SW, Sharma N, Arteiro A, Roy S, Rainsberger B. Composite double-double and grid/skin structures. 1st ed. Composites Design Group, Aeronautics & Astronautics Department, Stanford University; 2019.

[3] Arteiro A, Sharma N, Melo JDD, Ha SK, Miravete A, Miyano Y, et al. A case for Tsai's Modulus, an invariant-based approach to stiffness. Compos Struct 2020;252:112683. https://doi.org/10.1016/j.compstruct.2020.112683.

[4] Melo JDD, Bi J, Tsai SW. A novel invariant-based design approach to carbon fiber reinforced laminates. Compos Struct 2017;159:44–52. https://doi.org/10.1016/j.compstruct.2016.09.055.

[5] Dantas R, Targino TG, Cardoso C, Paiva E, Carlos R, Freire S, et al. Low velocity impact response of non-traditional double-double laminates. J Compos Mater 2023. https://doi.org/10.1177/00219983231163513.

[6] Tsai SW. Double–double: New family of composite laminates. AIAA J 2021;59:4293–305. https://doi.org/10.2514/1.J060659.

[7] Nielsen MWD, Butler R, Rhead AT. Minimum mass laminate design for uncertain in-plane loading. Compos Part A Appl Sci Manuf 2018;115:348–59. https://doi.org/10.1016/j.compositesa.2018.09.028.

[8] Vermes B, Tsai SW, Massard T, Springer GS, Czigany T. Design of laminates by a novel '"double-double"' layup. Thin-Walled Struct 2021;165. https://doi.org/https://doi.org/10.1016/j.tws.2021.107954.

[9] Millen SLJ, Falzon BG, Aravand MA. Invariant Based Approaches in the Design of Composite Laminates. Compos Sci Technol 2021;202. https://doi.org/10.1016/j.compscitech.2020.108526.

[10] Millen SLJ, Aravand MA. Assessing the relative conservatism of the unit circle failure envelope. Compos Commun 2021;24. https://doi.org/10.1016/j.coco.2021.100690.

[11] AITM Airbus Test Method. Fiber Reinforced plastics determination of compression strength

after impact. 2010.

[12] Tan W, Falzon BG, Chiu LNS, Price M. Predicting low velocity impact damage and Compression-After-Impact (CAI) behaviour of composite laminates. Compos Part A Appl Sci Manuf 2015;71:212–26. https://doi.org/10.1016/j.compositesa.2015.01.025.

[13] Liu H, Falzon BG, Tan W. Experimental and numerical studies on the impact response of damage-tolerant hybrid unidirectional/woven carbon-fibre reinforced composite laminates. Compos Part B Eng 2018;136:101–18. https://doi.org/10.1016/j.compositesb.2017.10.016.

[14] Millen SLJ, Ullah Z, Falzon BG. On the importance of finite element mesh alignment along the fibre direction for modelling damage in fibre-reinforced polymer composite laminates. Compos Struct 2021;278. https://doi.org/10.1016/j.compstruct.2021.114694.

[15] Catalanotti G, Camanho PP, Marques AT. Three-dimensional failure criteria for fiber-reinforced laminates. Compos Struct 2013;95:63–79. https://doi.org/10.1016/j.compstruct.2012.07.016.

[16] Chiu LNS, Falzon BG, Boman R. A continuum damage mechanics model for the analysis of the crashworthiness of composite structures: a work in progress. Proc. 15th Aust. Aeronaut. Conf., Melbourne, Australia: 2013.

[17] Tan W, Falzon BG. Modelling the nonlinear behaviour and fracture process of AS4/PEKK thermoplastic composite under shear loading. Compos Sci Technol 2016;126:60–77. https://doi.org/10.1016/j.compscitech.2016.02.008.

[18] Faggiani A, Falzon BG. Predicting low-velocity impact damage on a stiffened composite panel. Compos Part A Appl Sci Manuf 2010;41:737–49. https://doi.org/10.1016/j.compositesa.2010.02.005.

[19] Falzon BG, Apruzzese P. Numerical analysis of intralaminar failure mechanisms in composite structures. Part I: FE implementation. Compos Struct 2011;93:1039–46. https://doi.org/10.1016/j.compstruct.2010.06.028.

[20] Tan W, Falzon BG. Modelling the crush behaviour of thermoplastic composites. Compos Sci Technol 2016;134:57–71. https://doi.org/10.1016/j.compscitech.2016.07.015.

[21] Tan W, Falzon BG, Price M. Predicting the crushing behaviour of composite material using high-fidelity finite element modelling. Int J Crashworthiness 2015;20:60–77. https://doi.org/10.1080/13588265.2014.972122.

[22] Varandas LF, Catalanotti G, Melro AR, Falzon BG. On the importance of nesting considerations for accurate computational damage modelling in 2D woven composite materials. Comput Mater Sci 2020;172. https://doi.org/10.1016/j.commatsci.2019.109323.

[23] Hongkarnjanakul N, Bouvet C, Rivallant S. Validation of low velocity impact modelling on different stacking sequences of CFRP laminates and influence of fibre failure. Compos Struct 2013;106:549–59. https://doi.org/10.1016/j.compstruct.2013.07.008.

[24] Liu H, Falzon BG, Tan W. Predicting the Compression-After-Impact (CAI) strength of damage-tolerant hybrid unidirectional/woven carbon-fibre reinforced composite laminates. Compos Part A Appl Sci Manuf 2018;105:189–202. https://doi.org/10.1016/j.compositesa.2017.11.021.

CHAPTER 7

Interlaminar Shear Stress and Failure of Quad and Double-double Laminates

K.M. Yeoh[a], J.L. Liu[a], T.E. Tay[a], V.B.C Tan*[a]

a Department of Mechanical Engineering, National University of Singapore, Singapore

**Corresponding author, email address: mpetanbc@nus.edu.sg*

7.1 Introduction

Composites are gaining widespread use as structural materials for their high strength- and stiffness-to-weight ratios. However, their complex failure mechanisms, including interlaminar failures such as delamination and matrix cracking, need further study. Apart from stacking sequence (i.e., layup), the strength and stiffness of individual plies, the mechanical behaviour of composite laminates may also be influenced by interlaminar fracture toughness and strength.
Recently, we have shown that the interlaminar fracture toughness and strength are influenced by the difference in the fibre orientations in adjacent plies, also known as the interply mismatch angle [1]. This influence is discussed in this article in the context of quads and recently proposed double-double (DD) laminates.

Quads are a traditional group of laminates commonly used in the aerospace industry. They consist of 0°, 45°, -45° and 90° plies in various proportions and arrangements. Due to their relatively long history and frequency of use in the industry, they are fairly well studied and various rules have been devised by regulatory authorities. Examples include symmetric stacking and the 10% rule where the laminate must have at least 10% of each of the four base plies. Constrained by the restrictive nature of the rules, the design of quad laminates is less flexible and typically require more plies.

DDs are a recently proposed family of laminate configurations made by repeating four-ply subunits which consist of two pairs of angle-plies in various arrangements, e.g., [+Φ/-Φ/+Ψ/-Ψ] [2]. In contrast to quads, DDs offer a more flexible design approach due to the unrestricted ply angles. Manufacturing of DDs by repeatedly stacking the subunits would be simpler, faster and less prone to error compared to quads. Furthermore, ply dropping or tapering is also easier with DDs and can be achieved with fewer plies overall. Nevertheless, further studies are required to better understand their behaviour and design as well as to establish design codes for industrial adoption.

Here, a simple procedure to estimate the interlaminar shear (ILS) stress is presented, and the results discussed in the context of comparison between quads and DD laminates, with the implication of taking into account mismatch angles between plies.

7.2 Estimating interlaminar shear stresses

Consider the free body diagram of a laminated composite plate element with stresses acting along the x-direction as shown in Fig. 7.1 below.

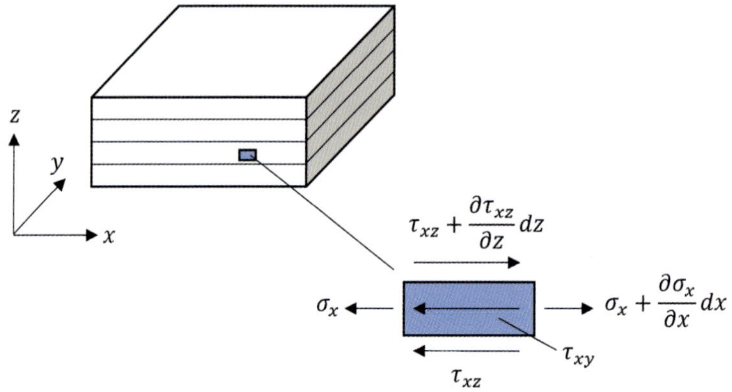

Fig. 7.1 Free body diagram of a laminated composite plate element with stresses in the x-direction

From the free body diagram, the gradient of the ILS stress τ_{xz} along the height of the element can be obtained from the equilibrium equation,

$$\frac{\partial \tau_{xz}}{\partial z} = -\frac{\partial \sigma_x}{\partial x} - \frac{\partial \tau_{xy}}{\partial y} \tag{7.1}$$

The expression for τ_{yz} is similarly obtained,

$$\frac{\partial \tau_{yz}}{\partial z} = -\frac{\partial \sigma_y}{\partial y} - \frac{\partial \tau_{xy}}{\partial x} \tag{7.2}$$

Combining Eqs. (7.1) and (7.2),

$$\frac{\partial}{\partial z}\begin{Bmatrix} \tau_{xz} \\ \tau_{yz} \end{Bmatrix} = -\partial \begin{Bmatrix} \sigma_x \\ \sigma_y \\ \tau_{xy} \end{Bmatrix} \tag{7.3}$$

where

$$\partial = \begin{bmatrix} \frac{\partial}{\partial x} & 0 & \frac{\partial}{\partial y} \\ 0 & \frac{\partial}{\partial y} & \frac{\partial}{\partial x} \end{bmatrix} \tag{7.4}$$

Therefore, the ILS stresses at interface K can be expressed as

$$\begin{Bmatrix}\tau_{xz}\\ \tau_{yz}\end{Bmatrix}_K = \begin{Bmatrix}\tau_{xz}\\ \tau_{yz}\end{Bmatrix}_{K-1} - \partial\left(\int_{z_{K-1}}^{z_K} \boldsymbol{\sigma}_K dz\right) \tag{7.5}$$

where $\{\boldsymbol{\sigma}\}_k = \begin{Bmatrix}\sigma_x\\ \sigma_y\\ \tau_{xy}\end{Bmatrix}_K$ denotes the in-plane stresses of ply K between interfaces K and $K-1$. The interfaces have positions z_k and z_{K-1} respectively.

The in-plane stresses of each ply in a composite plate can be obtained from Classical Laminate Theory (CLT). In CLT, the mechanical response of a composite plate is governed by the ABD matrix.

$$\begin{Bmatrix}\boldsymbol{N}\\ \boldsymbol{M}\end{Bmatrix} = \begin{bmatrix}\boldsymbol{A} & \boldsymbol{B}\\ \boldsymbol{B} & \boldsymbol{D}\end{bmatrix}\begin{Bmatrix}\boldsymbol{\varepsilon}^{0}\\ \boldsymbol{\kappa}\end{Bmatrix} \tag{7.6}$$

where $\boldsymbol{N}$ is the in-plane load vector, $\boldsymbol{M}$ is the moment vector, $\boldsymbol{\varepsilon}^{0}$ is the mid-plane strain vector and $\boldsymbol{\kappa}$ is the curvature vector. Details on the ABD matrix and CLT can be found in many textbooks [3].

The mid-plane strains and curvatures can be obtained from Eq. (7.6) by inverting the ABD matrix.

$$\begin{Bmatrix}\boldsymbol{\varepsilon}^{0}\\ \boldsymbol{\kappa}\end{Bmatrix} = \begin{bmatrix}\boldsymbol{A}' & \boldsymbol{B}'\\ \boldsymbol{C}' & \boldsymbol{D}'\end{bmatrix}\begin{Bmatrix}\boldsymbol{N}\\ \boldsymbol{M}\end{Bmatrix} \tag{7.7}$$

The in-plane stresses at each ply can then be obtained.

$$\boldsymbol{\sigma}_K = \overline{\boldsymbol{Q}}_K \boldsymbol{\varepsilon}_K \tag{7.8}$$

where

$$\boldsymbol{\varepsilon} = \boldsymbol{\varepsilon}^{0} + z\boldsymbol{\kappa} \tag{7.9}$$

Combining Eqs. (7.7) – (7.9),

$$\boldsymbol{\sigma}_K = \overline{\boldsymbol{Q}}_K\left[\left(\boldsymbol{A}' + z\boldsymbol{C}'\right)\boldsymbol{N} + \left(\boldsymbol{B}' + z\boldsymbol{D}'\right)\boldsymbol{M}\right] \tag{7.10}$$

The expressions for $\boldsymbol{\sigma}$ of the plies involved can then be substituted into Eq. (7.5) to obtain the ILS stress at the interface of interest.

The equations above have been coded into a simple programme as well as a spreadsheet calculator to calculate the ILS stresses of composites with any given layup. Full details, including access to the spreadsheet, are provided in Appendix C.

7.3 Interlaminar shear stresses of quad laminates

The proposed procedure is first used to analyse the ILS stresses of two different quad laminates.

a) Quad 1 (Q1) – $[0/45/90/-45]_{3s}$
b) Quad 2 (Q2) – $[90/90/90/45/-45/45/-45/45/-45/0/0/0]_s$

For this analysis, the properties of a T700/2510 carbon-epoxy lamina will be used and are given in Table 7.1 below.

Table 7.1 Geometric and material properties of the T700/2510 carbon-epoxy lamina

Property	Value
Thickness per lamina, t	$63\mu m$
Longitudinal modulus, E_1	107GPa
Transverse modulus, E_2	8GPa
Shear modulus, G_{12}	4GPa
Major Poisson's ratio, ν_{12}	0.32

We consider beams that are subjected to a transverse shear stress as shown in

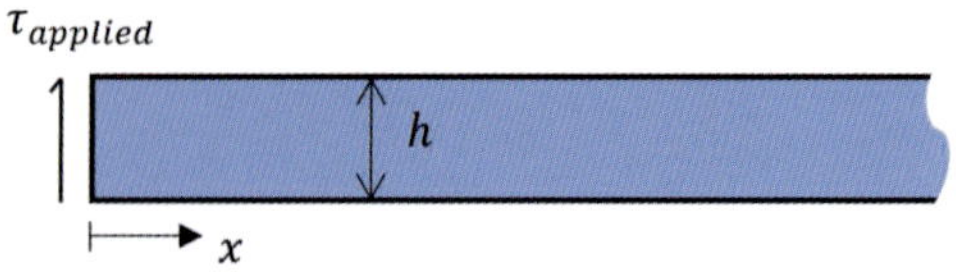

Fig. 7.2 below.

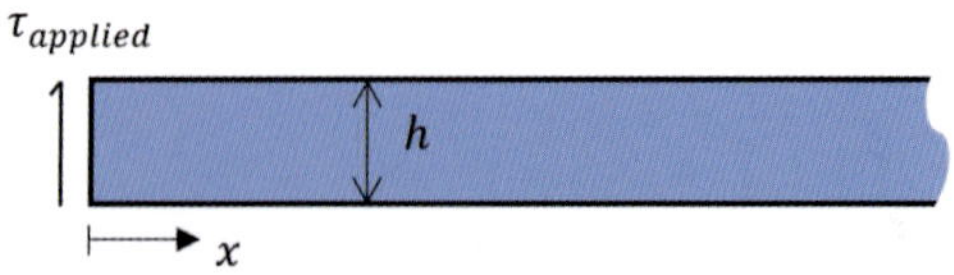

Fig. 7.2 24-ply composite beam subjected to a transverse shear stress

For this simple transverse shear stress, the load vectors are given as

$$\boldsymbol{N} = \begin{Bmatrix} 0 \\ 0 \\ 0 \end{Bmatrix} \quad \boldsymbol{M} = \begin{Bmatrix} -Px \\ 0 \\ 0 \end{Bmatrix} \tag{7.11}$$

where $P = h\tau_{applied}$.

The stresses in each ply can then be obtained by substituting Eq. (7.11) into Eq. (7.10).

$$\boldsymbol{\sigma}_k = -\overline{\boldsymbol{Q}}_k \left(\begin{Bmatrix} B'_{11} \\ B'_{21} \\ B'_{61} \end{Bmatrix} + z \begin{Bmatrix} D'_{11} \\ D'_{21} \\ D'_{61} \end{Bmatrix} \right) (Px) \tag{7.12}$$

Ch.7

From Eq. (7.12), it can be seen that all the ply stresses are merely dependent on x and z. As such, the expression for the dominant ILS stress τ_{xz} will only consist of the first term from Eq. (7.1).

$$\begin{aligned}
(\tau_{xz})_K &= (\tau_{xz})_{K-1} - \int_{z_{K-1}}^{z_K} \left(\frac{\partial(\sigma_x)}{\partial x} \right)_K dz \\
&= (\tau_{xz})_{K-1} + P[\overline{Q}_{11} \quad \overline{Q}_{12} \quad \overline{Q}_{16}]_K \int_{z_{K-1}}^{z_K} \left(\begin{Bmatrix} B'_{11} \\ B'_{21} \\ B'_{61} \end{Bmatrix} + z \begin{Bmatrix} D'_{11} \\ D'_{21} \\ D'_{61} \end{Bmatrix} \right) dz \\
&= (\tau_{xz})_{K-1} + Pt \left([\overline{Q}_{11} \quad \overline{Q}_{12} \quad \overline{Q}_{16}]_K \left(\begin{Bmatrix} B'_{11} \\ B'_{21} \\ B'_{61} \end{Bmatrix} + \frac{z_k + z_{K-1}}{2} \begin{Bmatrix} D'_{11} \\ D'_{21} \\ D'_{61} \end{Bmatrix} \right) \right)
\end{aligned} \tag{7.13}$$

where $t = z_K - z_{K-1}$ is the ply thickness.

From Eq. (7.13) above, the ILS stress at any interface of interest in the composite beam can then be obtained. Fig. 7.3 below shows the predicted ILS stresses at each interface of the beams, in terms of a scaling factor, $\tau_{xz}/\tau_{applied}$. The ILS stress is obtained by multiplying the applied transverse shear stress by this scaling factor. The detailed results are shown in Table 7.6 in Appendix A.

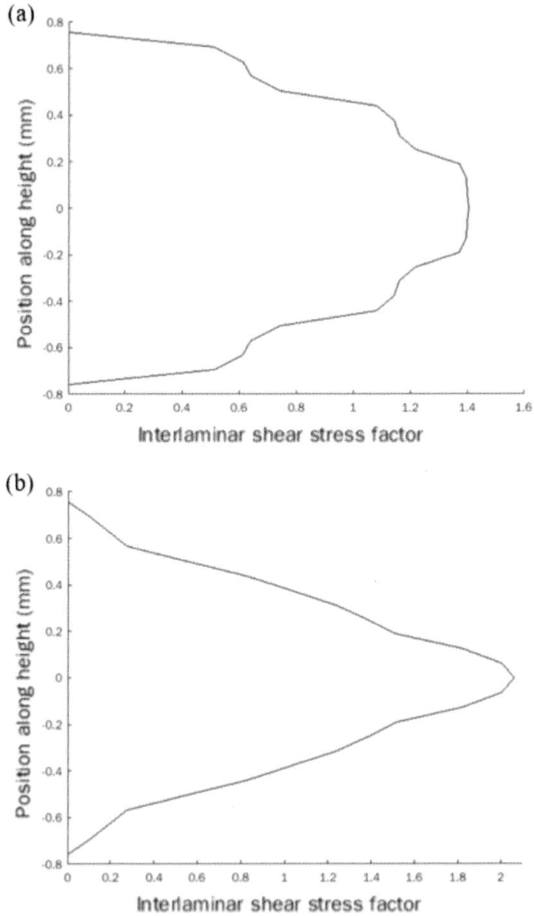

Fig. 7.3 ILS stress factor distribution along the thickness of the beams (a) Q1; (b) Q2

From the plots in Fig. 7.3 above, the ILS stresses of both beams show a generally increasing trend that peaks at the mid-plane of the beam. In contrast to the parabolic distribution expected in homogeneous beams, the ILS stresses in these beams show piece-wise distributions. This is attributed to the laminated nature of composite beams. The stiffness changes between adjacent plies of different orientations result in a piece-wise distribution of the normal bending stresses across the thickness of the laminate although the strains are assumed to vary linearly across the thickness in CLT. As a result, the ILS stresses, which is the sum of the gradient of the normal stresses across the plies involved, would vary in a piece-wise manner as well.

More importantly, it is noted that variations in the layup configuration can result in significant differences in the ILS stresses. Most notably, although both the Q1 and Q2 beams have the same number of 0°, 45°, -45° and 90° plies, the differences in their respective ply arrangements have resulted in significantly different stress magnitudes, with the peak stress factors of the Q1 and Q2 beams being 1.406 and 2.063 respectively. This is because the Q2 beam has its stiffer plies closer to the mid-plane of the beam, while the Q1 beam has its plies of varying stiffness distributed more evenly along its thickness, resulting in a reduced flexure stiffness and larger flexure deformation in the Q2 beam for the same applied load.

The calculated ILS stresses were compared to finite element (FE) predictions for validation in Appendix B.

7.4 Interlaminar shear stresses of DD laminates

The analysis is then repeated for two different DD laminates. The same material properties for a T700/2510 carbon-epoxy lamina, ply thickness and loads as shown in Table 7.1 and Fig. 7.2 will be used.

a) DD 1 (DD1) – $[10/60/-60/-10]_6$
b) DD 2 (DD2) – $[20/80/-20/-80]_6$

The predicted ILS stress factors for both DD beams are shown in Fig. 7.4 below, and the detailed results are shown in Table 7.6 in Appendix A.

(a)

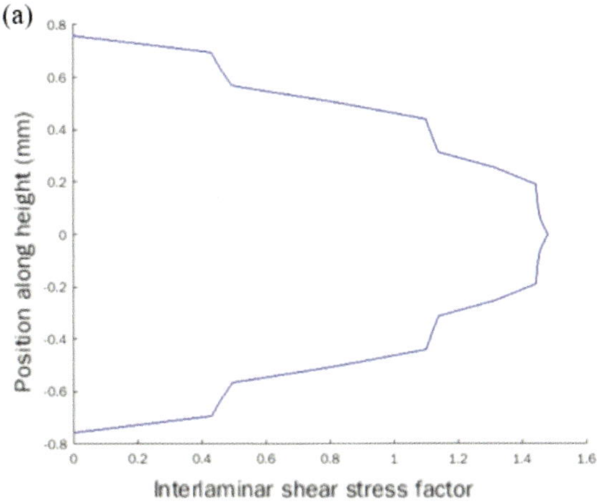

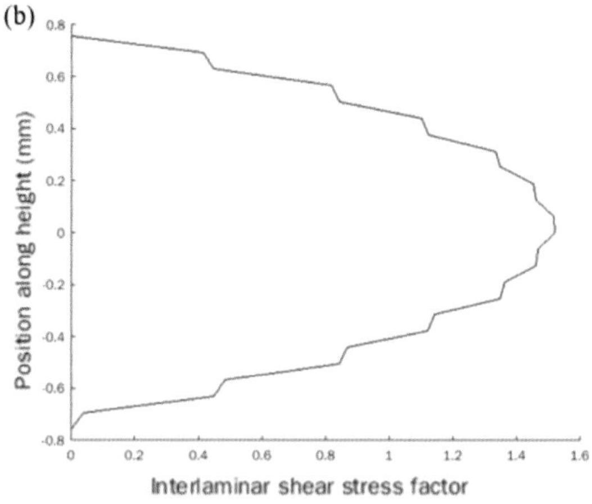

Fig. 7.4 ILS stress factor distribution along the thickness of the beams (a) DD1; (b) DD2

As seen from the plots, the ILS stress factors in both DD beams also increase in a piece-wise manner and peak at the mid-plane of the beams. The ILS stress factor increments along the thickness of the DD1 beam is much larger than the increments in the DD2 beam. This is due to the angle-plies being adjacent to each other in the DD1 beam subunit. These angle-ply pairs have similar stiffness when subjected to bending along the y-axis as considered in this example problem, effectively behaving as a single ply with twice the thickness. In the DD2 beam, the angle-ply pairs are instead separated from each other.

More interestingly, both the DD1 and DD2 beams show similar peak ILS stress factors, with values of 1.481 and 1.518 respectively. This is in stark contrast to the quad beams, especially since the DD beams consist of different angle-ply pairs and different four-ply subunit arrangements. These values suggest that DD beams with similar geometries and loads may show similar ILS stress distributions at their interfaces.

The same analysis is then extended to all possible 24-ply DD configurations. The peak ILS stress factors of each configuration are shown in Fig. 7.5, colour coded by magnitude with yellow being the smallest, followed by green, and blue the largest.

Paired 1 $[\Phi/-\Phi/\Psi/-\Psi]_6$

	0	10	20	30	40	50	60	70	80	90	Φ
0	1.500	1.501	1.503	1.509	1.516	1.521	1.524	1.525	1.525	1.525	
10	1.501	1.503	1.505	1.510	1.517	1.521	1.524	1.525	1.526	1.526	
20	1.503	1.505	1.505	1.508	1.515	1.523	1.525	1.525	1.526	1.527	
30	1.509	1.510	1.508	1.505	1.509	1.522	1.526	1.524	1.524	1.525	
40	1.516	1.517	1.515	1.509	1.504	1.507	1.518	1.518	1.519	1.520	
50	1.521	1.521	1.523	1.522	1.507	1.502	1.504	1.508	1.511	1.512	
60	1.524	1.524	1.525	1.526	1.518	1.504	1.501	1.502	1.504	1.506	
70	1.525	1.525	1.525	1.524	1.518	1.508	1.502	1.500	1.501	1.501	
80	1.525	1.526	1.526	1.524	1.519	1.511	1.504	1.501	1.500	1.500	
90	1.525	1.526	1.527	1.525	1.520	1.512	1.506	1.501	1.500	1.500	
Ψ											

Paired 2 $[\Phi/-\Phi/-\Psi/\Psi]_6$

	0	10	20	30	40	50	60	70	80	90	Φ
0	1.500	1.501	1.503	1.509	1.516	1.521	1.524	1.525	1.525	1.525	
10	1.501	1.500	1.502	1.507	1.516	1.523	1.525	1.526	1.526	1.526	
20	1.503	1.502	1.500	1.503	1.516	1.525	1.527	1.527	1.527	1.527	
30	1.509	1.507	1.503	1.500	1.509	1.523	1.526	1.526	1.525	1.525	
40	1.516	1.516	1.516	1.509	1.500	1.512	1.520	1.521	1.520	1.520	
50	1.521	1.523	1.525	1.523	1.512	1.500	1.506	1.511	1.512	1.512	
60	1.524	1.525	1.527	1.526	1.520	1.506	1.500	1.502	1.505	1.506	
70	1.525	1.526	1.527	1.526	1.521	1.511	1.502	1.500	1.501	1.501	
80	1.525	1.526	1.527	1.525	1.520	1.512	1.505	1.501	1.500	1.500	
90	1.525	1.526	1.527	1.525	1.520	1.512	1.506	1.501	1.500	1.500	
Ψ											

Staggered 1 $[\Phi/-\Psi/-\Phi/\Psi]_6$

	0	10	20	30	40	50	60	70	80	90	Φ
0	1.500	1.504	1.507	1.507	1.507	1.506	1.506	1.506	1.506	1.506	
10	1.504	1.500	1.501	1.502	1.503	1.507	1.509	1.510	1.512	1.513	
20	1.507	1.501	1.500	1.501	1.507	1.513	1.515	1.517	1.519	1.520	
30	1.507	1.502	1.501	1.500	1.506	1.516	1.519	1.520	1.521	1.522	
40	1.507	1.503	1.507	1.506	1.500	1.509	1.517	1.519	1.519	1.519	
50	1.506	1.507	1.513	1.516	1.509	1.500	1.505	1.510	1.513	1.513	
60	1.506	1.509	1.515	1.519	1.517	1.505	1.500	1.502	1.506	1.506	
70	1.506	1.510	1.517	1.520	1.519	1.510	1.502	1.500	1.501	1.501	
80	1.506	1.512	1.519	1.521	1.519	1.513	1.506	1.501	1.500	1.500	
90	1.506	1.513	1.520	1.522	1.519	1.513	1.506	1.501	1.500	1.500	
Ψ											

Ch.7

Staggered 2 [Φ/Ψ/-Φ/-Ψ]$_6$

	0	10	20	30	40	50	60	70	80	90	Φ
0	1.500	1.504	1.507	1.507	1.507	1.506	1.506	1.506	1.506	1.506	
10	1.504	1.512	1.515	1.515	1.513	1.511	1.510	1.510	1.511	1.513	
20	1.507	1.515	1.520	1.520	1.517	1.515	1.514	1.515	1.518	1.520	
30	1.507	1.515	1.520	1.520	1.518	1.516	1.515	1.516	1.519	1.522	
40	1.507	1.513	1.517	1.518	1.516	1.513	1.512	1.514	1.517	1.519	
50	1.506	1.511	1.515	1.516	1.513	1.508	1.507	1.508	1.511	1.513	
60	1.506	1.510	1.514	1.515	1.512	1.507	1.503	1.503	1.504	1.506	
70	1.506	1.510	1.515	1.516	1.514	1.508	1.503	1.500	1.501	1.501	
80	1.506	1.511	1.518	1.519	1.517	1.511	1.504	1.501	1.500	1.500	
90	1.506	1.513	1.520	1.522	1.519	1.513	1.506	1.501	1.500	1.500	
Ψ											

Staggered 3 [Φ/-Ψ/Ψ/-Φ]$_6$

	0	10	20	30	40	50	60	70	80	90	Φ
0	1.500	1.509	1.520	1.525	1.527	1.527	1.527	1.527	1.527	1.527	
10	1.500	1.503	1.512	1.519	1.523	1.525	1.526	1.528	1.529	1.529	
20	1.497	1.497	1.505	1.513	1.520	1.525	1.528	1.529	1.530	1.530	
30	1.489	1.489	1.496	1.505	1.515	1.524	1.528	1.529	1.530	1.529	
40	1.481	1.481	1.488	1.495	1.504	1.515	1.523	1.525	1.526	1.525	
50	1.476	1.478	1.485	1.491	1.495	1.502	1.511	1.517	1.519	1.518	
60	1.474	1.478	1.488	1.495	1.497	1.497	1.501	1.506	1.509	1.509	
70	1.474	1.481	1.494	1.502	1.504	1.502	1.499	1.500	1.502	1.502	
80	1.474	1.486	1.502	1.510	1.511	1.508	1.502	1.500	1.500	1.500	
90	1.474	1.489	1.506	1.513	1.514	1.509	1.503	1.500	1.500	1.500	
Ψ											

Staggered 4 [Φ/Ψ/-Ψ/-Φ]$_6$

	0	10	20	30	40	50	60	70	80	90	Φ
0	1.500	1.509	1.520	1.525	1.527	1.527	1.527	1.527	1.527	1.527	
10	1.500	1.512	1.523	1.529	1.530	1.530	1.529	1.529	1.529	1.529	
20	1.497	1.508	1.520	1.528	1.531	1.531	1.530	1.529	1.530	1.530	
30	1.489	1.498	1.510	1.520	1.527	1.529	1.529	1.528	1.528	1.529	
40	1.481	1.488	1.499	1.507	1.516	1.522	1.524	1.524	1.524	1.525	
50	1.476	1.483	1.492	1.498	1.503	1.508	1.514	1.516	1.517	1.518	
60	1.474	1.481	1.491	1.498	1.499	1.500	1.503	1.506	1.508	1.509	
70	1.474	1.482	1.495	1.502	1.503	1.501	1.499	1.500	1.502	1.502	
80	1.474	1.486	1.501	1.509	1.509	1.506	1.501	1.500	1.500	1.500	
90	1.474	1.489	1.506	1.513	1.514	1.509	1.503	1.500	1.500	1.500	
Ψ											

Fig. 7.5 Peak ILS stress factors for all possible 24-ply DD laminates

From the results in Fig. 7.5 above, it is noted that all the 24-ply DD laminates have similar peak ILS stress factors, with values of 1.5 ± 1.8%. This similarity indicates a uniformity across the DD laminates in terms of their ILS stress distributions, which can be attributed to the homogenised behaviour of DD laminates. Given enough repetitions of the four-ply subunits, the behaviour of DD laminates are expected to tend towards that of a homogeneous beam, commonly termed 'homogenised'. In this case, six repeats appear to be sufficient for the DD laminates to be homogenised. This is also observed to some extent in the Q1 beam above, where the plies of various stiffness are distributed evenly across the height of the beam.

Given that the transverse shear stress along the thickness of a homogeneous beam is known to have a parabolic distribution that peaks at the mid-plane and is independent of the material properties, it can potentially be used as another measure of homogenisation. This will be further discussed in the subsequent section. For a homogeneous beam with a rectangular cross-section subjected to a simple shear load, the peak ILS stress factor can be shown to be equal to 1.5, which is close to that of the DD beams.

The uniformity in the ILS stress distributions of DD laminates enables a simplified consideration regarding its design against interlaminar failure. Given sufficient repeats, the ILS stress distribution in DD laminates is expected to be similar to that of a homogeneous beam, which can be obtained from simple well-known equations. This means that only the interply mismatch angles, which result from the choice of angle-plies in the subunit, now need to be considered to predict and design against the onset of interlaminar failure in homogenised DD laminates.

7.5 A proposed measure of homogenisation

The transverse shear stresses of homogeneous beams are known to have a parabolic distribution that peaks at the mid-plane of the beam. This stress distribution is solely dependent on the cross-section of the beam. In other words, it is independent of the material properties for homogeneous beams.

As DD laminates approach a homogenised state with sufficient repeats of the four-ply subunit, the distribution of the transverse shear stresses should approach that of a homogeneous beam of the same geometry and load. This can be observed by the plot of ILS stress factors for the DD beams in Fig. 7.4, which approximate a kinked parabola. As such, how closely the distribution of ILS stresses in a DD laminate (or any other composite layup) resembles the transverse shear stress distribution of a homogeneous beam could be used as a measure of homogenisation.

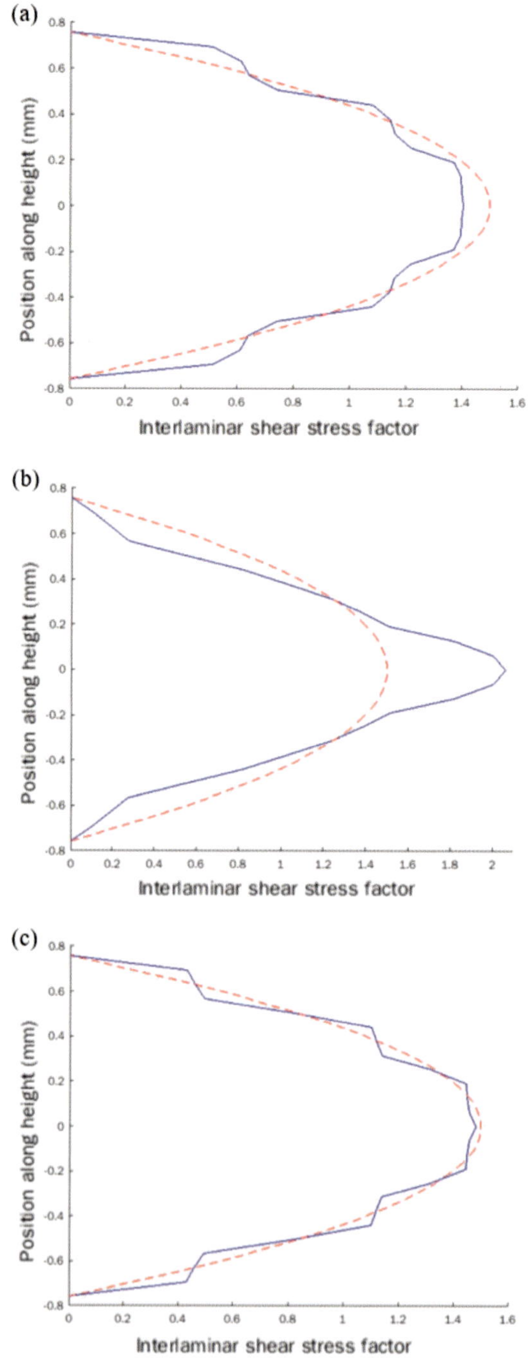
(a)
Position along height (mm)
Interlaminar shear stress factor
(b)
Position along height (mm)
Interlaminar shear stress factor
(c)
Position along height (mm)
Interlaminar shear stress factor

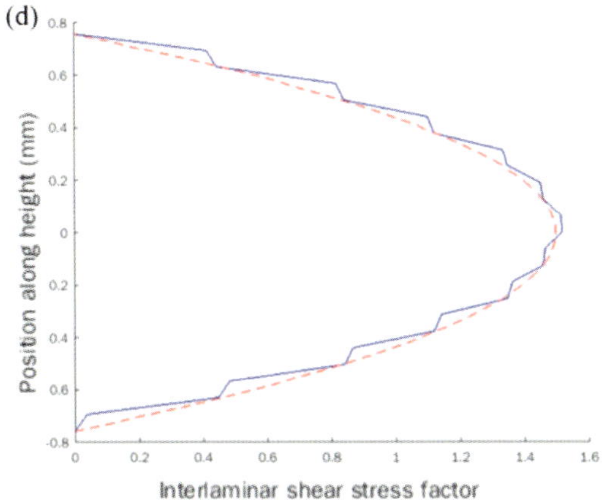

Fig. 7.6 ILS stress factor distribution along the thickness of the beams superimposed with that of a homogeneous beam (red dashed line) (a) Q1; (b) Q2; (c) DD1; (d) DD2

As can be seen from the superimposed plots above, the ILS stress distributions of the Q1, DD1 and DD2 beams resemble that of a true homogeneous beam (red dashed line) fairly closely, while that of the Q2 beam deviates quite significantly. This is once again due to the uniform distribution of plies with varying stiffness in the former three beams, which allow the difference in stiffness between plies of different orientations to be smeared out across the thickness of the beams.

One possible method to quantify the difference between the calculated ILS stresses in a composite beam and that of an equivalent homogeneous beam is to consider the root-mean-square (RMS) difference between the values at the interface positions. This RMS difference can then be normalised with respect to the applied shear stress value to obtain a dimensionless homogenisation parameter for comparison. A smaller parameter indicates that the composite beam is closer to being homogenised, with a value 0 indicating perfect homogenisation.

Table 7.2 Root mean square difference in ILS stress factors from homogeneous beams

Beam	Normalised RMS difference in ILS stress
Q1	0.1107
Q2	0.2706
DD1	0.0845
DD2	0.0839

Table 7.2 above shows the proposed homogenisation parameter for all four beams calculated using the ILS stress factors obtained in the earlier analyses. As expected, the Q2 beam has a parameter value that is significantly larger than the other three beams, while the two DD beams are the most homogenised. Such a parameter could be used as a guide in determining the number of repeats required in the design of DD laminates against ILS stresses. It can also be similarly used for any other laminate, although the design variables may require further efforts to determine and control.

7.6 Effect of inter-ply mismatch angle on interlaminar shear strength

In practice, ILS stresses are often used to predict the onset of interlaminar failure in composites. This is based on the assumption that all interlaminar interfaces of the same material system have the same interlaminar properties, which are typically obtained via testing unidirectional specimens.

Numerous studies however, have shown that the interlaminar properties of the same material system, such as ILS strength and fracture toughness, varies with the interply mismatch angle, $\theta_{mismatch}$ [4-7]. This means that for laminates with varying $\theta_{mismatch}$, the ILS stresses alone are insufficient to predict which interface will fail first, and the interlaminar properties that are dependent on the $\theta_{mismatch}$ must be taken into account. The ratio of ILS stress to strength should be considered instead. A higher ratio indicates that a given interface is more likely to fail, whereas the critical transverse shear stress or load that causes the first interlaminar failure in a laminate can be estimated by equating the highest ratio to one.

The previous example of composite beams subjected to a unit transverse shear stress is now extended by taking into account the ILS strengths at each interface. The variation in ILS strength with $\theta_{mismatch}$ is obtained from our previous study using the same T700/2510 carbon-epoxy material system as shown in Table 7.3 [1]. For $\theta_{mismatch}$ that lie in between the given data points, a linear interpolation is used to estimate the strength value.

Table 7.3 Variation of ILS strength with $\theta_{mismatch}$

$\theta_{mismatch}$ (°)	ILS strength (MPa)
0	89.32
10	88.53
20	86.49
30	84.90
40	79.47
45	75.41
50	81.14
60	82.15
70	83.17
80	80.44
90	78.52

From Table 7.3, the ILS strength shows a general decreasing trend as the $\theta_{mismatch}$ increases, with the minimum at 45°, followed by a slight increase and then it decreases again. This means that predictions made using a uniform ILS strength obtained from testing unidirectional specimens would likely over-predict the load bearing capacity of the composite structure.

Returning to the four beam example problems, the ILS strengths of all the interfaces are first assumed to be the same and correspond to that of $\theta_{mismatch} = 0°$. Naturally, the interface of each beam that has the highest predicted ILS stress would then have the highest ILS stress to strength ratio. Therefore, the first instance of interlaminar failure will most likely occur at those interfaces, which are located at the mid-plane of all the beams.

Table 7.4 ILS stress to strength ratio of all four beams (uniform ILS strength)

	Q1	Q2	DD1	DD2
Peak ILS stress to strength ratio	0.01574	0.02310	0.01658	0.01699
Interface no.	0	0	0	0
Interface angles	-45°/-45°	0°/0°	-10°/10°	-80°/20°
Critical load to cause interlaminar failure	63.53MPa	43.29MPa	60.31MPa	58.86MPa

From the Table 7.4 above, it can be seen that all four beams show different peak ILS stress to strength ratios and critical applied loads that would lead to interlaminar failure. This is a natural consequence of the different ILS stresses in each beam due to their individual layups while assuming the same ILS strength for all interfaces. It is also interesting to note that the Q1 beam turns out to be the best performing (lowest peak ratio) while the Q2 beam is the worst performing (highest peak ratio) among the four with a difference of about 20MPa in their critical loads, even though they have the same number of each ply. This once again underscores the importance of ply arrangement. The example problem is then reconsidered by using the ILS strengths of each interface based on Table 7.3, taking into account the corresponding $\theta_{mismatch}$.

Table 7.5 ILS stress to strength ratio of all four beams (varying ILS strength)

	Q1	Q2	DD1	DD2
Peak ILS stress to strength ratio	0.01856	0.02310	0.01795	0.01887
Interface no.	-1/+1	0	-1/+1	0
Interface angles	90°/-45° -45°/90°	0°/0°	-60°/-10° 10°/60°	80°/20°
Critical load to cause interlaminar failure	53.88MPa	43.29MPa	55.71MPa	52.99MPa

From Table 7.5 above, it is noted that the critical load to cause interlaminar failure has changed for most of the beams once the mismatch-angle adjusted ILS strength of the interfaces are taken into account. The only exception is the Q2 beam, due to the group of 0° plies at the mid-plane of the beam, where the peak ILS stress is located. For the Q1 and DD1 beams, it is also noted that the interface with the largest ILS stress to strength ratio has shifted as well. This shift can be attributed to the difference in $\theta_{mismatch}$ between the mid-plane interface as well as the interfaces adjacent to it. In both beams, the mid-plane interface has a much smaller $\theta_{mismatch}$ than both adjacent interfaces, resulting in an increase in ILS strength that is larger than the increase in ILS stress.

It is also interesting to note that now the DD1 beam is predicted to be the best performing beam in the context of interlaminar failure, outperforming the originally superior Q1 beam. This can be seen from the now lower peak ILS stress to strength ratio as well as the higher critical load required to cause interlaminar failure for the DD1 beam. This change can be attributed to the ply arrangement in the Q1 beam, which comprises $\theta_{mismatch}$ of 45° and 90° at almost all its interfaces, resulting in the lowest ILS strength among all the beams.

These changes in the results would have led to different decisions made in the context of interlaminar failure resistance or critical applied load. Thus, the ILS strength of interfaces that vary with $\theta_{mismatch}$ should be adopted in practice for a more accurate analysis of interlaminar failure in composites.

7.7 Conclusion

A simple procedure to estimate the ILS stresses at the ply interfaces of composite laminates, which can be easily coded into a computer programme, has been proposed. FE simulations were used to validate the proposed procedure, with the predicted ILS stress values deviating about 4% in the worst case. Analyses of two quad and two DD laminated beams subjected to simple transverse shear using the proposed procedure showed that the ILS stress distributions can be sensitive to layup. However, the analyses also showed that given enough repeats of the subunits (six in this case), the ILS stress distribution in DD beams tend towards that of a homogeneous beam, which simplifies the design of DD beams against interlaminar failure. A dimensionless homogenisation parameter based on the ILS stress distribution is proposed to quantify the degree of homogeneity of a given composite.

The article also discussed the importance of taking into account the variation of ILS strengths of interfaces with the interply mismatch angle. Further analyses using the same beam problems showed that calculations performed using an assumed uniform ILS strength obtained from solely testing unidirectional specimens are likely to over-predict the load bearing capacity of the composite structure. Using the ILS strength values that have been adjusted for the interply mismatch angles would change the final comparisons between different layups and alter any decisions made in the context of interlaminar failure resistance or critical applied loads. As such, when determining the ILS strength values of composite laminates through experiments, specimens with varying interply mismatch angles should be tested to account for this mismatch angle dependence.

7.8 References

[1] J.L. Liu, T.E. Tay, S.W. Ng, and V.B.C. Tan, "Effects of inter-ply mismatch angle on interlaminar properties and their influence in numerical simulations," *Compos. Part A Appl. Sci. Manuf.*, vol. 154, p. 106795, 2022.

[2] S.W. Tsai, "Double–double: New family of composite laminates," *AIAA J.*, vol. 59, pp. 4293-4305, 2021.

[3] S.W. Tsai and H.T. Hahn, *Introduction to composite materials*: Routledge, 2018.

[4] I. Chou, K. Kageyama, I. Kimpara, and I. Ohsawa, *Mode i and mode ii fracture toughness measured between differently oriented plies in graphite/epoxy composites*: ASTM International, 1995.

[5] B.W. Kim and A.H. Mayer, "Influence of fiber direction and mixed-mode ratio on delamination fracture toughness of carbon/epoxy laminates," *Compos. Sci. Technol.*, vol. 63, pp. 695-713, 2003.

[6] J. Polaha, B. Davidson, R. Hudson, and A. Pieracci, "Effects of mode ratio, ply orientation and precracking on the delamination toughness of a laminated composite," *J. Reinf. Plast. Compos.*, vol. 15, pp. 141-173, 1996.

[7] J. Tao and C.T. Sun, "Influence of ply orientation on delamination in composite laminates," *J. Compos. Mater.*, vol. 32, pp. 1933-1947, 1998.

Appendix A. Output data from ILS stress and strength calculations using Python programme

Table 7.6 ILS stress scaling factor at each interface

Interface	Q1	Q2	DD1	DD2
Top (free)	0	0	0	0
-11	0.513	0.101	0.432	0.412
-10	0.613	0.193	0.461	0.446
-9	0.642	0.276	0.494	0.815
-8	0.747	0.546	0.823	0.843
-7	1.081	0.825	1.103	1.101
-6	1.143	1.032	1.119	1.121
-5	1.160	1.237	1.140	1.335
-4	1.215	1.379	1.316	1.349
-3	1.371	1.510	1.444	1.452
-2	1.395	1.817	1.449	1.459
-1	1.400	2.002	1.457	1.517
0 (mid-plane)	1.406	2.063	1.481	1.518
1	1.400	2.002	1.457	1.467
2	1.395	1.817	1.449	1.460
3	1.371	1.510	1.444	1.362
4	1.215	1.379	1.316	1.349
5	1.160	1.237	1.140	1.144
6	1.143	1.032	1.119	1.123
7	1.081	0.825	1.103	0.870
8	0.747	0.546	0.823	0.843
9	0.642	0.276	0.494	0.484
10	0.613	0.193	0.461	0.449
11	0.513	0.101	0.432	0.041
Bottom (free)	0	0	0	0

Table 7.7 Ratio of ILS stress to strength at each interface (uniform ILS strength)

Interface	Q1	Q2	DD1	DD2
Top (free)	0	0	0	0
-11	0.00575	0.00113	0.00484	0.00461
-10	0.00686	0.00216	0.00516	0.00499
-9	0.00718	0.00309	0.00553	0.00912
-8	0.00836	0.00611	0.00921	0.00944
-7	0.01211	0.00924	0.01235	0.01232
-6	0.01280	0.01155	0.01253	0.01255
-5	0.01298	0.01384	0.01276	0.01494
-4	0.01361	0.01544	0.01474	0.01510
-3	0.01535	0.01690	0.01617	0.01626
-2	0.01562	0.02035	0.01622	0.01633

-1	0.01567	0.02241	0.01631	0.01698
0 (mid-plane)	0.01574	0.02310	0.01658	0.01699
1	0.01567	0.02241	0.01631	0.01642
2	0.01562	0.02035	0.01622	0.01634
3	0.01535	0.01690	0.01617	0.01525
4	0.01361	0.01544	0.01474	0.01510
5	0.01298	0.01384	0.01276	0.01281
6	0.01280	0.01155	0.01253	0.01257
7	0.01211	0.00924	0.01235	0.00974
8	0.00836	0.00611	0.00921	0.00944
9	0.00718	0.00309	0.00553	0.00542
10	0.00686	0.00216	0.00516	0.00503
11	0.00575	0.00113	0.00484	0.00045
Bottom (free)	0	0	0	0

Table 7.8 Ratio of ILS stress to strength at each interface (varying ILS strength)

Interface	Q1	Q2	DD1	DD2
Top (free)	0	0	0	0
-11	0.00681	0.00113	0.00533	0.00501
-10	0.00813	0.00216	0.00561	0.00554
-9	0.00851	0.00366	0.00609	0.00992
-8	0.00990	0.00695	0.00951	0.01048
-7	0.01434	0.01051	0.01359	0.01340
-6	0.01516	0.01314	0.01363	0.01394
-5	0.01538	0.01575	0.01405	0.01625
-4	0.01612	0.01757	0.01522	0.01677
-3	0.01819	0.02002	0.01780	0.01768
-2	0.01850	0.02035	0.01763	0.01814
-1	0.01856	0.02241	0.01795	0.01847
0 (mid-plane)	0.01574	0.02310	0.01712	0.01887
1	0.01856	0.02241	0.01795	0.01785
2	0.01850	0.02035	0.01763	0.01814
3	0.01819	0.02002	0.01780	0.01658
4	0.01612	0.01757	0.01522	0.01677
5	0.01538	0.01575	0.01405	0.01392
6	0.01516	0.01314	0.01363	0.01396
7	0.01434	0.01051	0.01359	0.01059
8	0.00990	0.00695	0.00951	0.01048
9	0.00851	0.00366	0.00609	0.00589
10	0.00813	0.00216	0.00561	0.00558
11	0.00681	0.00113	0.00533	0.00049
Bottom (free)	0	0	0	0

Appendix B. FE validation

To verify the accuracy of the proposed approach, FE simulations are conducted on the Q1, Q2, DD1 and DD2 beams subjected to transverse shear loads. Cantilever beams (DD1 and DD2) and 3-point bending (Q1 and Q2) are simulated as they are common situations where shear loads are induced.

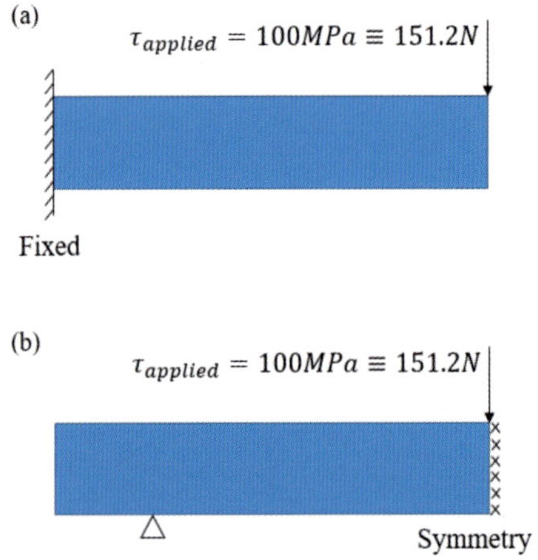

Fig. 7.7 Simple schematic of FE models (a) Cantilever; (b) 3-point bending

In the FE simulations, all four beams are modelled as 24 plies which are perfectly bonded. Each of the plies correspond to ten elements in thickness and are assigned the corresponding orientations based on the beam's layup. The in-plane dimension is 1mm. A downward force of 151.2N is then applied to the loading edge (for 3-point bending) or face (for cantilever) of the beam, which corresponds to an applied transverse shear stress of 100MPa.

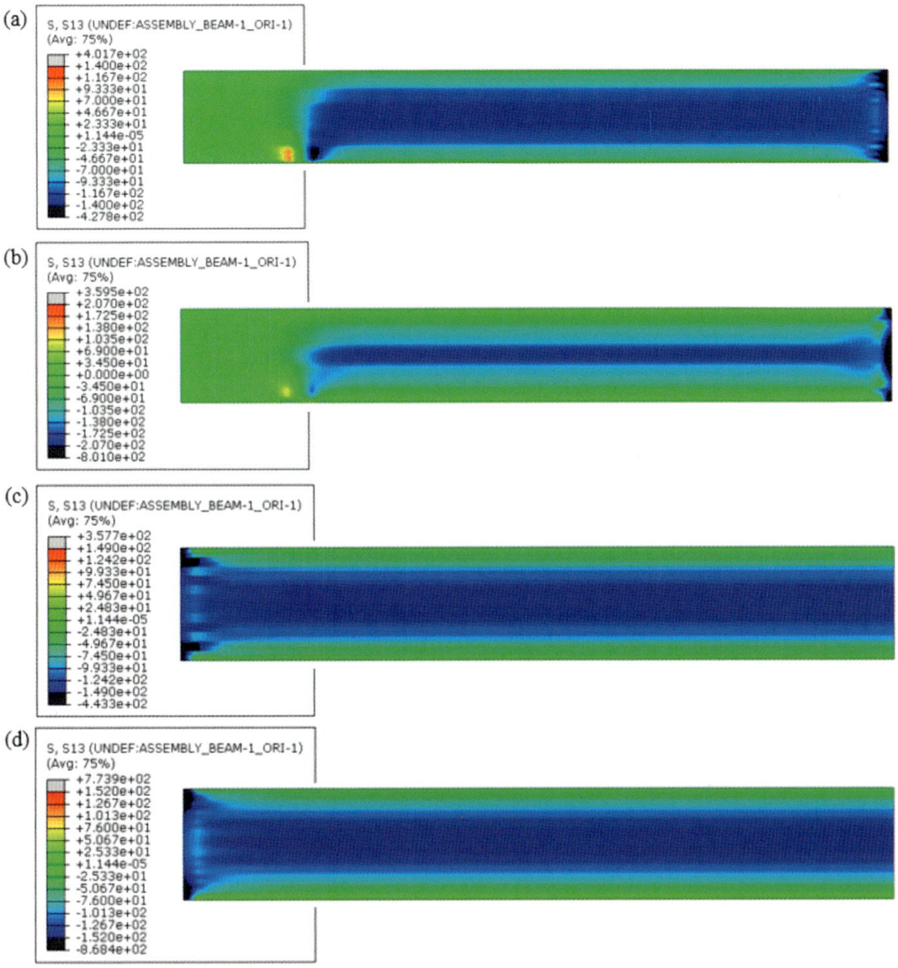

Fig. 7.8 Shear stresses predicted from FE models at mid-sections of each beam (a) Q1; (b) Q2; (c) DD1; (d) DD2

Table 7.9 Peak ILS stress scale factors predicted from CLT and FE models

Beam	CLT	FE	Deviation
Q1	1.406	1.385	1.52%
Q2	2.063	1.981	4.14%
DD1	1.481	1.460	1.44%
DD2	1.518	1.507	0.73%

Fig. 7.8 shows the shear stresses predicted from the FE simulations for all four beams, sectioned halfway along the y-direction (into plane). All four models show shear stresses that increases and peaks towards the mid-plane of the beam. This is similar to the predictions made using CLT. Furthermore, the peak shear stress factors predicted from the FE models, as shown in Table 7.9, also match closely to those from CLT, with deviations of 1.5% or less for the Q1, DD1 and DD2 beams, and 4.14% for the Q2 beam. The larger deviations in the Q2 beam can once again be attributed to the difference in ply arrangements. In the Q2 beam, plies with

similar stiffness are grouped together, resulting in a deformed cross-section that is more accurately described using a zigzag. On the other hand, the deformation of the Q1, DD1 and DD2 beams with a more uniform distribution of plies with varying stiffness will resemble the linear variation of strains assumed in CLT more closely.

Towards the free edges of the beams, the predicted shear stresses in the beams vary from the CLT predictions. This can be seen from the sectioned plan view (xy-plane) of the beams in Fig. 7.9 below. This variation is due to the free edge effects that were not taken into account as the simple load vectors in Eq. (7.11) in the CLT have a dependence along the x-direction only.

Fig. 7.9 Sectioned plan view (xy-plane) of each beam showing stress variations near the free edges (a) Q1; (b) Q2; (c) DD1; (d) DD2

Nevertheless, even with a simple load representation, this proposed procedure is a quick way to predict and compare the interlaminar shear stresses of different layups, which can vary significantly from solely a change in the ply stacking sequence, as seen in the example problem presented.

Appendix C. Relating interlaminar shear stress to laminate moments

As shown in Eq. (7.5), shear stresses arise only when there is a gradient in $\{\boldsymbol{\sigma}\}$. In most cases, $\boldsymbol{N}$ is uniform and hence can be omitted when inserting Eq. (7.10) into Eq. (7.5) to give,

$$\begin{aligned}\begin{Bmatrix}\tau_{xz}\\ \tau_{yz}\end{Bmatrix}_K &= \begin{Bmatrix}\tau_{xz}\\ \tau_{yz}\end{Bmatrix}_{K-1} - \partial\overline{\boldsymbol{Q}}_K\left(\int_{z_{K-1}}^{z_K}\left(\boldsymbol{B}' + z\boldsymbol{D}'\right)dz\right)\boldsymbol{M}\\ &= \begin{Bmatrix}\tau_{xz}\\ \tau_{yz}\end{Bmatrix}_{K-1} - \partial\left((z_K - z_{k-1})\overline{\boldsymbol{Q}}_K\boldsymbol{B}' + \frac{1}{2}(z_K^2 - z_{K-1}^2)\overline{\boldsymbol{Q}}_K\boldsymbol{D}'\right)\boldsymbol{M}\end{aligned} \quad (7.14)$$

Note that the sub-matrices of the ABD matrix in CLT are constructed from

$$\begin{aligned}\boldsymbol{A} &= \sum_{n=1}^{N}(z_K - z_{k-1})\overline{\boldsymbol{Q}}_K\\ \boldsymbol{B} &= \frac{1}{2}\sum_{n=1}^{N}(z_K^2 - z_{K-1}^2)\overline{\boldsymbol{Q}}_K\end{aligned} \quad (7.15)$$

Hence, Eq. (7.14) can be written as,

$$\begin{Bmatrix}\tau_{xz}\\ \tau_{yz}\end{Bmatrix}_K = \begin{Bmatrix}\tau_{xz}\\ \tau_{yz}\end{Bmatrix}_{K-1} - \partial\left(\overline{\boldsymbol{A}}_K\boldsymbol{B}' + \overline{\boldsymbol{B}}_K\boldsymbol{D}'\right)\boldsymbol{M} \quad (7.16)$$

where $\overline{\boldsymbol{A}}_K = (z_K - z_{k-1})\overline{\boldsymbol{Q}}_K$ and $\overline{\boldsymbol{B}}_K = \frac{1}{2}(z_K^2 - z_{K-1}^2)\overline{\boldsymbol{Q}}_K$

Expanding Eq. (7.16) gives,

$$\begin{Bmatrix}\tau_{xz}\\ \tau_{yz}\end{Bmatrix}_K = \begin{Bmatrix}\tau_{xz}\\ \tau_{yz}\end{Bmatrix}_{K-1} - \begin{bmatrix}E_{11} & E_{12} & E_{13} & E_{31} & E_{32} & E_{33}\\ E_{31} & E_{32} & E_{33} & E_{21} & E_{22} & E_{23}\end{bmatrix}_K \begin{bmatrix}\frac{\partial M_x}{\partial x}\\ \frac{\partial M_y}{\partial x}\\ \frac{\partial M_{xy}}{\partial x}\\ \frac{\partial M_x}{\partial y}\\ \frac{\partial M_y}{\partial y}\\ \frac{\partial M_{xy}}{\partial y}\end{bmatrix} \quad (7.17)$$

where $\boldsymbol{E}_K = \overline{\boldsymbol{A}}_K \boldsymbol{B}' + \overline{\boldsymbol{B}}_K \boldsymbol{D}'$.

Eq. (7.17) shows that interlaminar shear stresses are dependent on gradients of moments, which are in turn related to the transverse loads acting on the laminate through the usual plate kinetics,

$$\begin{aligned} \frac{\partial M_x}{\partial x} + \frac{\partial M_{xy}}{\partial y} &= N_{xz} \\ \frac{\partial M_{xy}}{\partial x} + \frac{\partial M_y}{\partial y} &= N_{yz} \\ \frac{\partial N_{xz}}{\partial x} + \frac{\partial N_{yz}}{\partial y} &= -q \end{aligned} \tag{7.18}$$

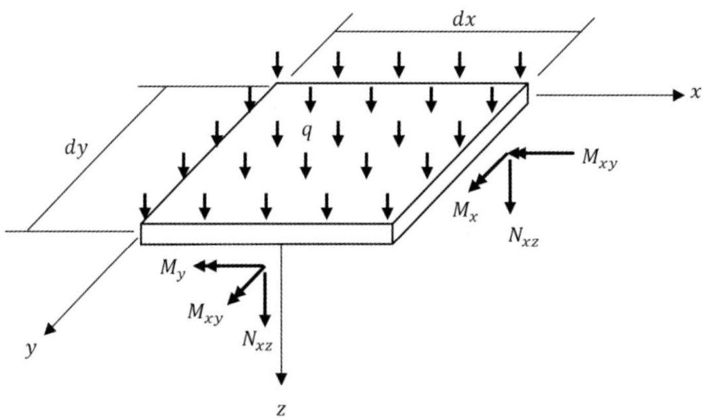

Fig. 7.10 Laminate load conventions

The equations presented have been packaged into an Excel worksheet that can be downloaded at https://scholarbank.nus.edu.sg/handle/10635/217796.

A screenshot of the worksheet is shown in Fig. 7.11. The required inputs are indicated by the coloured cells. The laminate inputs are ply thickness, E_1, E_2, G_{12} and ν_{12}, and the orientation of each ply. The load input is the vector of gradients of the moments in Eq. (7.17). Plies can be added or removed by simply adding or removing rows to the worksheet.

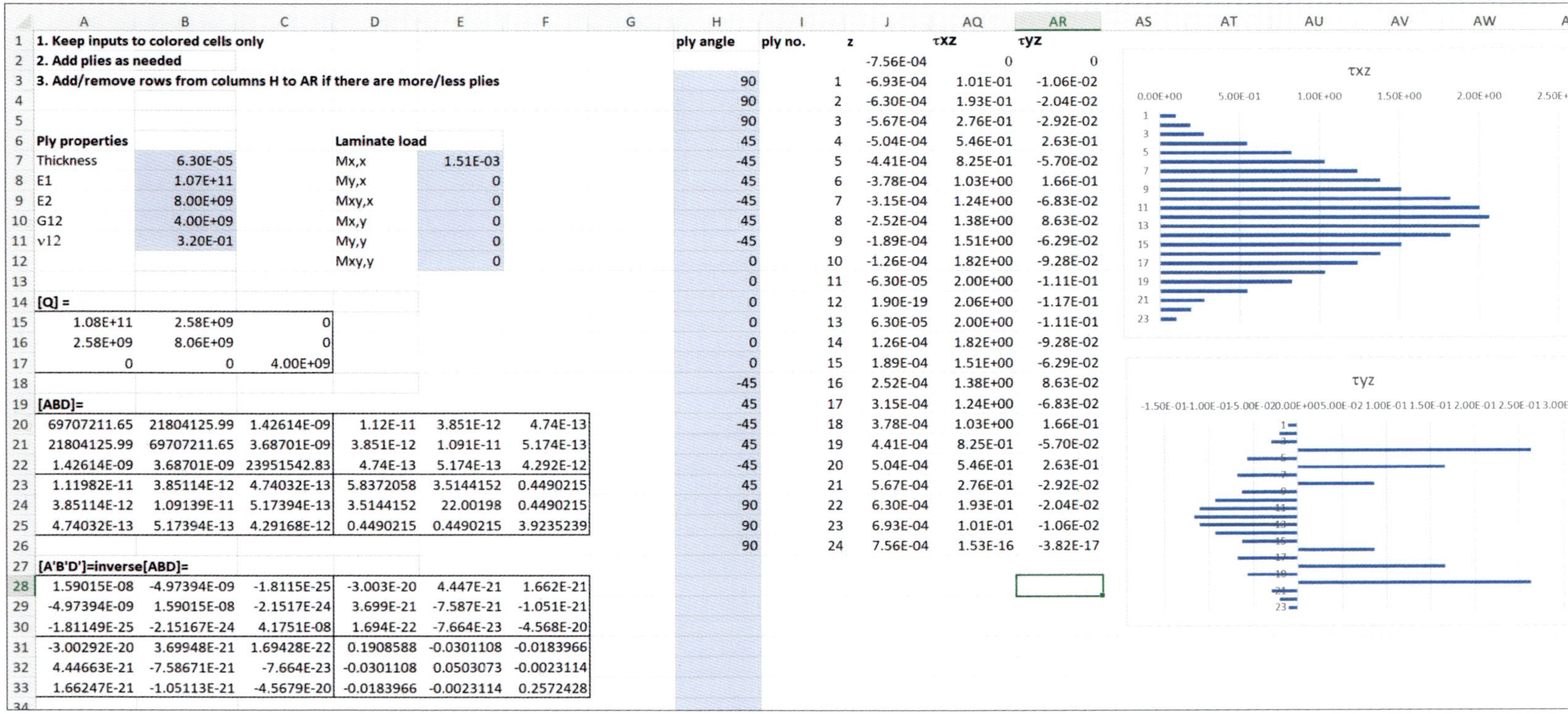

	A	B	C	D	E	F	G	H	I	J	AQ	AR
1	1. Keep inputs to colored cells only							ply angle	ply no.	z	τxz	τyz
2	2. Add plies as needed									-7.56E-04	0	0
3	3. Add/remove rows from columns H to AR if there are more/less plies							90	1	-6.93E-04	1.01E-01	-1.06E-02
4								90	2	-6.30E-04	1.93E-01	-2.04E-02
5								90	3	-5.67E-04	2.76E-01	-2.92E-02
6	Ply properties			Laminate load				45	4	-5.04E-04	5.46E-01	2.63E-01
7	Thickness	6.30E-05		Mx,x	1.51E-03			-45	5	-4.41E-04	8.25E-01	-5.70E-02
8	E1	1.07E+11		My,x	0			45	6	-3.78E-04	1.03E+00	1.66E-01
9	E2	8.00E+09		Mxy,x	0			-45	7	-3.15E-04	1.24E+00	-6.83E-02
10	G12	4.00E+09		Mx,y	0			45	8	-2.52E-04	1.38E+00	8.63E-02
11	ν12	3.20E-01		My,y	0			-45	9	-1.89E-04	1.51E+00	-6.29E-02
12				Mxy,y	0			0	10	-1.26E-04	1.82E+00	-9.28E-02
13								0	11	-6.30E-05	2.00E+00	-1.11E-01
14	[Q] =							0	12	1.90E-19	2.06E+00	-1.17E-01
15	1.08E+11	2.58E+09	0					0	13	6.30E-05	2.00E+00	-1.11E-01
16	2.58E+09	8.06E+09	0					0	14	1.26E-04	1.82E+00	-9.28E-02
17	0	0	4.00E+09					0	15	1.89E-04	1.51E+00	-6.29E-02
18								-45	16	2.52E-04	1.38E+00	8.63E-02
19	[ABD]=							45	17	3.15E-04	1.24E+00	-6.83E-02
20	69707211.65	21804125.99	1.42614E-09	1.12E-11	3.851E-12	4.74E-13		-45	18	3.78E-04	1.03E+00	1.66E-01
21	21804125.99	69707211.65	3.68701E-09	3.851E-12	1.091E-11	5.174E-13		45	19	4.41E-04	8.25E-01	-5.70E-02
22	1.42614E-09	3.68701E-09	23951542.83	4.74E-13	5.174E-13	4.292E-12		-45	20	5.04E-04	5.46E-01	2.63E-01
23	1.11982E-11	3.85114E-12	4.74032E-13	5.8372058	3.5144152	0.4490215		45	21	5.67E-04	2.76E-01	-2.92E-02
24	3.85114E-12	1.09139E-11	5.17394E-13	3.5144152	22.00198	0.4490215		90	22	6.30E-04	1.93E-01	-2.04E-02
25	4.74032E-13	5.17394E-13	4.29168E-12	0.4490215	0.4490215	3.9235239		90	23	6.93E-04	1.01E-01	-1.06E-02
26								90	24	7.56E-04	1.53E-16	-3.82E-17
27	[A'B'D']=inverse[ABD]=											
28	1.59015E-08	-4.97394E-09	-1.8115E-25	-3.003E-20	4.447E-21	1.662E-21						
29	-4.97394E-09	1.59015E-08	-2.1517E-24	3.699E-21	-7.587E-21	-1.051E-21						
30	-1.81149E-25	-2.15167E-24	4.1751E-08	1.694E-22	-7.664E-23	-4.568E-20						
31	-3.00292E-20	3.69948E-21	1.69428E-22	0.1908588	-0.0301108	-0.0183966						
32	4.44663E-21	-7.58671E-21	-7.664E-23	-0.0301108	0.0503073	-0.0023114						
33	1.66247E-21	-1.05113E-21	-4.5679E-20	-0.0183966	-0.0023114	0.2572428						

Fig. 7.11 Screenshot of Excel interlaminar shear stress calculator

Ch.7

With Michel Cognet at JEC World, the first believer and supporter of bi-angle non crimp fabric

With Thierry Massard at JEC World. Collaborated over 40 years, and developed MicMac and Lam search

CHAPTER 8

Buckling Optimization of DD

Wang Dan

Institute of High Performance Computing (IHPC), Agency for Science, Technology and Research (A*STAR), 1 Fusionopolis Way, 138632, Singapore

Multiple design space of DD laminates

DD is a novel laminate layup proposed by Prof Stephen Tsai as a repeat of a sub-laminate made up of two groups of balanced ply orientations. From the perspective of structural optimization, DD can provide better results naturally by expanding the design space largely with the tailored in-plane orientations. The simplicity of DD by compressing the redundancy of out-of-plane permutations makes it possible for easy and cheap manufacturing, which is the advantage that most other optimization results can't achieve. More importantly, the repeat patten of DD sub-laminate makes it straightforward for thickness tapering, which is significant in reducing the structural weight for a light-weight laminate for aerospace applications.

To obtain the best DD under one or multiply specific loading cases and customized structural requirements, both the DD ply orientations and the laminate thickness profile in 1D or 2D, as illustrated in Figure 8.1, can be varying to obtain the optimal solutions.

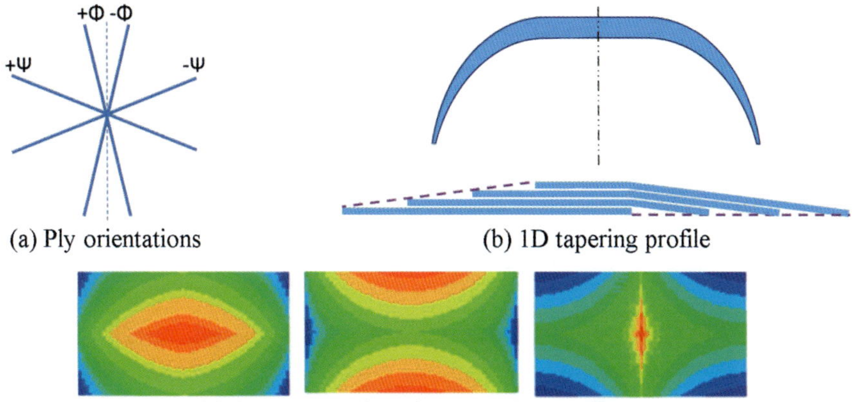

(a) Ply orientations (b) 1D tapering profile

(c) Some examples of 2D tapering profiles

Figure 8.1. Multiple design space of DD laminates

It is noticeable that the buckling performances is very critical in optimization of the DD for the following two reasons: 1) DD aims to be applied in thin-walled structure; 2) the risk of buckling is significantly increased by thickness thinning due to tapering. So integrated buckling consideration into the optimization procedure is very important for the success for the industry adoption of the developed tapered DD.

8.1 Buckling optimization of DD laminates by varying ply orientations

The mathematical optimization model of DD with varying ply orientations is formulated as follows.

$$\begin{cases} \text{Find:} x = [\Phi, \quad \Psi] \\ \text{Max:} \lambda_1 \\ \qquad 0 \leq x_i \leq \pi/2, \ i = 1,2 \end{cases} \tag{8.1}$$

The optimization problem is unconstraint to maximize the minimum buckling load λ_1 of the first mode. The optimization is performed for a rectangular panel with two different aspect ratios, as illustrated in Fig. 8.2, under three different loading cases, as illustrated in Fig. 8.3.

(a) b/a=1 (b) b/a=2

Figure 8.2. Sketch of a rectangular DD panel under investigation

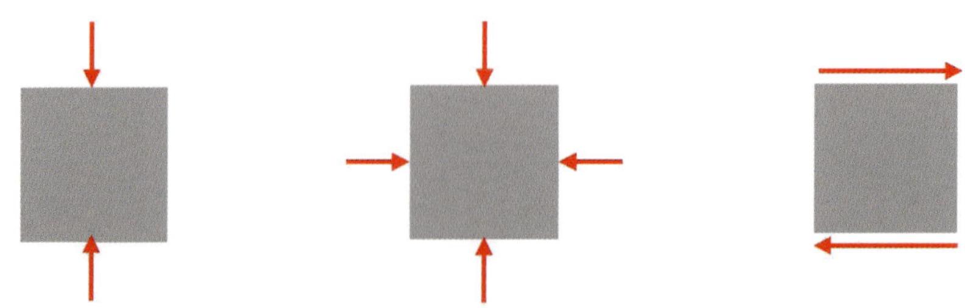

(a) Uniaxial compression (b) Biaxial compression (c) Shear

Figure 8.3. Three loading cases of a rectangular DD laminate panel.

Since there are only two design variables of two independent ply orientations Φ and Ψ, the optimal DD can be obtained using Lam Search by listing Φ and Ψ with an equal spacing, as given in Fig.s 8.4 and 8.5.

Uniaxial compression

Φ (°) \ Ψ (°)	0	15	30	45	60	75	90
0	3.63	Symm					
15	4.50	5.44					
30	6.69	7.67	9.99				
45	8.97	9.64	10.85	11.47			
60	8.43	9.07	10.22	10.85	10.34		
75	7.37	7.97	9.08	9.66	9.16	8.05	7.41
90	6.85	7.38	8.44	8.99	8.47	7.41	6.89

Biaxial compression

Φ (°) \ Ψ (°)	0	15	30	45	60	75	90
0	2.91	Symm					
15	3.61	4.02					
30	4.23	4.58	5.17				
45	4.49	4.83	5.43	5.74			
60	4.22	4.54	5.11	5.43	5.17		
75	3.69	3.98	4.54	4.83	4.58	4.02	
90	3.42	3.69	4.22	4.49	4.23	3.61	2.91

Simple shear

Φ (°) \ Ψ (°)	0	15	30	45	60	75	90
0	5.69	Symm					
15	7.51	8.51					
30	8.81	9.26	10.18				
45	9.26	9.77	10.75	11.30			
60	8.70	9.25	10.28	10.80	10.27		
75	7.54	8.08	9.15	9.68	9.13	8.00	
90	6.93	7.43	8.50	9.04	8.49	7.38	6.83

Figure 8.4. Buckling load distribution with respect to two independent ply orientations of DD for b/a=1

Uniaxial compression

Φ (°) \ Ψ (°)	0	15	30	45	60	75	90
0	3.63	Symm					
15	4.51	5.44					
30	6.67	7.55	9.40				
45	8.24	9.11	10.88	11.50			
60	8.45	9.09	10.26	10.88	10.00		
75	7.39	7.99	9.10	9.69	7.68	5.43	
90	6.86	7.40	8.47	9.02	6.69	4.49	3.62

Biaxial compression

Φ (°) \ Ψ (°)	0	15	30	45	60	75	90
0	2.90	Symm					
15	3.43	4.00					
30	4.23	4.20	3.96				
45	3.79	3.75	3.51	3.07			
60	3.27	3.22	2.96	2.53	2.00		
75	2.84	2.76	2.49	2.05	1.53	1.09	
90	2.67	2.57	2.29	1.85	1.34	1.73	1.43

Simple shear

Φ (°) \ Ψ (°)	0	15	30	45	60	75	90
0	5.65	Symm					
15	7.17	8.18					
30	8.52	8.09	7.32				
45	8.84	8.16	7.19	6.71			
60	8.65	8.90	7.45	6.52	5.90		
75	7.59	8.11	7.81	6.40	5.36	4.45	
90	7.03	7.52	7.95	6.31	5.08	3.98	3.46

Figure 8.5. Buckling load distribution with respect to two independent ply orientations of DD for b/a=2

It is found that for a square DD panel, the largest buckling load is obtained at $[\pm 45^{o}]_{rN}$ for all the three loading cases. The same conclusion only maintains for uniaxial compression. The optimal DD for the largest critical buckling load is $[\pm 30/0_2]_{rN}$ and $[\pm 60/\pm 15]_{rN}$ for biaxial compression and shear, respectively. Another important conclusion is that DD laminate provides a convex design space with only one global optimal, which is superior than the conventional laminate design with Quad or arbitrary ply orientations.

8.2 Buckling optimization of tapered DD laminates by varying laminate thickness profile

When the DD is tapered, its thickness profile will be varied. In that case, the number of design variables is largely increased, and the specific number depends on the mesh discretization if the local thickness is related to the element number. As illustrated in Fig. 8.6, the laminate thickness profile is defined by creating a non-uniform distribution of thicknesses. It is realized by assigning elemental thickness on a discretized mesh.

(a) Mesh discretization (b) Elemental thickness distribution

Figure 8.6. Thickness profile of the investigated panel

The mathematical optimization model of tapered DD with a varying thickness profile is formulated as follows.

$$\begin{cases} \text{Find}: x = \begin{bmatrix} h_1 & h_2 & \cdots & h_{n_e} \end{bmatrix} \\ \text{Min}: \max\left(r_i = \frac{1}{\lambda_i}\right), \quad i = 1,2,3 \\ s.t.\ \underline{h} \leq h_j \leq \overline{h}, \quad j = 1,2,\ldots,n_e \\ V \leq V_0 \end{cases} \tag{8.2}$$

where r_i is the reciprocal of the ith buckling load λ_i, that is to create a minimum objective function. h_j is the elemental thickness of the jth element and bounded by the lower and upper bounds of $\underline{h}$ and $\overline{h}$. V and V_0 are the structural volume and its upper bound.

Here, Lam Search is not sufficient to deal with the optimization of tapered DD with many design variables involved. Instead, advance structural optimization techniques are needed to efficiently solve this problem. The optimization procedure is provided in Fig. 8.7, including simulation and sensitivity calculation on a thickness varying model, and the integration with a highly efficient gradient-based optimizer. The details of a similar optimization procedure can be found in curved grid-stiffener design using streamline stiffener path optimization [1-3].

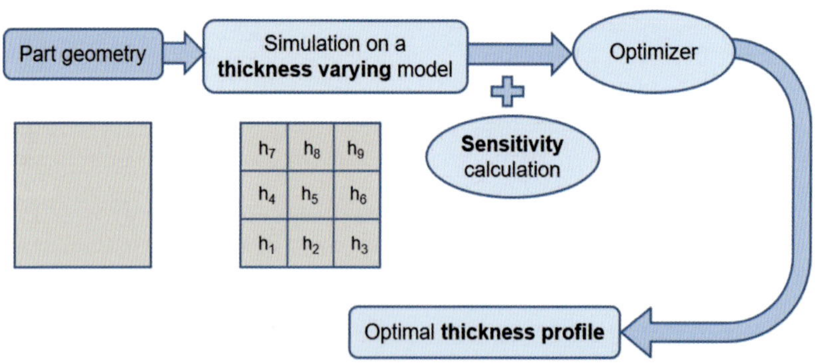

Figure 8.7. Optimization procedure for varying thicknesses of a tapered DD.

For a square DD with the initial layup of [±45°]7N, the tapering can help to increase the critical buckling load by 46%, 70% and 23%, under uniaxial compression, biaxial compression, and shear, respectively, compared with the flat panel with a uniform thickness under the same structural weight. The optimal thickness profiles for the tapered DD under different loading cases are provided in Fig. 8.8 and the corresponding buckling modes are provided in Fig. 8.9. It is shown that the thickness distribution of the laminate not only largely effect the critical buckling load but also the buckling modes.

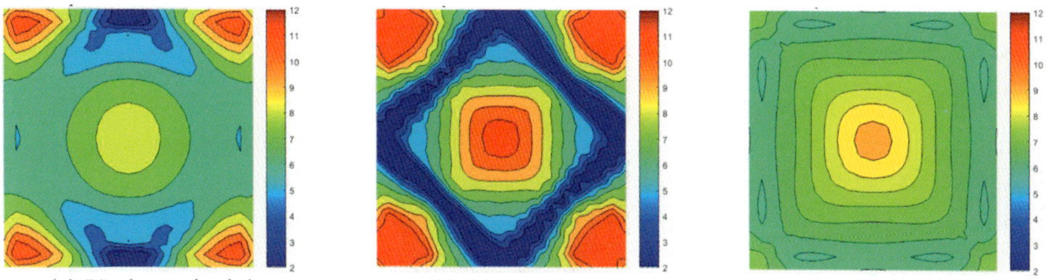

(a) Under uniaxial compression (b) Under biaxial compression (c) Under shear

Figure 8.8. Optimal thickness profiles for the tapered DD under different loading cases.

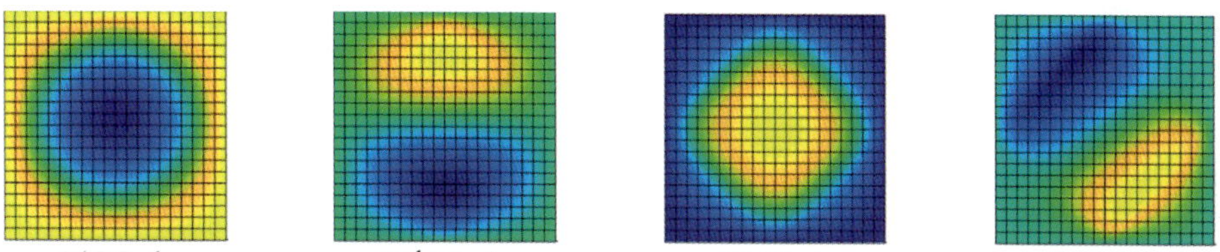

1st Mode 2nd Mode 1st Mode 2nd Mode

(a) Under uniaxial compression (b) Under biaxial compression

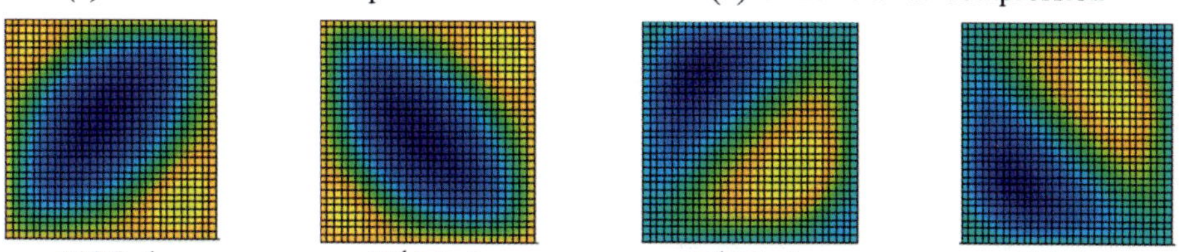

1st Mode 2nd Mode 3rd Mode 4th Mode

(c) Under shear

Figure 8.9. Buckling modes of the tapered DD with optimal thickness profiles under different loading cases.

8.3 **Buckling optimization of tapered DD laminates with spacing constraints**

As shown in Section 8.2, the optimal thickness distribution can largely increase the critical buckling load of DD laminates with the same structural weight. However, the lumped thickness distribution will may bring material failure forward and lead to delamination between different thicknesses.

To solve this problem, we can add spacing constraints into the optimization procedure and use the minimum spacing between different thicknesses to smooth the stress redistribution and postpone the occurrence of the delamination. With the added spacing constraints, the mathematical optimization model of a tapered DD with a varying thickness profile is reformulated as follows.

$$\begin{cases} \text{Find:} x = \begin{bmatrix} h_1 & h_2 & \cdots & h_{n_e} \end{bmatrix} \\ \text{Min:} \max\left(r_i = \frac{1}{\lambda_i}\right), \quad i = 1,2,3 \\ s.t. \ \underline{h} \le h_j \le \overline{h}, \quad j = 1,2,\ldots,n_e \\ \quad s_k \ge s_0, \quad k = 1,2,\ldots,n_e \\ \quad V \le V_0 \end{cases} \tag{8.3}$$

where s_k is the spacing of the elemental thickness and bounded by the lower bound of s_0. The spacing is calculated following the same equation provided in our previous work for curved grid-stiffener design [1-3].

After including the spacing constraints, the optimal thickness profiles for the tapered DD under uniaxial and biaxial compression are provided in Fig. 8.10. It is shown the optimal thickness profile is more reliable for yielding higher strength by controlling the spacing between different thicknesses. This advantage is achieved at the cost of a largely reduced optimization effect. The buckling increases are decreased from 46% and 70% to 23% and 21% for uniaxial and biaxial compression, respectively.

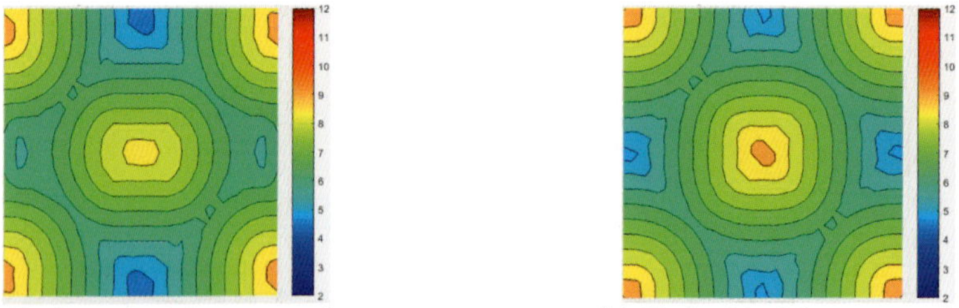

(a) Under uniaxial compression (b) Under biaxial compression

Figure 8.10. Optimal thickness profiles for the tapered DD including spacing constraints.

8.4 Conclusions and recommendations

We have discussed buckling optimization of DD laminates and revealed the big potential of buckling load increase by varying ply orientations and the thickness profile of DD. It is shown that the two independent ply orientations of DD provide a convex design space, which leads to the only optimal with the maximum critical buckling load. That can help to simplify the optimization process in advanced laminate design of a more general DD with variable stiffness.

Optimization of the thickness profile for tapered DD shows that other failure modes should also be considered in the optimization process to avoid the delamination of a thicken patch from the base panel due to the non-uniform thickness distribution. The delamination resistance can be included in the optimization model by adding the spacing constraints between different thicknesses.

The addition of spacing constraints provides the structure with a tapering profile more delamination resistance. However, more structural requirements, such as strength and stiffness also need to be considered in the optimization process in future research to provide reliable optimal configurations for the applications under multiple loading conditions. Moreover, introducing the manufacturing constraints into the optimization process is also very important to accelerate the industry adoption of DD.

Acknowledgement

The author would like to acknowledge the financial support from the Agency for Science, Technology and Research (A*STAR) - Career Development Fund (C210812010).

References:

[1] Dan Wang, Mostafa M. Abdalla, Zhen-Pei Wang, and Zhoucheng Su. Streamline stiffener path optimization (SSPO) for embedded stiffener layout design of non-uniform curved grid-stiffened composite (NCGC) structures. Computer Methods in Applied Mechanics and Engineering (2019): 1021-1050.

[2] Dan Wang, Mostafa M. Abdalla, and Weihong Zhang. Sensitivity analysis for optimization design of non-uniform curved grid-stiffened composite (NCGC) structures. Composite Structures (2018): 224-236.

[3] Dan Wang, Si-Yong Yeo, Zhoucheng Su, Zhen-Pei Wang, and Mostafa M. Abdalla. Data-driven streamline stiffener path optimization (SSPO) for sparse stiffener layout design of non-uniform curved grid-stiffened composite (NCGC) structures. Computer Methods in Applied Mechanics and Engineering 365 (2020): 113001

With Iris and Yasushi and Noriko Miyano. Together we organized Composites Durability Workshops for 20 years held over a dozen cities around the world

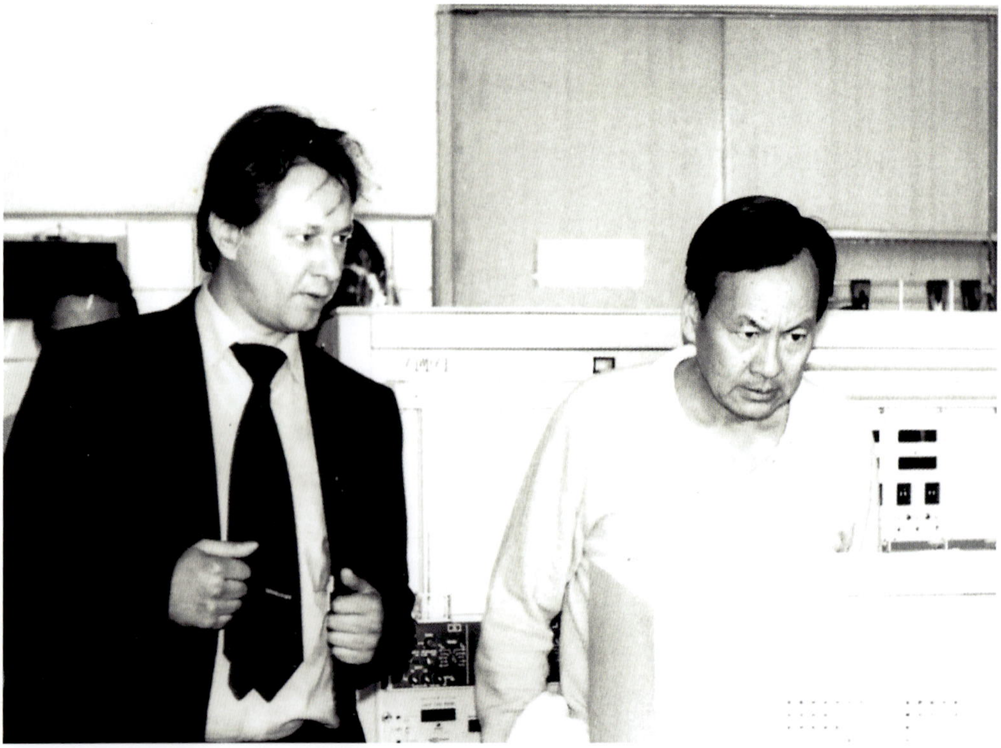

With Valery Vasiliev, the first proponent of isogrid, in Riga, 1975

CHAPTER 9

Fully Rationalised Tsai-Wu Failure Criterion for Transversely Isotropic Materials

Shuguang Li†, Mingming Xu and Elena Sitnikova

Faculty of Engineering, University of Nottingham, Nottingham NG8 1BB, UK

†Corresponding author: shuguang.li@nottingham.ac.uk

Abstract

By examining the quadratic function of the original Tsai-Wu failure criterion for transversely isotropic materials from the perspective of analytic geometry, two of the troublesome coefficients associated with the interactive terms, one describing interaction between in-plane direct stresses and one between transverse direct stresses, can be determined based on mathematical and logical considerations. An appropriate exploration of the quadratic failure function, whilst enhancing the mathematical rigour of the subject, reveals some useful physical relationships as intrinsic properties of a quadratic failure function. Two clear statements can be drawn: (i) To maintain a basic degree of consistency, a failure criterion based on a single quadratic failure function for transversely isotropic materials can only accommodate five independent strength properties, *viz.* the tensile and compressive strengths in directions along fibres and transverse to fibres, and the in-plane shear strength, and (ii) Amongst the transverse tensile, compressive and shear strengths, only two are independent. In presence of experimentally measured transverse shear strengths, they should be utilised to validate the fully rationalised Tsai-Wu criterion or to recalibrate the remaining transverse strength properties. A practical scheme to this effect has been proposed. To complete the development in the rationalisation of the Tsai-Wu criterion, the criterion has also been extended to offer its fully rationalised strain-based counterpart, unifying the approaches based on stresses and strains in a consistent manner.

Keywords: A. Quadratic failure function; B. Tsai-Wu criterion; C. Failure envelope; D. Strength; E. Transverse shear strengths; F. Transversely isotropic material; G. Strain-based criterion

9.1 Introduction

Failure prediction has always been a crucial consideration for structural applications of materials to ensure the safety of the structures under design. With the emergence of composites, which are heterogeneous and anisotropic in general, the formulation of failure criteria became challenging, which underlined the series of World Wide Failure Exercises (WWFEs) [1-3]. The Tsai-Wu failure criterion is one of the earliest failure criteria proposed for composites [4]. Since its publication, the criterion has enjoyed remarkable success as it has been employed by researchers and designers all over the world. It has been included in most textbooks on the subject of composites, e.g. [5]. It has also been incorporated in commercial FE codes, such as Abaqus [6]. On the other hand, it has been subjected to criticisms for not being failure mode related, because it employs a single quadratic function to account for all possible failure modes observed in experiments [7]. More recent attempts of appraising it can be found in [8,9]. The debate on its suitability for engineering applications is likely to continue, whilst its performances have been compared with other theories in the WWFE-I and II [1,2].

No matter how the Tsai-Wu criterion is perceived by its users, it seemed to have set a precedent that the polynomials employed to formulate failure criteria were to be kept to the second order, although one could legitimately argue for improvements by using higher orders of polynomials along the line as was suggested in [10], or by partitioning the failure function according to the failure modes as was attempted in [11]. There was one issue in the Tsai-Wu criterion which often casts doubt on its integrity. It is the determination of the interactive coefficients F_{12}. Given the uncertainties associated with this coefficient, Tsai and his co-workers tended to modestly refer to the criterion as a kind of empirical approach [12-14].

Another bitterly disputed issue on the subject of failure criteria for composites is between phenomenological approaches based on macroscopic stresses or strains and micromechanical approaches or even further down in the length scale. The former are not meant to be an ultimate solution and therefore are not expected to give a complete understanding with perfect accuracy. However, blaming the phenomenological nature alone for the deficiencies in the existing criteria seems to be little more than a quick and easy escape. In theories attempting to incorporate micromechanics of composites into their considerations, more plausible assumptions will have to be made, explicitly or implicitly, which are even more difficult to justify. Sometimes, the sheer complexity of such a theory could easily be turned into the safe haven for irrational treatments. Remarkably accurate predictions are often the consequence of fudging. As an example, representative volume elements and unit cells as basic and also necessary means of any micromechanical study have been subjected to a serious and systematic examination in [15]. It was revealed that their formulation and implementation had been in such a state that erroneous treatments were a common place. Expecting such micromechanical theories to bring truthful understanding of the failure of composites was rather wishful thinking. The point to make is that if the rationality cannot even be established at a macroscopic level, where there are only six macroscopic stresses to deal with, it is hard to imagine that the study of failure of composites can be accomplished at a microscopic scale, where one can easily be overwhelmed by new variables or constants, sometimes of little physical meaning. Whilst researchers should not be discouraged from explorations into micromechanics, it is always helpful to have a conventional and phenomenological failure criterion rationally established, either as a starting point and a firm basis for more sophisticated explorations, or as a reference to compare with before the new theory establishes itself. Given the recent development as pursued by the authors, it is believed that the Tsai-Wu criterion could become the first and the only fully rationally established interactive failure criterion for unidirectionally fibre reinforced composites (transversely isotropic), based on stresses or strains at macroscopic scale.

The Tsai-Wu criterion was formulated based on a quadratic failure function. Quadratic functions are one of the best understood family of functions in analytic geometry. However, the quadratic failure function as the basis of most existing failure criteria has not been appreciated appropriately until very recently. Systematic outcomes from mathematics can be taken advantage of in the formulation of composite failure criteria based on quadratic functions [16]. In this chapter, it will be demonstrated how such knowledge will help the rationalisation of the Tsai-Wu criterion as a representative composite failure criterion based on a quadratic function. Out of this exercise, the criterion will re-emerge as a fully rationalised, and hence consistent, composite failure criterion. Any disagreements with the reliable experimental data will originate from the assumptions, nothing else. These assumptions will be clearly stated.

As a rationally formulated failure criterion, it will not contradict itself in any way. Such basic self-consistency in composites failure criteria has often been taken for granted. However, in reality, none of the available popular failure criteria for composites meets such a basic requirement. As an illustration of the inconsistencies in popular existing failure criteria for composites that might not always be obvious to readers, their critical review has been reported in [9]. It is therefore fair to state that whilst the predictions of these theories fell short of the expectations of their users in accuracy, the formulations of them lack a basic level of self-consistency.

9.2 The original Tsai-Wu criterion and a critical examination

With conventional assumptions, such as homogeneity and linear elasticity up to failure, the Tsai-Wu criterion [4] was originally proposed in the context generally orthotropic materials by using a quadratic polynomial expression of stresses with tensorial coefficients.

$$F_1\sigma_1 + F_2\sigma_2 + F_3\sigma_3 + F_{11}\sigma_1^2 + F_{22}\sigma_2^2 + F_{33}\sigma_3^2 + 2F_{23}\sigma_2\sigma_3 + 2F_{13}\sigma_1\sigma_3 + 2F_{12}\sigma_1\sigma_2 + F_{44}\tau_{23}^2 + F_{55}\tau_{13}^2 + F_{66}\tau_{12}^2 = 1 \quad (1)$$

where the material symmetries as present in orthotropic materials have been duly reflected. In this function, coefficients F_{ij} and F_i $(i,j=1,2,\ldots,6)$ are contracted forms of 4th and 2nd ranked tensors, whereas the components not appearing in the function, such as F_{16} and F_4, are deemed to vanish due to the material symmetries. When the function on the left-hand side of (1) is less than 1, the material is safe. Otherwise, failure is either imminent or has taken place. The tensorial expressions employed enable its general applicability in terms of coordinate systems to be adopted to describe the problem, provided that they are aligned with the materials' principal axes so that symmetry considerations can be taken advantage of.

However, there have not been many genuinely generally orthotropic materials in engineering until the recent years. It is generally unsatisfactory to consider laminated composites, e.g. cross-ply, or quasi-isotropic layup, as orthotropic materials as far as their strength predictions are concerned. Even if they might exhibit orthotropic elastic behaviour macroscopically, they cannot be considered as homogeneous materials, given the distinct interlaminar interfaces. Unlike elastic properties that are dominated by the bulk of the material at a macroscale, strengths are more sensitive to localised features at a meso/micro scale. Fortunately, most composites are made of UD building blocks. There is always sufficient interest in understanding the behaviour of such building blocks. The present chapter will be devoted to the Tsai-Wu criterion within this category of composites aiming at its full rationalisation.

Given the random distribution of fibres over the cross-section of most UD composites, transverse isotropy can usually be justified as a sufficiently satisfactory description of behaviour of UD composites, for which one has

$$F_{33} = F_{22}, \quad (2a)$$

$$F_{13} = F_{12}, \quad (2b)$$

$$F_3 = F_2, \quad (2c)$$

$$F_{55} = F_{66}, \quad (2d)$$

and $$F_{23} = F_{22} - \frac{1}{2} F_{44}.$$ (2e)

The Tsai-Wu failure function can then be reduced to [4]

$$\begin{aligned} &F_1\sigma_1 + F_2(\sigma_2 + \sigma_3) + F_{11}\sigma_1^2 + F_{22}(\sigma_2^2 + \sigma_3^2) + 2F_{23}\sigma_2\sigma_3 + 2F_{12}\sigma_1(\sigma_3 + \sigma_2) \\ &+F_{44}\tau_{23}^2 + F_{66}\tau_{13}^2 + F_{66}\tau_{12}^2 = 1 \end{aligned} \quad (3)$$

where most of the coefficients involved can be determined from the conventional strengths of UD composite as follows [4]

$$F_{11} = \frac{1}{X_t X_c}, \qquad F_{22} = \frac{1}{Y_t Y_c}, \quad (4a)$$

$$F_1 = \frac{1}{X_t} - \frac{1}{X_c}, \qquad F_2 = \frac{1}{Y_t} - \frac{1}{Y_c}, \quad (4b)$$

$$F_{44} = \frac{1}{S_T^2} \quad \text{and} \quad F_{66} = \frac{1}{S_L^2} \quad (4c)$$

with X_t and X_c being the tensile and compressive strengths of the material along fibres, Y_t and Y_c those in the direction transverse to the fibres, and S_L and S_T the shear strengths along and transverse to fibres. They are measured when the material is subjected to uniaxial tension or compression, or pure shear in its principal axes. There are standards available for the experimental measurement of them, e.g. [17-19], although the authors have their reservations on the suitability of these standards for the measurement of the transverse shear strength S_T.

Coefficient F_{12} as the only missing item from (4) has caused great concern and confusion to the Tsai-Wu criterion over the past half century as can be found in the literature. It should ideally be determined by fitting systematic biaxial test data at different stress ratios to equation (3) before obtaining F_{12} for the material. Given the difficulties in conducting biaxial tests and the lack of standard procedure, an acceptable experimental approach is simply unavailable to date. Available accounts on the determination of F_{12} were reviewed in [8], and the experimental techniques employed will not be repeated here as this chapter is to address the theoretical aspect of the problem. The most representative and relevant attempts on the theoretical determinations of the coefficients to the interactive terms involved in the Tsai-Wu criterion are briefly reviewed as follows, confining the discussion to transversely isotropic materials.

Narayanaswami and Adelman [20] argued for $F_{12}=0$, i.e. the absence of this particular interaction between the longitudinal and the transverse stresses. This has been resorted to by many who attempted to propose failure criteria without such an interaction. In particular, in the Hashin criterion [11] another argument was made to eliminate the contribution from this interactive term by introducing the concept of failure of a plane in the material, as opposed to failure of the material at a point (hence in all planes through the point), based on the assumption that failure in a plane was produced by the stresses on that plane. Whilst this particular assumption was originally made convincingly by Mohr [21] when he proposed his failure criterion, it was pointed out in a recent review of the Mohr criterion [22] that this assumption could only be justified for isotropic materials, where the stress state had absolute freedom to determine the orientation of the failure plane. Unfortunately, this freedom was overshadowed, if not overridden, by the material anisotropy when applied to composites. It is clear that to support such a choice of the value of F_{12}, more appropriate argument would be required.

The most popular expression of F_{12} was undoubtedly given by Tsai and Hahn [23] as

$$F_{12} = -\frac{1}{2}\sqrt{F_{11}F_{22}}\ . \tag{5}$$

However, apart from the fact that it allowed to reduce the Tsai-Wu criterion to the von Mises criterion for isotropic materials of equal tensile and compressive strength, no other justification could be offered.

By investigating the diametral compression, Kallas and Hahn [24] managed to come up an expression

$$F_{23} = -F_{22}/2. \tag{6}$$

The complicated stress distribution involved did not seem to have convinced many readers, although the outcome turned out to be right as will be demonstrated later in this chapter through a more rigorous analysis.

In [25], DeTeresa and Larsen assumed infinite strength under hydrostatic compression, as was the case in the von Mises criterion for isotropic materials of equal tensile and compressive strength. They obtained $F_{12} = -F_{11}/4$ as a result. Given the anisotropy of the material, an infinite strength under an isotropic stress state is certainly questionable. Interestingly, in the same account, DeTeresa and Larsen also assumed infinite strength under equal biaxial compression transverse to fibres, which was also one of the assumptions in Hashin criterion [11]. They obtained an expression for the other coefficient as $F_{23} = -F_{22}$. These two coefficients had been dealt with in completely unrelated ways.

An attempt has been made in [8] to reconcile the issues described above. It was suggested that an elliptic paraboloidal envelope offers a reasonable compromise for logical consistency. Mathematically, between the ellipsoids and elliptic hyperboloids there exists a unique elliptic paraboloid sitting on the borderline between them. Employing an elliptic paraboloid as the failure envelope, given its open appearance, it allowed infinite strength, but only under a specific stress ratio [8] in triaxial compression. F_{12} was obtained as [8]

$$F_{12} = -\frac{1}{2}\sqrt{\delta}\sqrt{F_{11}F_{22}} \tag{7}$$

where $\delta = 4 - \frac{F_{44}}{F_{22}} = 4 - \frac{Y_T Y_C}{S_T^2}$. (8)

However, a critical drawback of this approach was that δ as defined in (8) could be negative for some materials as shown in [8], based on which it was concluded that the Tsai-Wu criterion was inapplicable to these materials. This remained an unsatisfactory aspect of that development.

The complete and rational determination of F_{12} have been established in [16]. Its expression was found to be the same as (5), i.e. its most popular form. It was shown that it could not come through alone but had to be accompanied by a specifically determined value of F_{23}, which happens to be the same as what was suggested by Kallas and Hahn in [24]. The rationalisation was in fact to draw a logical relationship between these two interactive coefficients. This forms a major part of the present chapter where a detailed derivation will be provided along with a number of other relevant considerations.

It has been established that determining F_{12} is not enough. How it should be used in the Tsai-Wu criterion is just as important for keeping the self-consistency of the criterion. A similar example in elasticity can be cited here. The minor Poisson's ratio ν_{21} is a measurable elastic property. If its measured value was employed in the elastic constitutive relationship directly along with the experimentally measured ν_{12}, it is almost certain that the symmetry of the stiffness and compliance matrices would be compromised. Such symmetry, shown as a relationship amongst Young's moduli and Poisson's ratios,

$$\frac{\nu_{21}}{E_2} = \frac{\nu_{12}}{E_1} \tag{9}$$

results from a mathematical consideration of the existence of strain energy density in the material, which must be obeyed. The point to make here is that an independently measured property is not necessarily an independent property physically. If there exists a certain physical relationship amongst the experimental data, such as (9), the satisfaction of such a relationship is one of the most important considerations in processing the experimental data. An appropriate approach to take would be to fit the raw test data to the designated relationship if systematic test data sets were available.

Without determining the coefficients for the interactive terms in general, Tsai and Wu suggested that the failure envelope should be an ellipsoid and hence should remain closed to avoid the possibility of having infinite strengths [4]. Restrictions were introduced on the ranges of the coefficients to the interactive terms of direct stresses in the context of orthotropic materials as

$$F_{23}^2 < F_{22}F_{33} \quad \text{or} \quad F_{23}^2 < F_{22}^2 \quad \text{under the condition of transverse isotropy,} \tag{10a}$$

$$F_{13}^2 < F_{11}F_{33} \quad \text{identical to the next condition for transversely isotropic materials,} \tag{10b}$$

$$F_{12}^2 < F_{11}F_{22}\,. \tag{10c}$$

Given the inequality form of (10), they are insufficient to determine F_{23}, F_{13} and F_{12}. However, confusions seemed to have risen from these conditions, and their appropriate clarification actually helps the rational formulation of the Tsai-Wu criterion as will be described in detail in this chapter.

The confusion can be revealed in two aspects. Firstly, conditions (10) do not ensure the closeness of the failure envelope. According to analytic geometry [26], each of the three conditions guarantees that the intersection of the failure envelope in the subspace of direct stresses with the respective coordinate plane is an ellipse. Take coordinate plane 1-2 for example. In this case, in the subspace of direct stresses, the intersection is obtained when $\sigma_3 = 0$ as

$$F_1\sigma_1 + F_2\sigma_2 + F_{11}\sigma_1^2 + F_{22}\sigma_2^2 + 2F_{12}\sigma_1\sigma_2 = 1\,. \tag{11}$$

Condition (10c) ensures that (11) is an ellipse in the 1-2 plane [26]. The same argument can be made for the remaining two conditions in (10). The precise physical interpretation of (10) is that the strength is finite under any plane stress condition in each of the principal planes of the material, which is a reasonable observation from all the available experimental data. However, intersecting three coordinate planes in ellipses does not exclude the possibility of the failure envelope being open in the 3D space. An elliptic paraboloid, as illustrated in Fig. 9.1, is apparently an open surface that satisfies (10) and therefore intersects three coordinate planes in ellipses.

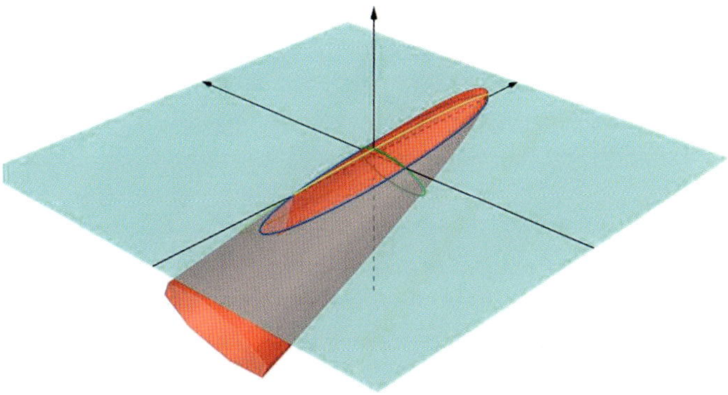

Fig. 9.1 Elliptic paraboloid in 3D space

Consider the T300/PR-319 composite as an example, which was one of the composites involved in WWFE-II [2] and also employed in [8,16] as one of the cases studied. The measured transverse shear strength is S_T=45 MPa. If F_{12} is determined using (5), whilst treating S_T as an independent strength property from which F_{44} can be obtained according to (4c) before F_{23} is determined using (2c), then in absence of shear stresses, the failure envelope (3) can be proven to be an ellipsoid. It is well-known that none of strength properties can be practically obtained without experimental error, although clear indications of such errors are not usually available as a part of the published data. Imagine that there was a 10% experimental error in S_T, and its value drops to 40.5 MPa. Then the failure envelope would turn into an elliptic hyperboloid. An elliptic paraboloid is obtained when S_T=40.82 MPa. In each of these cases, conditions (10) are satisfied.

The second aspect is that prohibition of open failure envelopes, or infinite strength, contradicts the claim that the Tsai-Wu criterion reproduces the von Mises criterion for isotropic materials of equal tensile and compressive strengths, because the failure envelope for the von Mises criterion is open [27]. In fact, it is a cylinder in the space of three principal stresses intersecting coordinate planes in closed ellipses. Its axis corresponds to the hydrostatic stress under which the material is assumed to have infinite strength. This was supported by Bridgman's experimental observations [28] and actually laid the foundation to the study of theory of plasticity [29].

Practically, an infinite strength requires an infinitely high stress, which can never be achieved physically. On the other hand, if a material exhibits a much higher strength under a specific stress state than its strength under any uniaxial stress state, it would not be unreasonable to assume the strength under that special condition was practically infinite. Scientifically, a claim of an infinite strength under a specific condition can be equally as unfounded as a claim of a finite strength under all conditions. A more meaningful point to argue therefore should not be whether the claim of finite or infinite strength is correct, but whether a claim has been made consistently within the theoretical framework.

9.3 A logical choice of the quadric surface resulting from a quadratic failure criterion

The failure envelope is defined in a 6-dimensional stress space in general. If the failure criterion for transversely isotropic materials (3) is re-written as

$$F_{11}\sigma_1^2 + F_{22}\left(\sigma_2^2 + \sigma_3^2\right) + 2F_{23}\sigma_2\sigma_3 + 2F_{12}\sigma_1\left(\sigma_3 + \sigma_2\right) + F_1\sigma_1 + F_2\left(\sigma_2 + \sigma_3\right) = K \tag{12}$$

where $K = 1 - \left(F_{44}\tau_{23}^2 + F_{66}\tau_{13}^2 + F_{66}\tau_{12}^2\right)$ (13)

it can be considered that the presence of any non-vanishing shear stress can be equivalently considered as a reduction of the value of the constant term in (3) from 1 to a smaller value, denoted as K, which could be negative under some stress states (K=0 implies that failure is due to shear stresses alone). The nature of the failure function can be sufficiently comprehensively investigated within the 3-dimensional subspace of direct stresses.

According to analytic geometry [26], there are a number of invariants associated with the quadratic equation (12), namely

$$A = \begin{vmatrix} F_{11} & F_{12} & F_{12} & \frac{1}{2}F_1 \\ F_{12} & F_{22} & F_{23} & \frac{1}{2}F_2 \\ F_{12} & F_{23} & F_{22} & \frac{1}{2}F_2 \\ \frac{1}{2}F_1 & \frac{1}{2}F_2 & \frac{1}{2}F_2 & -K \end{vmatrix}, \tag{14}$$

$$A' = \begin{vmatrix} F_{22} & F_{23} & \frac{1}{2}F_2 \\ F_{23} & F_{22} & \frac{1}{2}F_2 \\ \frac{1}{2}F_2 & \frac{1}{2}F_2 & -K \end{vmatrix} + 2\begin{vmatrix} F_{11} & F_{12} & \frac{1}{2}F_1 \\ F_{12} & F_{22} & \frac{1}{2}F_2 \\ \frac{1}{2}F_1 & \frac{1}{2}F_2 & -K \end{vmatrix} + \begin{vmatrix} F_{11} & F_{12} & F_{12} \\ F_{12} & F_{22} & F_{23} \\ F_{12} & F_{23} & F_{22} \end{vmatrix}, \tag{15}$$

$$I = F_{11} + 2F_{22}, \tag{16}$$

$$J = \begin{vmatrix} F_{11} & F_{12} \\ F_{12} & F_{22} \end{vmatrix} + \begin{vmatrix} F_{22} & F_{23} \\ F_{23} & F_{22} \end{vmatrix} + \begin{vmatrix} F_{22} & F_{12} \\ F_{12} & F_{11} \end{vmatrix} = 2\left(F_{11}F_{22} - F_{12}^2\right) + F_{22}^2 - F_{23}^2, \tag{17}$$

$$D = \begin{vmatrix} F_{11} & F_{12} & F_{12} \\ F_{12} & F_{22} & F_{23} \\ F_{12} & F_{23} & F_{22} \end{vmatrix} = F_{11}F_{22}{}^2 + 2F_{23}F_{12}{}^2 - 2F_{22}F_{12}{}^2 - F_{11}F_{23}{}^2 = \left(F_{22} - F_{23}\right)\Delta, \tag{18}$$

where $\Delta = F_{11}\left(F_{22} + F_{23}\right) - 2F_{12}^2$. (19)

There are in fact a few more invariants for complete classification of quadric surfaces, but they are not useful in the present discussion and are therefore waived here.

Given expressions of the coefficients in (4a), which mean that the failure envelope intersects the coordinate axes at the points corresponding to the strengths of the material obtained under uniaxial tension or compression, one always has

$I > 0$, because $F_{11} > 0$ and $F_{22} > 0$. (20)

According to the constraints on the coefficient for the interactive terms in (10a) and (10c), which mean that the intersections of the failure envelope with coordinate planes are closed loci, one also has

$$J > 0, \qquad \text{because} \qquad F_{11}F_{22} - F_{12}^2 > 0 \quad \text{and} \quad F_{22}^2 - F_{23}^2 > 0. \tag{21}$$

According to analytic geometry, quadratic equation (12) defines a quadric surface in the three-dimensional $\sigma_1 - \sigma_2 - \sigma_3$ space. The relevant characteristics of this quadric surface are dictated by the invariants as defined in (14)-(18). In presence of conditions (20) and (21), the following classification can be made [26]:

if $A<0$ (any A') and $D>0$, it is an ellipsoid; (22a)
if $A<0$ (any A') and $D=0$, it is an elliptic paraboloid; (22b)
if $A<0$ (any A') and $D<0$, it is a two-sheets hyperboloid; (22c)
if $A>0$, $A'<0$ and $D<0$; it is a one-sheet hyperboloid; (22d)
if $A=0$, $A'<0$ and $D=0$, it is an elliptic cylinder. (22e)

For the sake of completeness, there is in fact another proper surface, that is an elliptic hyperboloid, often called a saddle surface. However, it is irrelevant to the present discussion, because it requires $J<0$, which cannot be met by the physical problem under consideration. Other combinations of the values of the invariants correspond to surfaces degenerated to some lines, points or even imaginary surfaces.

It can be seen that condition of $D=0$ is only shared by the elliptic paraboloid and the elliptic cylinder. Under condition $D=0$, an elliptic cylinder can be obtained when the coefficients in (12) take a specific combination satisfying $A=0$ whilst $A'<0$ ($A'\geq 0$ will lead to an imaginary surface or a straight line). Therefore, for the failure criterion (12) to reproduce the von Mises criterion, i.e. to degenerate to an elliptic cylinder from another quadric surface, $D=0$ is a necessary condition. Under this condition, for (12) to produce a real (as opposed to imaginary) and proper (as opposed to a straight line) surface other than an elliptic cylinder, an elliptic paraboloid is the only possibility.

The logical consequence of the argument based on analytic geometry is that for a composite failure criterion based on a quadratic function to reproduce the von Mises criterion for isotropic materials of equal tensile and compressive strength, the only possible shape of the failure envelope is an elliptic paraboloid. There is no alternative.

Interestingly, for isotropic materials, if the tensile and compressive strengths are different, a logical extension of the von Mises criterion is the Raghava-Caddell-Yeh criterion [30], which has been more rationally formulated in [31] using invariants of the first and second orders. The failure envelope resulting from the Raghava-Caddell-Yeh criterion is a paraboloid, and it is a circular paraboloid due to the isotropy of the material. With this criterion placed between the von Mises and the Tsai-Wu, the logical transition between them becomes much smoother. If the Raghava-Caddell-Yeh criterion is considered as the generalisation of the von Mises criterion [27] to accommodate the disparity between the tensile and compressive strength of isotropic materials due to their brittleness, then the Tsai-Wu criterion is the generalisation of the Raghava-Caddell-Yeh criterion in order to accommodate the differences of the strengths of composites in different directions due to their anisotropy.

After achieving the above understanding, there is one more aspect that should be settled. It is the orientation of the elliptic paraboloid. As it has only one opening, corresponding to infinite strength, it is conceivable that the opening should be towards the direction of triaxial compression, i.e. in the octant where each of the three direct stresses are negative.

9.4 The consequence of the failure envelope being an elliptic paraboloid

For the failure envelope to be an elliptic paraboloid, one of the necessary conditions is D=0, which gives an equation and therefore can be employed to determine one of the coefficients in (12). Determining a coefficient this way not only reduces the demand on necessary experiments, but, more importantly, it will help keep the consistency of the formulation. Since condition (10a) rules out the possibility of $F_{22}-F_{23}=0$, given the expression of D in (18), condition D=0 is equivalent to Δ=0. From the expression of Δ as given in (19), one has

$$F_{12}^2=\frac{1}{2}F_{11}\left(F_{22}+F_{23}\right) \tag{23}$$

where F_{11} and F_{22} are considered to be known for a given material, as determined in (4a). Relationship (23) agrees with what was obtained in [8]. It means that F_{23} and F_{12} are not both independent. If F_{12} is considered as a function of F_{23}, it can be plotted in Fig. 9.2 as a parabola. The domain of variable F_{23} can be obtained from condition (10a) as ($-F_{22}$, F_{22}). Corresponding to the domain, the range of the function (23) happens to coincide with condition in (10c), i.e. $\left(-\sqrt{F_{11}F_{22}},\sqrt{F_{11}F_{22}}\right)$. The permissible values of F_{23} and F_{12} can only be within the green rectangle as marked in Fig. 9.2, with the parabola passing its two corners on the right and touching its side on the left.

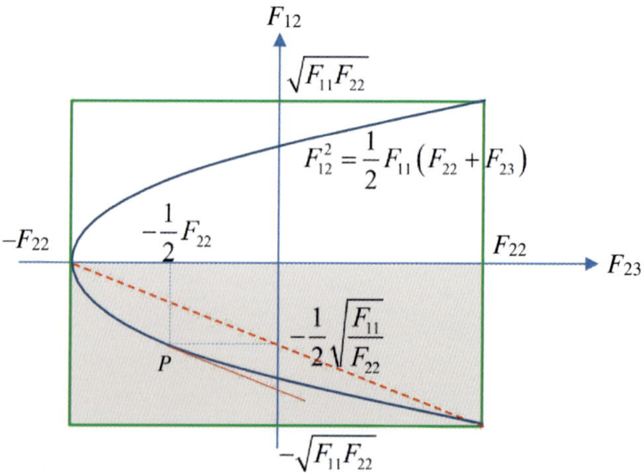

Fig. 9.2 The relationship between F_{12} and F_{23} and their permissible ranges

F_{12} as a function of F_{23} is apparently not single-valued. Each of the top and bottom halves of the parabola provides a single valued branch, corresponding to the positive and negative values of F_{12}, respectively. The physical interpretation of F_{12} is that the reinforcing effect brought in through the interactive term is observed only when the stresses involved are of the same sense, so that the term produces a negative value as its contribution to the failure function on the right-hand side of (12). Therefore, F_{12} will have to be negative. Then, the practical range for F_{12} of function (23) is only the negative half, i.e. $\left(-\sqrt{F_{11}F_{22}},0\right)$, and the meaningful single valued branch of the function is the bottom half of the parabola. Thus

$$F_{12}=-\frac{1}{\sqrt{2}}\sqrt{F_{11}\left(F_{22}+F_{23}\right)}\,. \tag{24}$$

Permissible values of F_{12} and F_{23} should therefore be taken from the shaded box as shown in Fig. 9.2.

9.5 The sensitivity of the quadratic failure criterion to the coefficients involved

Any point on the permissible branch of the parabola in Fig. 9.2 satisfies (24). To determine F_{12} and F_{23} completely means to fix a point, referred to as point P, on the parabola. Given a relationship between F_{23} and F_{12} as shown in (24), one could be tempted to shed the remaining responsibility of the full determination of F_{23} and F_{12} to experiments, i.e. to have one of them measured experimentally. Unfortunately, there are no reliable experimental means to directly determine either of them. The best one could do is to determine the transverse shear strength, hence F_{44}, and then obtain F_{23} using (2e). On the other hand, the available resources in terms of mathematics and logic have not been exhausted. The consideration of mathematical and logical consistency will offer another much-desired condition as will be pursued below.

Ch.9

Coefficients F_{12} and F_{23} are expressed one way or another in terms of strength properties, which are obtained experimentally, and hence errors are inevitable in their values. Any perturbation due to an experimental error in one of them will have to be met by a variation in the other to keep (24) satisfied. If P is selected close to the tip of the parabola, where dF_{12}/dF_{23} approaches infinity, it would require a large variation in F_{12} to correct a small perturbation in F_{23}, making the failure criterion extremely sensitive to small errors in the evaluation of F_{23}. Similarly, if P is placed close to the open end of the parabola, although the derivative is neither infinite nor zero, the failure criterion is bound to be more sensitive to F_{12} than to F_{23}. To the same effects of (24) as the relationship between F_{23} and F_{12}, such sensitivities can also be seen from (7) and (8) from the previous attempt to rationalise the Tsai-Wu criterion, although it was not completely satisfactory as reviewed earlier. Such sensitivities can be illustrated numerically as follows. Consider the same composite T300/PR-319 as cited earlier. Some numerical values have been generated as shown in Table 9.1 for F_{23} and F_{12} with respect to small variations of the transverse shear strength at different magnitude within its permissible range. The trend described above can be clearly seen, where the percentage variations were calculated with reference to their respective nominal values obtained at S_T=45 MPa as shaded in Table 9.1.

The parabola shown in Fig. 9.2 will be subject to further examination, as its intricacy is yet to be unravelled. In order to balance F_{12} and F_{23} on an unbiased basis, they can now be normalised with respect to the breadths of their single-valued ranges, i.e. within the shaded box in Fig. 9.2, respectively, as follows:

$$\tilde{F}_{12}=\frac{F_{12}}{\sqrt{F_{11}F_{22}}}\quad\text{and}\quad\tilde{F}_{23}=\frac{F_{23}}{2F_{22}}\,. \tag{25}$$

Thus, $\tilde{F}_{12}$ varies from 0 to –1 over a non-dimensional unit distance as $\tilde{F}_{23}$ takes its value from –1/2 to 1/2, also covering a non-dimensional unit distance. Both F_{12} and F_{23} are now normalised in a consistent and unbiased manner. Accordingly, the expression of Δ as given in (19) can also be normalised as

$$\tilde{\Delta}=\frac{\Delta}{2F_{11}F_{22}}=\frac{1}{2}+\tilde{F}_{23}-\tilde{F}_{12}^2. \tag{26}$$

Table 9.1 Sensitivities of F_{23} and F_{12} with respect to the variability of S_T

(T300/PR-319 composite with Y_t=40 MPa and Y_c=125 MPa)

S_T (MPa)	ΔS_T	$F_{23}(\times 10^{-6}/\text{MPa}^2)$	ΔF_{23}	$F_{12}(\times 10^{-6}/\text{MPa}^2)$	ΔF_{12}
46.0	0.22%	-36.29	2.2%	7.907	0.33%
45.9	N/A	-37.33	N/A	7.882	N/A
45.0		-46.91		-7.467	
35.5		-196.7		-5.027	
35.4	0.22%	-199.0	4.9%	-6.207	15.8%

The mathematical interpretation of equal sensitivity of $\tilde{\Delta}$ with respect to $\tilde{F}_{12}$ and $\tilde{F}_{23}$ is

$$\frac{\partial\tilde{\Delta}}{\partial\tilde{F}_{12}}=\frac{\partial\tilde{\Delta}}{\partial\tilde{F}_{23}}. \tag{27}$$

From (26), one has

$$\frac{\partial\tilde{\Delta}}{\partial\tilde{F}_{12}}=-2\tilde{F}_{12} \qquad \text{and} \qquad \frac{\partial\tilde{\Delta}}{\partial\tilde{F}_{23}}=1 \tag{28}$$

and thus, given (28), condition (27) leads to

$$\tilde{F}_{12}=-\frac{1}{2} \qquad \text{or} \qquad F_{12}=-\frac{1}{2}\sqrt{F_{11}F_{22}} \quad \text{as also given in (5).} \tag{29}$$

Since Δ=0, its nondimensionalised form $\tilde{\Delta}$ as defined in (26) should also vanish, i.e.

$$\tilde{\Delta}=\frac{1}{2}+\tilde{F}_{23}-\tilde{F}_{12}^2=0. \tag{30}$$

Given (29), one has

$$\tilde{F}_{23}=-\frac{1}{4} \qquad \text{or} \qquad F_{23}=-\frac{1}{2}F_{22} \quad \text{as also given in (6).} \tag{31}$$

Both F_{12} and F_{23} have been determined collectively as shown above. Whilst relationship (23) represents a rigorous consideration from analytic geometry, condition (27) results from the logical consideration of unbiased sensitivity of the failure criterion to F_{12} and F_{23}. The normalisations made to F_{12} and F_{23} in (25) ensure that they can be compared like with like.

From (31), given (2e) and the expressions of F_{44} and F_{22} in (4c) and (4a), respectively, a natural consequence is that

$$F_{44} = 2(F_{22} - F_{23}) = 3F_{22} \qquad \text{or} \qquad S_T = \sqrt{Y_t Y_c / 3}\,. \tag{32}$$

This suggests that amongst three transverse strengths, Y_t, Y_c and S_T, of a transversely isotropic material, only two can be independent if the failure criterion is based on a single quadratic function.

9.6 The fully rationalised Tsai-Wu failure criterion

With the coefficients to the interactive terms determined as in the previous section, the fully rationalised Tsai-Wu criterion can be delivered as:

$$F_1\sigma_1 + F_2(\sigma_2+\sigma_3) + F_{11}\sigma_1^2 + F_{22}\left(\sigma_2^2+\sigma_3^2-\sigma_2\sigma_3+3\tau_{23}^2\right) - \sqrt{F_{11}F_{22}}\sigma_1(\sigma_3+\sigma_2) + F_{66}\left(\tau_{13}^2+\tau_{12}^2\right) = 1\,. \tag{33}$$

It involves only 5 independent coefficients F_1, F_2, F_{11}, F_{22} and F_{66} which can be evaluated from the conventional strength properties, *viz*. tensile and compressive strengths in directions along and transverse to fibres, and the in-plane shear strength, according to (4), out of which S_{44} has been made redundant. The nature of the failure envelope defined by (33) can be confirmed with full mathematical certainty as follows.

With F_{12} and F_{23} as determined in (29) and (31), one can verify

$$A = -\frac{3}{16}F_{22}\left(\sqrt{F_{22}}F_1 + 2\sqrt{F_{11}}F_2\right)^2 < 0 \tag{34}$$

provided

$$-F_1\big/\sqrt{F_{11}} \neq 2F_2\big/\sqrt{F_{22}}\,. \tag{35}$$

Conditions $J>0$ and $D=0$ have been satisfied as shown in (21) and (23). Therefore, conditions for an elliptic paraboloid in the subspace of direct stresses as given in (22b) are all met and (33) is indeed an elliptic paraboloid in the 3-dimensional subspace of the direct stresses. It should be noted that because of the anisotropy, its failure envelope is a genuine elliptic paraboloid, not a circular paraboloid, in general.

Condition (35) is met by most of practical composites and any exception means $-F_1\big/\sqrt{F_{11}} = 2F_2\big/\sqrt{F_{22}}$, which is very rare. Even if it is met, it only turns $A<0$ to $A=0$, as $A>0$ will never be possible. A special case of such an exception when $F_1=F_2=0$ will be discussed later.

For 2D stress states (plane stress) in the σ_1-σ_2 plane, (33) reduces to

$$F_1\sigma_1 + F_2\sigma_2 + F_{11}\sigma_1^2 + F_{22}\sigma_2^2 - \sqrt{F_{11}F_{22}}\sigma_1\sigma_2 + F_{66}\tau_{12}^2 = 1 . \tag{36}$$

An important observation to make is that to predict the failure of a material under a 3D stress state from (33), the same set of coefficients is involved and therefore the same set of strength properties is required as that for a 2D stress state from (36). According to analytic geometry, it can be easily proven that (36) defines an ellipse in the σ_1-σ_2 plane in absence of the shear stress, as expected.

It should be emphasised here that rationalisation above has to be carried out to the criterion in its 3D form before it is reduced to 2D. Although (36) is a well-known form [23], its rational justification can only come through the route of manipulating the criterion in its general form (33) for 3D stress states.

9.7 A practical aspect of the Tsai-Wu criterion and its implications

The quadratic form of the failure function of the Tsai-Wu criterion can cause certain inconvenience when applied in design practices. Specifically, one is not always lucky enough to obtain a value 1 directly under a given stress condition. If the criterion returns a value of 1.2, it does not mean that the material is 20% overloaded because of the nonlinear nature of the failure function. An appropriate measure is therefore desirable for application of the criterion in practical engineering design of composite structures.

Assume a loading conditions where all stress components increase or decrease proportionally. In this case, a loading parameter λ can be introduced as the proportion factor, so that the stress state can be expressed as

$$\{\sigma\} = \lambda\{\bar{\sigma}\} \tag{37}$$

where $\{\bar{\sigma}\}$ signifies the relative magnitudes of the stress components and can be set as a unit vector in 6-dimensional space. The Tsai-Wu criterion can be expressed as

$$\lambda\bar{L} + \lambda^2\bar{Q} = 1 \tag{38}$$

where

$$\bar{L} = F_1\bar{\sigma}_1 + F_2\left(\bar{\sigma}_2 + \bar{\sigma}_3\right), \tag{39a}$$

$$\bar{Q} = F_{11}\bar{\sigma}_1^2 + F_{22}\left(\bar{\sigma}_2^2 + \bar{\sigma}_3^2\right) - F_{22}\bar{\sigma}_2\bar{\sigma}_3 - \sqrt{F_{11}F_{22}}\bar{\sigma}_1\left(\bar{\sigma}_3 + \bar{\sigma}_2\right) + 3F_{22}\bar{\tau}_{23}^2 + F_{66}\left(\bar{\tau}_{13}^2 + \bar{\tau}_{12}^2\right). \tag{39b}$$

After a re-arrangement, the non-negative sense of $\bar{Q}$ can be shown as

$$\bar{Q} = \frac{3}{4}F_{22}\left(\bar{\sigma}_2 - \bar{\sigma}_3\right)^2 + \frac{1}{4}\left(2\sqrt{F_{11}}\bar{\sigma}_1 - \sqrt{F_{22}}\left(\bar{\sigma}_2 + \bar{\sigma}_3\right)\right)^2 + 3F_{22}\bar{\tau}_{23}^2 + F_{66}\left(\bar{\tau}_{13}^2 + \bar{\tau}_{12}^2\right) \geq 0 . \tag{39c}$$

Solving (38) as a quadratic equation for λ, one obtains two roots and, given the non-negative sense of $\bar{Q}$ as shown in (39c), their senses are also determined as follows:

$$\lambda_1 = \frac{-\bar{L}+\sqrt{\bar{L}^2+4\bar{Q}}}{2\bar{Q}} \geq 0 \quad \& \quad \lambda_2 = \frac{-\bar{L}-\sqrt{\bar{L}^2+4\bar{Q}}}{2\bar{Q}} \leq 0. \tag{40}$$

They represent the two distinct scenarios respectively: λ_1 gives the proportion factor by which $\{\bar{\sigma}\}$ should be increased or reduced to bring the stress level to the critical state of failure whilst maintaining the sense of the stress state; and λ_2 is the proportion factor that should be applied to hit to the critical state of failure whilst the sense of the stress state is reversed, as illustrated in Fig. 9.3. Usually, if one is sure about the sense of the stress state, i.e. $\{\bar{\sigma}\}$, then λ_1 should be used. However, in problems where the sense of the stress state is not certain, for instance, during an optimisation process, the user is advised to take both into consideration.

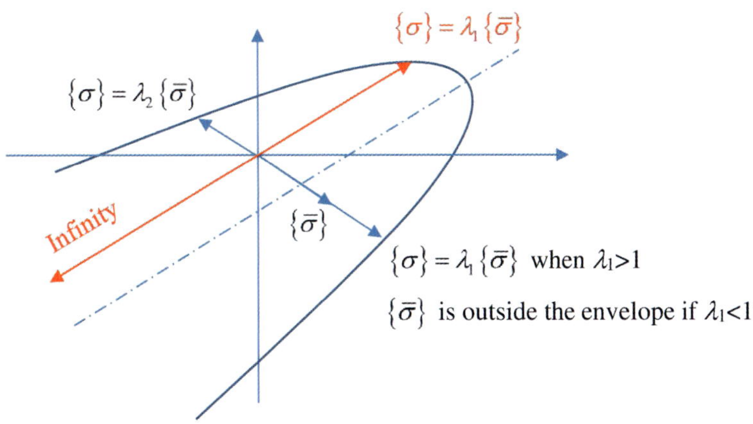

Fig. 9.3 Failure envelope in σ_I-σ_T plane

A special but also important case is $\bar{Q}=0$. In this case,

$$\lambda_1 = \lim_{\bar{Q}\to 0} \frac{-\bar{L}+\sqrt{\bar{L}^2+4\bar{Q}}}{2\bar{Q}} = \lim_{\bar{Q}\to 0} \frac{4}{4\sqrt{\bar{L}^2+4\bar{Q}}} = \frac{1}{|\bar{L}|} \quad \& \quad \lambda_2 = \frac{-\bar{L}-\sqrt{\bar{L}^2+4\bar{Q}}}{2\bar{Q}} \to -\infty. \tag{41}$$

In the subspace of direct stresses, given (39c), $\bar{Q}=0$ implies

$$\bar{\sigma}_2 = \bar{\sigma}_3 \qquad \text{and} \qquad \sqrt{F_{11}}\bar{\sigma}_1 = \sqrt{F_{22}}\bar{\sigma}_2. \tag{42}$$

In other words, under triaxial compression of the following stress ratio

$$\sigma_1 : \sigma_2 : \sigma_3 = -\sqrt{F_{22}/F_{11}} : -1 : -1 \tag{43}$$

the material exhibits infinite strength. In this case, the axis corresponding to this stress ratio is parallel to the axis of the elliptic paraboloid (usually they do not coincide with each other unless the tensile and compressive strengths of the material happen to be equal), as shown in Fig. 9.3 in red. According to the focus principle of parabola, all such axes intersect the parabola only once corresponding to λ_1. Imagine that a light source was placed at the focus of parabola and the light beams from the source reflect at the parabola. The loading axis along the red line in Fig. 9.3 will be the beam of light reflected at the parabolic surface, and it would remain parallel to the axis of the parabola tending to infinity on the open side of the parabola.

It can be stated clearly now that the assumption of infinite strength under hydrostatic compression [27] is applicable only to isotropic materials when $F_{11}=F_{22}$. Assuming this for anisotropic composites as in [25] would certainly be misleading. For a transversely isotropic material, an infinite strength can be obtained, but not under the hydrostatic pressure anymore. It should be under a triaxial compression state at the stress ratio as given by (43), associated with the anisotropy of the strength of the material.

The same consideration can be applied to the 2D version of the Tsai-Wu criterion, although the mathematical manipulation associated with $\bar{Q}$ and the outcomes will be a little different.

$$\bar{L} = F_1\bar{\sigma}_1 + F_2\bar{\sigma}_2, \tag{44a}$$

$$\bar{Q} = F_{11}\bar{\sigma}_1^2 + F_{22}\bar{\sigma}_2^2 - \sqrt{F_{11}F_{22}}\bar{\sigma}_1\bar{\sigma}_2 + F_{66}\bar{\tau}_{12}^2 = \frac{1}{2}\left(F_{11}\bar{\sigma}_1^2 + F_{22}\bar{\sigma}_2^2\right) + \frac{1}{2}\left(\sqrt{F_{11}}\bar{\sigma}_1 - \sqrt{F_{22}}\bar{\sigma}_2\right)^2 + F_{66}\bar{\tau}_{12}^2 \geq 0. \tag{44b}$$

In this case, $\bar{Q}=0$ implies

$$\bar{\sigma}_1 = \bar{\sigma}_2 = \bar{\tau}_{12} = 0. \tag{45}$$

In other words, $\bar{Q}=0$ is possible only if the material is not stressed at all. Otherwise, there will always be a positive and a negative root for λ. This also agree with the condition that the intersection of the failure envelope with σ_1-σ_2 coordinate plane, which is effectively what the Tsai-Wu criterion for plane stresses predicts, is closed. In fact, it is an ellipse as has been mentioned earlier.

9.8 Special cases of the fully rationalised Tsai-Wu failure criterion as necessary verifications

The point on marked the parabola in Fig. 9.2, as defined by (29) for F_{12} and (31) for F_{23}, is special in multiple ways. It has been argued above that it renders failure criterion (33) to be equally sensitive to F_{12} and F_{23} under the condition that the failure envelope is an elliptic paraboloid. The slope to the parabola at this point happens to be equal to the slope of the secant between the two extremes of the single valued branch concerned, i.e. the tip and the parabola and the bottom right end. The tangent and secant have been marked in Fig. 9.1 in solid and dashed red lines, respectively. It is therefore the point whose existence has been asserted by the mean-value theorem [26].

To appreciate the outcomes above and to justify the normalisations introduced in (25), all the terms in the failure criterion (33) are normalised as follows, which is similar to procedure used in [23]:

$$\tilde{F}_1\tilde{\sigma}_1 + \tilde{F}_2\left(\tilde{\sigma}_2 + \tilde{\sigma}_3\right) + \tilde{\sigma}_1^2 + \tilde{\sigma}_2^2 + \tilde{\sigma}_3^2 - \tilde{\sigma}_2\tilde{\sigma}_3 - \tilde{\sigma}_1\left(\tilde{\sigma}_3 + \tilde{\sigma}_2\right) + \tilde{\tau}_{23}^2 + \tilde{\tau}_{13}^2 + \tilde{\tau}_{12}^2 = 1 \tag{46}$$

where

$$\tilde{\sigma}_1 = \sqrt{F_{11}}\sigma_1, \quad \tilde{\sigma}_2 = \sqrt{F_{22}}\sigma_2, \quad \tilde{\sigma}_3 = \sqrt{F_{22}}\sigma_3,$$
$$\tilde{\tau}_{23} = \sqrt{F_{44}}\tau_{23} = \sqrt{3F_{22}}\tau_{23}, \quad \tilde{\tau}_{13} = \sqrt{F_{66}}\tau_{13}, \quad \tilde{\tau}_{12} = \sqrt{F_{66}}\tau_{12}, \tag{47}$$
$$\tilde{F}_1 = F_1/\sqrt{F_{11}}, \quad \tilde{F}_2 = F_2/\sqrt{F_{22}}$$

and $\tilde{F}_{12}$ and $\tilde{F}_{23}$ obtained in (29) and (31) have also been incorporated. All terms of the similar nature tend to contribute to F in a similar way, whilst the differences due to the anisotropy of the material in all second order terms have been absorbed in the normalisation scheme. The contributions of F_{12} and F_{23} as obtained in (29) and (31) have indeed been equalised in their normalised form.

As a further verification, reducing the materials to isotropic ones of unequal tensile and compressive strengths, for which $F_1=F_2$ and $F_{11}=F_{22}$, whilst conditions $J>0$ and $D=0$ remain satisfied, given (21) and (23), the expression of A in (34) reduces to

$$A = -\frac{27}{16}F_{11}^2F_1^2 < 0 \quad \text{since} \quad F_1 \neq 0 \text{ for unequal tensile and compressive strengths.} \tag{48}$$

In this case, equation (33) reduces to

$$\left(\frac{1}{X_t} - \frac{1}{X_c}\right)(\sigma_1 + \sigma_2 + \sigma_3) + \frac{1}{X_t X_c}\left(\sigma_1^2 + \sigma_2^2 + \sigma_3^2 - \sigma_2\sigma_3 - \sigma_1\sigma_3 - \sigma_1\sigma_2 + 3\tau_{23}^2 + 3\tau_{13}^2 + 3\tau_{12}^2\right) = 1 \tag{49}$$

where X_t and X_c are the tensile and compressive strengths of the isotropic material, respectively. This reproduces the Raghava-Caddell-Yeh criterion [30], for which the failure envelope is a circular paraboloid demonstrating infinite strength against hydrostatic compression, but not hydrostatic tension.

The existence of a relationship amongst transverse strengths as shown in (32) can also be argued as follows. For an isotropic material under plane stresses in the σ_2-σ_3 plane, a proper failure criterion for transversely isotropic materials should reproduce the result of a failure criterion for isotropic materials. The Raghava-Caddell-Yeh criterion is for isotropic materials, which requires only two independent strength properties. For any failure criterion for transversely isotropic materials to reproduce the Raghava-Caddell-Yeh criterion in this respect, only two strength properties in this plane can be accommodated. The third one has to be related to the other two.

As an even more special case, for isotropic materials of equal tensile and compressive strengths, i.e. $X_t=X_c=X$, condition (35) cannot be met, because $-F_1/\sqrt{F_{11}} = 2F_2/\sqrt{F_{22}} = 0$. As a result, $A=0$. The failure envelope as a quadric surface become a cylinder and the Raghava-Caddell-Yeh criterion (49) further reduces to the von Mises criterion [27]

$$\sigma_1^2 + \sigma_2^2 + \sigma_3^2 - \sigma_2\sigma_3 - \sigma_1\sigma_3 - \sigma_1\sigma_2 + 3\tau_{23}^2 + 3\tau_{13}^2 + 3\tau_{12}^2 = X^2. \tag{50}$$

It can be seen that the cylindrical failure envelope can be considered as an extreme case of an elliptic paraboloid mathematically as tensile and compressive strengths approach each other, allowing the von Mises criterion to be seen as a special case of the fully rationalised Tsai-Wu criterion with the Raghava-Caddell-Yeh criterion as an intermediate position.

As a further point to discuss, relationship (32) amongst transverse strengths means that δ=1 according to (8). Equation (7) as obtained in [8] naturally leads to (29), reproducing (5) as Tsai and Hahn suggested in [23]. In [8], it was suggested that δ was a material constant which could vary from material to material if the measured data were used directly. Relationship (32) asserts further that δ is a universal constant, i.e. δ=1, for all transversely isotropic materials, including isotropic materials as a special case, brittle or ductile. This is apparently observed by isotropic materials obeying the Raghava-Caddell-Yeh criterion [30] or the von Mises criterion [27], as some special cases. Any variation in δ as those shown in [8] over the range of composites involved there should be attributed to the variability in experimentally measured transverse strength properties, in the same way as was argued for elastic properties associated with relationship (9).

As a concluding remark on the exploration above, relationship (29) had been suggested by Tsai and Hahn [23] whilst (31) obtained by Kallas and Hahn [24]. There is nothing new from the present exploration in this respect as both were known individually. The effort made above merely establishes a logical relationship between the two based on analytic geometry of quadratic functions. As an analogy, apples falling off the trees and the Moon hanging in the sky were both common sense but unrelated phenomena until the law of universal gravitation came to light.

9.9 Transverse shear strength: its true significance and appropriate role

Equation (32) gives a relationship amongst transverse strengths. If two of them are known, the third can be predicted using the relationship. Practically, amongst the three, the transverse tensile is the easiest to obtain, and the method employed to obtain it is the most reliable, since the test can be performed under the uniaxial stress state accurately, as long as the specimens are made sufficiently straight and aligned properly in the test machine. However, this does not eliminate the variability in the test data as it is inevitable due to the brittle nature of the material. The transverse compressive strength is also relatively easy to obtain. One might sometimes find less variability in these tests than in the tensile tests. However, its accuracy is undermined by the lack of uniaxial stress state. To avoid premature failure due to buckling, the specimens usually have to be made wide and short. The failure does not actually take place under uniaxial compression as this strength property is defined. Relatively, the measurement method for the transverse shear strength is least established. In fact, there has hardly been any specific standard for this. Available data were likely obtained following the procedure designed for the in-plane shear strength instead. Any method adopted is bound to be more complicated than its tensile and compressive counterparts, adding to the uncertainty. It can be stated without any reservation that this strength property is the least reliable amongst all experimentally measured transverse strength properties. Therefore, when they are employed in practical failure analysis, users ought to be aware of the difference in reliability of these transverse strength properties and place appropriate levels of confidence on each of them accordingly.

Given what was described above, direct use of transverse shear strength in a failure analysis in the same way as other strength properties is a questionable approach. In fact, for most composites, transverse shear strengths are usually unavailable. In this case, relationship (32) can be conveniently adapted to avoid the need for such a property, as it is not independent, very much like the minor Poisson's ratio ν_{21} as an elastic property, which also tends to show more variability than other elastic properties, if measured.

In the rare circumstances where the transverse shear strength is available, a question arises of how it should be used. Ignoring it seems wasteful, bearing in mind that it is often obtained with a lot more efforts than the other properties. The following two ways of taking advantage of it can be proposed.

9.9.1 As a means of validation of the theory

Available transverse shear strengths can be employed to validate relationship (32). Given that if (32) is considered as the strength of the material under pure transverse shear predicted from failure criterion (33), the degree of its agreement with experimentally obtained transverse shear strength can serve as a basic level of validation of criterion (33). In this respect, users have to bear in mind two generic considerations. One is that any failure criterion is a theoretical idealisation based on many simplifications. Its qualitative outcomes should be given sufficient emphasis before quantitative information becomes relevant. Perfect quantitative agreement with experiments, whilst desirable, is a luxury not always available. Ample room will have to be allowed for possible disagreements. Another consideration is that experimental data should never be taken for granted as the truth for theories to meet. So-called experimental values are not always obtained through strict and scientific processes. Given the lack of reliability in the transverse shear strength as described above in particular, due diligence is essential.

As an attempt along this line, (32) has been applied to the range of materials involved in WWFE-I~III as was referred to in [1-3]. The results are shown in Table 9.2 with the measured transverse strengths provided in row 1 to 3 and the transverse shear strength predicted from (32) listed in row 4.

On the prediction of transverse shear strengths, it would be interesting to cite a couples of independent accounts in the literature. In Puck's theory [32,33] the transverse shear strength is a derived quantity and it is equal to the transverse tensile strength, i.e.

$$S_T = Y_t \,. \tag{51}$$

It corresponds to the special case of (32) when Y_c=$3Y_t$. The predictions according to the Puck criterion, i.e. from (51), have been shown in row 5 in Table 9.2, which are identical to those in row 1, as expected. Relative to the measured values in row 3, those from (32) outperform the predictions from the Puck theory for all except S-2 glass/Epoxy and G40-800/5026 as highlighted in Table 9.2. Even for these two cases, one only marginally underperforms, whilst the other happens to be identical to the Puck theory prediction. As far as the prediction of the transverse shear strength is concerned, the fully rationalised Tsai-Wu criterion seems to be superior to the Puck criterion. When applying the fully rationalised Tsai-Wu criterion (33) to problems involving general stress states, one should be prepared for discrepancies of a magnitude comparable with those between the figures in rows 3 and 4 in Table 9.2 as an indication of the level of accuracy this criterion is capable of offering. The same statement can be made to the Puck criterion where level of accuracy is indicated by discrepancies between the figures in rows 3 and 5 of Table 9.2.

Table 9.2 Experimentally measured transverse strengths and predicted transverse shear strengths according to different theories, all in MPa

	Row no.	Strength	Type of composite considered								
			AS4 3501-6	T300 BSL914C	E-glass LY556	E-Glass MY750	IM7 8551-7	T300 PR-319	A-S Epoxy 1	S-2 glass Epoxy 2	G40-800 5026
Experimentally measured strengths	1	Y_t	48	27	35	40	68	40	38	56.5	70
	2	Y_c	200	200	114	145	185	125	150	180	210
	3	S_T	55	N/A	50	50	57	45	50	40	57
Predicted S_T from Y_t & Y_c as the input	4	Present prediction according to Eq.(32)	56.57	42.43	36.47	43.97	64.76	40.82	43.59	58.22	70.00
	5	Puck's prediction according to Eq. (51)[32,33]	48	27	35	40	68	40	38	56.5	70
	6	Christensen's prediction according to Eq.(52)[34]	50.93	43.82	33.50	42.56	58.59	36.91	41.90	54.48	56.12

Another relationship amongst transverse strengths can be found in [34], given as follows

$$S_T^2 = \left(\frac{Y_c + Y_t}{3Y_c + 5Y_t}\right) Y_t Y_c \quad \text{or} \quad \frac{S_T^2}{Y_t Y_c} = \frac{Y_c + Y_t}{3Y_c + 5Y_t}. \tag{52}$$

It is apparently a more complicated expression than that given in (32). The derivation in [34] was based on the following failure criterion stemming from Hashin's matrix failure criterion [11] before it was split into tensile and compressive modes.

$$F_2(\sigma_2 + \sigma_3) + F_{22}(\sigma_2 + \sigma_3)^2 + F_{44}\left(\tau_{23}^2 - \sigma_2\sigma_3\right) = 1. \tag{53}$$

This can be equally obtained from (12) under a plane stress condition in the plane transverse to the fibres, given relationship (2e) due to the transverse isotopy. The contributions from fibre direction as well as the interactive term associated with F_{12} have been excluded *a priori*. Therefore, (53) only represents the restricted view of the behaviour of the material from a specific perspective under the plane stress condition, and therefore it is not expected to be any more comprehensive than the fully rationalised Tsai-Wu criterion (33).

The interaction between stresses represented by F_{12} resembles the role of the Poisson's ratios in the generalised Hooke's law in the theory of elasticity. It might be true that, in the solutions to many elastic problems, the contributions from the Poisson's ratios are limited in terms of magnitude. Whilst ignoring such contributions would not lead to excessive error, the absence of Poisson's ratios would prevent appropriate understanding of many important mechanical behaviours of materials, such as the anticlastic behaviour, free edge effects, etc.

Similarly, to reveal the special relationships amongst strength properties, it is important to incorporate the interactive terms in the failure criterion and leave the mathematics to determine the magnitude of their contributions. In the well-known Hashin criterion [11], the interactive term F_{12} was ignored based on an assumption that failure is determined by stresses on the failure plane, which was inherited from the Mohr criterion [21] for isotropic materials, but this particular assumption is inapplicable to anisotropic composites as was argued in [22]. Whilst the numerical errors could be insignificant in many cases, as Hashin also stated in [11], the exclusion of such interactions would not allow the intricate logical relationships (29) and (31) to be revealed. The way of treating the interactive terms can also be viewed as a meaningful indicator of consistency of the formulation. Including one interactive coefficients, F_{23}, in the failure function, whilst excluding another, F_{12}, as was the practice in [11], can at least be viewed as a biased approach deviating from the spirit of being objective and consistent. The numerical closeness between the results from a theory and the experimental data in one aspect or another is important, but the authors believe that the mathematical and logical consistency is even more important.

The numerical values obtained from (52) according to [34] over the same range of materials in the previous subsection are shown in row 6 in Table 9.2. Relative to the experimental data as in row 3, the predictions from (32) as listed in row 4 outperform those from (52) for 5 out of 8 materials. The trend can be further shown graphically in Fig. 9.4 where ratio S_T^2/Y_tY_c is plotted as a function of ratio Y_t/Y_c. The curve corresponding to (52) does not seem to be more representative than the straight line obtained according to (32), understanding the variability in measured transverse strengths as argued in [34]. The predictions are not more relevant to carbon composites (black symbols) than to glass composites (green symbols) either as was claimed in [34].

Logically, the major premise underlying (52) is failure criterion (53), which is at its best a special form of the criterion based on a full quadratic failure function from which (32) is obtained. There is no reason to expect (52) to be more representative than (32). In [34], efforts were made to eliminate one of the strength properties from (53) whilst keeping the manipulations all within the transverse plane, whilst the (32) was obtained from the fully rationalised Tsai-Wu criterion which is formulated for 3D stresses and can also be consistently degenerated to 2D stresses. In contrast to (53), the consistent criterion for plane stresses in the σ_2-σ_3 plane from the fully rationalised Tsai-Wu criterion gives

$$F_2\left(\sigma_2+\sigma_3\right)+F_{22}\left(\sigma_2^2+\sigma_3^2-\sigma_2\sigma_3+3\tau_{23}^2\right)=1. \tag{54}$$

Apparently, for mathematical and logical consistency, the coefficients to both second order stress invariants must be related, as was the case in the Raghava-Caddell-Yeh criterion, as well as the von Mises criterion, in the plane of isotropy.

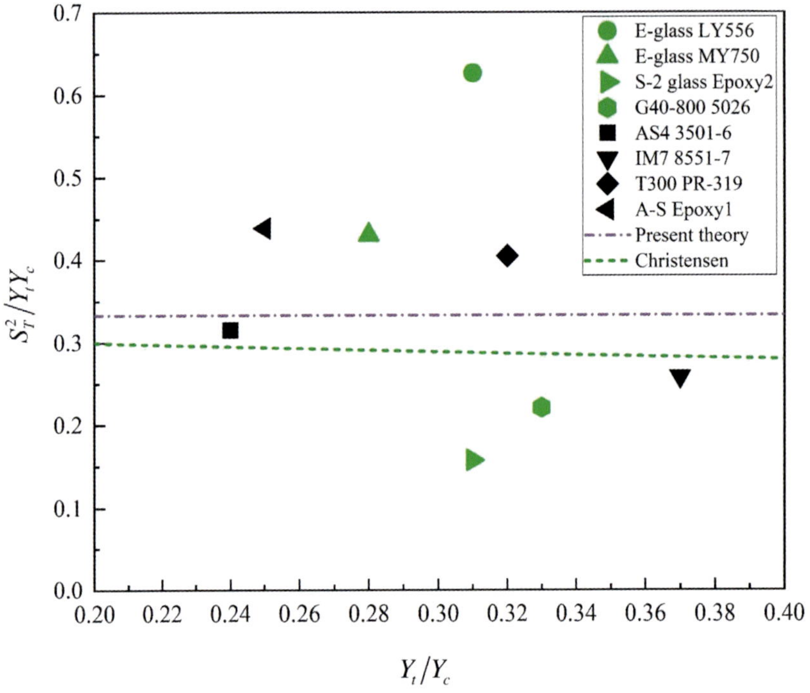

Fig. 9.4 S_T^2/Y_tY_c versus Y_t/Y_c

9.9.2 As a means of calibrating other transverse strength properties

There is genuine lack of complete sets of experimental data in the public domain. A widely publicised document [35] prepared to meet the FAA (Federal Aviation Administration) certification requirements contains detailed testing data for transverse tensile and compressive strengths of Hexcel 8552 AS4 UD composite. Taking room temperature dry condition as an example, results for transverse tensile strength were obtained from two batches of specimens. Relative to a mean value of 9.27 ksi, the variabilities were 19% and 30% for the two batches respectively. In the case of the second batch, the measured values fluctuate between about 7.4 and 10.3 ksi. The compressive strengths as obtained in the same report seemed more consistent, but this should not be assumed to have higher reliability, given the compromised uniaxial stress condition as mentioned earlier. Unfortunately, no transverse shear strengths were provided. The range of variabilities as reported in [35] was by no means the extreme case, but should be sufficiently representative. If one is prepared to accept this level of variability, then corrections to the test data within this range would not be unacceptable as a recalibrated presentation of the data. For the same set of composites as shown in Table 9.2, the measured data points are shown in Fig. 9.5 as solid symbols. To a different degree, they are off the 1:1 straight line representing equation (32). If the transverse shear strengths predicted from (32) is employed instead of the measured ones, it will draw all points vertically onto the 1:1 straight line, shown as hollow symbols of the same shape and colour. The magnitudes of the changes in the transverse shear strengths can be measured as the distance between the solid and hollow symbols of the same shape and colour in Fig. 9.5. The numerical values of these changes are provided in row 4 in Table 9.3 as were also shown in Table 9.2 and in percentage in row 5 of Table 9.3 relative to the measured values. In most cases, the magnitudes of the changes are not unacceptable. However, making such changes would mean that the measured transverse shear strengths should not be used at all in the failure criterion, e.g. in (33), even when they are available. This would be a waste of resources.

Given the above discussion, as the second way of making use of available transverse shear strength data, they can be employed to moderate or recalibrate the remaining transverse strengths. A simple way of achieving this is to fit the data of all three transverse strengths into relationship (32) to eliminate systematic errors arising from test methods and data processing. Given the fact that these strengths carry different degrees of reliability, it is therefore advisable to incorporate such reliability measure into the fitting process. As a first attempt, the following scheme can be proposed.

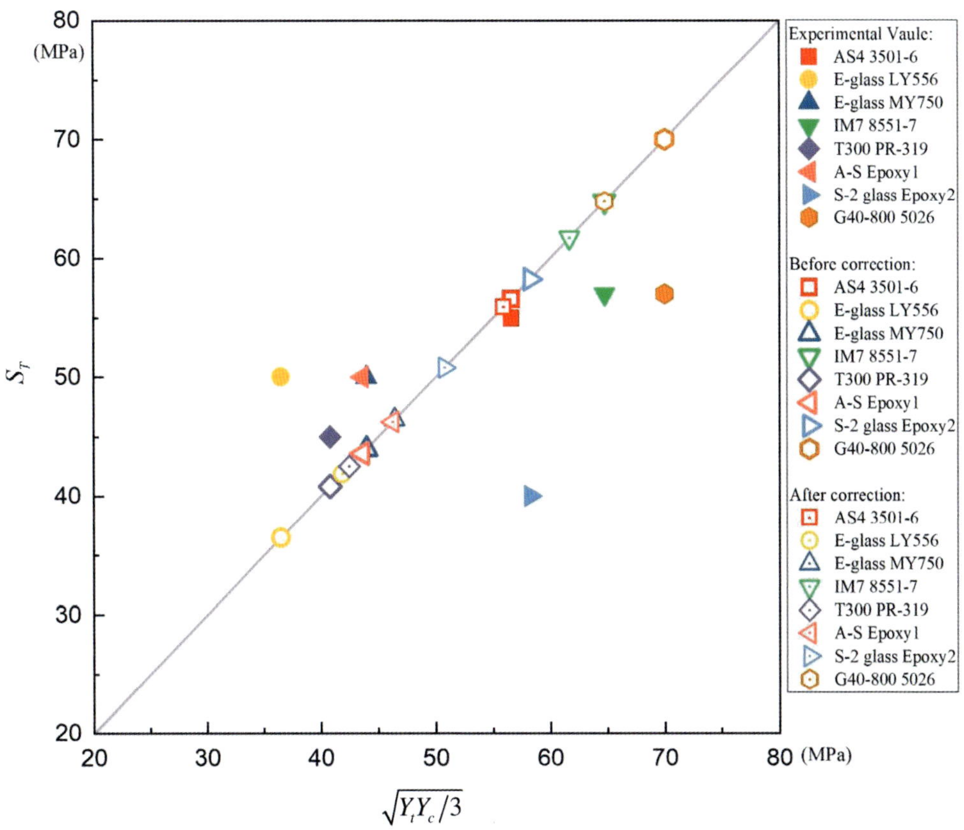

Fig. 9.5 Transverse strength data for a range of materials plotted on the $\sqrt{Y_t Y_c / 3}$ - S_T plane

Assume a systematic source of error ω in all measured strengths. Because of the different reliability in the measured strengths, the correction to each strength property is weighed by a measure of its variability. Thus

$$\begin{aligned} S_T^c &= S_T + \alpha\omega \\ Y_t^c &= Y_t \pm \beta\omega \\ Y_c^c &= Y_c \pm \chi\omega \end{aligned} \tag{55}$$

where superscript c stands for corrected value, α, β and χ are the weights as a measure of the variability of the respective strength property employed. The higher the variability, the higher the value of α, β or χ. In absence of such a measure, one might consider the standard deviation from the respective data set as substitute, if available. A further concession will be for the users to exercise their own judgement in the absence of any measure, which is probably the most likely position in reality. It would be reasonable to expect χ to have a higher value than α and β based on the discussion above about the lack of reliability of the measured transverse shear strength. Since ω can be positive as well as negative, there is no need to have ± sign in the first equation in (55). The corrections aim at (32), i.e.

$$3\left(S_T^c\right)^2 = Y_t^c Y_c^c \qquad \text{or} \qquad 3\left(S_T + \alpha\omega\right)^2 = \left(Y_t \pm \beta\omega\right)\left(Y_c \pm \chi\omega\right). \tag{56}$$

With all combinations of + and – in front of the ω terms, (56) gives rise 4 quadratic equations for ω. There can be eight different solutions for ω in general. It is the real solution having the minimum absolute value that will be taken as the desirable solution, since one is interested in a minimum correction to the experimental data, as long as the condition (56) is satisfied.

The original values and their corrected counterparts for the range of materials as listed in Table 9.2 are presented in Table 9.3 (row 7, 9 and 11) along with the magnitude of the corrections ω. This time, all transverse strengths have been modified a bit. The results were obtained with $\alpha=\beta=\chi=1$, for the purpose of illustration. The moderation scheme is basically to spread systematic errors amongst all transverse strengths, and hopefully some of the errors might cancel each other. After the moderation, the magnitudes of corrections are modest, with the maximum of 27% relative to the original value in the case of high strength S-2 glass fibre composite, as shown Table 9.3. For readers' ease of evaluation, all corrections above 10% in Table 9.3 have been highlighted and the remaining can be considered as insignificant.

Given the variability of the measured transverse strengths and the brittle nature of the behaviour of UD composites in this respect, even the highest correction as obtained in Table 9.3 should not be excessive comparing with the scatters one normally has for these strengths from experiments. The corrected strengths are also plotted in Fig. 9.5 in hollow symbols with a dot in the middle, and, as expected, they also fall exactly on the 1:1 straight line. If one accepts relationship (32), then the corrected values should be considered as a set of more representative transverse strengths as given in rows 7, 9 and 11 in Table 9.3 for the materials concerned than the original ones in rows 1 to 3, respectively. One may observe that the corrections are not entirely insignificant but certainly comparable with the scatter of measured transverse strengths. The corrected values (11th row) show improved transverse shear strengths in all cases as compared with those uncorrected (4th row) relative to the original experimental values (3rd row). For instance, the measured S_T for the AS4 3501-6 composite is 55 MPa whilst using (32) directly gives 56.57MPa. The correction results in 55.91 MPa which falls closer to 55MPa than 56.57MPa.

Table 9.3 Transverse strengths and their corrected values

Row No.	Material	AS4 3501-6	T300 BSL914C	E-glass LY556	E-Glass MY750	IM7 8551-7	T300 PR-319	A-S Epoxy 1	S-2 glass Epoxy 2	G40-800 5026
1	Y_t (MPa)	48	27	35	40	68	40	38	56.5	70
2	Y_c (MPa)	200	200	114	145	185	125	150	180	210
3	S_T (MPa)	55	N/A	50	50	57	45	50	40	57
4	S_T from (32) (MPa)	56.57	42.43	36.47	43.97	64.76	40.82	43.59	58.22	70.00
5	Correction	2.9%	N/A	-27%	-12%	14%	9.3%	13%	46%	23%
6	ω (MPa)	0.91	N/A	-8.11	-3.56	4.69	-2.50	-3.75	10.62	7.77
7	Y_t^c (MPa)	47.09	N/A	43.11	43.56	63.44	42.50	41.75	45.88	62.23
8	Correction	-1.9%	N/A	23%	8.9%	-6.9%	6.3%	9.9%	-19%	-11%
9	Y_c^c (MPa)	199.09	N/A	122.11	148.56	180.44	127.50	153.75	169.38	202.23
10	Correction	-0.5%	N/A	7.1%	2.5%	-2.5%	2.0%	2.5%	-6.0%	-3.7%
11	S_T^c (MPa)	55.91	N/A	41.89	46.44	61.69	42.50	46.25	50.62	64.77
12	Correction	1.7%	N/A	-16%	-7.1%	8.2%	-5.6%	-7.5%	27%	14%

The proposed correction can be considered as a part of experimental data processing for their more representative values, if all three transverse strengths are available. However, one can use relationship (32) to obtain the third strength if only two of them are known, e.g. in the case of T300/BSL914C as included in Tables 9.2 and 9.3. Usually, the tests for transverse shear strength are relatively more sophisticated and demanding to conduct than the other two. Using (32), the need of such test can be waived. However, when deriving the third one from the other two, one will have to accept the fact that experimental errors from the two known strengths will be accumulated onto the third and none of them can be corrected for their representativeness. Nevertheless, one should not be discouraged too much from taking advantage of this approach to minimise the experimental efforts. Given the sophistications involved in the transverse shear testing, experimental errors could be even more substantial in the measured strength than in the calculated ones. For instance, it would be much harder to achieve the pure shear stress state in the material where failure takes place. This undermines the applicability of available test standards, e.g. [19], to this particular strength property.

9.10 The strain-based Tsai-Wu failure criterion and its rationalised presentation

Majority of the failure criteria are expressed in terms of stresses. For various reasons, some designers seem to favour strain-based failure criteria. Two classic failure criteria in the composites field are the maximum stress criterion and the maximum strain criterion. Their major recognised drawback is insufficient accuracy of failure predictions due to the lack of interactions amongst stress or strain components. Their logical inconsistencies have rarely been examined objectively until fairly recently [9,36]. If the maximum stress criterion is considered as stress-based, the maximum strain is its strain-based counterpart. For every stress-based criterion, one might expect that its strain-based counterpart may be simply reproduced by replacing stresses and strengths by corresponding strains and strains at failure. However, this has not quite been the case.

The lack of symmetry between these two approaches has rarely been explained. The underlying reason rests on the nature of the input data. The input data to a stress-based failure criterion are a set of strengths which should be measured and obtained from tests conducted under uniaxial stress states or pure shear stress states. The input data to a strain-based failure criterion are a set of strains at failure. If they had been measured and obtained from testing the material under uniaxial strain states or pure shear strain states, then the strain-based criterion and the stress-based criterion would complement each other as two parallel and symmetric approaches. The predictions would not be identical as the input data were not the same, but they should shadow each other in terms of failure prediction. They would be of equal level of consistency. However, practicality prohibits the strains at failure to be obtained under uniaxial strain states or pure shear strain states. They are obtained under uniaxial stress states or pure shear stress states, instead, thus spoiling the symmetry between the two approaches. It is worth noting that the claim that a structural design and/or analysis could be carried out entirely based on strains without involving stresses at all is at most a fantasy, because at least the input data, including the elastic properties and strength properties, can only be measured under designated stress states, not strain states. As a result, a stress-based criterion and its strain-based counterpart no longer share equal level of consistency. In general, the latter will have one additional inconsistency on top of whatever it inherits from the former. The additional inconsistency can be easily demonstrated as its incapability of reproducing the input data, in other words, failure to pass a 'sanity check'. It is a deep-rooted confusion primarily due to a simple recognition of the differences between uniaxial stress states and uniaxial strain states [9,36]. Based on such understanding, the maximum stress criterion and the maximum strain criterion were unified for the first time in [36], where other rational considerations are also presented.

Stresses and strains are related through the generalised Hooke's law within the linear regime at least, a strain-based criterion can always be obtained simply by expressing the stresses in the criterion in terms of strains according to the generalised Hooke's law. Obviously, such a practice would not improve the consistency of the theory in any way but would make the determination of the coefficients involved in the failure function a formidable task [37] in general. They would not be as simple as (4) anymore. However, to meet the demand of some of the users who favour strains, the following can be proposed in a consistent manner to come up with a strain-based counterpart from any stress-based criterion rationally.

Strains are a measure of deformations. However, not all deformations are responsible for material failure. A typical example of such non-failure-related strains are the strains resulting from hygrothermal effects. They do not result in stresses, therefore, such strains are discounted from the total strains before being multiplied by the stiffness matrix $[C]$ of the material to evaluate stresses $\{\sigma\}$ in problems involving such effects [5,23], i.e.

$$\{\sigma\}=[C]\left(\{\varepsilon\}-\{\varepsilon\}_{\text{hygrothermal}}\right) \tag{57}$$

where $\{\varepsilon\}$ are the conventional or total strains. To distinguish them from their partitions that will be introduced later, they will be referred to as apparent strains and denoted as $\{\varepsilon\}_{\text{apparent}}$. In general, apparent strains can be partitioned into the 'force strains' and 'free strains' as follows:

$$\{\varepsilon\}_{\text{apparent}} = \{\varepsilon\}_{\text{forced}} + \{\varepsilon\}_{\text{free}} = \{\tilde{\varepsilon}\} + \{\varepsilon\}_{\text{free}} \tag{58}$$

where $\{\tilde{\varepsilon}\}$ are employed to denote 'forced strains'.

Similar to the strains caused by hygrothermal effects, those induced by the Poisson's effort are 'free strains', as far as material failure is concerned, as one does not expect such a strain to cause failure directly. Thus,

$$\{\varepsilon\}_{\text{free}} = \{\varepsilon\}_{\text{Poisson}} + \{\varepsilon\}_{\text{hygrothermal}}. \tag{59}$$

If all 'free strains' are removed from the consideration for material failure, what is left are 'forced strains' and they are responsible for the failure. Under a general 3D stress state, the 'force strains' result from their respective stresses σ. For transversely isotropic materials, they can be obtained as

$$\{\varepsilon\}_{\text{forced}} = \{\tilde{\varepsilon}\} = \begin{Bmatrix} \sigma_1/E_L \\ \sigma_2/E_T \\ \sigma_3/E_T \\ \tau_{23}/G_T \\ \tau_{13}/G_L \\ \tau_{12}/G_L \end{Bmatrix} = \begin{Bmatrix} \tilde{\varepsilon}_1 \\ \tilde{\varepsilon}_2 \\ \tilde{\varepsilon}_3 \\ \gamma_{23} \\ \gamma_{13} \\ \gamma_{12} \end{Bmatrix} \tag{60}$$

where E_L and E_T, and G_L and G_T are the longitudinal and transverse Young's moduli, and shear moduli of the transversely isotropic material, respectively. Apparently, they are not related to Poisson's ratios in any way. Each of 'forced stress' is associated with its corresponding uniaxial stress or pure shear stress in a one-to-one relationship. Given the transverse isotropy of the material, there is no 'free' part in the shear strains.

On the other hand, 'free strains' are induced by the 'forced strains' through Poisson's effect

$$\{\varepsilon\}_{\text{Poisson}} = \begin{Bmatrix} -\nu_L'(\tilde{\varepsilon}_2 + \tilde{\varepsilon}_3) \\ -\nu_L\tilde{\varepsilon}_1 - \nu_T\tilde{\varepsilon}_3 \\ -\nu_L\tilde{\varepsilon}_1 - \nu_T\tilde{\varepsilon}_2 \\ 0 \\ 0 \\ 0 \end{Bmatrix} \tag{61}$$

where ν_L and ν_T are the longitudinal and transverse Poisson's ratios of the transversely isotropic material, respectively, and the minor Poisson's ratio in the longitudinal direction can be obtained as

$$\nu_L' = \nu_L E_T / E_L \,. \tag{62}$$

The objective of the formulation is to evaluate failure based on the apparent strains, not stresses, and therefore it is desirable to express the direct components of the 'forced strains' in (60) in terms of apparent strains. This can be achieved after some algebraic manipulations [36] as

$$\begin{Bmatrix} \tilde{\varepsilon}_1 \\ \tilde{\varepsilon}_2 \\ \tilde{\varepsilon}_3 \end{Bmatrix} = \frac{1}{\left(1-\nu_T-2\nu_L\nu_L'\right)\left(1+\nu_T\right)} \begin{bmatrix} 1-\nu_T^2 & \nu_L'\left(1+\nu_T\right) & \nu_L'\left(1+\nu_T\right) \\ \nu_L\left(1+\nu_T\right) & 1-\nu_L\nu_L' & \nu_T+\nu_L\nu_L' \\ \nu_L\left(1+\nu_T\right) & \nu_T+\nu_L\nu_L' & 1-\nu_L\nu_L' \end{bmatrix} \begin{Bmatrix} \varepsilon_1 \\ \varepsilon_2 \\ \varepsilon_3 \end{Bmatrix}. \tag{63}$$

It is clear that the coefficient matrix above is not symmetric in general. Under a plane stress condition, its 2D version can be obtained as

$$\begin{Bmatrix} \tilde{\varepsilon}_1 \\ \tilde{\varepsilon}_2 \\ \tilde{\gamma}_{12} \end{Bmatrix} = \begin{bmatrix} 1/\left(1-\nu_L\nu_L'\right) & \nu_L'/\left(1-\nu_L\nu_L'\right) & 0 \\ \nu_L/\left(1-\nu_L\nu_L'\right) & 1/\left(1-\nu_L\nu_L'\right) & 0 \\ 0 & 0 & 1 \end{bmatrix} \begin{Bmatrix} \varepsilon_1 \\ \varepsilon_2 \\ \gamma_{12} \end{Bmatrix}. \tag{64}$$

It is regretted that (64) was incorrectly given in reference [36] by the first author of this chapter.

If the 'forced strains' are used in a strain-based failure criterion instead of the apparent strains, with the strains at failure obtained under uniaxial stress states as the input data, it will eliminate the inconsistency as described previously. A stress-based failure criterion can be easily converted into a strain-based one in a straightforward manner. Under the assumption of linear stress-strain relationship up to failure, such formulated strain-based criteria will result in exactly the same predictions as their stress-based counterpart. The two approaches are therefore unified. The only difference is that they are now expressed in terms of strains. It should be noted that although stresses do not appear in the strain-based criterion, the concept of stresses is indispensable.

In this way, the strain-based fully rationalised Tsai-Wu failure criterion can be given as

$$f_1\tilde{\varepsilon}_1 + f_2\left(\tilde{\varepsilon}_2+\tilde{\varepsilon}_3\right) + f_{11}\tilde{\varepsilon}_1^2 + f_{22}\left(\tilde{\varepsilon}_2^2+\tilde{\varepsilon}_3^2-\tilde{\varepsilon}_2\tilde{\varepsilon}_3+3\gamma_{23}^2\right) - \sqrt{f_{11}f_{22}}\,\tilde{\varepsilon}_1\left(\tilde{\varepsilon}_3+\tilde{\varepsilon}_2\right) + f_{66}\left(\gamma_{13}^2+\gamma_{12}^2\right) = 1 \tag{65}$$

where

$$f_{11} = \frac{1}{x_t x_c}, \quad f_{22} = \frac{1}{y_t y_c}, \quad f_1 = \frac{1}{x_t} - \frac{1}{x_c}, \quad f_2 = \frac{1}{y_t} - \frac{1}{y_c} \quad \text{and} \quad f_{66} = \frac{1}{s_L^2} \tag{66}$$

with x_t and x_c being the tensile and compressive strains at failure of the material along fibres, y_t and y_c those in the direction transverse to the fibres, and s_L the shear strain at failure along fibres. Readers are reminded that the direct components of the 'forced strains' in (65) can be readily expressed in terms of their apparent counterparts according to (63) so that the above criterion can be applied in terms of apparent strains effectively.

The counterpart of (65) under the plane stress condition (not plane strain!) can be given as follows, noticing the 'forced strains' in this case as in (64).

$$f_1\tilde{\varepsilon}_1 + f_2\tilde{\varepsilon}_2 + f_{11}\tilde{\varepsilon}_1^2 + f_{22}\tilde{\varepsilon}_2^2 - \sqrt{f_{11}f_{22}}\tilde{\varepsilon}_1\tilde{\varepsilon}_2 + f_{66}\gamma_{12}^2 = 1. \qquad (67)$$

It can be seen that the strain-based Tsai-Wu criterion shares the same features as its stress-based counterpart, and the same set of strength properties are required for both 2D and 3D stress states, which are expressed in terms of strengths and strains at failure, respectively.

As mentioned already, the strains at failure are measured under uniaxial tensile or compressive or pure shear stress states, not uniaxial tensile or compressive or pure shear strain states. When the material is under such a simple stress state (note that the corresponding strain state is not so simple), the above strain-based Tsai-Wu criterion will reproduce the input data as a minimum requirement of consistency, whilst maintaining the format of the conventional Tsai-Wu criterion. Using apparent strains directly in the above instead will be very wrong.

If the constitutive relationship of the material remains linearly elastic, the strain-based approach will reproduce the results from its stress-based counterpart. However, if the strains at failure are measured independently of their stresses at failure, they will provide different predictions of the failure of the material. For example, the relationship (63) between the 'forced strains' and the apparent strains does no longer hold, and one should also check if the Poisson's ratios still remain constant. Readers are therefore warned against any unjustified claim that strain-based criteria are more reliable. Individual coincidences should not be generalised blindly. What can be seen here is that the strain-based fully rationalised Tsai-Wu criterion can be as consistent as its stress-based counterpart, but not any more consistent than its stress-based counterpart.

9.11 Conclusions

The Tsai-Wu failure criterion has been fully rationalised. Any significant deficiency or discrepancy with genuine experimental observations can now be confidently attributed to the lack of representativeness of the basic assumptions underlying the criterion:

(1) The transverse isotropy and homogeneity of the material;
(2) Failure function being a single quadratic function.

Any deviation from these ideal positions will inevitably have an effect on the outcomes and should be anticipated in assessing the accuracy of the predictions. The use of these assumptions brings convenience but also restrictions. The obtained criterion can only be as accurate as its underlying assumptions allow. Consistent use of a quadratic function for a failure criterion helps to eliminate undue errors and anomalies resulting from mathematical and logical oversights, but not the generic deficiency associated with these assumptions.

When complete quadratic failure function is used, there is only one rational choice for the failure envelope — an elliptic paraboloid, a single-sided open surface that allows infinite strength under a unique stress ratio in triaxial compression. Any alternative shape for the failure envelope will lead to contradictions one way or another.

The quadratic failure function corresponding to the elliptic paraboloidal failure envelope can only accommodate five independent strength properties. Seven were initially introduced in the original Tsai-Wu criterion [4] after considerations of transverse isotropy. A relationship between the two interactive coefficients F_{12} and F_{23} has been revealed purely from requirements of analytic geometry to ensure that the failure envelope is an elliptic paraboloid. A further relationship between them can be obtained from a logical consideration that the sensitivity of the failure criterion to these two coefficients should be unbiased. Thus, both of them can be uniquely determined analytically as $F_{12} = -\frac{1}{2}\sqrt{F_{11}F_{22}}$ and $F_{23} = -\frac{1}{2}F_{22}$. Although they were previously suggested, they can only be fully justified as a single proposition. The second of the two relationships can be re-written as $3S_T = Y_t Y_c$, given the transverse isotropy of the material. This introduces a relationship amongst transverse strengths as a natural consequence of the rationalisation, and the relationship is an intrinsic property of the quadratic failure function for transversely isotropic materials. Failing to comply with this relationship compromises the consistency of the criterion based on a single quadratic failure function.

Due to the high variability in the measured transverse strengths, the obtained relationship amongst transverse strengths may not be perfectly satisfied by the available experimental data. In presence of all three transverse strengths independently obtained experimentally, the transverse shear strength can be employed to validate the relationship. The failure criterion can only be as accurate as the agreement between S_T and $\sqrt{Y_t Y_c/3}$. Alternatively, the available transverse shear strength can be employed to calibrate the remaining two transverse strengths. Calibration method that can potentially be used for this purpose has been proposed in this chapter.

The fully rationalised Tsai-Wu criterion is given in (33) for 3D stress states. Its 2D version in (36) looks identical to its familiar form. Although no difference has been made to (36), the rationalisation has offered a firm basis to (36), and the determination of F_{12} is no longer experimental data fitting or any other empiricism, but a rigorous matter of mathematical and logical consequence of the basic assumptions introduced.

The required five independent strength properties will be limited to conventional and widely available tensile and compressive strengths in directions along and transverse to fibres, and the in-plane shear strength for failure predictions of both 2D and 3D stress states.

The most significant outcome of the present work is that a thorough understanding of the nature of quadratic failure function has been achieved with its intrinsic relationships revealed. Obeying these relationships ensures the self-consistency of the failure criterion. This should conclude a phase of explorations on the subject of failure criteria based on a quadratic failure function reported in [8,34,38] as far as transversely isotropic materials are concerned. The failure criterion as given in (33) can be employed in design and analysis with confidence within its applicability defined by its assumptions.

The fully rationalised Tsai-Wu criterion has also been extended from its conventional stress-based form to its strain-based counterpart whilst maintaining full consistency, allowing the failure to be predicted in terms of strains. It should reproduce the predictions from its stress-based counterpart if the material behaviour is perfectly linear up to failure. Difference can be expected in the predictions if nonlinearity is present in the behaviour of the material. In this case, the consistency of the strain-based criterion can no longer be taken for granted.

Acknowledgement

The second author wishes to acknowledge the financial support from CSC, China and the scholarship from the Faculty of Engineering, the University of Nottingham.

References

[1] M.J. Hinton, P.D. Soden, and A.S. Kaddour, "Failure Criteria in Fibre-Reinforced-Polymer Composites", *Composites Science and Technology*, Part A, **58**(7), 1998; Part B, **62**(12-13), 2002 and Part C, **64**(3-4), 2004

[2] A.S. Kaddour, and M.J. Hinton, "Evaluation of Theories for Predicting Failure in Polymer Composite Laminates under 3-D States of Stress", *Journal of Composite Materials*, Part A, **46**(19-20), 2012, and Part B, **47**(6-7), 2013

[3] A.S. Kaddour, M.J. Hinton, P.A. Smith, and S. Li, "Benchmarking of matrix cracking, damage and failure models for composites: Comparison between theories", Part A, *Journal of Composite Materials*, **47**(20-21), 2013

[4] S.W. Tsai and E.M. Wu, "A general theory of strength for anisotropic materials", *Journal of Composite Materials*, **5**:58-80, 1971

[5] R.M. Jones, *Mechanics of Composite materials*, 2nd edn. Taylor & Francis, London, 1999

[6] Abaqus, *ABAQUS User's Manual*, Version 6.14, 2014

[7] D. Hull, and T.W. Clyne, *An Introduction to Composite Materials*, 2nd edn., Section 8.2, Cambridge University Press, Cambridge, 1996

[8] S. Li, E. Sitnikova, Y. Liang and A.-S. Kaddour, "The Tsai-Wu failure criterion rationalised in the context of UD composites", *Composites A*, **102**:207-217, 2017

[9] S. Li and E. Sitnikova, "A critical review on the rationality of popular failure criteria for composites", *Composites Communications*, **8**:7-13, 2018

[10] I.I. Gol'denblat and V.A. Kopnov, "Strength of glass-reinforced plastics in the complex stress state", *Mekhanika Polimerov*, 7:70, 1965

[11] Z. Hashin, "Failure criteria for unidirectional fiber composites", *Journal of Applied Mechanics*, **47**:329–334, 1980

[12] S.W. Tsai and J.D.D. Melo, "An invariant-based theory of composites", *Composites Science and Technology*, **100**:237–243, 2014

[13] S.W. Tsai and J.D.D. Melo, "A unit circle failure criterion for carbon fiber reinforced polymer composites", *Composites Science and Technology*, **123**:71-78, 2016

[14] S.W. Tsai, "A survey of macroscopic failure criteria for composite materials", *J. Reinforced Plastics and Composites*, **3**:40-62, 1984

[15] S. Li and E. Sitnikova (2019) *Representative Volume Elements and Unit Cells — Concepts, Theory, Applications and Implementation*, Woodhead Publishing Series in Composites Science and Engineering, Elsevier, Duxford, UK

[16] S. Li, M. Xu and E. Sitnikova, The formulation of the quadratic failure criterion for transversely isotropic materials: mathematical and logical considerations, *J Composites*

Sci, **6**:82, 2022

[17] ASTM D3039/D3039M-17, *Standard Test Method for Tensile Properties of Polymer Matrix Composite* Materials, ASTM International, West Conshohocken, PA, 2017

[18] ASTM D3410 / D3410M-16, *Standard Test Method for Compressive Properties of Polymer Matrix Composite Materials with Unsupported Gage Section by Shear Loading*, ASTM International, West Conshohocken, PA, 2016

[19] D5379/D5379M-12, *Standard Test Method for Shear Properties of Composite Materials by the V-Notched Beam Method*, ASTM International, West Conshohocken, PA, 2012

[20] R. Narayanaswami, and H.M. Adelman, "Evaluation of the tensor polynomial and Hoffman strength theories for composite materials", *Journal of Composite Materials*, **11**:366, 1977

[21] O. Mohr, "Welche Umstände bedingen die Elastizitätsgrenze und den Bruch eines Materials?" *Civilingenieur*, **44**(45):1524-1530 & **44**(46):1572-1577, 1900

[22] S. Li, "A reflection on the Mohr failure criterion", *Mechanics of Materials*, **148**:103442, 2020

[23] S.W. Tsai and H.T. Hahn, *Introduction to Composite materials*, Technomic Publishing Company, Westport, CT06880, USA, 1980

[24] M.N. Kallas, and H.T. Hahn, "A failure criterion for composites under dimensional state of stress", *35th International SAMPE Symposium*, 674-686, 1990

[25] S.J. Deteresa and G.J. Larsen, "Reduction in the number of independent parameters for the Tsai–Wu tensor polynomial theory of strength for composite materials", *Journal of Composite Materials*, **37**:1769-1785, 2003

[26] G.A. Korn and T.M. Korn, *Mathematical Handbook for Scientists and Engineers*, Sections 2.4, 3.5 & 4.7, McGraw-Hill, New York, USA, 1968

[27] R. von Mises, "Mechanik der festen Körper im plastisch-deformablen Zustand, *Nachrichten von der Gesellschaft der Wissenschaften zu Göttingen*", *Mathematisch-Physikalische Klasse.* **1913**(1):582–592, 1913

[28] P.W. Bridgman, "Fracture and flow in metals", *Metals Technology*, T.P. 1782, The American Institute of Mining and Metallurgical Engineers, New York, Dec. 1944

[29] R. Hill, *The Mathematical Theory of Plasticity*, Oxford University Press, Oxford, 1998

[30] Raghava, R.M. Caddell and G.S.Y. Yeh, "The macroscopic yield behaviour of polymers", *J. Mater. Sci.*, **8**:225-232, 1973

[31] R.M. Christensen, *The Theory of Materials Failure*, Oxford University Press, Oxford, 2013

based phenomenological models", *Composites Sci. Tech.*, **58:**1045-1067, 1998

[33] M. Knops, *Analysis of Failure in Fiber Polymer Laminates, The Theory of Alfred Puck*, Springer, Berlin, 2008

[34] R.M. Christensen, "Completion and closure on failure criteria for unidirectional fiber composite materials", *ASME J. Appl. Mech.*, **81**:011011-1-6, 2013

[35] E. Clarkson, "Hexcel 8552 AS4 unidirectional prepreg qualification statistical analysis report", FAA special project no SP4614WI-Q, Report No NCP-RP-2010-008 Rev D, Wichita State University, 2011

[36] S. Li, "The maximum stress failure criterion and the maximum strain failure criterion: their unification and rationalization", *J. Composites Science*, **4**:157, 2020

[37] W.W. Feng, "A failure criterion for composite materials", *J. Compos. Mater.*, **25**:88-100, 1991

[38] R.M. Christensen, "Failure criteria for fiber composite materials, the astonishing sixty year search, definitive usable results", *Composites Sci and Tech*, **182**:107718, 2019

With Becky and Edward Wu of the Tsai-Wu failure criterion.
Ed passed away 10 years ago.

With Manuel Torres at JEC World, a self-made innovator,
passed away in November 2020

CHAPTER 10

Feasible Region of Lamination Parameters for Double-Double Laminates

Kai Zhao[a], David Kennedy[a], Antonio Miravete[b], Stephen W. Tsai[c], Carol A. Featherston[a], Xiaoyang Liu[d]

[a] *School of Engineering, Cardiff University, Queen's Buildings, The Parade, Cardiff CF24 3AA, UK*

[b] *Department of Aeronautics and Astronautics, Massachusetts Institute of Technology, Cambridge, MA 02139, United States*

[c] *Department of Aeronautics & Astronautics, Stanford University, Durand Building, 496 Lomita Mall, Stanford, CA 94305-4035, United States*

[d] *School of Civil Engineering, Qingdao University of Technology, Qingdao 266033, China*

10.1. Introduction

Several studies have shown that DD laminates are able to replace most quad-axial (QUAD) laminates by taking advantage of their larger design space. However, the exact design space of DD laminates remains unknown. York[1] concluded that the lower bound of the design space of lamination parameters ξ_1^A, ξ_2^A and ξ_1^D, ξ_2^D for general two-angle laminates can be represented by parabolas. However, the upper bound and design spaces for other lamination parameters, especially for $\xi_{1,2,3,4}^B$ which relate to the in-plane-out-of-plane coupling and also the homogenization, have not been discussed. Therefore, the extension of the homogenization of DD laminates is hard to define mathematically, resulting in sacrificing design space for the near-homogenized laminates.

In this chapter, taking advantage of repetition, formulations for obtaining the lamination parameters of DD laminates are derived, based on which lamination parameters can be expressed in simple forms. The formulations are not limited to the most commonly used DD laminate sequence, i.e. staggered 1, staggered 2, staggered 3 and paired. They are also applicable to DD laminates with rigid body rotation - unbalanced DD laminates which have four different angles. With the aid of these formulations, a method based on Lagrange multipliers is presented to derive feasible regions of any combination of two lamination parameters. Benefiting from the simple expressions of lamination parameters, all feasible regions can be obtained analytically, providing mathematical support for DD families. The obtained result shows that their design space is larger than that of conventional QUAD laminates in terms of industrial practices. A homogenization criterion is implemented into the design space, based on which a tailored DD laminate is proposed, expanding design possibilities and enabling homogenization to be achieved using only 16 plies/4 repeats.

The content of this chapter is largely based on a previously published paper in the AIAA [2]. The study described in that paper provides the foundation for the research presented here, and many of the findings and conclusions have been adapted and expanded upon for this chapter.

10.2. Lamination parameters

Lamination parameters were first introduced by Tsai et al. in 1968 [3], enabling the stiffness matrix to be represented by a linear function, and also significantly reducing the number of design variables for optimization problems. For the design of either flat or curved composite laminates, lamination parameters can be used as design variables instead of the large number of ply angles to meet the mechanical requirements of engineering scenarios with regard to buckling, postbuckling and vibration, etc. By using lamination parameters as design variables, the number of design variables can be reduced to a maximum of twelve, and the stiffness matrix can then be expressed as a linear function of these lamination parameters instead of the conventional set of equations with a large number of ply orientations.

Based on classical laminate theory, the stress-strain relationship for a composite laminate can be expressed as

$$\begin{bmatrix} \mathbf{n} \\ \mathbf{m} \end{bmatrix} = \begin{bmatrix} \mathbf{A} & \mathbf{B} \\ \mathbf{B} & \mathbf{D} \end{bmatrix} \begin{bmatrix} \boldsymbol{\varepsilon}^0 \\ \boldsymbol{\kappa} \end{bmatrix} \tag{10.1}$$

where $\mathbf{n}$ and $\mathbf{m}$ are vectors of in-plane forces and moments per unit width, $\mathbf{A}, \mathbf{B}$ and $\mathbf{D}$ are in-plane, coupling, and out-of-plane stiffness matrices, $\boldsymbol{\varepsilon}^0$ is a vector of in-plane strains and $\boldsymbol{\kappa}$ is a vector of mid plane curvatures.

The stiffness matrices $\mathbf{A}, \mathbf{B}$ and $\mathbf{D}$ can be expressed in terms of 12 lamination parameters $\xi_j^k (j = 1,2,3,4; k = A, B, D)$ and material stiffness invariants $\mathbf{u}$ as shown below:

$$\begin{bmatrix} A_{11} \\ A_{22} \\ A_{12} \\ A_{66} \\ A_{16} \\ A_{26} \end{bmatrix} = h \begin{bmatrix} 1 & \xi_1^A & \xi_2^A & 0 & 0 \\ 1 & -\xi_1^A & \xi_2^A & 0 & 0 \\ 0 & 0 & -\xi_2^A & 1 & 0 \\ 0 & 0 & -\xi_2^A & 0 & 1 \\ 0 & \xi_3^A/2 & \xi_4^A & 0 & 0 \\ 0 & \xi_3^A/2 & -\xi_4^A & 0 & 0 \end{bmatrix} \begin{bmatrix} U_1 \\ U_2 \\ U_3 \\ U_4 \\ U_5 \end{bmatrix} \tag{10.2}$$

$$\begin{bmatrix} B_{11} \\ B_{22} \\ B_{12} \\ B_{66} \\ B_{16} \\ B_{26} \end{bmatrix} = \frac{h^2}{2} \begin{bmatrix} 0 & \xi_1^B & \xi_2^B & 0 & 0 \\ 0 & -\xi_1^B & \xi_2^B & 0 & 0 \\ 0 & 0 & -\xi_2^B & 0 & 0 \\ 0 & 0 & -\xi_2^B & 0 & 0 \\ 0 & \xi_3^B/2 & \xi_4^B & 0 & 0 \\ 0 & \xi_3^B/2 & -\xi_4^B & 0 & 0 \end{bmatrix} \begin{bmatrix} U_1 \\ U_2 \\ U_3 \\ U_4 \\ U_5 \end{bmatrix} \tag{10.3}$$

$$\begin{bmatrix} D_{11} \\ D_{22} \\ D_{12} \\ D_{66} \\ D_{16} \\ D_{26} \end{bmatrix} = \frac{h^3}{12} \begin{bmatrix} 1 & \xi_1^D & \xi_2^D & 0 & 0 \\ 1 & -\xi_1^D & \xi_2^D & 0 & 0 \\ 0 & 0 & -\xi_2^D & 1 & 0 \\ 0 & 0 & -\xi_2^D & 0 & 1 \\ 0 & \xi_3^D/2 & \xi_4^D & 0 & 0 \\ 0 & \xi_3^D/2 & -\xi_4^D & 0 & 0 \end{bmatrix} \begin{bmatrix} U_1 \\ U_2 \\ U_3 \\ U_4 \\ U_5 \end{bmatrix} \tag{10.4}$$

where the material stiffness invariants **u** and the stiffness properties **Q** are:

$$\begin{bmatrix} U_1 \\ U_2 \\ U_3 \\ U_4 \\ U_5 \end{bmatrix} = \frac{1}{8} \begin{bmatrix} 3 & 3 & 2 & 4 \\ 4 & -4 & 0 & 0 \\ 1 & 1 & -2 & -4 \\ 1 & 1 & 6 & -4 \\ 1 & 1 & -2 & 4 \end{bmatrix} \begin{bmatrix} Q_{11} \\ Q_{22} \\ Q_{12} \\ Q_{66} \end{bmatrix} \tag{10.5}$$

$$\begin{cases} Q_{11} = E_{11}/(1 - \nu_x \nu_y) \\ Q_{22} = E_{22}/(1 - \nu_x \nu_y) \\ Q_{12} = \upsilon_x Q_{22} \\ Q_{66} = G_{12} \end{cases} \tag{10.6}$$

Ch.10

Where E_{11} is the longitudinal Young's modulus, E_{22} is the transverse modulus, G_{12} is the shear modulus, υ_x, υ_y are the major, minor Poisson's ratio, respectively, and h is the thickness of the laminate.

The lamination parameters are obtained by the following integrals

$$\begin{bmatrix} \xi_1^k \\ \xi_2^k \\ \xi_3^k \\ \xi_4^k \end{bmatrix} = \int_{-h/2}^{h/2} Z^k \begin{bmatrix} \cos 2\theta \\ \cos 4\theta \\ \sin 2\theta \\ \sin 4\theta \end{bmatrix} dz, k = A, B, D, \begin{cases} Z^A = 1/h \\ Z^B = 2z/h^2 \\ Z^D = 12z^2/h^3 \end{cases} \tag{10.7}$$

where θ represents the ply angle at depth z below the mid-surface. For a general angle laminate, such an equation cannot be simplified further as the angle of each ply differs at different depths.

10.3. Lamination parameters of general DD laminates

As the sub-laminate of DD laminates has only four angles, namely two angles with opposite signs or four different angles [4] (i.e. unbalanced DD laminates), Eq.10.7 can be derived with respect to the four angles $\theta_1, \theta_2, \theta_3, \theta_4$.

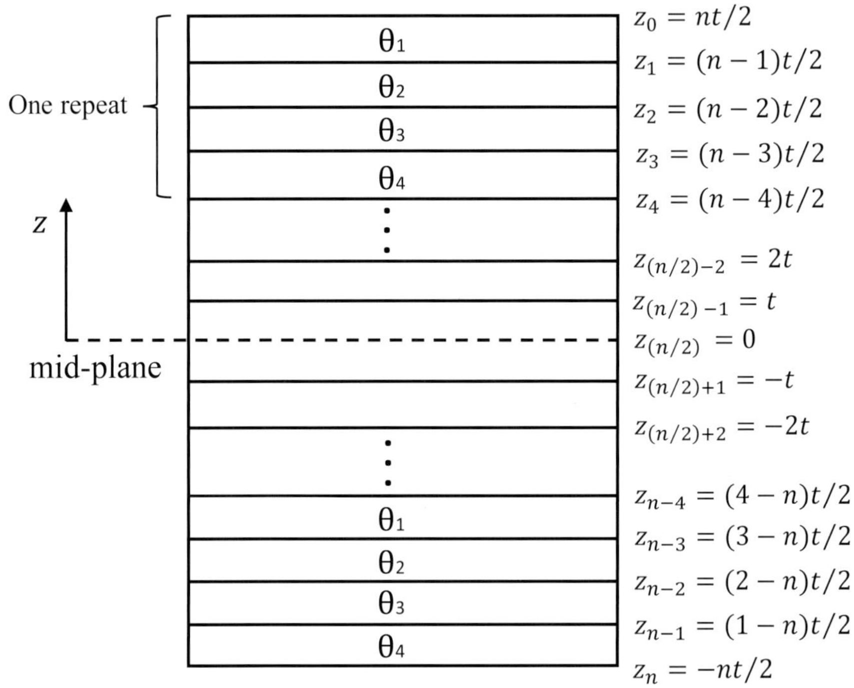

Assuming a laminate has n plies with thickness t, identified by $n+1$ nodes, comprising $n/4$ repetitions of sub-laminate $[\theta_1, \theta_2, \theta_3, \theta_4]$, whose locations are shown in Fig. 10.1, it should be noted that the total number of layers n is always an even number, and the mid-plane always lies on the boundary of two plies.

Then, for $k = 1,2,\ldots,n/4$, the kth repetition has its top surface at $z = t[-1/2 + (4k-4)/n]$ and its bottom surface at $z = t[-1/2 + (4k)/n]$.

Thus the kth θ_1 ply runs from $z = t[-1/2 + (4k-4)/n]$ to $z = t[-1/2 + (4k-3)/n]$.

The kth θ_2 ply runs from $z = t[-1/2 + (4k-3)/n]$ to $z = t[-1/2 + (4k-2)/n]$.

The kth θ_3 ply runs from $z = t[-1/2 + (4k-2)/n]$ to $z = t[-1/2 + (4k-1)/n]$.

And the kth θ_4 ply runs from $z = t[-1/2 + (4k-1)/n]$ to $z = t[-1/2 + (4k)/n]$.

Defining trigonometric functions $f_1(\theta) = \cos(2\theta), f_2(\theta) = \cos(4\theta), f_3(\theta) = \sin(2\theta), f_4(\theta) = \sin(4\theta)$, the lamination parameters $\xi_{1,2,3,4}^{A,B,D}$ can be calculated as follows

For $\xi_{1,2,3,4}^{A}$, Eq. 10.7 can be expressed as

$$\begin{bmatrix} \xi_1^A \\ \xi_2^A \\ \xi_3^A \\ \xi_4^A \end{bmatrix} = \int_{-nt/2}^{nt/2} \frac{1}{nt} \begin{bmatrix} f_1(\theta) \\ f_2(\theta) \\ f_3(\theta) \\ f_4(\theta) \end{bmatrix} dz \tag{10.8}$$

Integrating each ply through the thickness

$$\xi_j^A = \frac{1}{t}\sum_{k=1}^{n/4}\left\{\int_{t[-1/2+(4k-4)/n]}^{t[-1/2+(4k-3)/n]} f_j(\theta_1)dz + \int_{t[-1/2+(4k-3)/n]}^{t[-1/2+(4k-2)/n]} f_j(\theta_2)dz + \int_{t[-1/2+(4k-2)/n]}^{t[-1/2+(4k-1)/n]} f_j(\theta_3)dz + \int_{t[-1/2+(4k-1)/n]}^{t[-1/2+(4k)/n]} f_j(\theta_4)dz\right\} \tag{10.9}$$

Simplifying Eq. 10.9, we have

$$\xi_j^A = \left\{\frac{1}{4}f_j(\theta_1) + \frac{1}{4}f_j(\theta_2) + \frac{1}{4}f_j(\theta_3) + \frac{1}{4}f_j(\theta_4)\right\} \tag{10.10}$$

Eq. 10.10 shows that equal contributions are made by all four of the angles. So, for the DD laminates without rigid body rotation, the contributions based on cos functions will be equal and opposite, so that ξ_3^A and ξ_4^A are zero.

For $\xi_{1,2,3,4}^B$, similarly to Eq. 10.8, substituting number of plies n and thickness t into Eq.10.7, we have

$$\begin{bmatrix}\xi_1^B\\ \xi_2^B\\ \xi_3^B\\ \xi_4^B\end{bmatrix} = \int_{-nt/2}^{nt/2} \frac{2z}{(nt)^2}\begin{bmatrix}f_1(\theta)\\ f_2(\theta)\\ f_3(\theta)\\ f_4(\theta)\end{bmatrix} dz \tag{10.11}$$

Integrating each ply through the thickness

$$\xi_j^B = \frac{2}{t^2}\sum_{k=1}^{n/4}\left\{\int_{t[-1/2+(4k-4)/n]}^{t[-1/2+(4k-3)/n]} zf_j(\theta_1)dz + \int_{t[-1/2+(4k-3)/n]}^{t[-1/2+(4k-2)/n]} zf_j(\theta_2)dz + \int_{t[-1/2+(4k-2)/n]}^{t[-1/2+(4k-1)/n]} zf_j(\theta_3)dz + \int_{t[-1/2+(4k-1)/n]}^{t[-1/2+(4k)/n]} zf_j(\theta_4)dz\right\} \tag{10.12}$$

And simplifying

$$\xi_j^B = \left\{-\frac{3}{n}f_j(\theta_1) - \frac{1}{n}f_j(\theta_2) + \frac{1}{n}f_j(\theta_3) + \frac{3}{n}f_j(\theta_4)\right\} \tag{10.13}$$

It can be seen that the contributions are inversely proportional to the number of plies n, so that increasing the number of plies can restrict the magnitude of the ξ_j^B lamination parameters to any required limit, i.e. approximating a symmetric laminate.

For $\xi_{1,2,3,4}^D$, Eq. 10.7 can be rewritten as

$$\begin{bmatrix} \xi_1^D \\ \xi_2^D \\ \xi_3^D \\ \xi_4^D \end{bmatrix} = \int_{-nt/2}^{nt/2} \frac{12z^2}{(nt)^3} \begin{bmatrix} f_1(\theta) \\ f_2(\theta) \\ f_3(\theta) \\ f_4(\theta) \end{bmatrix} dz \tag{10.14}$$

Integrating each ply through the thickness

$$\xi_j^D = \frac{12}{t^3} \sum_{k=1}^{n/4} \left\{ \int_{t[-1/2+(4k-4)/n]}^{t[-1/2+(4k-3)/n]} z^2 f_j(\theta_1) dz + \int_{t[-1/2+(4k-3)/n]}^{t[-1/2+(4k-2)/n]} z^2 f_j(\theta_2) dz \right. \\ \left. + \int_{t[-1/2+(4k-2)/n]}^{t[-1/2+(4k-1)/n]} z^2 f_j(\theta_3) dz + \int_{t[-1/2+(4k-1)/n]}^{t[-1/2+(4k)/n]} z^2 f_j(\theta_4) dz \right\} \tag{10.15}$$

And simplifying

$$\xi_j^D = \left\{ \left(\frac{1}{4} + \frac{3}{n^2}\right) f_j(\theta_1) + \left(\frac{1}{4} - \frac{3}{n^2}\right) f_j(\theta_2) + \left(\frac{1}{4} - \frac{3}{n^2}\right) f_j(\theta_3) \right. \\ \left. + \left(\frac{1}{4} + \frac{3}{n^2}\right) f_j(\theta_4) \right\} \tag{10.16}$$

As can be seen from Eq. 10.10, Eq. 10.13, and Eq. 10.16, each set of lamination parameters ($\xi_{1,2,3,4}^A$, $\xi_{1,2,3,4}^B$, $\xi_{1,2,3,4}^D$) can be obtained through a single equation in contrast to the integral equations which need to be solved for QUAD layups, making the feasible regions easier to obtain.

It can be observed from the equations above that the values of $\xi_{1,2,3,4}^A$ are related only to the ply angles, hence, increasing the number of plies in DD laminates has no effect on their in-plane lamination parameters. Besides, $\xi_{1,2,3,4}^B$ are inversely proportional to the number of plies and $\xi_{1,2,3,4}^D$ are inversely proportional to the square of the number of plies. As a result, as the number of plies increases, $\xi_{1,2,3,4}^B$ decreases with a first order relation while $\xi_{1,2,3,4}^D$ decreases with a second order one. These relations are impossible to explicitly describe for QUAD layups due to the complexity of their stacking sequences. On the other hand, such explicit expressions can be very advantageous when deriving the boundaries of the design space in terms of lamination parameters, as will be shown in the following sections.

10.3.1. Lamination parameters of Staggered 1

In this section, one of the DD laminates presented by Tsai [4] which has sub-laminates with the stacking sequence $[+\phi/-\psi/-\phi/+\psi]$ will be taken as an example to validate the method proposed. Based on Eq.10.10, 10.13 and 10.16, $\xi_{1,2,3,4}^A$ can be written as:

$$\begin{aligned} \xi_1^A &= \frac{1}{2} cos(2\phi) + \frac{1}{2} cos(2\psi); \\ \xi_2^A &= \frac{1}{2} cos(4\phi) + \frac{1}{2} cos(4\psi); \\ \xi_3^A &= 0;\ \xi_4^A = 0 \end{aligned} \tag{10.17}$$

It can be seen from Eq. 10.17 that equal contributions are made by the four plies. Therefore, in a DD laminate comprising angles $(\pm\phi, \pm\varphi)$ the contributions based on sine functions will be equal and opposite, so that $\xi^A_{3,4}$ which are related to the shear-extension coupling stiffness are intrinsically zeros. It should be noted that DD laminates could include shear-extension coupling when the plies are rotated, however research on the DD laminates with ply rotation is out of the scope of this study. As for $\xi^B_{1,2,3,4}$, they can be expressed as:

$$\begin{aligned} \xi^B_1 &= -\frac{1}{2n}\cos(2\phi) + \frac{1}{2n}\cos(2\psi); \\ \xi^B_2 &= -\frac{1}{2n}\cos(4\phi) + \frac{1}{2n}\cos(4\psi); \\ \xi^B_3 &= -\frac{1}{n}\sin(2\phi) + \frac{1}{n}\sin(2\psi); \\ \xi^B_4 &= -\frac{1}{n}\sin(4\phi) + \frac{1}{n}\sin(4\psi) \end{aligned} \tag{10.18}$$

It can be observed that the magnitudes of these lamination parameters are inversely proportional to the number of plies n, so that increasing the number of plies will decrease the magnitude of the $\xi^B_{1,2,3,4}$ to any required limit, i.e. approximating a symmetric laminate. Hence homogeneity can be approximated to any required accuracy by increasing the number of plies, more detailed research is presented in the following sections.

Ch.10

Expressions for $\xi^D_{1,2,3,4}$ are given in Eq. 10.19.

$$\begin{aligned} \xi^D_1 &= \frac{1}{2}\cos(2\phi) + \frac{1}{2}\cos(2\psi) \\ \xi^D_2 &= \frac{1}{2}\cos(4\phi) + \frac{1}{2}\cos(4\psi) \\ \xi^D_3 &= \frac{6}{n^2}\sin(2\phi) + \frac{6}{n^2}\sin(2\psi) \\ \xi^D_4 &= \frac{6}{n^2}\sin(4\phi) + \frac{6}{n^2}\sin(4\psi) \end{aligned} \tag{10.19}$$

10.4. Feasible regions of DD laminates

The feasible regions of the lamination parameters are treated as constraints when employing lamination parameters as non-dimensional design variables. The earliest work on feasible regions of lamination parameters was by Miki [5] and [6] who defined the feasible region for in-plane or out-of-plane stiffness of an orthotropic laminate. Over the years, researchers have been pushing the boundary of feasible regions of lamination parameters of QUAD to exploit any possible stiffness tailoring of a laminate through either new methodologies or representations of feasible regions. However, research on the boundaries of the design space highly depends on replacing or adding constraints regardless of the methods used, leading to more complex design space. Compared to QUAD layups, the design complexity of DD laminates is significantly reduced due to the simple stacking concept. Furthermore, the design spaces corresponding to DD laminates are found to be larger than those corresponding to QUAD layups in industrial practice which will be proved later in this chapter.

This section describes a methodology to obtain the feasible regions of combinations of any two lamination parameters for DD laminates (i.e. staggered 1: $[+\phi/-\psi/-\phi/\psi]$, Staggered 2: $[+\phi/+\psi/-\phi/-\psi]$, Staggered 3: $[+\phi/-\psi/+\psi/-\phi]$ and Paired: $[\pm\phi/\pm\psi]$) without rigid body rotation, using Lagrange multipliers, in favour of simplifying composite design.

To obtain the feasible regions of any two lamination parameters for DD laminates, a two-step method has been proposed herein. In the first step, by using Lagrange multipliers, the feasible regions of the laminates with ϕ and ψ varying between -90° and 90°, which covers three DD laminates (i.e. staggered 1, staggered 2, and a part of paired) are obtained as base solutions. In the second step, the exact feasible regions of each type of DD laminate can be obtained by adding angle constraints on the top of the base solutions.

10.4.1. Method based on Lagrange multipliers

For the first step, we plot one of the parameters along the horizontal (x) axis, and find the minimum and maximum possible values of the other parameter (y). A constrained optimization problem can therefore be summarized in the form:

Minimise and maximize $y = f(\theta_1, \theta_2)$ subject to $x = g(\theta_1, \theta_2)$, where f and g denote the functions of θ_1, θ_2.

Solutions correspond to the stationary values of the augmented function

$$F(\theta_1, \theta_2, \lambda) = f(\theta_1, \theta_2) + \lambda\big(g(\theta_1, \theta_2) - x\big) \tag{10.20}$$

where λ is a Lagrange multiplier.
The conditions to be satisfied at each stationary point are

$$\frac{\partial F}{\partial \theta_1} = \frac{\partial f}{\partial \theta_1} + \lambda \frac{\partial g}{\partial \theta_1} = 0 \tag{10.21}$$

$$\frac{\partial F}{\partial \theta_2} = \frac{\partial f}{\partial \theta_2} + \lambda \frac{\partial g}{\partial \theta_2} = 0 \tag{10.22}$$

$$\frac{\partial F}{\partial \lambda} = g(\theta_1, \theta_2) - x = 0 \tag{10.23}$$

Solving Eq. 10.21-23, the outer boundaries of the feasible region for any combination of non-zero lamination parameters for any DD laminate can be found. The key to solving the above equations depends largely on pairs of $\cos 2\theta$, $\sin 2\theta$, $\cos 4\theta$ and $\sin 4\theta$ in Eq. 10.10, 10.13 and 10.16, from which it can be deduced that there are, by combination rule, C(4, 2) + 4 = 10 solution types in total, as shown in appendix A, where 4 are pairs of their own. Furthermore, the coefficients of Eq. 10.10, 10.13 and 10.16 can potentially reduce the complexity of solving the above equations depending on the type of DD laminates.

The feasible regions of all possible combinations of any two lamination parameters can be analytically obtained based on the above equations using the method of Lagrange multipliers. However, in this step, constraints on ϕ and ψ are impossible to be added as it will cause transcendental functions which have to be solved numerically. Such an approach will require a lot of computational effort and might not reach enough accuracy which is not the intention of this research. Therefore, the ϕ and ψ are supposed to vary between -90° and 90° herein, which covers three DD laminates (i.e. staggered 1, staggered 2, and a part of paired). Based on this range, the obtained feasible regions can be used as base solutions for each type of DD laminate, the results of staggered 1 which is the most promising DD laminate are presented in the following paragraphs. According to the combination rule, there are 66 solutions in total, i.e. C(12, 2) = 66. From Eq. 10.17 and 10.19, we see that $\xi_1^A = \xi_1^D$, $\xi_2^A = \xi_2^D$, $\xi_{3,4}^A = 0$, reducing the total number of combinations to 28, i.e. C(8,2). And whilst 28 solutions can be derived, there are only five basic solution types/unique region shapes, as shown in Fig 10.2, with the rest of the 28 solutions being obtained from these by either rotation or scaling.

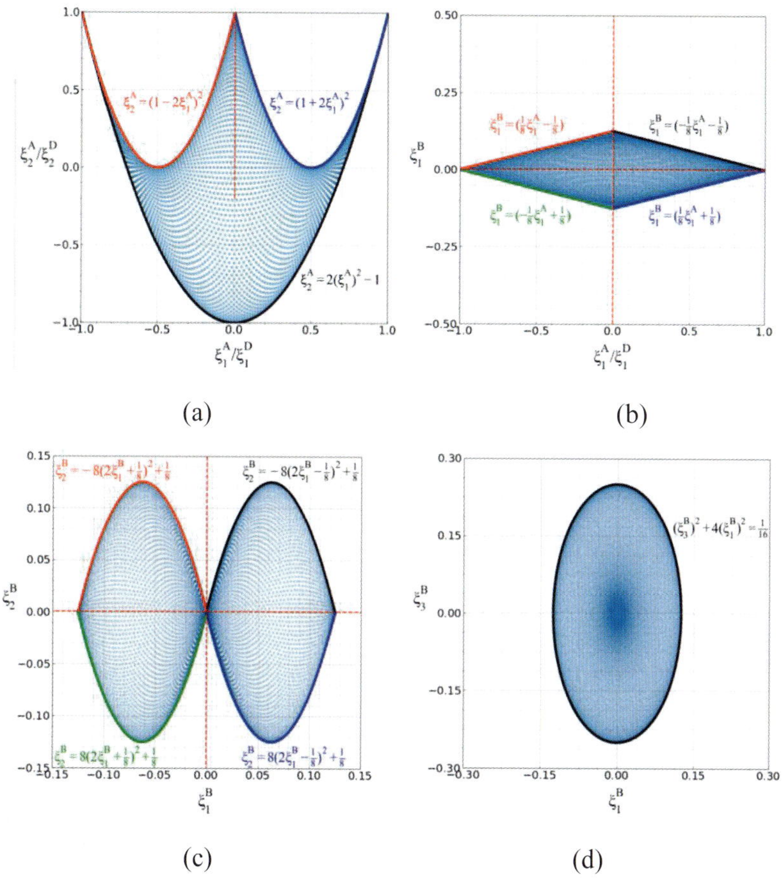

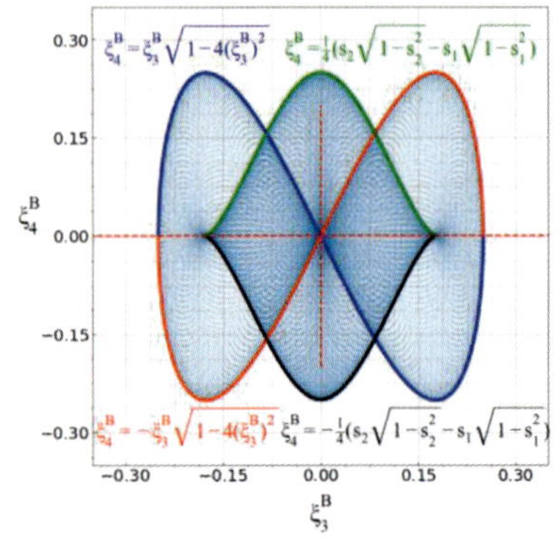

(e)

Fig. 10.2. Two dimensional feasible regions of 8-ply laminate: 6 unique shapes.

Fig 10.2 illustrates feasible regions of the five unique pairs of lamination parameters (i.e. ξ_1^A-ξ_2^A/ξ_1^D-ξ_2^D, ξ_1^A-ξ_1^B, ξ_1^B-ξ_2^B, ξ_1^B-ξ_3^B and ξ_3^B-ξ_4^B). Five different shapes of feasible region are obtained from six different types of solutions. The equations for the boundaries which include straight lines, quadratics, ellipses and quartics are also included in the figure with the equations being in the same colour as the corresponding boundaries. In order to validate the obtained feasible regions, the lamination parameters of all of the possible layups for 8-ply laminates with ply angle increments of 1 degree are plotted (the blue scatter points) and can be seen to lie within the feasible region boundaries.

10.4.2. Feasible regions of Staggered 1

Above solution covers larger design possibilities than each type of DD family. However, most of the types brings extra disadvantageous design space in coupling stiffness. It is found by [4] that the optimal sub-laminate is the staggered 1. Such a configuration retains the same design space of in-plane and out-of-plane stiffness and has the minimum space for coupling stiffness. To obtain its feasible regions, extra constraints restricting angles lying within 0° to 90° must be added on the top of base solutions. Those extra boundaries can be easily found by restricting one of the DD angles to be 0° or 90°. From Eq.10.17-20 it can be deduced that restricting ϕ and ψ to be positive only change lamination parameters $\xi_{3,4}^B, \xi_{3,4}^D$. Therefore, the feasible regions not related to these four lamination parameters are the same as the base solutions, shown in Appendix B. Such a two-step method can be applied to any sequence of DD families.

Fig 10.3 illustrates unique shapes of feasible regions for staggered 1 marked in yellow scatters with 8 plies in total, and their solutions are listed within plots. Since staggered 1 has different values of $\xi_{3,4}^B, \xi_{3,4}^D$ compared with the base solutions, 22 feasible regions related to them are changed. Among these feasible regions, 6 unique and basic shapes are identified, as shown in Fig 10.3 (a-f). It can be observed that restricting ϕ and ψ between 0° and 90° significantly reduces the feasible regions of coupling lamination parameters $\xi_{1,2,3,4}^B$ compared with base solutions marked in blue scatters, hence, the coupling effect on staggered 1 is reduced.

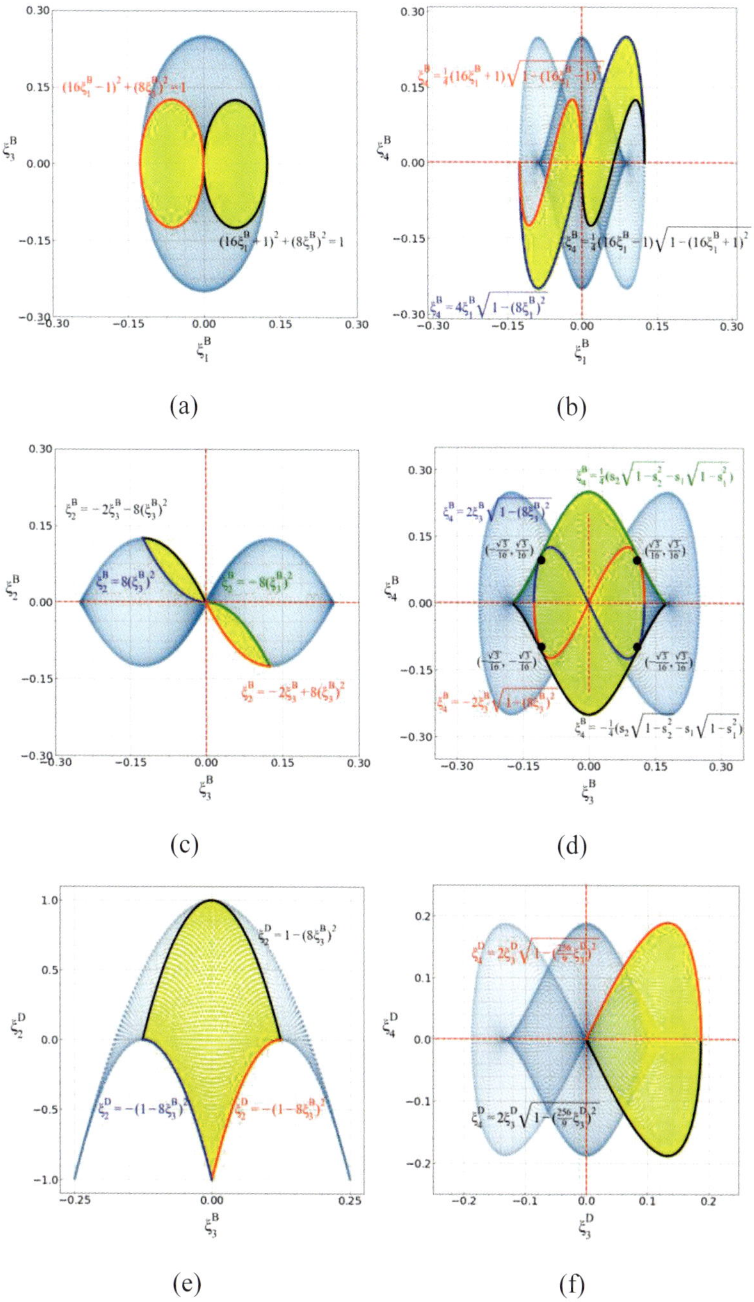

Fig. 10.3. Feasible regions for staggered 1 with 8 plies.

10.4.3. Staggered 1 vs QUAD

Comparison of the feasible region of ξ_1^A-ξ_2^A between QUAD laminates and DD laminates is shown in Fig.10.4. The solid green lines in Fig.10.4 (a), (b), and (c) represent the feasible regions of 6-ply, 12-ply and 20-ply QUAD laminates, respectively. As can be seen that the lower boundaries of the feasible region of DD laminates are lower than those of the three QUAD laminates, achieving more design possibilities. However, DD laminates cannot reach the lamination parameters located in the top left and top right of the feasible region of QUAD laminates, especially when the number of plies is large. The red parabola represents the boundary above which the percentage of 90° plies should be more than 50% (hard laminates), while the blue parabola is the same for the 0° plies (hard laminates). Due to the nature of DD laminates, the maximum percentage of 90° or 0° is 50%. As shown in Fig.10.4 (b), for QUAD laminates with ply number less than 12, which are mostly adopted in the industry, most of its feasible region can be covered by DD laminates and the rest can be approximated by the nearest point of DD laminates. For QUAD laminates with a large number of plies, the regions of soft laminates (±45° plies are more than 50%) and neutral laminates can still be fully covered. The hard laminates of QUAD, however, cannot be covered by DD laminates. To cover this area, some strategies can be applied to DD laminates e.g. replacing the QUAD laminate with a DD laminate with the same A11*, A22* or A66*. Besides, the failure envelopes of the hard QUAD laminates and approximated DD laminates are close. Hence, this reduction of the feasible region has only a small effect on DD laminates in terms of practical design.

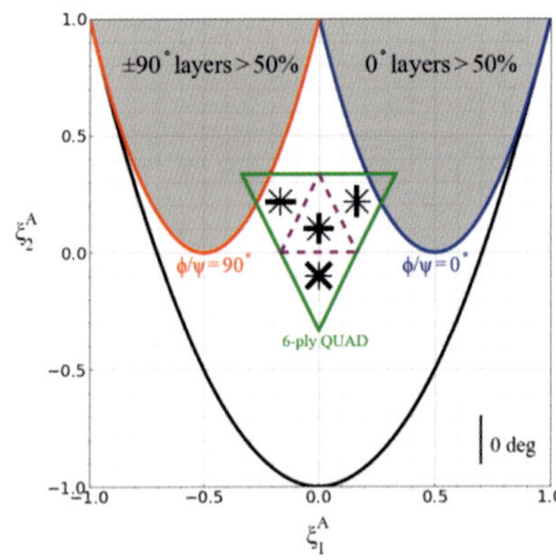

(a)

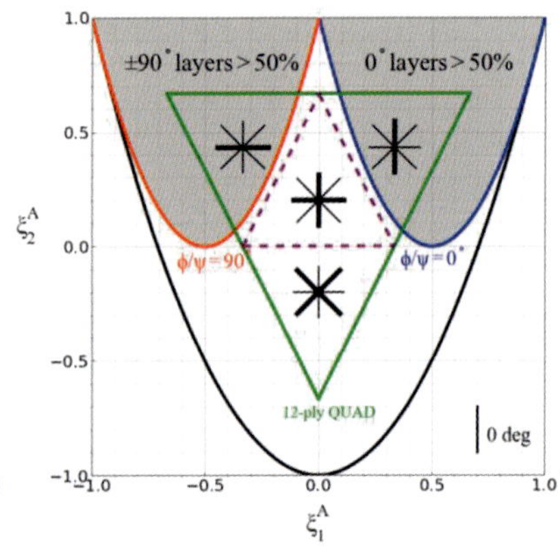

(b)

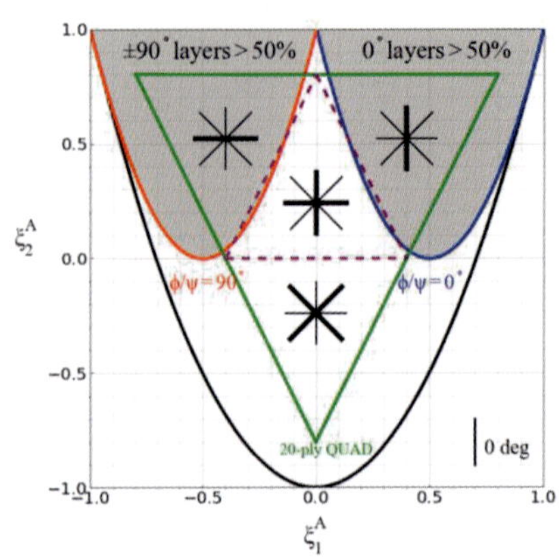

(c)

Fig. 10.4. Feasible region of ξ_1^A-ξ_2^A for DD laminates and QUAD laminates, (a) 6-ply; (b) 12-ply; (c) 20-ply.

10.4.4. The effect on the number of plies/repeats

For practical design, the coupling stiffness **B** is an important concern as it causes warping within a laminate. For QUAD laminates, the coupling stiffness is forced to be zero by stacking laminas symmetrically, resulting in high-thickness laminates. Due to the nature of DD laminates, their coupling stiffness **B** cannot be eliminated. However, the coupling effect can be significantly reduced by increasing the number of plies. Such a phenomenon is known as homogenization [7]. Fig. 10.5 shows the effect of the number of plies on the feasible regions of $\xi^B_{1,2,3,4}$. Because the boundaries of the feasible regions of DD laminates are inversely proportional to the number of plies, see table 10.1, a DD laminate is able to achieve homogenization with only a few repeats. As can be seen from Fig. 10.5, the values of $\xi^B_{1,2,3,4}$ are halved when the number of plies is doubled. The maximum absolute values of $\xi^B_{3,4}$ for a DD laminate with 8 plies are both 0.5, while those of a DD laminate with 32 plies are reduced to 0.125, which means the effects of B_{16} and B_{26} could be neglected. Compared to $\xi^B_{3,4}$, $\xi^B_{1,2}$ have even smaller absolute values. As can be seen from Fig. 10.5 (a), the maximum absolute values of $\xi^B_{1,2}$ for a DD laminate with 32 plies are less than 0.1.

On the other hand, homogenization is also seen for ξ^D_3 and ξ^D_4, as the contributions made by each angle are equal, with discrepancies that are inversely proportional to the square of the number of plies. Thus, the coupling parameters ξ^D_3 and ξ^D_4 are approaching zeros when the number of plies increases, see Fig 10.5(g), while for QUAD laminates, these stiffnesses have to be tailored in order to avoid such coupling, which limits their design space.

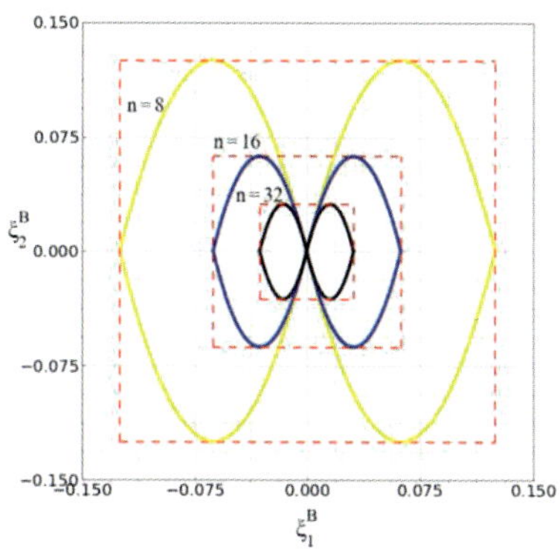

(a)

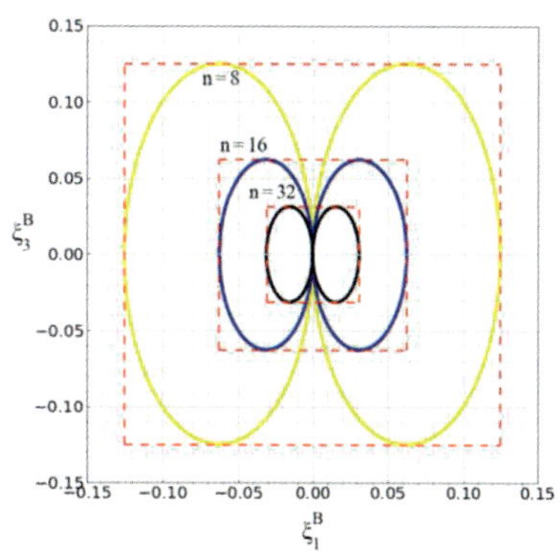

(b)

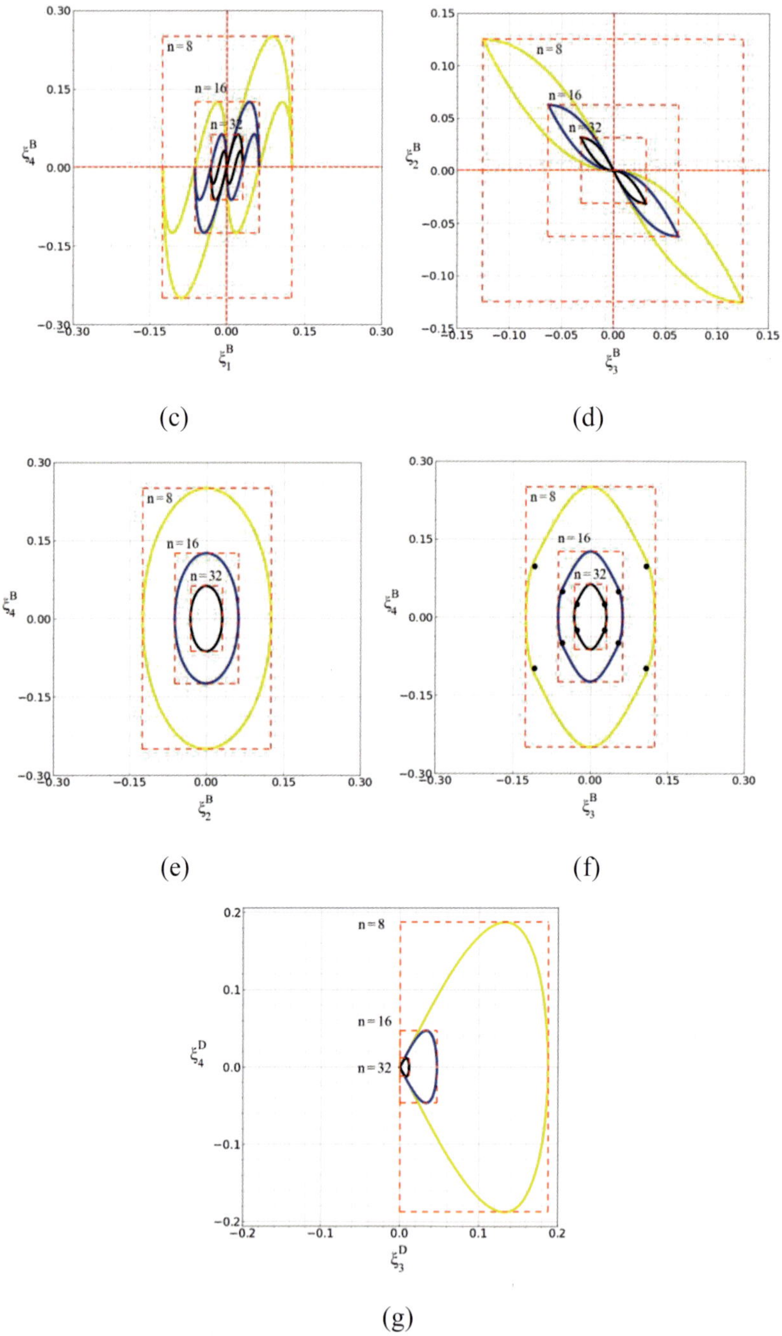

Fig. 10.5. Feasible regions of lamination parameters: (a) ξ_1^B-ξ_2^B; (b) ξ_1^B-ξ_3^B; (c) ξ_1^B-ξ_4^B; (d) ξ_3^B-ξ_2^B; (e) ξ_2^B-ξ_4^B; (f) ξ_3^B-ξ_4^B; (g) ξ_3^D-ξ_4^D.

Table 10.1. feasible region boundaries for staggered 1.

(x, y)	Feasible region	Range

ξ_1^B, ξ_2^B	$\lvert y\rvert = \frac{1}{n}[(1-2n\lvert x\rvert)^2 - 1]$	$-\frac{1}{n} \le x \le \frac{1}{n}; -\frac{1}{n} \le y \le \frac{1}{n}$
ξ_1^B, ξ_3^B	$n^2y^2 + (2nx \pm 1)^2 = 1$	$-\frac{1}{n} \le x \le \frac{1}{n}; -\frac{1}{n} \le y \le \frac{1}{n}$
ξ_1^B, ξ_4^B	$y = \frac{2}{n}(2nx \pm 1)\sqrt{1-(2nx \pm 1)^2}$ $y = 4x\sqrt{1-(nx)^2}$	$-\frac{1}{n} \le x \le \frac{1}{n}; -\frac{2}{n} \le y \le \frac{2}{n}$
ξ_3^B, ξ_2^B	$y = \pm nx^2$ $y = -2x \pm nx^2$	$-\frac{1}{n} \le x \le \frac{1}{n}; -\frac{1}{n} \le y \le \frac{1}{n}$
ξ_2^B, ξ_4^B	$(\frac{y}{2})^2 + x^2 = \frac{1}{n^2}$	$-\frac{1}{n} \le x \le \frac{1}{n}; -\frac{2}{n} \le y \le \frac{2}{n}$
ξ_3^B, ξ_4^B	$y = \pm 2x\sqrt{1-(nx)^2}$ $y = \frac{2}{n}\left(\left\lvert s_1\sqrt{1-s_1^2} - s_2\sqrt{1-s_2^2}\right\rvert\right)$ $s_1 = \frac{n}{2}x - \delta, s_2 = \frac{n}{2}x + \delta$ $\delta = \sqrt{\left(1 + \frac{n^2}{4}x^2\right) - \sqrt{n^2x^2 + \frac{1}{4}}}$	$-\frac{1}{n} \le x \le \frac{1}{n}; -\frac{2}{n} \le y \le \frac{2}{n}$
ξ_3^D, ξ_4^D	$y = 2\lvert x\rvert\sqrt{1 - \frac{n^4x^2}{144}}$	$-\frac{12}{n^2} \le x \le \frac{12}{n^2}; -\frac{12}{n^2} \le y \le \frac{12}{n^2}$

10.5. Homogenization criterion

According to [8], two criterions have to be met to reach homogenization, which require the largest components of $[\mathbf{B}^*]$ to be sufficient small and $\lvert[\mathbf{A}^*] - [\mathbf{D}^*]\rvert$ is approaching 0[50], where $[\mathbf{A}^*,\ \mathbf{B}^*, \mathbf{D}^*]$ are thickness normalized stiffness defined in Eq. 10.24 [9].

$$\mathbf{A}^*, \mathbf{B}^*, \mathbf{D}^* = [\frac{1}{h}A_{ij}, \frac{2}{h^2}B_{ij}, \frac{12}{h^3}D_{ij}] \tag{10.24}$$

Specific homogenization criterion is defined in [4] i.e. a percentage I of the absolute value of Tsai's modulus [10] which can be defined in the following equations:

$$\left\lvert \mathbf{B}_{ij}{}^* \right\rvert < I * Tsai's \tag{10.25}$$

$$\lvert[\mathbf{A}^*] - [\mathbf{D}^*]\rvert < I * Tsai's \tag{10.26}$$

As for the DD laminates with sub-laminate of $[+\phi/-\psi/-\phi/+\psi]$, as $\xi_{1,2}^A$=$\xi_{1,2}^D$ and $\xi_{3,4}^A = 0$, with Eq. 10.25, it can be easily deduced that A_{11}^*, A_{22}^*, A_{66}^*=D_{11}^*, D_{22}^*, D_{66}^* and $A_{16}^*, A_{26}^* = 0$. Hence, Eq.10.27 can be simplified as

$$|[\mathbf{D_{16}}^*]| < I * Tsai's \tag{10.27}$$

$$|[\mathbf{D_{26}}^*]| < I * Tsai's \tag{10.28}$$

Normalizing as in Eq. 10.4 of Eq.10.25, $\mathbf{B}^* = (2/h^2)\mathbf{B}$ so that

$$\begin{bmatrix} B_{11}^* \\ B_{22}^* \\ B_{12}^* \\ B_{66}^* \\ B_{16}^* \\ B_{26}^* \end{bmatrix} = \begin{bmatrix} 0 & \xi_1^B & \xi_2^B & 0 & 0 \\ 0 & -\xi_1^B & \xi_2^B & 0 & 0 \\ 0 & 0 & -\xi_2^B & 0 & 0 \\ 0 & 0 & -\xi_2^B & 0 & 0 \\ 0 & \xi_3^B/2 & \xi_4^B & 0 & 0 \\ 0 & \xi_3^B/2 & -\xi_4^B & 0 & 0 \end{bmatrix} \begin{bmatrix} U_1 \\ U_2 \\ U_3 \\ U_4 \\ U_5 \end{bmatrix} \tag{10.29}$$

As the values of components in the stiffness matrix $\mathbf{B}^*$ are set to be less than a predefined value of I to satisfy the homogenization criterion, the criterion can be expressed as follows:

$$-I * Tsai's \le (U_2 * \xi_1^B + U_3 * \xi_2^B) \le I * Tsai's \tag{10.30}$$

$$-I * Tsai's \le (-U_2 * \xi_1^B + U_3 * \xi_2^B) \le I * Tsai's \tag{10.31}$$

$$-I * Tsai's \le (-U_3 * \xi_2^B) \le I * Tsai's \tag{10.32}$$

$$-I * Tsai's \le (\frac{1}{2}U_2 * \xi_3^B + U_3 * \xi_4^B) \le I * Tsai's \tag{10.33}$$

$$-I * Tsai's \le (\frac{1}{2}U_2 * \xi_3^B - U_3 * \xi_4^B) \le I * Tsai's \tag{10.34}$$

From Eq. 10.30-32:

$$-\frac{I}{U_2^*} \le \xi_1^B \le \frac{I}{U_2^*} \text{ and } -\frac{I}{U_3^*} \le \xi_2^B \le \frac{I}{U_3^*} \tag{10.35a,35b}$$

Where U_2^* and $U_3^*$are normalized U_2 and U_3 by Tsai's modulus [11].
These are extreme values. The feasible region of (ξ_1^B, ξ_2^B) is a parallelogram with vertices at

$$\left(-\frac{I}{U_2^*}, 0\right), \left(0, -\frac{I}{U_3^*}\right), \left(\frac{I}{U_2^*}, 0\right) \text{ and } \left(0, \frac{I}{U_3^*}\right)$$

Combining Eq. 10.33 and Eq. 10.34

$$-\frac{2I}{U_2^*} \le \xi_3^B \le \frac{2I}{U_2^*} \text{ and } -\frac{I}{U_3^*} \le \xi_4^B \le \frac{I}{U_3^*} \tag{10.36a,36b}$$

These are extreme values. The feasible region of (ξ_3^B, ξ_4^B) is a parallelogram with vertices at

$$\left(-\frac{2I}{U_2^*}, 0\right), \left(0, -\frac{I}{U_3^*}\right), \left(\frac{2I}{U_2^*}, 0\right) \text{ and } \left(0, \frac{I}{U_3^*}\right)$$

As for the homogenization criterion defined in Eq.10.32, $\xi_{3,4}^{D}$ are inversely proportional to the second order of number of plies, hence ξ_3^D and ξ_4^D are always smaller than ξ_3^B and ξ_4^B, respectively. Therefore, for staggered 1, when $[\mathbf{B}^*]$ has met the criterion of 2%, $||[\mathbf{A}^*]-[\mathbf{D}^*]||$ is naturally less than 2% of Tsai's. Fig.6 presents the feasible regions in terms of $\xi_{1,2,3,4}^{B}$ for DD laminates with 8, 16 and 32 plies with a homogenization criterion (red dashed line) of 2% of Tsai's modulus, replacing by master ply values [11] i.e. $U_2^* = 0.417$ and $U_3^* = 0.098$ leading to homogenization criterion at $(-0.048,0),(0,-0.204),(0.048,0)$ and $(0,0.204)$ for the feasible region of (ξ_1^B,ξ_2^B) and $(-0.096,0),(0,-0.204),(0.096,0)$ and $(0,0.204)$ for the feasible region of (ξ_3^B,ξ_4^B).

As discussed in Section 10.4.3, the DD laminate with a larger number of plies has smaller design spaces in terms of $\xi_{1,2,3,4}^{B}$, hence, the homogenization criterion can be relatively easily achieved by the laminates with a larger number of plies/repeats. As can be seen that for staggered 1 with 16 plies (4 repeats), most of the design space is located inside the region defined by the homogenization criterion, while for DD laminates with 32 plies (8 repeats), the design space is fully covered by the homogenization criterion.

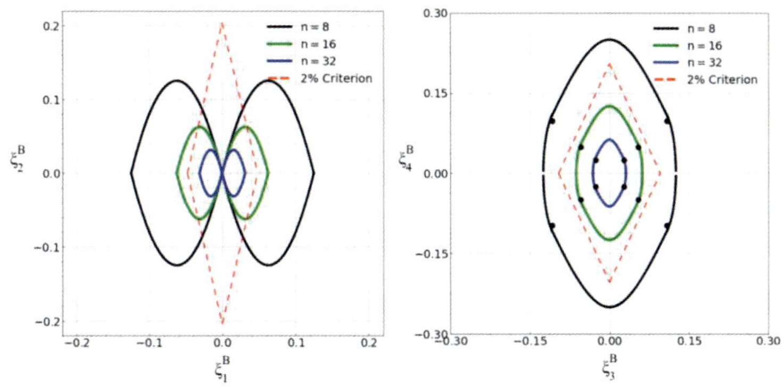

(a) (b)

Fig. 10.6. Feasible region of ξ_j^B with 2% homogenization criterion.

For the feasible regions out of this homogenization criterion envelope, other types of DD laminates can be employed instead to cover this area, and 16 plies/4 repeats are enough for DD families to reach homogenization. Thus, DD laminates can still retain the maximum design space in bending and extension stiffness while having the minimum or no effect on coupling stiffness. By such means, staggered 1 can reach homogenization at 16 plies/4 repeats using staggered 3 as replacements on the unreachable area.

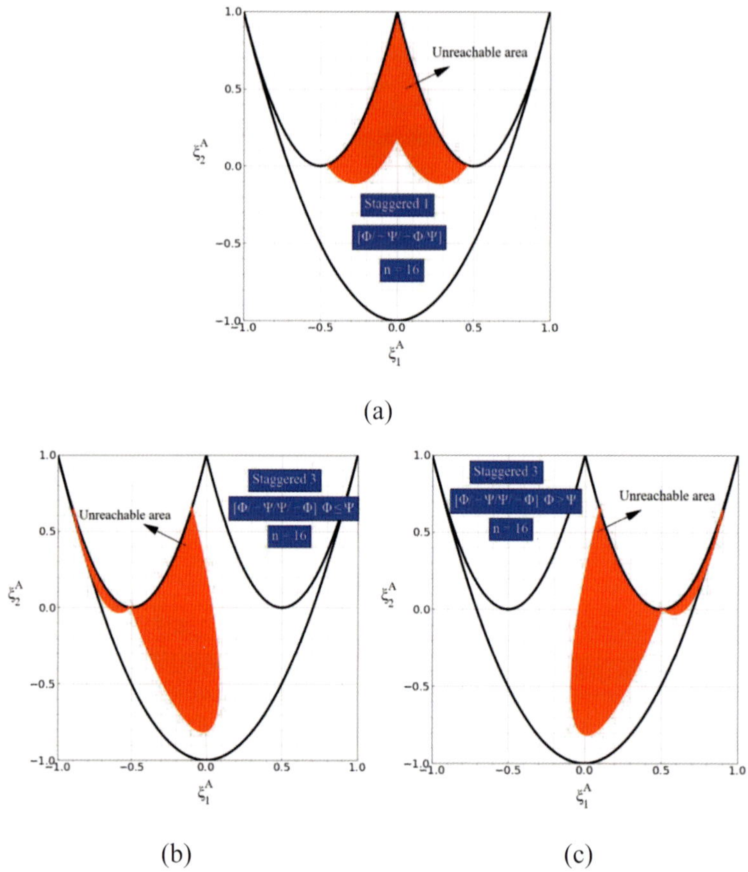

Fig. 10.7. unreachable feasible region under 2% homogenization criterion for 4-repeat DD laminates on ξ_1^A vs ξ_2^A.

Fig 10.7 illustrates the unreachable feasible region under the 2% homogenization criterion for 4-repeat staggered 1 and 3. For staggered 1, this area in feasible region ξ_1^B and ξ_2^B is reflected to the feasible region ξ_1^A and ξ_2^A shown in Fig.10.7(a).Unlike staggered 1, the unreachable area for staggered 3 is caused by the criterion on ξ_3^B and ξ_4^B rather than ξ_1^B and ξ_2^B. It can be observed that the red area for staggered 1, see Fig 10.7(a), can be covered by a combination of staggered 3 when $\Phi \leq \Psi$, see Fig 10.7(b), and $\Phi > \Psi$, see Fig 10.7(c). Thus, this combination of DD sequences shown in Fig 10.8 allows DD laminates to reach all the possible design space covered by the 2% homogenization criterion in just 16 plies/4 repeats. As for the conventional QUAD laminates which are characterized by heterogeneous, they would require at least 120 plies to reach homogenization, the large number of plies makes their optimization design hard to perform[4]. Great benefits can be, therefore, obtained by replacing QUAD laminates with DD laminates.

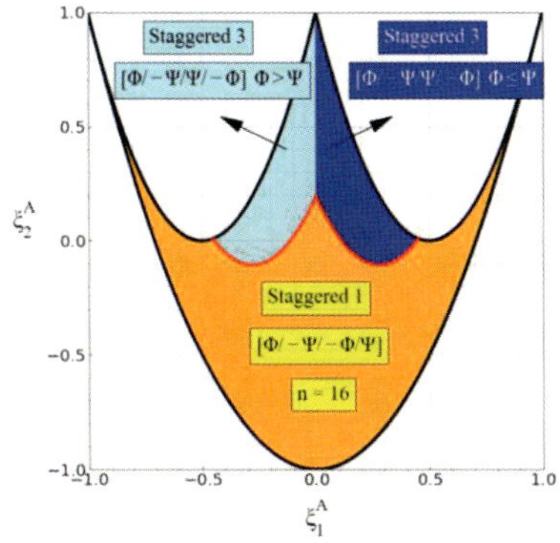

Fig. 10.8. A tailored feasible region for a combination of staggered 1 and 3.

10.6. Conclusions

In this chapter, novel formulations for obtaining the lamination parameters of DD laminates are presented, based on which the feasible regions of the combinations of any two lamination parameters are analytically obtained using Lagrange multipliers. All the feasible regions with regards to two lamination parameters have been obtained in this research, and the six typical feasible regions have been presented, while the rest are either rotations or scaling of these basic shapes. These novel formulations are also applicable to the DD laminates with rigid body rotation, which will be discussed in future work. The proposed method is further validated against one of the popular DD laminates with the stacking sequence of the sub-laminate as $[+\phi/-\psi/-\phi/+\psi]$. The obtained results indicate that the lamination parameters ξ_1^A and ξ_2^A are always equal to lamination parameters ξ_1^D and ξ_2^D, respectively, meanwhile ξ_3^A and ξ_4^A always remain zeros. The feasible regions of ξ_{1234}^B have first order negative correlations with the number of plies. Such a feature is significantly beneficial for DD laminates to achieve homogenization. Besides, based on the simple expressions of lamination parameters, feasible regions for any two lamination parameters can be obtained analytically using the method of Lagrange multipliers. Finally, the homogenization criterion of 2% has been drawn in the feasible region of DD laminates, and combination sequences of staggered 1 and 3 have been proven to be able to reach all the design space covered by the 2% homogenization criterion with only 16 plies/4 repeats.

References

[1] York CB. Laminate stiffness tailoring for improved buckling performance. Thin-Walled Structures 2021;161:107482. https://doi.org/10.1016/j.tws.2021.107482.

[2] Zhao K, Kennedy D, Miravete A, Tsai SW, Featherston CA, Liu X. Defining the Design Space for Double–Double Laminates by Considering Homogenization Criterion. AIAA Journal 2023:1–14. https://doi.org/10.2514/1.j062639.

[3] Tsai SW, Pagano NJ. Invariant properties of composite materials. AIR FORCE MATERIALS LAB WRIGHT-PATTERSON AFB OH WRIGHT-PATTERSON AFB; 1968.

[4] Tsai SW. Double–double: New family of composite laminates. AIAA Journal 2021;59:4293–305. https://doi.org/10.2514/1.J060659.

[5] Miki M. Material design of composite laminates with required in-plane elastic properties. Progress in Science and Engineering of Composites 1982;2:1725–31.

[6] Miki M. Design of laminated fibrous composite plates with required flexural stiffness. ASTM International; 1985.

[7] Shrivastava S, Sharma N, Tsai SW, Mohite PM. D and DD-drop layup optimization of aircraft wing panels under multi-load case design environment. Compos Struct 2020;248:112518. https://doi.org/10.1016/j.compstruct.2020.112518.

[8] Vermes B, Tsai SW, Riccio A, di Caprio F, Roy S. Application of the Tsai's modulus and double-double concepts to the definition of a new affordable design approach for composite laminates. Compos Struct 2021;259:113246. https://doi.org/10.1016/j.compstruct.2020.113246.

[9] Tsai SW, Melo JDD. An invariant-based theory of composites. Compos Sci Technol 2014;100:237–43. https://doi.org/10.1016/j.compscitech.2014.06.017.

[10] Tsai SW, Melo JDD. An invariant-based theory of composites. Compos Sci Technol 2014;100:237–43. https://doi.org/10.1016/j.compscitech.2014.06.017.

[11] Tsai SW, Melo JDD, Sihn S, Arteiro A, Rainsberger R. Composite Laminates: Theory and practice of analysis, design and automated layup. Stanford Aeronautics & Astronautics; 2017.

Appendix A

There are 10 possible types, as given in the following list where superscripts x and y equal A, B or D and denote the lamination parameters to be plotted on the x and y axes, respectively.

Type (1,1) compares ξ_1^x and ξ_1^y: $f(\theta_1,\theta_2) = p_1c_1 + p_2c_2\ ;\ g(\theta_1,\theta_2) = q_1c_1 + q_2c_2$
Type (2,2) compares ξ_2^x and ξ_2^y: $f(\theta_1,\theta_2) = p_1C_1 + p_2C_2\ ;\ g(\theta_1,\theta_2) = q_1C_1 + q_2C_2$
Type (3,3) compares ξ_3^x and ξ_3^y: $f(\theta_1,\theta_2) = p_1s_1 + p_2s_2\ ;\ g(\theta_1,\theta_2) = q_1s_1 + q_2s_2$
Type (4,4) compares ξ_4^x and ξ_4^y: $f(\theta_1,\theta_2) = p_1S_1 + p_2S_2\ ;\ g(\theta_1,\theta_2) = q_1S_1 + q_2S_2$
Type (1,2) compares ξ_1^x and ξ_2^y: $f(\theta_1,\theta_2) = p_1C_1 + p_2C_2\ ;\ g(\theta_1,\theta_2) = q_1c_1 + q_2c_2$
Type (1,3) compares ξ_1^x and ξ_3^y: $f(\theta_1,\theta_2) = p_1s_1 + p_2s_2\ ;\ g(\theta_1,\theta_2) = q_1c_1 + q_2c_2$
Type (1,4) compares ξ_1^x and ξ_4^y: $f(\theta_1,\theta_2) = p_1S_1 + p_2S_2\ ;\ g(\theta_1,\theta_2) = q_1c_1 + q_2c_2$
Type (3,2) compares ξ_3^x and ξ_2^y: $f(\theta_1,\theta_2) = p_1C_1 + p_2C_2\ ;\ g(\theta_1,\theta_2) = q_1s_1 + q_2s_2$
Type (2,4) compares ξ_2^x and ξ_4^y: $f(\theta_1,\theta_2) = p_1S_1 + p_2S_2\ ;\ g(\theta_1,\theta_2) = q_1C_1 + q_2C_2$
Type (3,4) compares ξ_2^x and ξ_4^y: $f(\theta_1,\theta_2) = p_1S_1 + p_2S_2\ ;\ g(\theta_1,\theta_2) = q_1s_1 + q_2s_2$

Where

$c_1 = cos(2\theta_1)$ $\quad$ $c_2 = cos(2\theta_2)$,
$C_1 = cos(4\theta_1) = 2c_1^2 - 1 = 1 - 2s_1^2$ $\quad$ $C_2 = cos(4\theta_2) = 2c_2^2 - 1 = 1 - 2s_2^2$
$s_1 = sin(2\theta_1)$ $\quad$ $s_2 = sin(2\theta_2)$
$S_1 = sin(4\theta_1) = 2c_1s_1$ $\quad$ $S_2 = sin(4\theta_2) = 2c_2s_2$

And (p_1, p_2) and (q_1, q_2) are the coefficients of the trigonometric functions in the objective function and constraint, respectively.

Appendix B

Feasible regions of lamination parameters: specific solutions for DD laminates: $-90 \leq \phi, \psi \leq 90$

There is no need to compare a parameter with itself. Also $\xi_1^A = \xi_1^D$, $\xi_2^A = \xi_2^D$ and $\xi_3^A = \xi_4^A = 0$. So, there are 28 comparisons, i.e. C(8,2)=28, between pairs of the 4 $\xi_{1,2,3,4}^D$parameters and the 4 $\xi_{1,2,3,4}^B$ parameters.

x	q_1	q_2	y	p_1	p_2	Type	$-90 \leq \phi, \psi \leq 90$
ξ_1^D	$\frac{1}{2}$	$\frac{1}{2}$	ξ_1^B	$-\frac{1}{2n}$	$\frac{1}{2n}$	(1,1)	$\lvert y\rvert \leq \frac{1}{n}(1 - \lvert x\rvert)$
ξ_2^D	$\frac{1}{2}$	$\frac{1}{2}$	ξ_2^B	$-\frac{1}{2n}$	$\frac{1}{2n}$	(2,2)	$\lvert y\rvert \leq \frac{1}{n}(1 - \lvert x\rvert)$
ξ_3^D	$\frac{6}{n^2}$	$\frac{6}{n^2}$	ξ_3^B	$-\frac{1}{n}$	$\frac{1}{n}$	(3,3)	$\lvert y\rvert \leq \left(\frac{2}{n} - \frac{n}{6}\lvert x\rvert\right)$
ξ_4^D	$\frac{6}{n^2}$	$\frac{6}{n^2}$	ξ_4^B	$-\frac{1}{n}$	$\frac{1}{n}$	(4,4)	$\lvert y\rvert \leq \left(\frac{2}{n} - \frac{n}{6}\lvert x\rvert\right)$
ξ_1^D	$\frac{1}{2}$	$\frac{1}{2}$	ξ_2^D	$\frac{1}{2}$	$\frac{1}{2}$	(1,2)	$2x^2 - 1 \leq y \leq (1 - 2\lvert x\rvert)^2$
ξ_1^D	$\frac{1}{2}$	$\frac{1}{2}$	ξ_2^B	$-\frac{1}{2n}$	$\frac{1}{2n}$	(1,2)	$\lvert y\rvert \leq \frac{1}{n}[(1 - 2\lvert x\rvert)^2 - 1]$
ξ_1^B	$-\frac{1}{2n}$	$\frac{1}{2n}$	ξ_2^D	$\frac{1}{2}$	$\frac{1}{2}$	(1,2)	$2n^2x^2 - 1 \leq y \leq (1 - 2n\lvert x\rvert)^2$
ξ_1^B	$-\frac{1}{2n}$	$\frac{1}{2n}$	ξ_2^B	$-\frac{1}{2n}$	$\frac{1}{2n}$	(1,2)	$\lvert y\rvert \leq \frac{1}{n}[(1 - 2n\lvert x\rvert)^2 - 1]$
ξ_1^D	$\frac{1}{2}$	$\frac{1}{2}$	ξ_3^D	$\frac{6}{n^2}$	$\frac{6}{n^2}$	(1,3)	$\lvert y\rvert \leq \frac{12}{n^2}\sqrt{1 - x^2}$
ξ_1^D	$\frac{1}{2}$	$\frac{1}{2}$	ξ_3^B	$-\frac{1}{n}$	$\frac{1}{n}$	(1,3)	$\lvert y\rvert \leq \frac{2}{n}\sqrt{1 - x^2}$

ξ_1^B	$-\frac{1}{2n}$	$\frac{1}{2n}$	ξ_3^D	$\frac{6}{n^2}$	$\frac{6}{n^2}$	(1,3)	$\lvert y\rvert \le \frac{12}{n^2}\sqrt{1-n^2x^2}$
ξ_1^B	$-\frac{1}{2n}$	$\frac{1}{2n}$	ξ_3^B	$-\frac{1}{n}$	$\frac{1}{n}$	(1,3)	$\lvert y\rvert \le \frac{2}{n}\sqrt{1-n^2x^2}$
ξ_1^D	$\frac{1}{2}$	$\frac{1}{2}$	ξ_4^D	$\frac{6}{n^2}$	$\frac{6}{n^2}$	(1,4)	$\lvert y\rvert \le \max\left[\frac{24\lvert x\rvert}{n^2}\sqrt{1-x^2}, \frac{12}{n^2}\left(\left\lvert c_1\sqrt{1-c_1^2} - c_2\sqrt{1-c_2^2}\right\rvert\right)\right]$ $c_1 = x-\delta, c_2 = x+\delta$ $\delta = \sqrt{(1+x^2) - \sqrt{4x^2+\frac{1}{4}}}$
ξ_1^D	$\frac{1}{2}$	$\frac{1}{2}$	ξ_4^B	$-\frac{1}{n}$	$\frac{1}{n}$	(1,4)	$\lvert y\rvert \le \max\left[\frac{4\lvert x\rvert}{n}\sqrt{1-x^2}, \frac{2}{n}\left(\left\lvert c_1\sqrt{1-c_1^2} - c_2\sqrt{1-c_2^2}\right\rvert\right)\right]$ $c_1 = x-\delta, c_2 = x+\delta$ $\delta = \sqrt{(1+x^2) - \sqrt{4x^2+\frac{1}{4}}}$
ξ_1^B	$-\frac{1}{2n}$	$\frac{1}{2n}$	ξ_4^D	$\frac{6}{n^2}$	$\frac{6}{n^2}$	(1,4)	$\lvert y\rvert \le \max\left[\frac{24\lvert x\rvert}{n}\sqrt{1-n^2x^2}, \frac{12}{n^2}\left(\left\lvert c_1\sqrt{1-c_1^2} - c_2\sqrt{1-c_2^2}\right\rvert\right)\right]$ $c_1 = nx-\delta, c_2 = nx+\delta$ $\delta = \sqrt{(1+(nx)^2) - \sqrt{4n^2x^2+\frac{1}{4}}}$
ξ_1^B	$-\frac{1}{2n}$	$\frac{1}{2n}$	ξ_4^B	$-\frac{1}{n}$	$\frac{1}{n}$	(1,4)	$\lvert y\rvert \le \max\left[4\lvert x\rvert\sqrt{1-n^2x^2}, \frac{2}{n}\left(\left\lvert c_1\sqrt{1-c_1^2} - c_2\sqrt{1-c_2^2}\right\rvert\right)\right]$ $c_1 = nx-\delta, c_2 = nx+\delta$ $\delta = \sqrt{(1+(nx)^2) - \sqrt{4n^2x^2+\frac{1}{4}}}$

ξ_3^D	$\frac{6}{n^2}$	$\frac{6}{n^2}$	ξ_2^D	$\frac{1}{2}$	$\frac{1}{2}$	(3,2)	$-\left(1-\frac{n^2\lvert x\rvert}{6}\right)^2 \leq y \leq 1-\frac{n^4x^2}{72}$
ξ_3^D	$\frac{6}{n^2}$	$\frac{6}{n^2}$	ξ_2^B	$-\frac{1}{2n}$	$\frac{1}{2n}$	(3,2)	$\lvert y\rvert \leq \frac{1}{n}\left[1-\left(1-\frac{n^2\lvert x\rvert}{6}\right)^2\right]$
ξ_3^B	$-\frac{1}{n}$	$\frac{1}{n}$	ξ_2^D	$\frac{1}{2}$	$\frac{1}{2}$	(3,2)	$-(1-n\lvert x\rvert)^2 \leq y \leq 1-\frac{n^2}{2}x^2$
ξ_3^B	$-\frac{1}{n}$	$\frac{1}{n}$	ξ_2^B	$-\frac{1}{2n}$	$\frac{1}{2n}$	(3.2)	$\lvert y\rvert \leq \frac{1}{n}(1-n\lvert x\rvert)^2$
ξ_2^D	$\frac{1}{2}$	$\frac{1}{2}$	ξ_4^D	$\frac{6}{n^2}$	$\frac{6}{n^2}$	(2,4)	$\lvert y\rvert \leq \frac{12}{n^2}\sqrt{1-x^2}$
ξ_2^D	$\frac{1}{2}$	$\frac{1}{2}$	ξ_4^B	$-\frac{1}{n}$	$\frac{1}{n}$	(2,4)	$\lvert y\rvert \leq \frac{2}{n}\sqrt{1-x^2}$
ξ_2^B	$-\frac{1}{2n}$	$\frac{1}{2n}$	ξ_4^D	$\frac{6}{n^2}$	$\frac{6}{n^2}$	(2,4)	$\lvert y\rvert \leq \frac{12}{n^2}\sqrt{1-n^2x^2}$
ξ_2^B	$-\frac{1}{2n}$	$\frac{1}{2n}$	ξ_4^B	$-\frac{1}{n}$	$\frac{1}{n}$	(2,4)	$\lvert y\rvert \leq \frac{2}{n}\sqrt{1-n^2x^2}$
ξ_3^D	$\frac{6}{n^2}$	$\frac{6}{n^2}$	ξ_4^D	$\frac{6}{n^2}$	$\frac{6}{n^2}$	(3,4)	$\lvert y\rvert \leq \max\left[2\lvert x\rvert\sqrt{1-\frac{n^4x^2}{144}}, \frac{12}{n^2}\left(\left\lvert s_1\sqrt{1-s_1^2}-s_2\sqrt{1-s_2^2}\right\rvert\right)\right]$ $s_1=\frac{n^2}{12}x-\delta, s_2=\frac{n^2}{12}x+\delta$ $\delta=\sqrt{\left(1+\frac{n^4x^2}{144}\right)-\sqrt{\frac{n^4x^2}{36}+\frac{1}{4}}}$
ξ_3^D	$\frac{6}{n^2}$	$\frac{6}{n^2}$	ξ_4^B	$-\frac{1}{n}$	$\frac{1}{n}$	(3,4)	$\lvert y\rvert \leq \max\left[\frac{n\lvert x\rvert}{3}\sqrt{1-\frac{n^4x^2}{144}}, \frac{2}{n}\left(\left\lvert s_1\sqrt{1-s_1^2}-s_2\sqrt{1-s_2^2}\right\rvert\right)\right]$ $s_1=\frac{n^2}{12}x-\delta, s_2=\frac{n^2}{12}x+\delta$ $\delta=\sqrt{\left(1+\frac{n^4x^2}{144}\right)-\sqrt{\frac{n^4x^2}{36}+\frac{1}{4}}}$

Ch.10

ξ_3^B	$-\frac{1}{n}$	$\frac{1}{n}$	ξ_4^D	$\frac{6}{n^2}$	$\frac{6}{n^2}$	(3,4)	$\frac{12}{n^2}\left(\left\lvert s_1\sqrt{1-s_1^2}-s_2\sqrt{1-s_2^2}\right\rvert\right)\Big]$ $s_1=\frac{n}{2}x-\delta, s_2=\frac{n}{2}x+\delta$ $\delta=\sqrt{\left(1+\frac{n^2}{4}x^2\right)-\sqrt{n^2x^2+\frac{1}{4}}}$
ξ_3^B	$-\frac{1}{n}$	$\frac{1}{n}$	ξ_4^B	$-\frac{1}{n}$	$\frac{1}{n}$	(3,4)	$\lvert y\rvert \leq \max\Big[2\lvert x\rvert\sqrt{1-\frac{n^2}{4}x^2},$ $\frac{2}{n}\left(\left\lvert s_1\sqrt{1-s_1^2}-s_2\sqrt{1-s_2^2}\right\rvert\right)\Big]$ $s_1=\frac{n}{2}x-\delta, s_2=\frac{n}{2}x+\delta$ $\delta=\sqrt{\left(1+\frac{n^2}{4}x^2\right)-\sqrt{n^2x^2+\frac{1}{4}}}$

Noted that above solutions are for DD laminates with angles ranging from -90° to 90°. To obtain each feasible region for staggered 1, 2, 3 and paired, it can be simply achieved by adding angle constraints (0° or 90°.) onto above boundaries.

CHAPTER 11

Notched Response of DD Laminates for Robust Characterization of Design Allowables

Albertino Arteiro
DEMec, Faculdade de Engenharia, Universidade do Porto
Rua Dr. Roberto Frias, s/n, 4200-465 Porto, Portugal
E-mail: aarteiro@fe.up.pt

11.1 Introduction

Structural components are typically designed taking into account the **effects of stress concentrations** resulting from **geometrical discontinuities** or **in-service damage**. Therefore, aircraft design demands the generation of notched strength allowables, which ensure safe design of access holes for wiring and other system functions and contribute to the damage tolerance-based design of through-penetration type of damage (Figure 11.1).

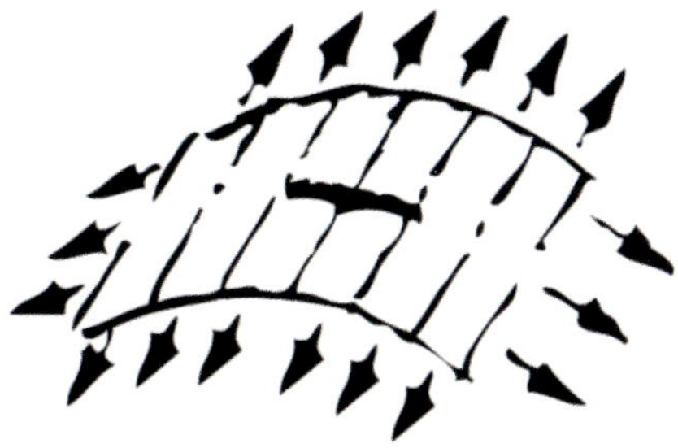

Figure 11.1 Illustration of through-penetration damage in a stiffened composite structure (DOD, 2002).

On the other hand, the **replacement of metallic materials by polymer composites** is dependent on the **capability to predict the strength of composite materials accurately and quickly**. Throughout the years, many analytical models have been proposed to predict the notched response of composite laminate coupons (Figure 11.2), ranging from classical linear elastic fracture mechanics (LEFM, Figure 11.3) to semi-empirical inherent flaw (Figure 11.4), point stress and average stress models (Figure 11.5). The latter, although usually providing good predictions, require that a baseline coupon, specific to a given notch geometry (e.g., open hole) and laminate family (e.g., quasi-isotropic, hard or soft) is tested to determine an empirical parameter, usually referred to as *characteristic length* or *distance*. This means that new baseline coupons need to be tested and new parameters **determined every time lay-up** and **notch geometry change**, making analysis and, consequently, optimization less robust.

Therefore, **analysis models** should not rely on **empirical parameters** but, instead, take the **laminate layers** as the **basic building blocks** and use the orthotropic material properties related to the directions of the fibers to predict the response of composite laminates. This is key to save time and cost in the design process, and to extend the analysis to a broader range of composite laminates, including unconventional designs.

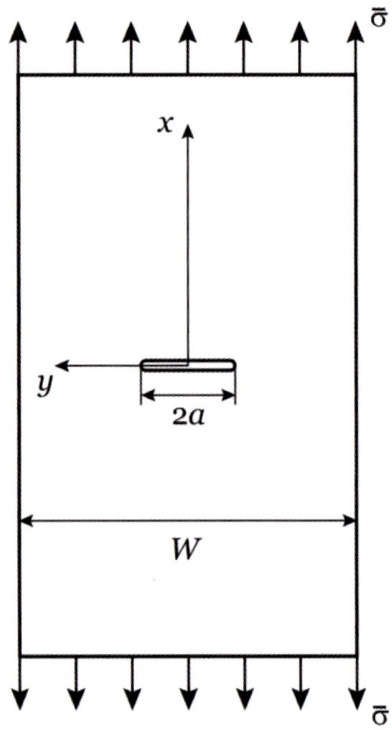

Figure 11.2 Centered notch plate with crack-like notch of length 2*a*.

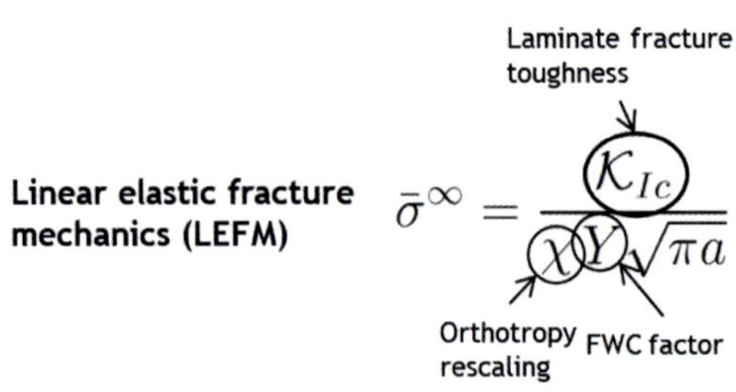

Figure 11.3 LEFM model for notched strength of laminates with crack-like notch length 2*a* (Suo, Bao, Fan, & Wang, 1991) (Chen, Weiss, Stickler, & Wang, 1992).

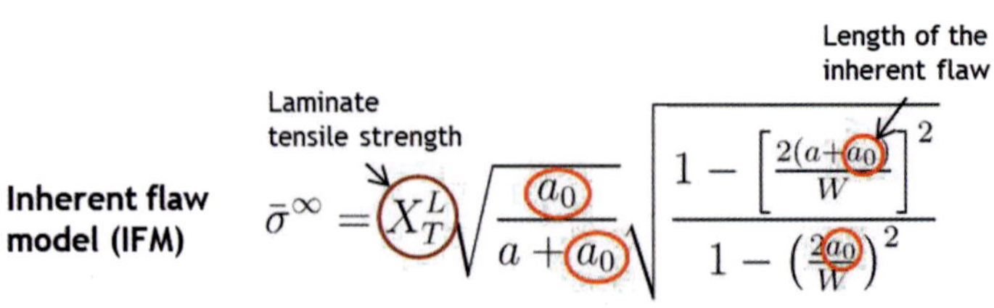

Figure 11.4 Inherent flaw model for notched strength of laminates with crack-like notch length 2*a* (Waddoups, Eisenmann, & Kaminski, 1971).

Characteristic distance

Point stress (PS) model $\sigma_{yy}(x,0)\big|_{x=a+r_{ot}} = X_T^L$

Average stress (AS) model $\dfrac{1}{r_{ot}}\displaystyle\int_a^{a+r_{ot}} \sigma_{yy}(x,0)\,\mathrm{d}x = X_T^L$

Figure 11.5 Point stress and average stress models for notched strength of laminates with crack-like notch length $2a$ (Whitney & Nuismer, 1974).

11.2 Notched strength of composite laminates

A more physically sound model for the notched strength of composite laminates resides in the application of a **Finite Fracture Mechanics (FFMs)** coupled criterion (Camanho, Erçin, Catalanotti, Mahdi, & Linde, 2012). This model is based on the simultaneous fulfilment of a stress-based criterion and an energy-based criterion (Figure 11.6). Only **three ingredients** are required to solve this system of equations:

1. The laminate elastic properties to obtain the stress field and Energy Release Rate (ERR) at the notch boundaries;
2. Laminate unnotched strength; and
3. Laminate fracture toughness.

The laminate elastic properties, the laminate strength and the laminate fracture toughness required in the system of equations of Figure 11.6 can be directly measured from testing the laminate of interest. However, design and optimization of composite structures requires additional flexibility in obtaining these laminate properties, and it is highly desirable that they are obtained from the properties measured at the ply (or lamina) level. In particular, it is of great interest to be able to scale between unidirectional ply (or lamina) and laminate properties, or between the properties of different laminates. This is possible through:

1. Tsai's modulus and Master Ply to scale the elastic properties between UD plies and laminates, and between laminates;
2. Omni envelopes to scale the plain strengths between UD plies and laminates, and between laminates; and
3. Fracture mechanics to scale the fracture toughness between UD plies and laminates, and between laminates.

11.3 Laminate elastic properties, stress field and ERR

Tsai's modulus – the trace of the stiffness matrices of composite laminates, $Tr = \mathrm{tr}(\mathbf{Q}) = \mathrm{tr}(\mathbf{A}^*) = \mathrm{tr}(\mathbf{D}^*)$ (Arteiro, et al., 2020), an invariant with respect to coordinate transformations – enables the definition of normalized elastic moduli of carbon-fiber reinforced polymer (CFRP) laminates (among other continuous fiber-reinforced composite systems) leading to a set of materially invariant elastic properties (Arteiro,

Pereira, Bessa, Furtado, & Camanho, 2019) – the **Master Ply** (Tsai & Melo, 2014), for example:

	E_1/Tr	E_2/Tr	G_{12}/Tr	ν_{12}
Master Ply	0.880	0.052	0.031	0.320

or, written as Tsai-normalized stiffness components:

	Q_{11}/Tr	Q_{12}/Tr	Q_{22}/Tr	Q_{66}/Tr
Master Ply	0.885	0.017	0.052	0.031

$$\begin{cases} \frac{1}{l}\int_R^{R+l} \sigma_x(0,y)\,\mathrm{d}y = X^L \\ \frac{1}{l}\int_R^{R+l} \mathcal{G}_I(a)\,\mathrm{d}a = \mathcal{G}_{Ic} \end{cases}$$

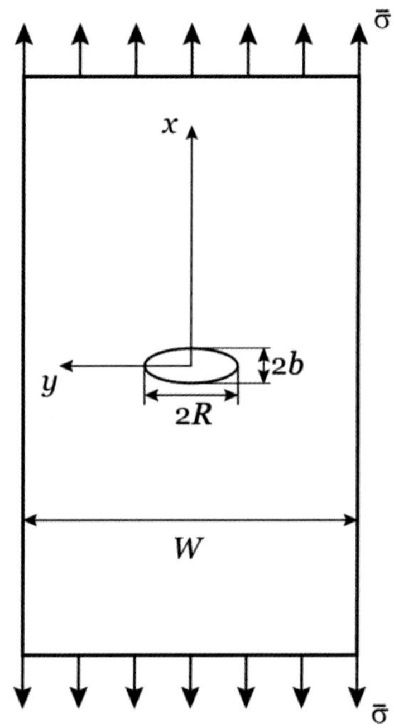

Figure 11.6 FFMs coupled criterion (left) for notched strength of laminates with an elliptical notch (right) of major axis $2R$ (Arteiro A. , Catalanotti, Xavier, & Camanho, 2013).

Using the Classical Laminated Plate Theory (CLPT), **Tsai-normalized laminate factors** can be generated using the Master Ply as ply properties. Whether considering conventional QUAD laminates or **double-double** (**DD**) laminates, the laminate in-plane engineering constants (or its matrix components A_{ij}*) can be obtained from the corresponding (materially invariant) laminate factors and from just one stiffness property – Tsai's modulus, Tr. Therefore, knowing that the stress distribution can be determined from the applied stress (Figure 11.6), elastic properties of the laminate and geometry (Tan, 1994):

$$\sigma_x(0,y) = \bar{\sigma}_x f(y, R, b, W, E_x, E_y, G_{xy}, \nu_{xy})$$

the only elastic property that is required is the Tsai's modulus, which can be used to determine the laminate elastic properties E_x, E_y, G_{xy} and ν_{xy} as a function of lay-up. This is required for the solution of the stress concentration factor for an infinitely large orthotropic plate with an elliptical notch (Tan, 1994):

$$K_T^{\infty} = 1 + \frac{1}{\lambda}\sqrt{2\left(\sqrt{\frac{E_x}{E_y}} - \nu_{xy}\right) + \frac{E_x}{G_{xy}}}$$

to obtain the stress field from the solution of an infinitely large plate and from a geometry correction factor (Tan, 1994):

$$\frac{K_T}{K_T^{\infty}}\sigma_x^{\infty}(0,y) = \sigma_x(0,y)$$

where the former can be determined as (Tan, 1994):

$$\frac{\sigma_x^{\infty}(0,y)}{\bar{\sigma}_x} = \frac{\lambda^2}{(1-\lambda)^2} + \frac{(1-2\lambda)\gamma}{(1-\lambda)^2\sqrt{\gamma^2-1+\lambda^2}} + \frac{\lambda^2\gamma}{(1-\lambda)(\gamma^2-1+\lambda^2)^{3/2}} - \frac{\lambda^7}{2}\left(K_T^{\infty}-1-\frac{2}{\lambda}\right)\left[\frac{5\gamma}{(\gamma^2-1+\lambda^2)^{7/2}} - \frac{7\lambda^2\gamma}{(\gamma^2-1+\lambda^2)^{9/2}}\right]$$

Ch.11

and the geometry correction factor is (Tan, 1994):

$$\frac{K_T^{\infty}}{K_T} = \frac{\lambda^2}{(1-\lambda)^2} + \frac{(1-2\lambda)}{(1-\lambda)^2}\sqrt{1+(\lambda^2-1)(2R/W)^2} - \frac{\lambda^2}{(1-\lambda)}\frac{(2R/W)^2}{\sqrt{1+(\lambda^2-1)(2R/W)^2}} + \frac{\lambda^7}{2}\left(\frac{2R}{W}\right)^6\left(K_T^{\infty}-1-\frac{2}{\lambda}\right)\left\{\left[1+(\lambda^2-1)\left(\frac{2R}{W}\right)^2\right]^{-5/2} - \left(\frac{2R}{W}\right)^2\left[1+(\lambda^2-1)\left(\frac{2R}{W}\right)^2\right]^{-7/2}\right\}$$

with:

$$\lambda = \frac{b}{R}$$

$$\gamma = \frac{y}{R}$$

Likewise, the ERR for two cracks, each of length a, emanating symmetrically from the notch boundaries can be determined from the applied stress (Figure 11.6), elastic properties of the laminate and geometry (Catalanotti, Salgado, & Camanho, 2021). In this case, the stress intensity factor can be obtained as:

$$\mathcal{K}_I(a) = \sqrt{R}\bar{\sigma}_x \Psi(a, R, b, W, L, E_x, E_y, G_{xy}, \nu_{xy})$$

where **Ψ** is a geometry and orthotropy correction factor (Catalanotti, Salgado, & Camanho, 2021):

$$\Psi = \left(\tan\frac{\pi\xi}{2}\right)^{\frac{1}{4}} \sum_{i,j,k,l} P_{ijkl}\xi^i\rho^j\left(\frac{\vartheta}{\lambda^4}\right)^{-\frac{1}{4}k}\omega^l$$

where ξ and α are crack geometry parameters, ρ and ϑ are orthotropy parameters, and ω is a notched plate geometry parameter:

$$\xi = \frac{\alpha}{\omega - 1} = \frac{2a}{W - 2R}$$

$$\alpha = \frac{a}{R}$$

$$\rho = \frac{E_y - 2\nu_{yx}G_{xy}}{2E_yG_{xy}\sqrt{\frac{1}{E_yE_x}}}$$

$$\vartheta = \frac{E_x}{E_y}$$

$$\omega = \frac{W}{2R}$$

The orthotropy parameters ρ and ϑ can be obtained from Tsai's modulus and CLPT as a function of lay-up. The solutions of the stress field and ERR are both written as functions of the applied stress, which, together with the characteristic length l comprise the two unknows of the system of equations of the FFMs coupled criterion (Figure 11.6).

11.4 Laminate strength

To obtain the **laminate strength** required to solve the stress equilibrium of the FFMs coupled criterion (Figure 11.6), the laminate in-plane engineering constants obtained with Tsai's modulus and the Master Ply can be used to determine the laminate strengths using the strains-to-failure obtained from omni-strain failure envelopes (Tsai & Melo, 2014). Ply failure criteria, such as Tsai-Wu (Tsai & Melo, 2014) or other phenomenological criteria (Corrado, et al., 2022), can be used to generate ply failure envelopes of any ply orientation. When written in strain space, these ply failure envelopes can be superposed to generate laminate failure envelopes by selecting the controlling plies (inner envelopes). This is possible because, in strain space, failure envelopes are independent of the presence of other plies. Omni strain failure envelopes can be generated from superposing the ply failure envelopes for any ply orientation, hence defining an in-plane failure criterion that is independent of laminate stacking sequence or orientation.

If the ply failure envelopes are generated for last-ply failure (LPF), e.g., considering degraded transverse and longitudinal compressive properties (Tsai & Melo, 2014) or netting analysis, simplified last-ply-failure criteria can be generated for composite laminates. Generating these criteria in strain space, it is straightforward to obtain laminate strains-to-failure, which can then be used to determine the laminate strengths using the stress-strain relationships, the Tsai-normalized laminate factors, Tsai's modulus and the ply strengths.

In the case of LPF omni-strain envelopes (Figure 11.7), the strengths are invariably controlled by longitudinal failure of the 0° or 90° plies, which is accurate for *well-behaved* laminates, i.e., laminates that do not exhibit gauge delamination or massive matrix cracking (gauge longitudinal splitting or transverse cracking) before ultimate failure – e.g. *thin-ply laminates* (Arteiro, Catalanotti, Reinoso, Linde, & Camanho, 2019) (Arteiro, Furtado, Catalanotti, Linde, & Camanho, 2020). Therefore, in these cases, ply characterization can consist of just two tests of a 0° coupon to obtain the longitudinal tensile strength, X_T, and the longitudinal compressive strength, X_C.

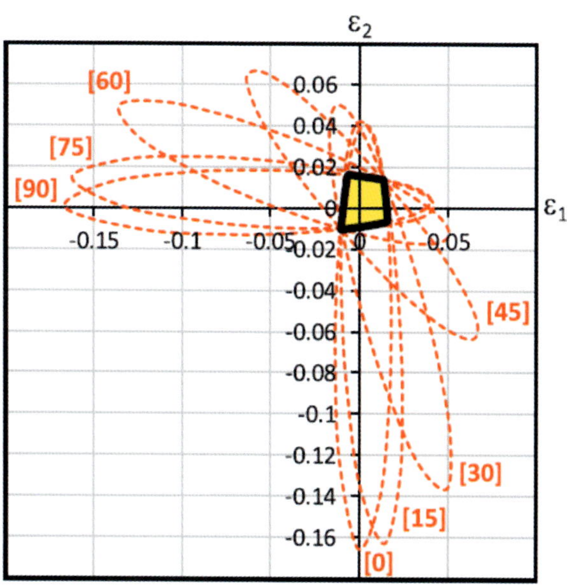

Figure 11.7 LPF omni-strain envelope (bold line) for IM7/8552 obtained from superposing ply failure envelopes for any ply orientation in strain space (examples with 15° steps in red dashed lines).

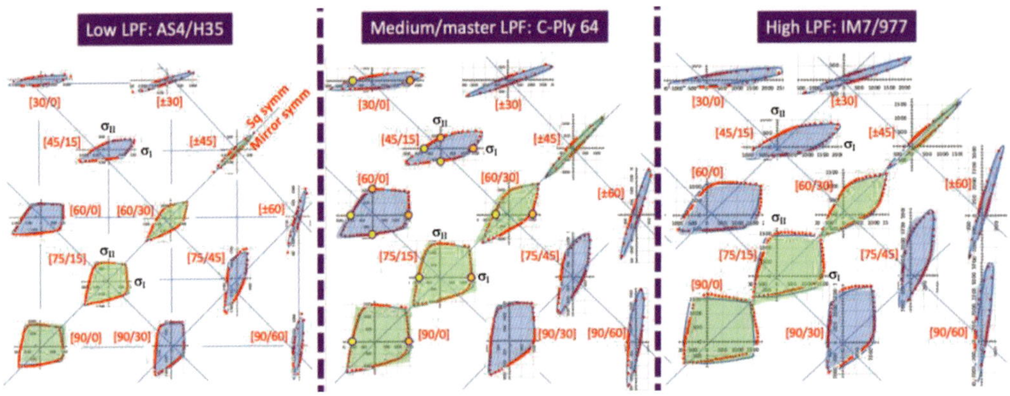

Figure 11.8 Master LPF envelopes and scaling for different materials and laminates – see **Chapter 2**.

Based on these observations, for conventional QUAD laminates and **longitudinal uniaxial loading**, a **maximum allowable strain failure criterion** is sufficient to determine the laminate tensile strength, X_T^L, and the laminate compressive strength, X_C^L, from the strain-to-failure of the 0° ply (Furtado, Arteiro, Bessa, Wardle, & Camanho, 2017). On the other hand, since DD laminates do not necessarily have 0° plies (unlike QUAD laminates with 10% rule), a maximum allowable strain failure criterion, for longitudinal uniaxial loading, is generally conservative. However, it still provides reasonable predictions without the need of recomputing each time the laminate failure envelope for the angle-plies.

Therefore, in this study, for the sake of simplicity, the maximum allowable strain failure criterion will be considered to determine the laminate strength required to solve the stress equilibrium of the FFMs coupled criterion (Figure 11.6). Alternatively, especially when considering multiaxial stress states, **Master Strength Envelopes** (Figure 11.8) can be used to determine X_T^L and X_C^L.

11.5 Laminate fracture toughness

Finally, the **laminate fracture toughness** is required to solve the energy equilibrium of the FFMs coupled criterion (Figure 11.6). Fracture mechanics solutions (Camanho & Catalanotti, 2011) can be used to establish the relationships between the mode-I ply fracture toughness and the mode-I fracture toughness of *well-behaved* balanced laminates. This is possible since the mode-I fracture toughness, $\mathcal{G}_{Ic}$, of a balanced laminate can be obtained from the weighted sum of the fracture toughness of its balanced sub-laminates (i):

$$\mathcal{G}_{Ic} = \frac{\sum_{(i)}^{N} \mathcal{G}_{Ic}^{(i)} t^{(i)}}{t}, \quad (i) \neq 90^\circ$$

The mode-I fracture toughness of each sub-laminate (i), $G_{Ic}^{(i)}$, can be related to its critical stress intensity factor, $K_{Ic}^{(i)}$, through:

$$\mathcal{G}_{Ic}^{(i)} = \frac{\left(\mathcal{K}_{Ic}^{(i)}\right)^2}{\acute{E}^{(i)}}$$

where $\acute{E}^{(i)}$ is an equivalent modulus obtained from the engineering constants of each sub-laminate (i), or, alternatively, directly from their Tsai-normalized laminate factors and the (sub-laminate independent) Tsai's modulus:

$$\acute{E}^{(i)} = \left(\frac{1+\rho^{(i)}}{2E_x^{(i)}E_y^{(i)}}\right)^{-1/2}\left(\frac{E_y^{(i)}}{E_x^{(i)}}\right)^{-1/4}$$

$$\rho^{(i)} = \frac{\left(E_x^{(i)}E_y^{(i)}\right)^{1/2}}{2G_{xy}^{(i)}} - \left(\nu_{xy}^{(i)}\nu_{yx}^{(i)}\right)^{1/2}$$

On the other hand, the remote failure stresses of sub-laminate (i) and of the 0° ply can be related, through LEFM, to the corresponding critical stress intensity factors, $K_{Ic}^{(i)}$ and K_{Ic}^{0}, respectively:

$$\bar{\sigma}^{(i)} = \frac{\mathcal{K}_{Ic}^{(i)}}{\chi^{(i)}Y\sqrt{\pi a}}$$

$$\bar{\sigma}^{0} = \frac{\mathcal{K}_{Ic}^{0}}{\chi^{0}Y\sqrt{\pi a}}$$

where $\chi^{(i)}$ and χ^{0} are the orthotropy correction factors of sub-laminate (i) and of the 0° ply respectively (Bao, Ho, Suo, & Fan, 1992):

$$\chi^{(i)} = \frac{1+0.1(\rho^{(i)}-1)-0.016(\rho^{(i)}-1)^2+0.002(\rho^{(i)}-1)^3}{[(1+\rho^{(i)})/2]^{1/4}}$$

and Y is a geometrical correction factor. Dividing the equations for the remote failure stresses of sub-laminate (i) and of the 0° ply leads to:

$$\mathcal{K}_{Ic}^{(i)} = \frac{\chi^{(i)}\Omega_0^{(i)}}{\chi^{0}}\mathcal{K}_{Ic}^{0}$$

where $\Omega_0^{(i)}$ is the stress ratio:

$$\Omega_0^{(i)} = \frac{\bar{\sigma}^{(i)}}{\bar{\sigma}^{0}}$$

which can be determined from CLPT assuming equal strain applied to sub-laminate (i) and to the 0° ply (Camanho & Catalanotti, 2011).

Using this fracture mechanics solution, it is possible to determine the mode-I fracture toughness, G_{Ic}, of any homogenized (or symmetric) balanced laminate using only **(a)** the Master Ply, to obtain the elastic relationships from the Tsai-normalized sub-laminate factors (*i*), **(b)** the Tsai's modulus to determine the equivalent modulus, and **(c)** the fracture toughness of the 0° ply, K_{Ic}^{0} (or G_{Ic}^{0}), to determine each sub-laminate fracture toughness, $K_{Ic}^{(i)}$ and $G_{Ic}^{(i)}$, whose weighted sum gives the mode-I fracture toughness, G_{Ic}.

To measure the fracture toughness of the 0° ply, robust fracture toughness characterization of the 0°plies (and of multidirectional laminates, in tension and in compression) are not standardized, and test methods derived from metals characterization, such as Compact Tension (CT), or its Over-height Compact Tension (OCT) variant (Figure 11.9), tend to be unsuitable. It is, therefore, proposed to determine the fracture toughness from the identification of the size effect laws of *geometrically similar specimens*[1] of different sizes and *positive geometry*[2] (Catalanotti, Xavier, & Camanho, 2014) (Catalanotti, Arteiro, Hayati, & Camanho, 2014) – Figure 11.10.

When scaled notched specimens are not available, the fracture toughness can also be back calculated using a single notched configuration (e.g., an open-hole coupon) and solving the system of equations in Figure 11.6 for G_{Ic} (Camanho, Erçin, Catalanotti, Mahdi, & Linde, 2012). Knowing the mode-I laminate fracture toughness, G_{Ic}, the model proposed by (Camanho & Catalanotti, 2011) can be used to back calculate the fracture toughness of the 0° ply, K_{Ic}^{0} (or G_{Ic}^{0}), required to scale the fracture toughness to other laminates.

Figure 11.9 Overall buckling of the unnotched end of an OCT specimen (despite the use of an anti-buckling support).

[1] *Geometrically similar specimens*: specimens scaled proportionally, i.e., with constant crack length-to-width and length ratios.

[2] *Positive geometry*: the geometry correction factor (Y) is an increasing function of the crack length, i.e., the ERR at constant load increases with crack extension.

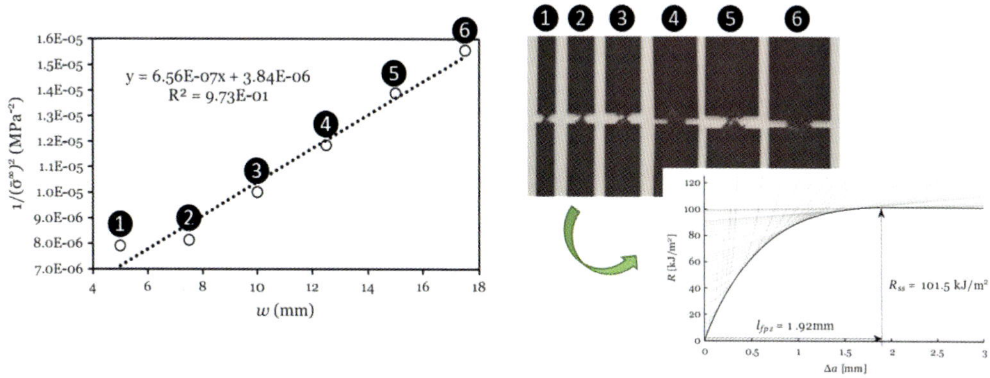

Figure 11.10 Measurement of intralaminar fracture resistance curves (*R*-curves) using the size effect method (Catalanotti, Arteiro, Hayati, & Camanho, 2014).

11.6 Material selection and characterization

Current practice includes obtaining notched strength allowables from **open-hole coupons** for several laminate configurations (*quasi-isotropic*, *hard* and *soft*). In a previous study (Tsai, Sharma, Arteiro, Roy, & Rainsberger, 2019), smooth and open-hole tension tests were conducted on DD [Φ/Ψ/-Ψ/-Φ] laminates, with Φ = 0°, Ψ = 50° and Φ = 15°, Ψ = 60°, i.e., [0/50/-50/0] and [15/60/-60/-15] laminates, respectively. All coupons were manufactured with NASHERO™ proprietary infusion process using Chomarat™ C-PLY™ [0/50] and [0/45] non-crimp fabrics (NCFs).

Ch.11

Plain tension tests were used to measure the unnotched strength of these laminates. In addition, digital image correlation was used to obtain full-field strain measurements from the 25-mm wide smooth specimens. Figure 11.11 shows representative stress-strain curves for the DD [0/50/-50/0] and [15/60/-60/-15] laminates. Averaging the stress-strain results from both DD laminates, an average Tsai's modulus of 120,006 MPa and a STDV of 3,019 MPa is obtained from the measured laminate Young's moduli, see Table 11.1.

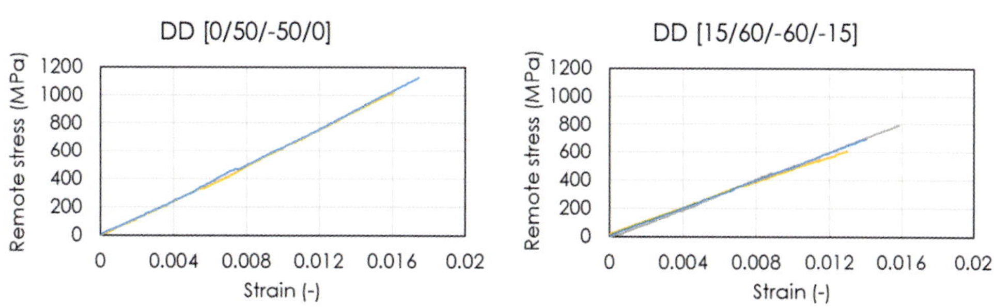

Figure 11.11 Stress-strain curves for smooth DD [0/50/-50/0] and [15/60/-60/-15] laminates (Tsai, Sharma, Arteiro, Roy, & Rainsberger, 2019).

The plain and open-hole tension data identified experimentally for the DD [0/50/-50/0] laminate (Table 11.2) is used to obtain the 0° ply strength and fracture toughness (Table 11.1). These are the only properties required as inputs to the omni failure criterion (Tsai & Melo, 2016), fracture mechanics model (Camanho & Catalanotti, 2011) and FFMs coupled criterion (Camanho, Erçin, Catalanotti, Mahdi, & Linde, 2012) to predict the open-hole tension (OHT) strength of multidirectional (incl. DD) laminates.

Tsai's modulus Tr (MPa)	**Ply strength** X (MPa)	**Ply fracture toughness** G_{Ic}^0 (N/mm)
120,006	1,963	177.7
STDV	**STDV**	**STDV**
3,019	148	8.9

Table 11.1 Tsai's modulus obtained from the DD [0/50/-50/0] and [15/60/-60/-15] laminates and ply strength and ply fracture toughness obtained from the DD [0/50/-50/0] laminate (Tsai, et al., 2022). STDV: standard deviation.

	Max. Stress (MPa)			
	DD [0/50/-50/0]		DD [15/60/-60/-15]	
$W = 25$ mm	**Average**	**STDV**	**Average**	**STDV**
Plain tension	1082	82	698	71
$W/D = 6$	**Average**	**STDV**	**Average**	**STDV**
OHT ($D = 6$ mm)	555	33	410	46
OHT ($D = 12$ mm)	469	23	342	30

Table 11.2 Unnotched and notched strength data of the DD [0/50/-50/0] and [15/60/-60/-15] laminates (Tsai, Sharma, Arteiro, Roy, & Rainsberger, 2019). STDV: standard deviation.

To validate the predictions of the FFMs coupled criterion (Camanho, Erçin, Catalanotti, Mahdi, & Linde, 2012), the analytical results are compared with the notched strength data obtained from the open-hole coupons with 6 mm and 12 mm of diameter ($D = 2R$) and a width-to-hole diameter ratio $W/D = 6$. The strength data is reported in Table 11.1 and Figure 11.12.

11.7 Notched strength predictions and analysis of DD laminates

Figure 11.13 shows the comparison with the experimental data, demonstrating the remarkable extrapolation capability of the FFMs coupled criterion (Tsai, Sharma, Arteiro, Roy, & Rainsberger, 2019). Benefiting from these known capabilities, the predictions are extended to other DD laminates, to understand the sensitivity of the notched strength of these laminates to changes in lay-up. Figure 11.14 shows the predicted OHT strength range when considering variations of ±3 STDVs around the average input values obtained from the DD [0/50/-50/0] laminate (Table 11.1). The experimental data of the two available DD laminates (Table 11.2) is also shown, including the error bar for 1 STDV on the measured OHT strengths.

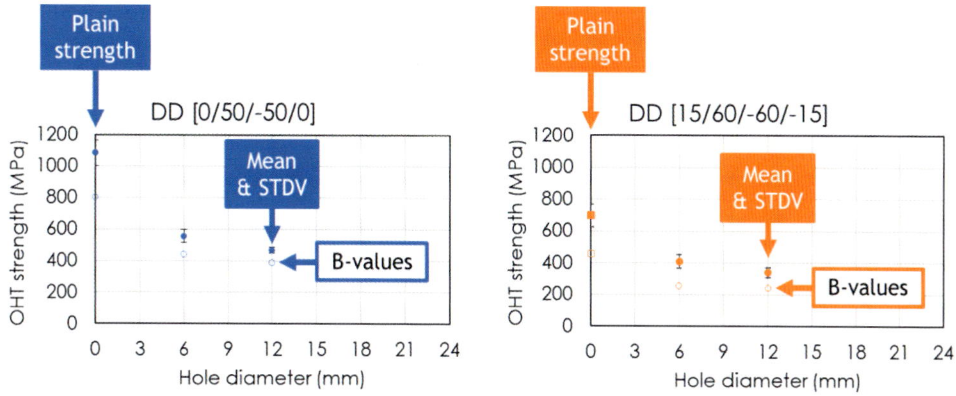

Figure 11.12 Plain and notched strength data for DD [0/50/-50/0] and [15/60/-60/-15] laminates (Tsai, Sharma, Arteiro, Roy, & Rainsberger, 2019).

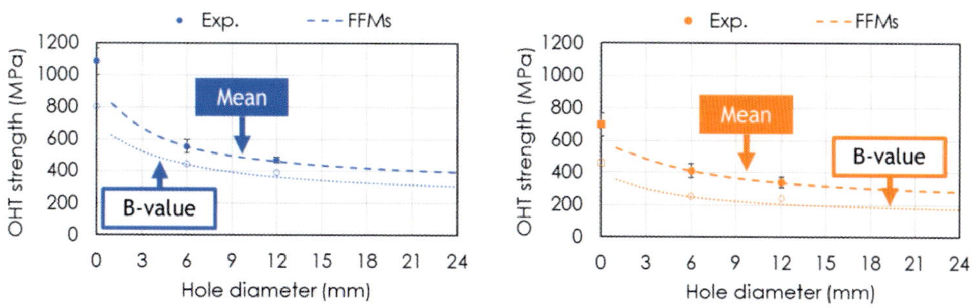

Figure 11.13 Comparison between test data and FFMs predictions for DD [0/50/-50/0] and [15/60/-60/-15] laminates (Tsai, Sharma, Arteiro, Roy, & Rainsberger, 2019).

As can be observed in Figure 11.14, softer laminates (with higher Φ and Ψ angles) show lower notched strength and, consequently, less variance (narrower bands) on the output notched strengths for the same variance (3 STDVs) on the material property (strengths and toughness) inputs. The same can be observed for conventional (QUAD) laminates (Figure 11.15), especially as the number of 90° plies increases (closer to the origin of the *x*-axis) and the number of 0° plies decreases. It is also interesting to note that the OHT strengths of DD laminates are not inferior to QUAD laminates. Moreover, unlike QUAD, whose OHT strengths lead to a discontinuous field, DD laminates show a continuous field from where the notched strengths can be evaluated and selected.

Figure 11.14 also shows that, within some ranges of Φ and Ψ, the notched strength remains insensitive to lay-up changes. It is the case of angle-ply (single-D) laminates with Φ = Ψ between 10° and 15°, or DD laminates with Φ between 15° and 25° and Ψ from 75° to 90°, see Figure 11.16.

In a previous publication (Tsai & Melo, 2015), it has already been shown that there is a minimum coefficient of variation in the elastic properties of composites, which could be used to define the preferred laminate for elastic properties identification, being the one less sensitive to intrinsic material variability and process variability. In addition, a "target laminate" has been identified – [23/75/-75/-23] (Φ = 23°, Ψ = 75°) – showing the lowest coefficient of variation (C.V.) for the Tsai-normalized E1* among all DD.

The results in Figure 11.16 confirm that this is a suitable laminate also for notched strength (design allowables) identification, as the notched strength for $\Phi = 23°$, $\Psi = 75°$ is almost insensitive to small lay-up changes, hence almost insensitive to process (lamination) variability, a type of variability that is extremely difficult to identify and control.

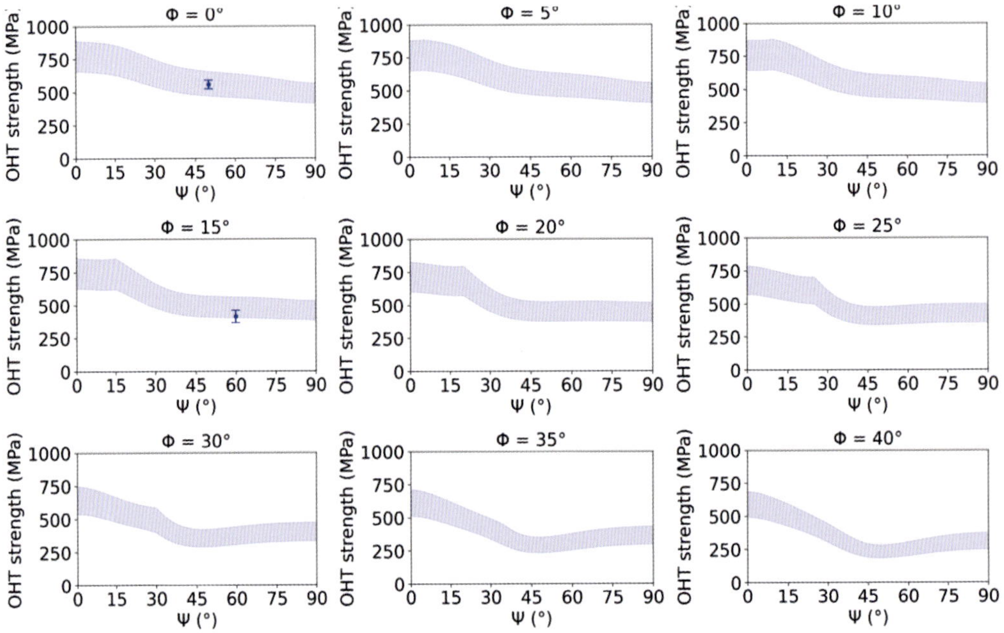

Figure 11.14 OHT strength predictions for a representative set of DD laminates.

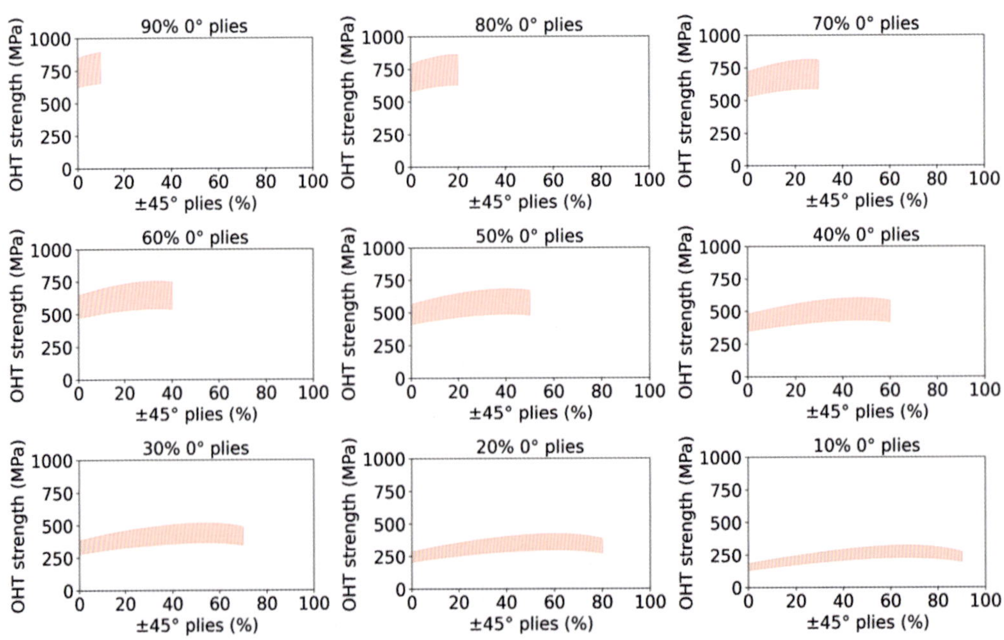

Figure 11.15 OHT strength predictions for a representative set of QUAD laminates.

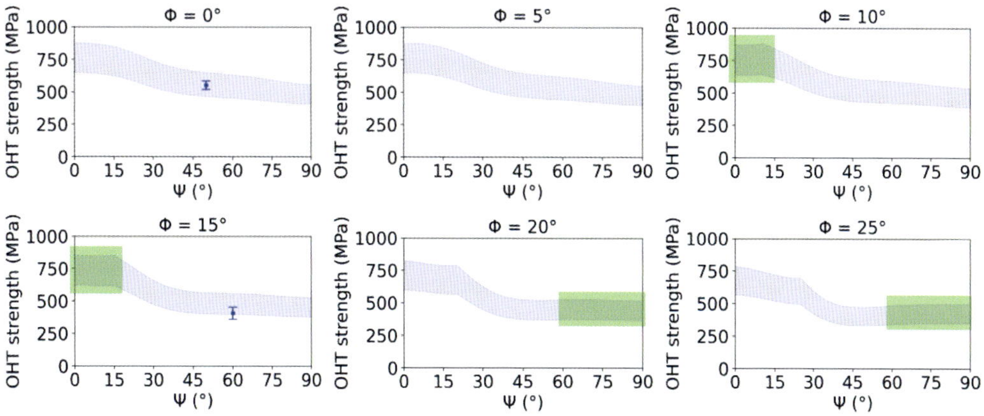

Figure 11.16 DD "*target laminates*".

11.8 Conclusions

An analytical FFMs coupled criterion has been proposed as a reliable tool to predict and assess the notched strengths of composite laminates. The extension of the predictions to a wider range of DD laminates shows that this new class of composites is not inferior, in any aspect, to conventional QUAD laminates. The range of notched strengths of DD, when moving from harder laminates to softer laminates, match the range of notched strengths of the QUAD laminates. This comes with the added flexibility of DD thanks to the continuous nature of the lay-up parameters Φ and Ψ, unlike the discrete nature of ply percentages in QUAD laminates.

In addition, it can be observed that the OHT strength of some DD configurations (within specific Φ and Ψ ranges) are practically insensitive to small changes in lay-up. This makes these configurations also less sensitive to process (lamination) variability. DD laminates with Φ between 15° and 25° and Ψ from 75° to 90° is one of such configurations. Considering that the [23/75/-75/-23] (Φ = 23°, Ψ = 75°) DD laminate shows the lowest C.V. for the Tsai-normalized E_1* among all DD, this is one possible "target laminate" for material and OHT characterization, due to lower sensitivity to both material and process variabilities.

Finally, it is important to stress that the combination of reliable analysis tools, incl. failure criteria and fracture mechanics, which make use of the invariant approaches to stiffness, through Tsai's modulus, and strength, through omni envelopes, allow robust identification of material properties even in the case of unconventional laminates. By understanding how these laminates respond to the load cases of interest (here, the notched response), it is possible to define a laminate configuration that is best suited for properties and allowables identification, which can then be reliably extended to other laminates (and even load cases). It is, therefore, possible to reduce the number of laminate configurations currently considered in common practice from at least four different lay-ups – unidirectional and quasi-isotropic, hard and soft QUAD – to one configuration only: a multidirectional laminate (Figure 11.17), which, while accounting for the specific lamination and manufacturing processes (something not possible with unidirectional laminates), also minimizes the sensitivity to material and process variabilities for robust material characterization.

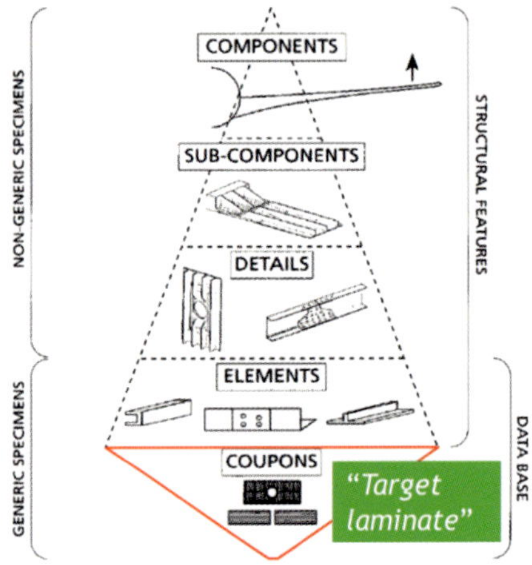

Figure 11.17 Approach to reduced characterization effort through Tsai's modulus and "*target laminates*" (DOD, 2002).

References

Arteiro, A., Catalanotti, G., Reinoso, J., Linde, P., & Camanho, P. P. (2019). Simulation of the Mechanical Response of Thin-Ply Composites: From Computational Micro-Mechanics to Structural Analysis. *Archives of Computational Methods in Engineering*, 1445-1487.

Arteiro, A., Catalanotti, G., Xavier, J., & Camanho, P. P. (2013). Notched response of non-crimp fabric thin-ply laminates: Analysis methods. *Composites Science and Technology*, 165-171.

Arteiro, A., Furtado, C., Catalanotti, G., Linde, P., & Camanho, P. P. (2020). Thin-ply polymer composite materials: A review. *Composites Part A*, 105777.

Arteiro, A., Pereira, L. F., Bessa, M. A., Furtado, C., & Camanho, P. P. (2019). A micro-mechanics perspective to the invariant-based approach to stiffness. *Composites Science and Technology*, 72-80.

Arteiro, A., Sharma, N., Melo, J. D., Ha, S. K., Miravete, A., Miyano, Y., . . . Hahn, H. T. (2020). A case for Tsai's Modulus, an invariant-based approach to stiffness. *Composite Structures*, 112683.

Bao, G., Ho, S., Suo, Z., & Fan, B. (1992). CORRIGENDA The role of material orthotropy in fracture specimens for composites. *Int. J. Solids Structures*, 2115.

Bao, G., Ho, S., Suo, Z., & Fan, B. (1992). The role of material orthotropy in fracture specimens for composites. *Int. J. Solids Structures*, 1105-1116.

Camanho, P. P., & Catalanotti, G. (2011). On the relation between the mode I fracture toughness of a composite laminate and that of a 0° ply: analytical model and experimental validation. *Eng Fract Mech*, 2535-2546.

Camanho, P. P., Erçin, G. H., Catalanotti, G., Mahdi, S., & Linde, P. (2012). A finite fracture mechanics model for the prediction of the open-hole strength of composite laminates. *Composites Part A*, 1219-1225.

Catalanotti, G., Arteiro, A., Hayati, M., & Camanho, P. (2014). Determination of the mode I crack resistance curve of polymer composites using the size-effect law. *Eng Fract Mech*, 49-65.

Catalanotti, G., Salgado, R. M., & Camanho, P. P. (2021). On the Stress Intensity Factor of cracks emanating from circular and elliptical holes in orthotropic plates. *Engineering Fracture Mechanics*, 107805.

Catalanotti, G., Xavier, J., & Camanho, P. (2014). Measurement of the compressive crack resistance curve of composites using the size effect law. *Composites Part A*, 300-307.

Chen, D. L., Weiss, B., Stickler, R., & Wang, Z. G. (1992). *Int J Fracture*, R3-R8.

Corrado, G., Arteiro, A., Marques, A., Reinoso, J., Daoud, F., & Glock, F. (2022). An extended invariant approach to laminate failure of fibre-reinforced polymer structures. *The Aeronautical Journal*, 1045-1068.

DOD. (2002). *COMPOSITE MATERIALS HANDBOOK, VOLUME 3. POLYMER MATRIX COMPOSITES* (Vols. MIL-HDBK-17-3F). DEPARTMENT OF DEFENSE.

Furtado, C., Arteiro, A., Bessa, M. A., Wardle, B. L., & Camanho, P. P. (2017). Prediction of size effects in open-hole laminates using only the Young's modulus, the strength, and the R-curve of the 0° ply. *Composites Part A*, 306-317.

Suo, Z., Bao, G., Fan, B., & Wang, T. C. (1991). Orthotropy rescaling and implications for fracture in composites. *Int. J. Solids Structures*, 235-248.

Tan, S. C. (1994). *Stress concentrations in laminated composites.* Lancaster: Technomic.

Tsai, S. W., & Melo, J. D. (2014). An invariant-based theory of composites. *Composites Science and Technology*, 237-243.

Tsai, S. W., & Melo, J. D. (2015). *Composite Materials Design and Testing, unlocking the mystery with invarients.* Stanford, CA, USA: Composites Design Group, Department of Aeronautics and Astronautics, Stanford University.

Tsai, S. W., & Melo, J. D. (2016). A unit circle failure criterion for carbon fiber reinforced polymer composites. *Composites Science and Technology*, 71-78.

Tsai, S. W., Kappel, E., Miravete, A., Riccio, A., Di Caprio, F., Roy, S., . . . Massard, T. (2022). *DOUBLE–DOUBLE: A New Perspective in The Manufacture and Design of Composites.* (S. W. Tsai, Ed.) Stanford: Department of Aeronautics and Astronautics, Stanford University.

Tsai, S., Sharma, N., Arteiro, A., Roy, S., & Rainsberger, B. (2019). *Composite Double-Double and Grid/Skin Structures – Low Weight/Low Cost Design and Manufacturing.* Stanford: Stanford University.

Waddoups, M. E., Eisenmann, J. R., & Kaminski, B. E. (1971). *J Compos Mater*, 446-454.

Whitney, J. M., & Nuismer, R. J. (1974). Stress Fracture Criteria for Laminated Composites Containing Stress Concentrations. *J Compos Mater*, 253-265.

Younger son, Ming, who cooked for the vice presidents
who became presidents soon after

Granddaughter Lauren Tsai, an artist, and TV and movie actress now
working in Tokyo

CHAPTER 12

Uniaxial Tensile Strength of Open-Hole Double-Double Laminates

J.L. Liu*[a], Gang Yang[a], T.E. Tay[b], V.B.C. Tan[b], D.Z. Jiang*[a], Jie Zhi[c]

a, School of Materials, Sun Yat-Sen University, China

b, Department of Mechanical Engineering, National University, Singapore

c, School of Aerospace Engineering and Applied Mechanics, Tongji University, China

*Corresponding authors, email addresses: liujlong9@mail.sysu.edu.cn (J.L.Liu), jiangdzhi@mail.sysu.edu.cn (D.Z. Jiang)

12.1 Introduction

Double-Double (DD) laminates have many advantages over Quad laminates because they are homogeneous, simple to design and can potentially improve manufacturing efficiency and reduce cost [1, 2]. Recently, the automotive industry has been deploying more composite components to achieve weight savings. However, current woven and Quad laminates struggle to meet some of the key requirements of automotive manufacturers. As shown in *Figure 1*: 1, fast and low-cost composite chassis designs are critical as new car models are introduced to the market regularly; 2, with increasing production volumes, composite components have to be produced as quickly as possible with simpler ply placement and improved manufacturing techniques; 3, lower direct material cost and lightweight are important. A promising approach is to improve the laminate through layup design [2]. As shown in *Figure 1*, woven CFRP laminates have little room for layup design, whereas the design and manufacturing cost of Quad laminate components are too high for the automotive industry. DD laminates, however, can mitigate these disadvantages.

This chapter investigates whether it is feasible to rate or scale the behaviour of Double-Double (DD) laminates under open-hole tension (OHT). A notch or open hole would invariably exist in actual parts, either as necessary notches such as design features or holes for joining parts, or unwanted notches such as defects or damage. Predicting the strength of laminates with open holes is critical as they weaken the structural strength of the component. Due to the stress concentration near the open hole, the rating of the laminates may be affected. Therefore, it is worth investigating if laminate scaling can also be applied to laminates with open hole.

The most reliable approach to investigate if scaling can be applied to laminates with open hole is through a systematic experimental study involving several materials and laminate layups. However, due to the high initial cost and difficulty to acquire several different materials, conducting a full-scale experimental study is challenging. On the other hand, conducting FEA simulation may be a good initial study.

This study conducts FEA simulation to investigate the tensile strength of open-hole Double-Double laminates. Finite element modelling techniques based on the methods that have shown reliable predictions in several of our previous studies [3-6] were adopted. Four different materials with 12 different laminate layups were investigated to compare their rating. The tensile strengths of the open-hole DD laminates were then compared.

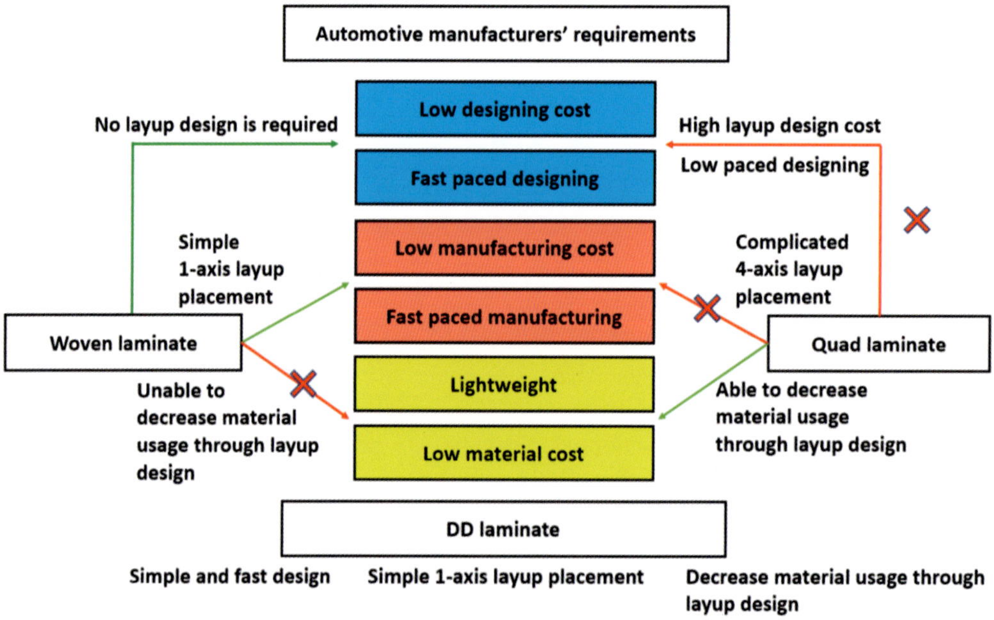

Figure 1, Automotive manufacturer's requirements, and the advantages/disadvantages of woven, Quad, and Double-Double laminates.

12.2 Finite element modelling for open-hole tension

The open-hole laminate as shown in *Figure 2* was discretized with full-integration solid elements (C3D8) for each ply and cohesive elements (COH3D8) for each interface between plies of different angles. With the left end fully clamped, the specimen was loaded by a uniform tensile displacement in X direction imposed at the right end. To model the progressive damage, two material subroutines (UMATs) were employed for the ply and the interface, respectively. The damage initiation criteria are given as follows:

Fiber compression $$\left(\frac{\sigma_{11}}{X_C}\right)^2 \geq 1 \quad (2)$$

Matrix tension $$\frac{1}{Y_t^2}(\sigma_{22}+\sigma_{33})^2 + \frac{1}{S_{23}^2}(\sigma_{23}^2 - \sigma_{22}\sigma_{33}) + \frac{1}{S_{12}^2}(\sigma_{12}^2+\sigma_{31}^2) - 1 \geq 0 \quad (3)$$

Matrix compression $$\frac{1}{Y_c}\left[\left(\frac{Y_c}{2S_{23}}\right)^2 - 1\right](\sigma_{22}+\sigma_{33}) + \frac{1}{4S_{23}^2}(\sigma_{22}+\sigma_{33})^2 + \frac{1}{S_{23}^2}(\sigma_{23}^2 - \sigma_{22}\sigma_{33}) + \frac{1}{S_{12}^2}(\sigma_{12}^2+\sigma_{31}^2) - 1 \geq 0 \quad (4)$$

Interface $$\left(\frac{\langle t_n\rangle}{t_n^0}\right)^2 + \left(\frac{t_s}{t_s^0}\right)^2 + \left(\frac{t_t}{t_t^0}\right)^2 \geq 1, \quad \langle * \rangle = \begin{cases} 0 & * < 0 \\ * & * \geq 0 \end{cases} \quad (5)$$

Ch.12

where X_T, X_C are longitudinal tensile and compressive strength; Y_T, Y_C are transverse tensile and compressive strength; S_{12} is in-plane shear strength and S_{23} is out-of-plane shear strength; t_n^0 is interface normal strength and t_s^0, t_t^0 are interface shear strength. Once the damage is initiated, a linear softening law is adopted to model the damage evolution [3-6].

The material properties of T3/N52, T4708/MR, T800/Cyt, and T700/2510 are shown in Table 1. The laminate layups are shown in Table 2. The tensile strength of the open hole laminates was then obtained from the FEA simulation result.

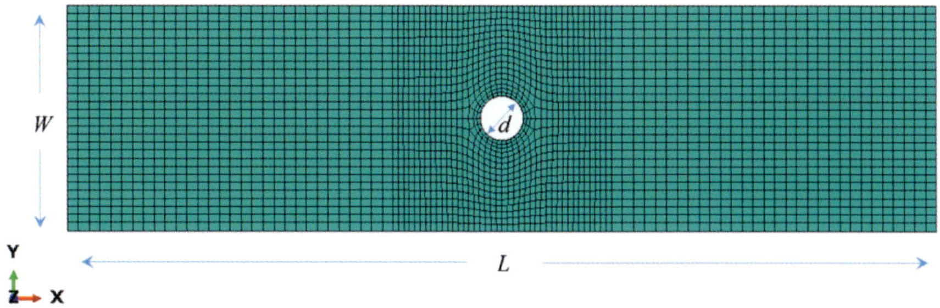

Figure 2, Finite element model of open-hole laminate, where L=80mm, W=20mm, d=4mm, and the thickness of the laminate is 1.6mm.

Table 1 material properties of T3/N52, T4708/MR, T800/Cyt (provided by Prof. S.W. Tsai) and T700/2510 (from [7]).

	T3/N52	T4708/MR	T700/2510	T800/Cyt
Modulus (fiber direction), GPa	181	142	107	162
Modulus (Transverse direction), GPa	10.3	7.72	8.0	9.0
In-plane Poisson's ratio	0.3	0.34	0.32	0.4
In-plane shear modulus, GPa	7.2	3.8	4.0	5.0
Out-of-plane shear modulus, GPa	3.4	3.4	2.7	3.4
Tensile strength (fiber direction), MPa	1500	2524	1650	3768
Compressive strength (fiber direction), MPa	1500	1700	1250	1656
Tensile strength (transverse direction), MPa	40	39.6	45.1	56
Compressive strength (transverse direction), MPa	246	165	165	150
In-plane shear strength, MPa	68	48	100	98

Table 2, laminate layups

Material	Designation	Laminate layup	Material	Designation	Laminate layup
T3/N52	T3_0/90	$[0/90]_8$	T700/2510	T7_0/90	$[0/90]_8$
	T3_0/75	$[75/0/-75/0]_4$		T7_0/75	$[75/0/-75/0]_4$
	T3_75/15	$[75/-15/15/-75]_4$		T7_75/15	$[75/-15/15/-75]_4$
	T3_0/60	$[60/0/-60/0]_4$		T7_0/60	$[60/0/-60/0]_4$
	T3_60/15	$[60/-15/15/-60]_4$		T7_60/15	$[60/-15/15/-60]_4$
	T3_0/48	$[48/0/-48/0]_4$		T7_0/48	$[48/0/-48/0]_4$
	T3_67.5/22.5	$[67.5/-22.5/22.5/-67.5]_4$		T7_67.5/22.5	$[67.5/-22.5/22.5/-67.5]_4$
	T3_48/38	$[48/-38/38/-48]_4$		T7_48/38	$[48/-38/38/-48]_4$
	T3_45/-45	$[45/-45]_8$		T7_45/-45	$[45/-45]_8$
	T3_60/30	$[60/-30/30/-60]_4$		T7_60/30	$[60/-30/30/-60]_4$
	T3_45/30	$[45/-30/30/-45]_4$		T7_45/30	$[45/-30/30/-45]_4$
	T3_30/15	$[30/-15/15/-30]_4$		T7_30/15	$[30/-15/15/-30]_4$
T4708/MR	T4_0/90	$[0/90]_8$	T800/Cyt	T8_0/90	$[0/90]_8$
	T4_0/75	$[75/0/-75/0]_4$		T8_0/75	$[75/0/-75/0]_4$
	T4_75/15	$[75/-15/15/-75]_4$		T8_75/15	$[75/-15/15/-75]_4$
	T4_0/60	$[60/0/-60/0]_4$		T8_0/60	$[60/0/-60/0]_4$
	T4_60/15	$[60/-15/15/-60]_4$		T8_60/15	$[60/-15/15/-60]_4$
	T4_0/48	$[48/0/-48/0]_4$		T8_0/48	$[48/0/-48/0]_4$
	T4_67.5/22.5	$[67.5/-22.5/22.5/-67.5]_4$		T8_67.5/22.5	$[67.5/-22.5/22.5/-67.5]_4$
	T4_48/38	$[48/-38/38/-48]_4$		T8_48/38	$[48/-38/38/-48]_4$
	T4_45/-45	$[45/-45]_8$		T8_45/-45	$[45/-45]_8$
	T4_60/30	$[60/-30/30/-60]_4$		T8_60/30	$[60/-30/30/-60]_4$
	T4_45/30	$[45/-30/30/-45]_4$		T8_45/30	$[45/-30/30/-45]_4$
	T4_30/15	$[30/-15/15/-30]_4$		T8_30/15	$[30/-15/15/-30]_4$

12.3 Tensile strength rating of open-hole laminates

Figure 3 and Table 3 show the open-hole tensile (OHT) strength of the laminates. The corresponding load vs displacement curves are given in Appendix A. The ratings of the laminates in Table 3 and *Figure 4* are obtained by normalizing the OHT strength of a laminate with the OHT strength of the corresponding T4708/MR laminate with the same layup. For example, the rating for T3_0/90 is its strength normalized with the strength of T4_0/90.

As shown in *Figure 4*, it appears that the laminates with their smaller mismatch angle (i.e. angle between fiber direction and the loading direction) equal to or below 22.5° have similar (almost constant) ratings. However, the ratings of the laminates with their smaller mismatch angles (i.e. greater than 22.5°) is not similar but show an upward trend from [45/**30**] and [60/**30**], to [48/**38**] and peaks at [**45**/-**45**].

This suggests that the open hole tensile strength of the laminates in other similar material systems can be predicted by just testing one reference material system with the layups [0/90] and [45/-45] laminates.

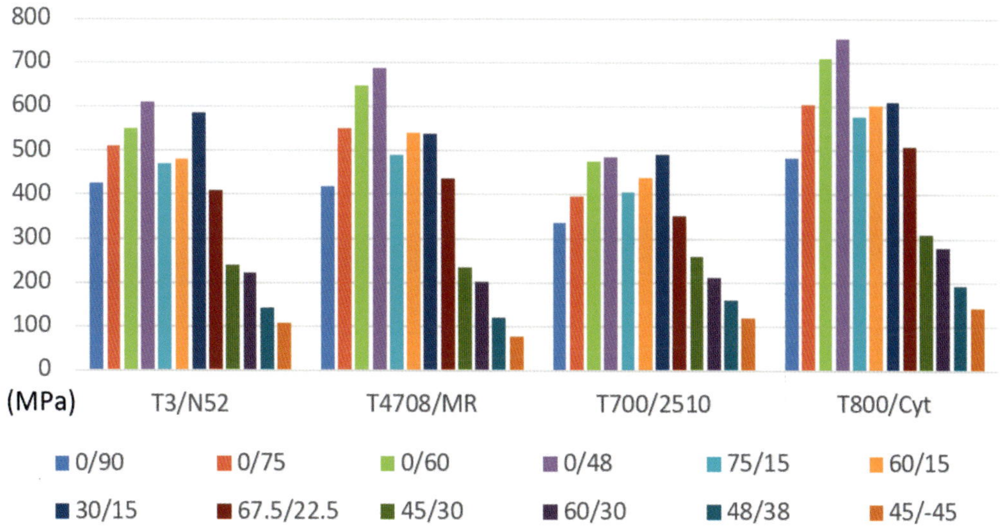

Figure 3, tensile strength of the open hole laminates

Table 3, tensile strength of the open hole laminates and their rating (ratio over T4708/MR) – the cases in blue colored boxes below are fiber-dominated, while the cases in yellow boxes are matrix-dominated.

Designation	Strength (MPa)	Rating	Designation	Strength (MPa)	Rating
T3_0/90	426.2	1.02	T7_0/90	335.2	0.8
T3_0/75	510.1	0.92	T7_0/75	395.4	0.72
T3_75/15	470.3	0.96	T7_75/15	406.6	0.83
T3_0/60	550.4	0.85	T7_0/60	476.3	0.74
T3_60/15	481.0	0.89	T7_60/15	439.5	0.81
T3_0/48	609.7	0.89	T7_0/48	484.8	0.71
T3_30/15	585.0	1.09	T7_30/15	491.0	0.91
T3_67.5/22.5	409.6	0.94	T7_67.5/22.5	352.2	0.81
T3_45/30	239.2	1.02	T7_45/30	258.2	1.10
T3_60/30	221.4	1.10	T7_60/30	210.6	1.05
T3_48/38	140.7	1.17	T7_48/38	159.0	1.32
T3_45/-45	107.8	1.38	T7_45/-45	118.6	1.52
T4_0/90	418.2	1	T8_0/90	483.7	1.16
T4_0/75	551.5	1	T8_0/75	605.1	1.10
T4_75/15	490.0	1	T8_75/15	578.9	1.18
T4_0/60	647.1	1	T8_0/60	710.5	1.10
T4_60/15	539.3	1	T8_60/15	603.1	1.12
T4_0/48	687.8	1	T8_0/48	754.0	1.10
T4_30/15	538.3	1	T8_30/15	609.3	1.13
T4_67.5/22.5	436.9	1	T8_67.5/22.5	508.2	1.16
T4_45/30	235.2	1	T8_45/30	308.0	1.31
T4_60/30	200.6	1	T8_60/30	278.0	1.39
T4_48/38	120.6	1	T8_48/38	191.0	1.58
T4_45/-45	78.1	1	T8_45/-45	141.1	1.81

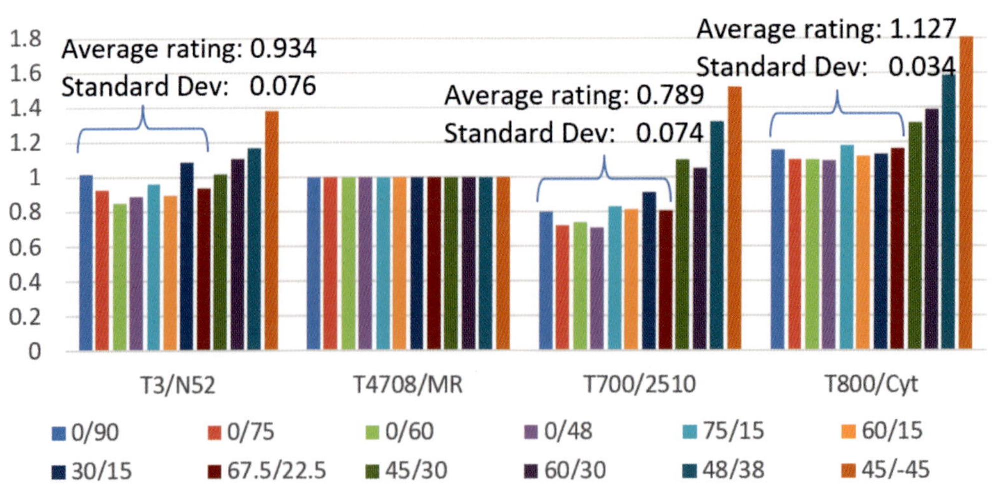

Figure 4, rating of the open hole laminates

12.4 Damage pattern of the open-hole laminates

The simulation result shows that failure of the laminates is dominated by either fiber damage or matrix damage. As shown in *Figure 5*, as T4_0/60 reaches its peak load, fiber damage occurs near the open hole and causes load drop, whereas its matrix damage remains local (*Figure 6*). It is observed that fiber damage dominated failure mostly occurs in DD laminates with their smaller mismatch angle below 22.5°. On the other hand, when T4_60/30 reaches its peak load, extensive matrix damage developed in the laminate (*Figure 7*). It is observed that matrix damage dominated failure mostly occurs in DD laminates with both mismatch angles greater than 22.5°.

For laminates with fiber damage dominated failure, the material's tensile strength along fiber direction might be more critical. The tensile strength along fiber direction for T700/2510, T3/N52, T800/Cyt, and T4708/MR are 1500 MPa, 1650 MPa, 3768 MPa, and 2524 MPa, respectively. The simulation results also showed that the rating is lowest for T700/2510, T3/N52, T800/Cyt, and T4708/MR, in the same order.

For laminates with matrix-dominated damage, the materials' in-plane shear strength might be more critical, as shown in Table 1, the in-plane shear strength of T700/2510, T3/N52, and T800/Cyt is significantly higher than T4708/MR, and simulation results showed that the ratings of the laminates with matrix-dominated damage are also higher than T4708/MR.

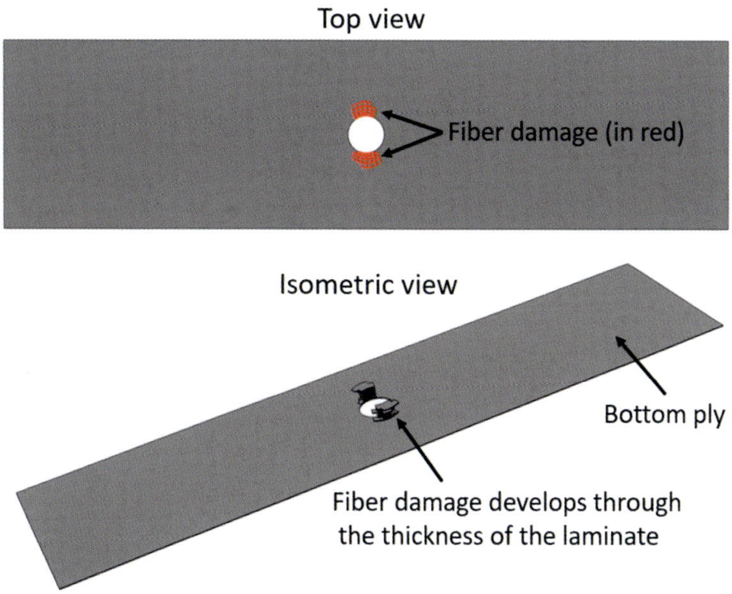

Figure 5, location of the fiber damage of T4_0/60 right after the laminate reaches its peak load. Bottom ply is also displayed to indicate the size of the damage.

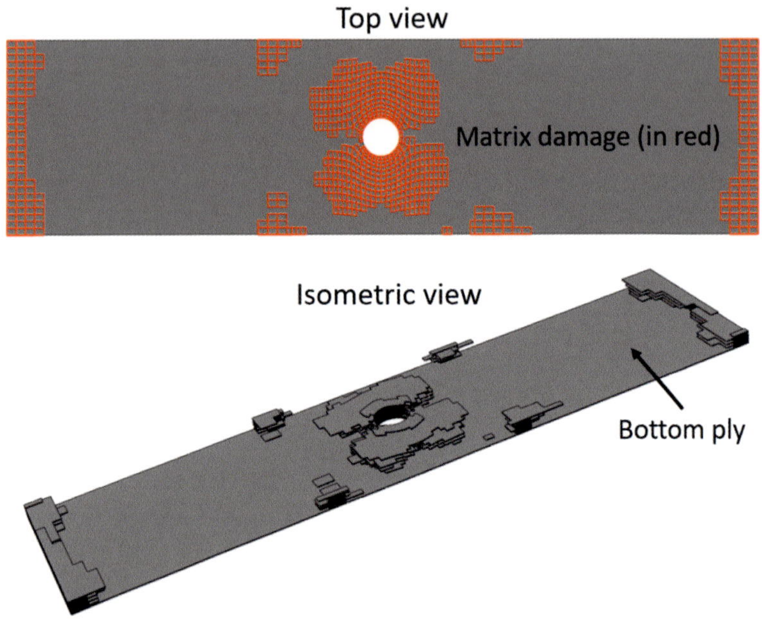

Figure 6, location of the matrix damage of T4_0/60 right after the laminate reaches its peak load. Bottom ply is also displayed to indicate the size of the damage.

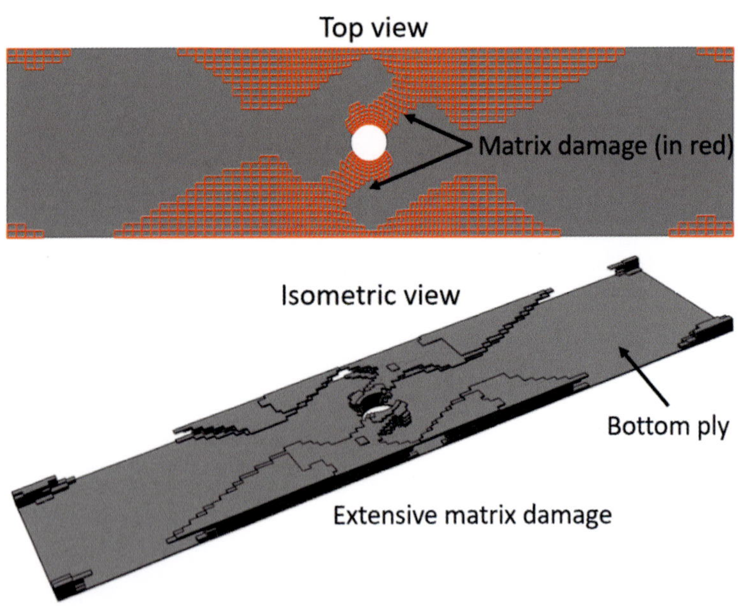

Figure 7, location of the matrix damage of T4_60/30 right after the laminate reaches its peak load. Bottom ply is also displayed to indicate the size of the damage.

12.5 Conclusion

Finite element analysis (FEA) was adopted to investigate the tensile strength of open hole DD laminates with 4 different material systems and 12 laminate layups. The simulation results showed that laminate scaling may also apply to open hole DD laminates but there might be 2 different scaling values depending on whether the failure is fiber damage dominated or matrix damage dominated. From the FEA simulation results, it appears that the strength of DD laminates could be predicted through laminate rating with marginal error. Strength found from a [0/90] and [45/-45] laminate may define comparable properties for all DD laminates in a material system. Although similar FEA models have been adopted by several of our previous publications [3-6]. Further investigations including experimental studies are needed to verify the claims made in this chapter.

References

[1] Tsai SW, Sharma N, Arteiro A, Roy S, Rainsberger B. Composite Double-Double and Grid/Skin structures, low weight/low cost design and manufacturing. International Paris Air Show 2019. Chapter 1, Page 18.

[2] Vermes B, Tsai SW, Riccio A, Di Caprio F, Roy S. Application of the Tsai's modulus and double-double concepts to the definition of a new affordable design approach for composite laminates. Composite Structures.2021;259:113246.

[3] Chen BY, Tay TE, Baiz PM, Pinho ST. Numerical analysis of size effects on open-hole tensile composite laminates.Composites Part A: Applied Science and Manufacturing.2013;47:52-62.

[4] Ridha M, Wang CH, Chen BY, Tay TE. Modelling complex progressive failure in notched composite laminates with varying sizes and stacking sequences.Composites Part A: Applied Science and Manufacturing.2014;58:16-23.

[5] Su ZC, Tay TE, Ridha M, Chen BY. Progressive damage modeling of open-hole composite laminates under compression.COMPOSITE STRUCTURES.2015;122:507-17.

[6] Zhi J, Tay TE. Interrogating failure mechanisms of notched composites through a discrete crack modeling approach.COMPOSITES SCIENCE AND TECHNOLOGY.2020;196:108203.

[7] Liu JL, Tay TE, Ng SW, Tan VBC. Effects of inter-ply mismatch angle on interlaminar properties and their influence in numerical simulations. COMPOSITES PART A. 2022;154:106795.

Appendix A. FEA Simulation result: Load vs displacement curve and tensile strength of the laminates.

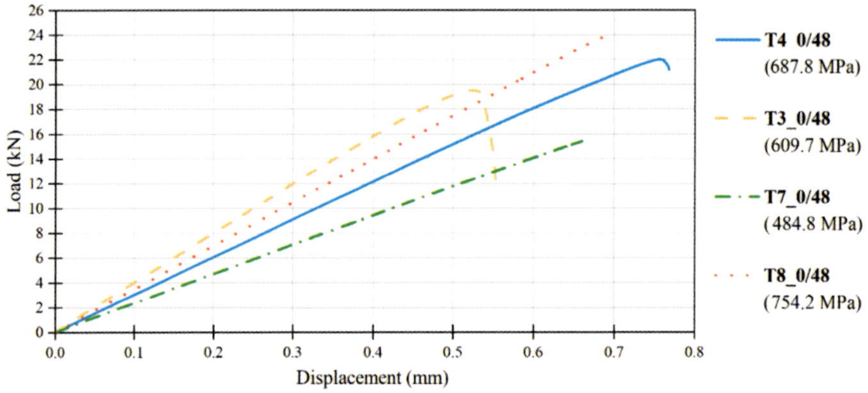

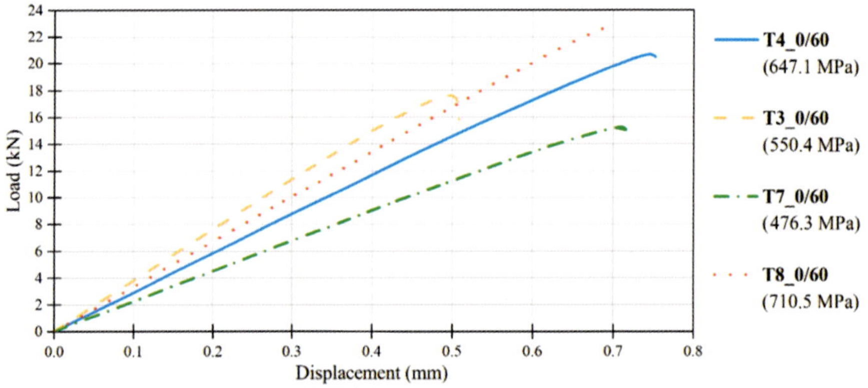

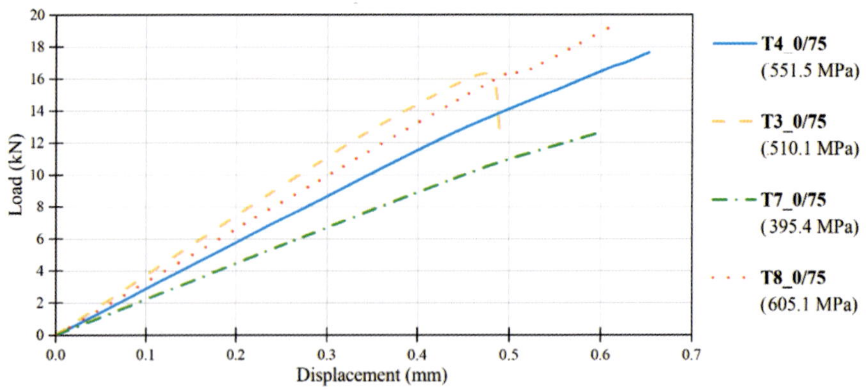

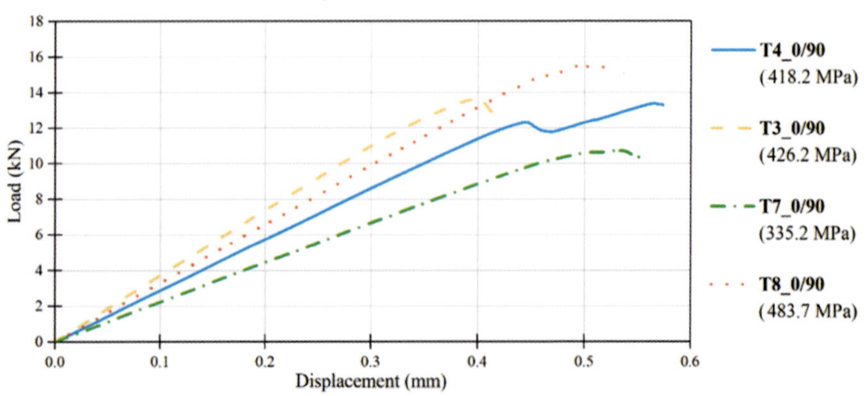

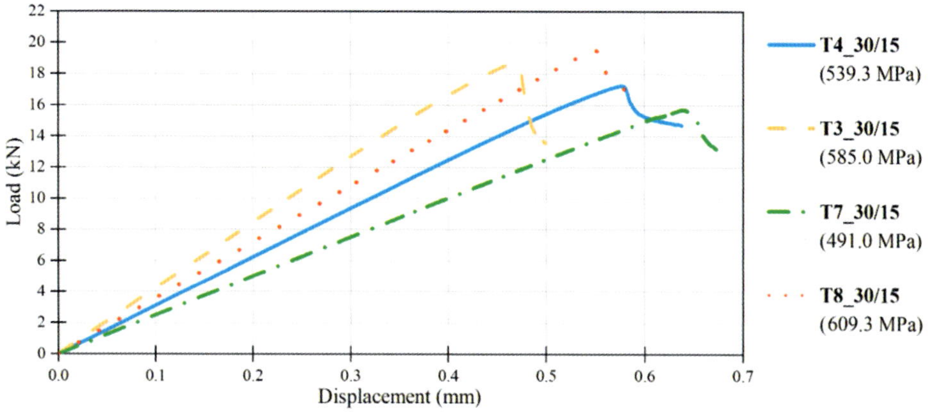
Load (kN)
Displacement (mm)
T4_30/15
(539.3 MPa)
T3_30/15
(585.0 MPa)
T7_30/15
(491.0 MPa)
T8_30/15
(609.3 MPa)

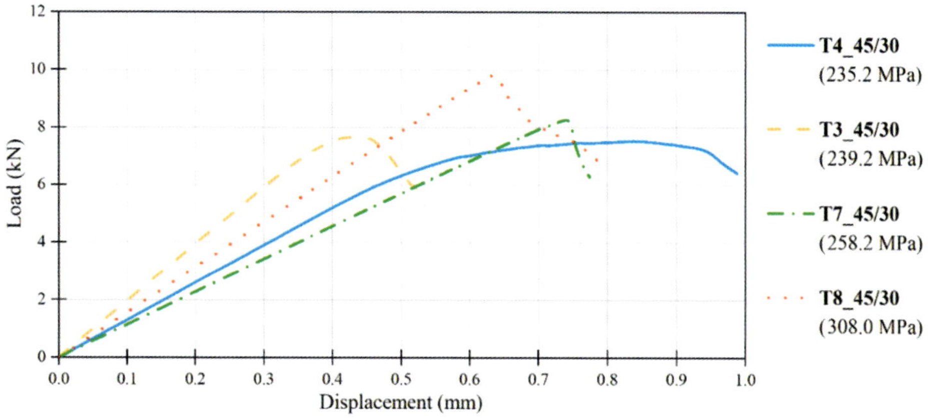
Load (kN)
Displacement (mm)
T4_45/30
(235.2 MPa)
T3_45/30
(239.2 MPa)
T7_45/30
(258.2 MPa)
T8_45/30
(308.0 MPa)

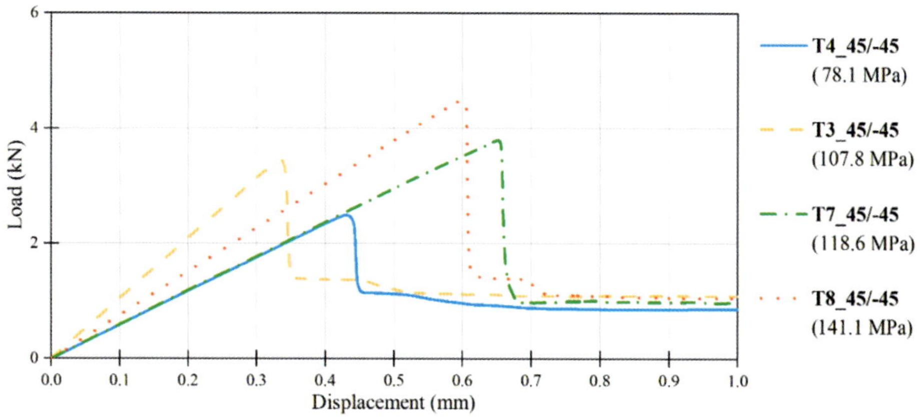
Load (kN)
Displacement (mm)
T4_45/-45
(78.1 MPa)
T3_45/-45
(107.8 MPa)
T7_45/-45
(118.6 MPa)
T8_45/-45
(141.1 MPa)

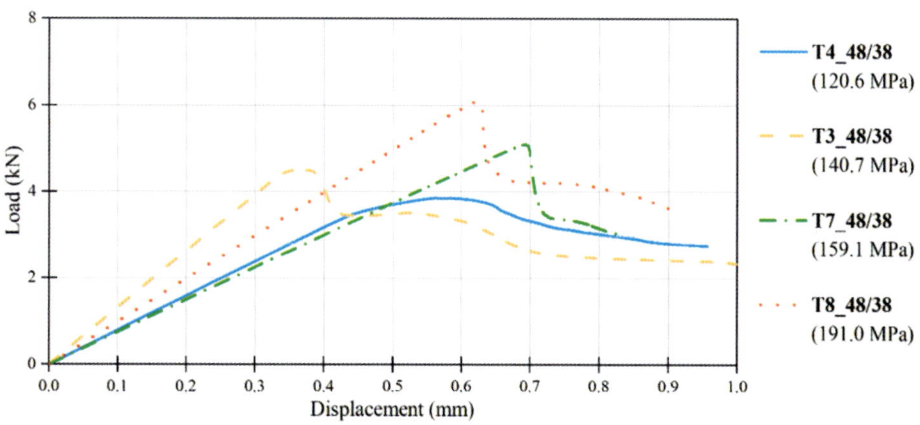
Load (kN)
Displacement (mm)
T4_48/38
(120.6 MPa)
T3_48/38
(140.7 MPa)
T7_48/38
(159.1 MPa)
T8_48/38
(191.0 MPa)

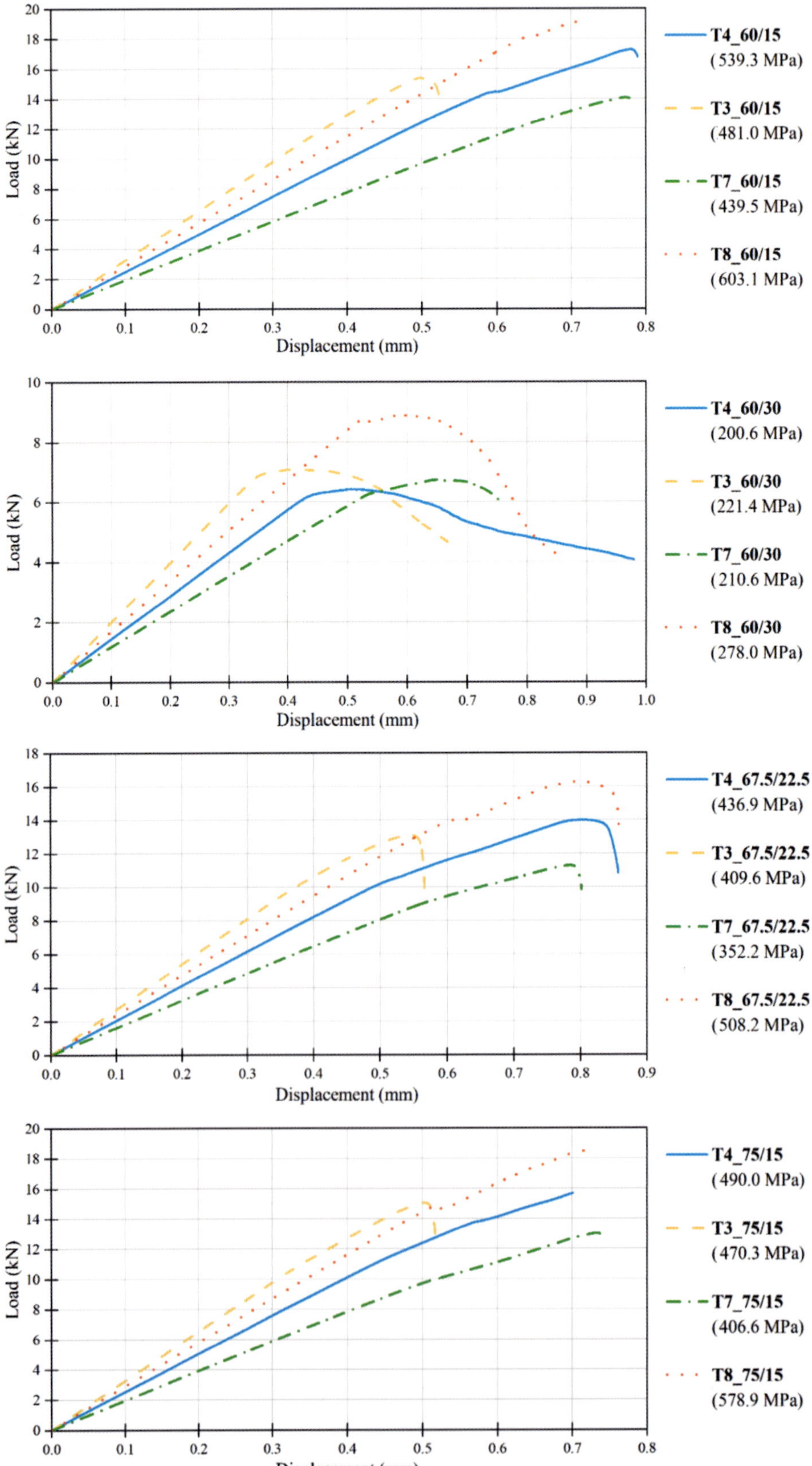
Load (kN)
Displacement (mm)
T4_60/15
(539.3 MPa)
T3_60/15
(481.0 MPa)
T7_60/15
(439.5 MPa)
T8_60/15
(603.1 MPa)
Load (kN)
Displacement (mm)
T4_60/30
(200.6 MPa)
T3_60/30
(221.4 MPa)
T7_60/30
(210.6 MPa)
T8_60/30
(278.0 MPa)
Load (kN)
Displacement (mm)
T4_67.5/22.5
(436.9 MPa)
T3_67.5/22.5
(409.6 MPa)
T7_67.5/22.5
(352.2 MPa)
T8_67.5/22.5
(508.2 MPa)
Load (kN)
Displacement (mm)
T4_75/15
(490.0 MPa)
T3_75/15
(470.3 MPa)
T7_75/15
(406.6 MPa)
T8_75/15
(578.9 MPa)

CHAPTER 13

A Double-Double 'Flex Panel' – How Tapering Can Be Utilized to Tailor Deflection Shapes

Authors:

Dr.-Ing. Erik Kappel*	https://orcid.org/0000-0002-8760-8451	Erik.kappel@dlr.de
Yannick Boose	https://orcid.org/0000-0002-7119-3335	Yannick.boose@dlr.de
Mirko Mißbach		Mirko.missbach@dlr.de

German Aerospace Center
Institute of Lightweight Systems (DLR-SY)
Composite Design Department
Lilienthalplatz 7
38108 Braunschweig, Germany

13.1 Motivation

This article reports on the designing, making and testing of prepreg-made CFRP panels, which aim to provide a specified deflection when subjected by a defined tip displacement. The task can be solved for isotropic materials by locally adapting the panel's local material thickness. However, it is challenging for CFRP aerospace laminates, which are typically composed of 0°, 45°, -45°, 90° plies. Established laminate guidelines for conventional laminates, as for example the symmetry requirement or the 10 % rule, constrict the composite laminate's design space drastically, which makes it difficult to optimize local stackings, zone dimensions and zone-to-zone transitions.

The family of Double-Double (DD) laminates is applied hereafter [1]. DD laminates offer unique opportunities for laminate-thickness tapering [2], which cannot be realized with conventional laminates. DD simplifies design with remarkable parallels to design processes for isotropic materials, as finally the DD laminate thickness is locally adapted in discrete steps.

The particular DD laminate architecture is characterized by a sequence of a four-ply building-block (BB), which is repeated r times to create a full $[\pm\Psi, \pm\Phi]_{rT}$ laminate. The DD-laminates' bending stiffness can be controlled by adjusting the number of building-block repeats r. The symmetry requirement as well as the limitation to pre-set ply orientations 0°, 45°, -45°, 90° cease.

13.2 Challenge 'aero surface'

Aerodynamic surfaces have to fulfill tight dimensional tolerances. The present study examines whether DD laminates allow for precisely adjusting a panel's shape for a specified loading scenario by locally adapting the number of BB repeats. Figure 13.1 shows the simplified object of investigation, which shall represent a section of a large-scale CFRP panel structure. It is the ambition to achieve a defined z-deflection curve w(x), which can for example be defined by aerodynamic requirements.

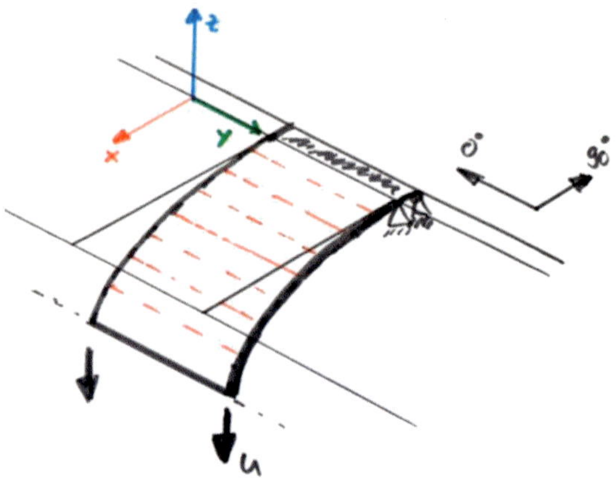

Figure 13.1: Object of investigation. Tip-displacement induced panel deflection

In the present study the panel's tip is positioned displacement (u in Figure 13.1) controlled. The study is executed to examine whether DD's tapering opportunities enable a precise design of panel's deflection shape.

13.3 DD background

The DD laminates investigated hereafter feature a $[\pm 0, \pm 50]$ BB. This is in line with a recent study of Cunha et al. [3], in which the authors define a DD substitute for a more-orthotropic $[0_3/90/\pm45/0_2/\pm45]_{2S}$ QUAD laminate ([50%,40%,10%], with Laminate parameters $V_1^* = 0.4$, $V_2^* = 0.2$). The $[\pm 0, \pm 50]$ BB is considered a potential candidate for a DD upper-wing cover laminate, with the 0° direction being aligned along the wing-span direction.

The unique tapering opportunities of DD have already been demonstrated for selected components [2], such as the tapered omega stiffened component shown in Figure 13.2.

Figure 13.2: Omega-shaped stringer with tapered DD laminate

The stacking sequence in a conventional laminate determines a panel's bending stiffness properties [4,5], as known from CLT basic equation, provided in Figure 13.3.

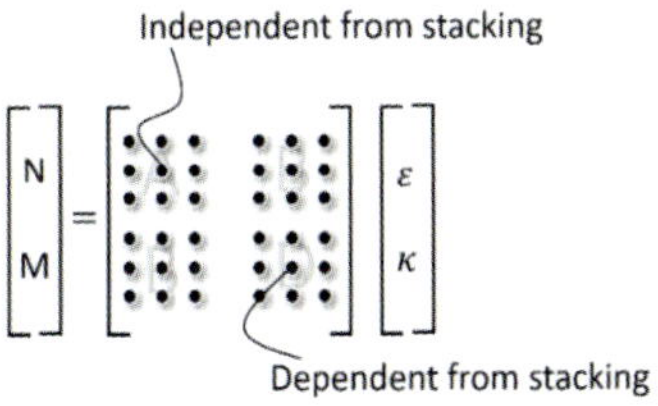

Figure 13.3: Basic relation of the Classical laminate theory (CLT)

With increasing number of plies, the number of stacking permutations grows nonlinearly, which makes optimization challenging. Zone-to-zone transitions need to be designed and ply-drop-offs need to be located (staggering) in transition zones. When conventional laminates shall be tapered, it is essential to keep the laminate symmetry, in order to assure the desired uncoupling between in-plane strain and out-of-plane curvature. This limits the design space!

For DD laminates, there is no laminate symmetry requirement. The DD laminate architecture is composed of a sequence of building blocks, all with a constant ply order within the BB. It is found that for those laminates the coupling matrix [B] as well as the bending-twist coupling terms in matrix [D] change proportional to $1/r$ and $1/r^2$, respectively.

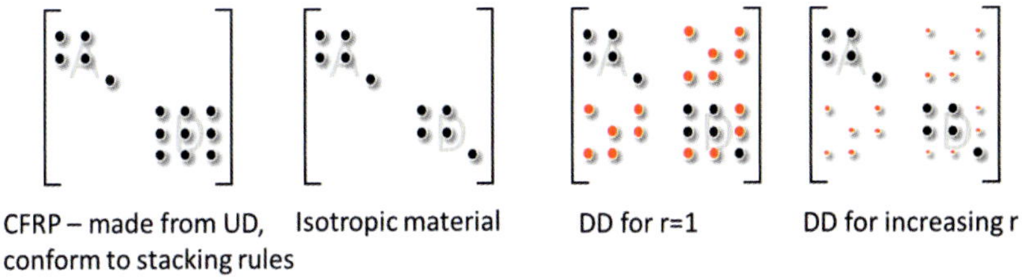

Figure 13.4: Homogenization in DD context. Red dots indication non-zero entries

Thus, the negative consequences of asymmetry diminish when the number of building block repeats is increased. This effect is denoted as homogenization. The effect allows for making warpage and twist-free structures and panels, even though the DD laminates remain asymmetric.

The ability to simply stack BBs promotes tapering. A base laminate can be created and local patches can be added without generating the necessity to re-shuffle the whole ply stack. Figure 13.5 shows how this option is utilized for the tapered 500 mm long panel examined in the present study.

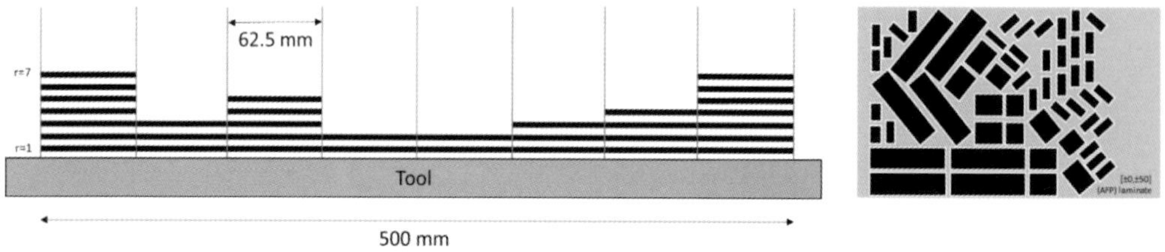

Figure 13.5: Schematic of the laminate stacking for the tapered sample and BB cut pattern for a pre-manufactured, 4-ply BB laminate

An eight-ply base laminate is locally thickened individually in eight different zones. Building-block run outs are all located on the bag-side of the laminate, leading to a clean inner laminate architecture, free of resin-rich pockets. Note that r>4 is a reasonable baseline [1,2], in order to minimize unwanted warping and twist. This recommendation was intentionally disregarded for the study at hand, to minimize actuation forces for the planned bending tests. When thin-laminate structures are in particular focus, it is recommended to use thin-ply prepregs to assure a reasonable number of BB repeats. For an isotropic material the [D] matrix in the CLT can be expressed as

$$[D]_{iso} = \frac{t^3}{12} \cdot [Q] \quad .$$

A similar relation is found for DD laminates [1,2,3,5].

$$[D] = \frac{{t_{lam}}^3}{12} \cdot [D^*] = r^3 \cdot \frac{16 \cdot t_{ply}^3}{3} \cdot [D^*]$$

The 11, 12, 22, 66 coefficients of the normalized [D*] matrix are constants, independent from r.

$$[D^*] = \begin{bmatrix} const & const & (1/r^2) \cdot (...) \\ const & const & (1/r^2) \cdot (...) \\ (1/r^2) \cdot (...) & (1/r^2) \cdot (...) & const \end{bmatrix}$$

For a sufficient high number of repeats, the entries $(1/r^2) \cdot (...)$approach zero and one finds that r is the dominant parameter to change the laminate curvature.

$$\kappa_y \approx \frac{1}{D_{22}} \cdot M_y \qquad \rightarrow \qquad \kappa_y \approx \frac{1}{r^3} \cdot \underbrace{\frac{3}{16 \cdot t_{ply}^3 \cdot D_{22}^*}}_{\substack{constant\ for \\ specific\ BB}} \cdot M_y$$

For the samples at hand the following scaling parameters are relevant.

r	2	3	4	5	6	7
$1/r^3$	1/8	1/27	1/64	1/125	1/216	1/343

The presented homogenization process creates simplicity in laminate design and optimization. For conventional QUAD laminates an optimization process needs to interrelate the number of plies, the specific ply orientations, the stacking sequence and constraints such as laminate symmetry for example. For DD, the optimization problem is simplified, as only a single integer-type parameter needs to be determined for each zone. Of course, engineering judgement is mandatory to define appropriate tapers, avoiding drastic laminate thickness steps. The samples presented hereafter already show courageous tapering.

13.4 Experimental study

Sample manufacturing

Two CFRP samples have been manufactured in a single autoclave process from Hexcel's M21E/IMA medium grade, unidirectionally reinforced (UD) prepreg material [6] (ply data in [7]). The samples are 500 mm long and 150 mm wide, with the laminates' 0° direction being aligned along the width direction (see Figure 13.6). The carbon-fiber epoxy prepreg has a nominal ply thickness of 0.184 mm, leading to a building-block thickness of 0.736 mm. One sample, denoted as 'homogeneous', is made with a constant thickness of 3.68 mm, which refers to five BB repeats or 20 plies. The second sample is denoted as 'tapered'. Its 500 mm length is sectioned into eight zones, each 62.5 mm long. The zone-specific number of BB repeats varies between 2 and 7, which is equivalent to 1.47mm to 5.15mm (see Figure 13.5).

The [±0, ±50] building block has been used for the study at hand. This BB is in context of potential DD laminate substitutes for upper wing covers, as indicated earlier.

Ch.13

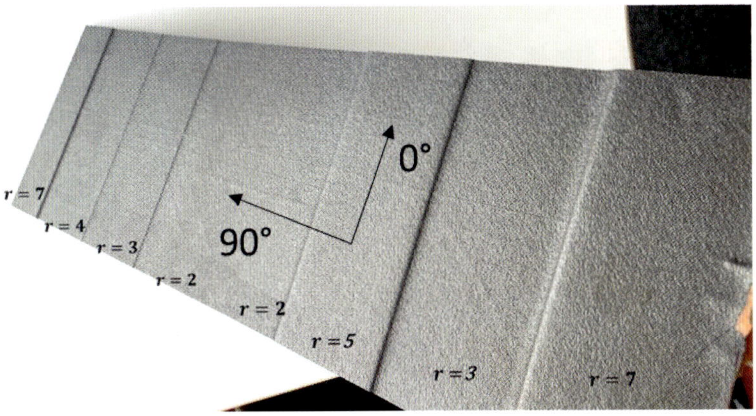

Figure 13.6: Tapered sample (vacuum bag side) with locally varying BB repeats

Both samples have been cured with a single-sided tool concept on a steel tool as indicated in Figure 13.5. The tool side creates the outer aero-relevant surface. The bag-side of the tapered sample is shown in Figure 13.6. All building block run-outs are located on the bag-side surface to avoid internal neat-resin pockets. The application of a covering BB is conceivable as well.

Creating the fixture and apply panel bending

The realized testing fixture mimics a conventional cantilever scenario, with a defined vertical tip displacement. Both samples are clamped individually using rollers on the right to prevent unwanted affectations of the individual slopes at the load-introduction side. Metal rollers were used to create a line-load scenario uncoupling tip displacement and slope. Figure 13.7 shows the schematic of the clamping scenario.

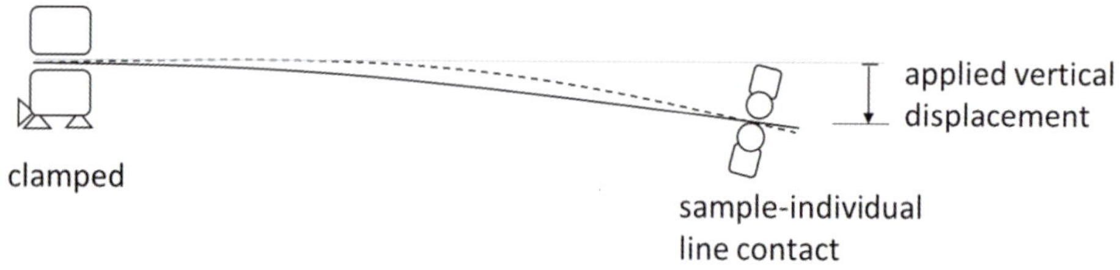

Figure 13.7: Fixture concept. Note that both samples were individually fixed in order not to constrain sample slope

Figure 13.8 shows the final fixture, made from aluminum profiles. Care was taken to define and assure a clean and twist-free bending condition to both samples. Two threaded rods are used to adjust the tip-displacement as desired. A tip displacement of 46.5 mm was applied in the experiment.

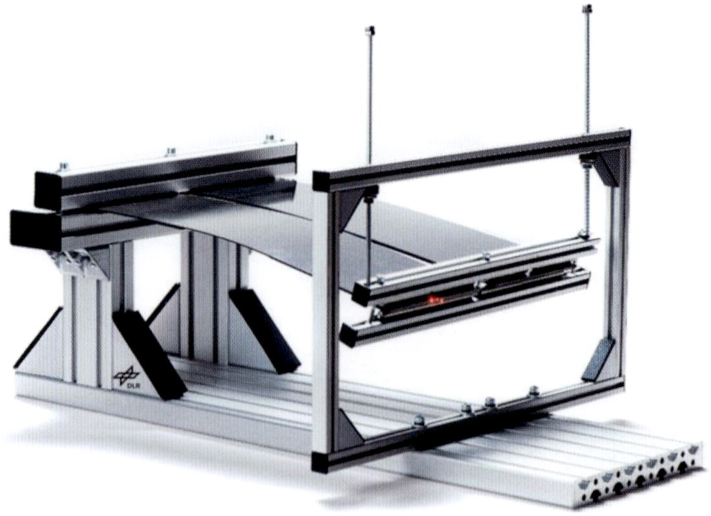

Figure 13.8: Fixture for applying a defined tip displacement. Red spot are integrated fiber-optic sensors for planned strain measurements

Evaluating deflection curves with GOM ATOS

A GOM ATOS 5 scanner has been used to digitize the outer surface of both bent samples, which is shown in Figure 13.9.

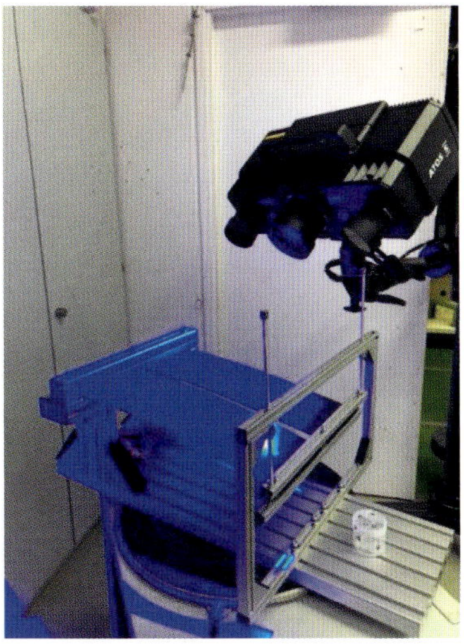

Figure 13.9: GOM ATOS 5 scanner used to measure the bent samples

The measurement data has been post-processed using GOM Inspect software. Figure 13.10 shows the determined tessellated mesh data, as well as the defined coordinate system. It is used to define section planes needed for quantifying the bending characteristics of both CFRP samples.

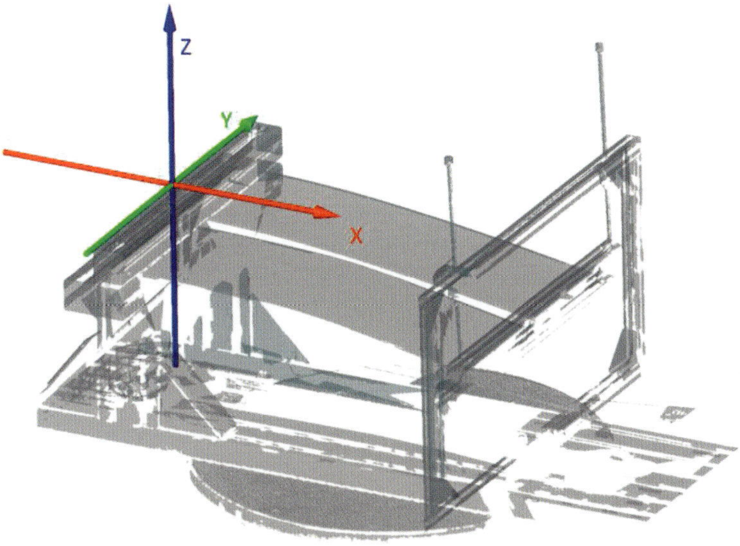

Figure 13.10: Measurement data and defined alignment

A contour plot has been created in a first step, which shows the z-displacement with respect to the illustrated xy-plane. As the local differences between both samples are rather small compared to considerable large total deflection of approximately 50 mm, the contour plot only indicates little differences. For more detailed analysis and for assessing differences on a 2D basis, two section-cuts were created. Each sample mesh was cut by a plane parallel to xz-plane, at the samples center of width location. The created sections consist of 713 and 760 points, respectively, which allows for meaningful evaluation of the deflection curves. Figure

13.11 summarizes the executed evaluation procedure. The illustrated sections of the tapered (green line) and the homogeneous sample (black line) are used for the comparison to FE data, as it is presented in the subsequent section. Figure 13.11 already shows a side view (along y-direction), which directly reveals the differences between both deflection curves.

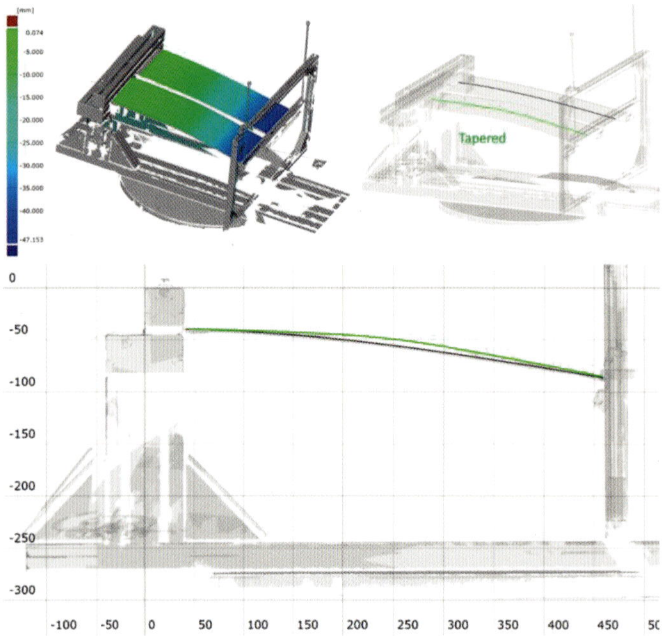

Figure 13.11: ATOS measurements for deformed state and extracted 2D deflection lines

The point information of the sections was exported for post processing in Python.

13.5 FE model setup

ABAQUS CAE has been used to create shell-element-based representations of both CFRP samples. Both are modeled in a common assembly. The rather simple FE model features a single static step. The samples are modeled using the composite shell section functionality. Laminates are defined locally for the 62.5 mm long zones. Linear-elastic Engineering constants for the prepreg at hand is provided in [7].

Figure 13.12 shows both samples in the FE model, while the local shell thickness is illustrated.

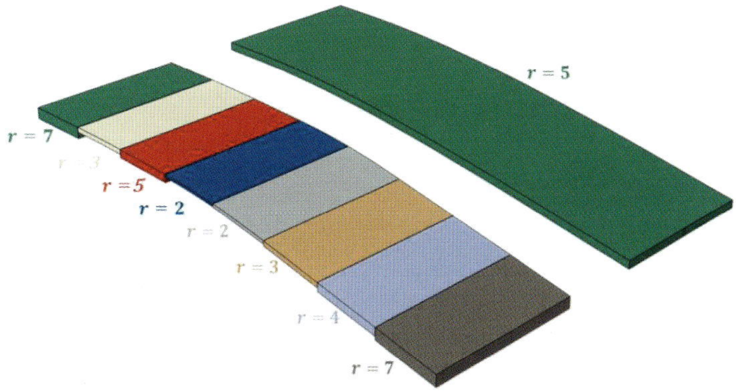

Figure 13.12: Shell-element models with illustrated local laminate thicknesses. Parameter r denotes the number of repeats.

Top-surface modelling option was executed to be in line with the aero surface condition and the experiment, in which the smooth outer surface is examined.

Individual node sets were defined along the samples' center lines. Each set consists of 240 nodes, as can be seen in Figure 13.13. The vertical displacements of those nodes were exported from the FE result files for post processing.

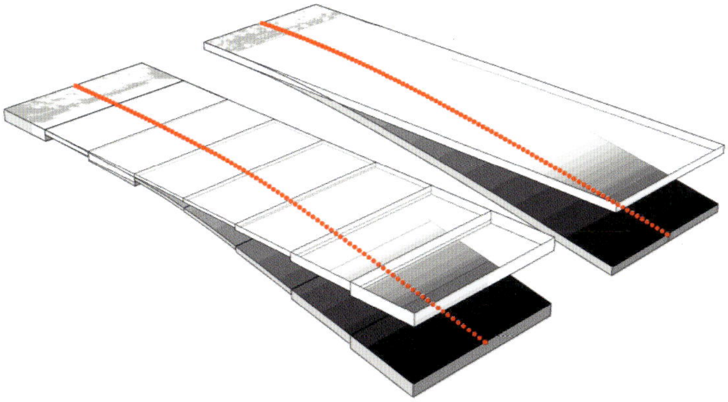

Figure 13.13: Evaluation node sets within FE models (100 nodes illustrated only)

Figure 13.14 shows a contour plot for a modeled vertical tip displacement of -46.5 mm, which is in line with the experimental setup. The displacement has been applied to two reference nodes, which were individually coupled with the samples, using the kinematic coupling feature in ABAQUS. Only the U2 degree of freedom has been coupled to not influence the samples' slope. Note that the y-direction in the ABAQUS model is equivalent to the z-direction in the measurement coordinate system, shown in Figure 13.1010.

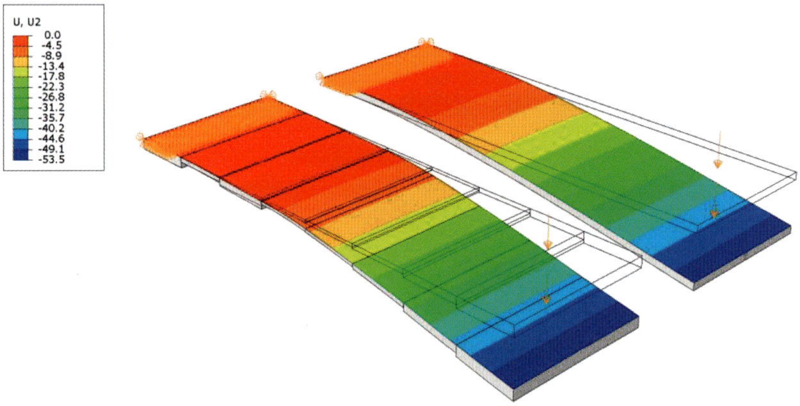

Figure 13.14: Contour plot for both samples (tapered left). Only little deviation can be seen as the differences are small compared to the overall tip displacement

The modeling strategy allows for evaluation reaction forces at both reference nodes. 6.2 N were determined for the tapered sample, while 39.1 N were determined for homogeneous sample.

13.6 Results

The deflection curves from the experiments and the numerical analysis are plotted in Figure 13.15. The qualitative and quantitative match between the experimental and the numerical data is considered satisfactory. Characteristic curvature changes, related to the local laminate thicknesses, are well met for both samples.

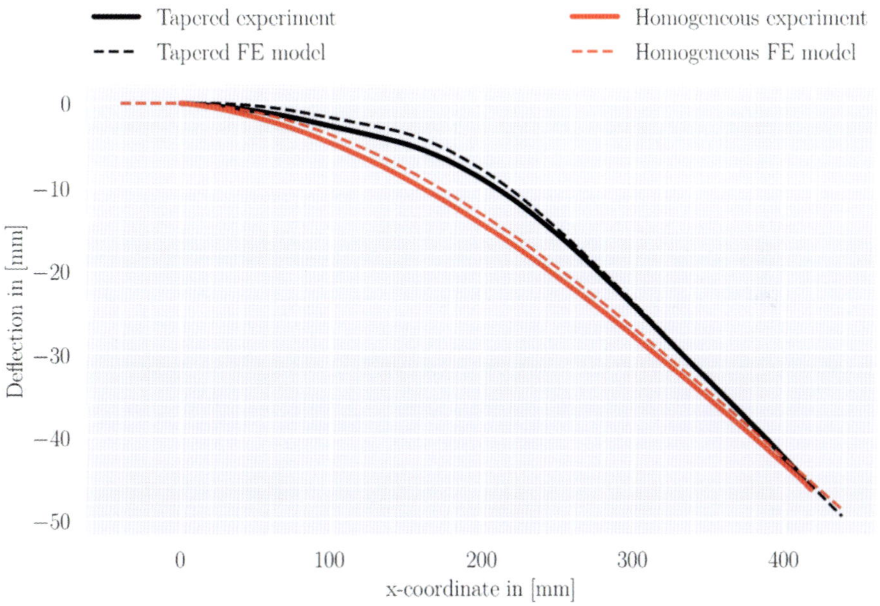

Figure 13.15: Deflection curves from the experiment and the FE simulation

The maximum deviations between the simulated and the measured shape do not exceed 2 mm. Figure 13.16 shows the deflection curve of the tapered sample. The illustrated colored offset indicates the corresponding local zones and the local laminate thickness. Note, that no data is available for the clamped areas, which explains the smaller length of the zones one and eight.

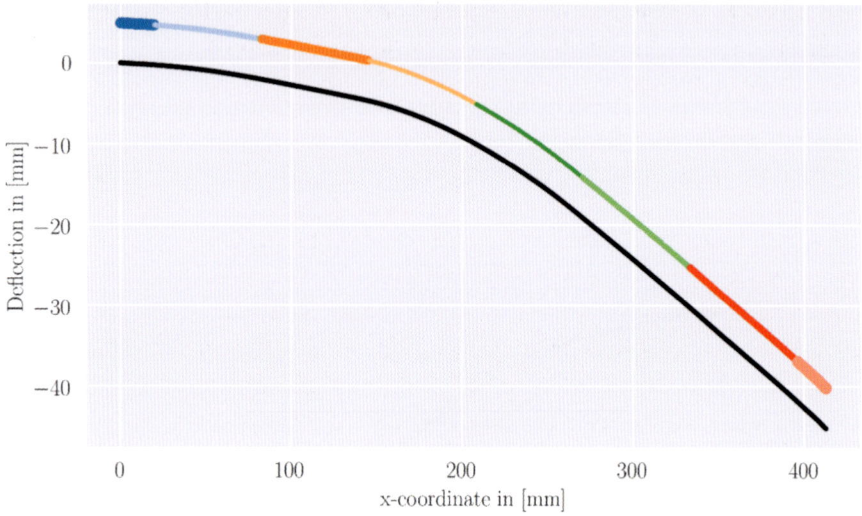

Figure 13.16: Deflection curve of the tapered sample. Zones and the local laminate thickness are indicated by the colored offset line. Zones 1-8 from left to right

The highest curvature is localized around the center area, as expected, where two zones with r=2 are located. Given the good match between FE results and the experiment, the numerical model was used to determine deflection curves for three additional taper configurations. All are subjected by the same tip displacement from above. Figure 13.17 shows the deflection curves of those configurations, which outlines the bandwidth of conceivable shapes.

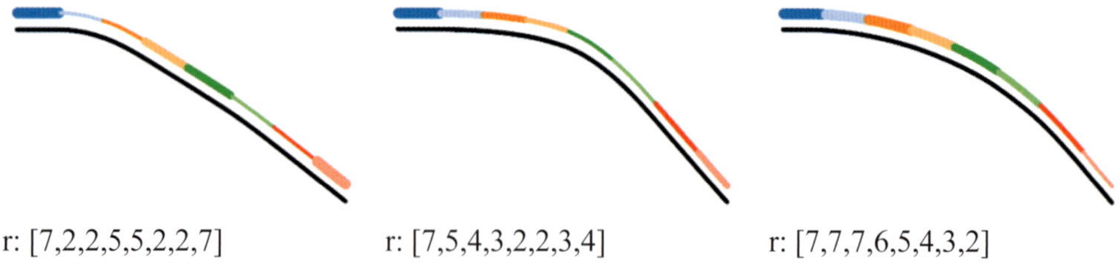

Figure 13.17: Deflection curves for selected r- distributions

Note, that r:[i,j,k,…] indicates the number of BB repeats in the eight laminate zones from left to right (root to tip). The examples show that rather smooth or more localized deflection curves can be realized. When it is considered that the number of zones and the specific zone lengths can be incorporated as design parameters in the design space, the results prospect a wide range of feasible deflection curves.

13.7 Conclusion

The unique tapering opportunities of Double-Double laminates have been utilized to tailor a CFRP panel's deflection shape, for a tip-displacement controlled bending scenario. A [±0, ±50] building block was used for the DD laminates. Two 500 mm long and 150 mm wide samples were made from UD prepreg. A 'homogeneous' and a 'tapered' sample were bent and the outer surface was captured with an optical GOM ATOS scanner. Deflection curves of the experiment and a corresponding numerical simulation were compared. The comparison shows a satisfying match, which verifies that tapered DD laminates are well suited to adjust panel deflection curves by adapting the number of building-block repeats locally. The numerical model was further used to predict deflection curves for additional laminate-thickness tapers, which underlined the wide range of conceivable deflection shapes feasible with DD laminates.

13.8 Outlook

The present study represents the experimental justification of a DD laminate optimization procedure, which shall be applied to a large-scale panel in a next step. The present study shows, that the numerically calculated deflection shape is well met by the manufactured samples. Within the next step it will be the challenge to implement additional aspects, such as aero loads within the design process. When panel size increases, it is expected that lateral stiffeners need to be attached to the panel in order to minimize corresponding deflection effects in lateral direction. Depending on the target shape, the equidistant sectioning approach of the panel, applied for the presented tests, can be challenged, to even improve the match to a prescribed panel shape or to further simplify manufacturing.

Deforming a tapered panel by mechanism attached to the panel's tip requires energy. The deformation procedure is reversible. An undeformed configuration has been manufactured in the present study. When manufacturing the panel according to the deformed contour, actuation forces change sign, which can lead to an overall reduction of the required energy on system level. Manufacturing a panel with a contour in between the undeformed and the deformed shape will reduce the maximum strain on the outer surfaces, when the panel is deformed.

Literature

[1] Tsai SW. Double–Double: New Family of Composite Laminates. AIAA Journal (https://doi.org/10.2514/1.J060659), 2021

[2] Tsai SW et al. DOUBLE–DOUBLE - A New Perspective in The Manufacture and Design of Composites. Composites Design Group. Stanford University, Stanford, CA 94305-4035, 2022

[3] Cunha RDd et al. Low velocity impact response of non-traditional double-double laminates. Journal of Composite Materials Vol. 0(0) 1–11, 2023

[4] Kappel E. Double–Double laminates for aerospace applications — Finding best laminates for given load sets. Composites Part C: Open Access 8, 100244, 2022

[5] Kappel E. Buckling of Double-Double laminates. Composites Part C: Open Access (submitted for publication) 2023

[6] HEXCEL M21 Data sheet. HexPly® M21 180°C (350°F) curing epoxy matrix ,2020

[7] Talreja R, Modeling Damage, Fatigue and Failure of Composite Materials, Cambridge: Elsevier Science. Woodhead Publishing Series in Composites Science and Engineering, 2015

Table of figures:

Section list

CHAPTER 14

Certifiable General Aviation Structures in DD

Naresh Sharma
naresh@nashero.com

The utilization of double angle plies, DD, and Tsai's Modulus allows for a precise analytical formulation of the mechanical behavior of composites. As demonstrated in previous chapters, these innovative technologies provide a unique perspective and simplify the design of composite structures, enhancing our understanding of them. Among these technologies, DD stands out as the sole method that enables seamless tapering of composite laminates, resulting in significant reductions in structural mass compared to optimized traditional quad-angle laminates. By homogenizing a laminate with just 8 ply repetitions, thin and structurally efficient laminates can be achieved. In this chapter, our focus is to showcase the immense potential impact of these technologies on reducing carbon emissions in the transportation industry. As thoroughly demonstrated in previous chapters, no other technology can achieve such substantial mass savings in already optimized structures.

The primary responsibility of a structural engineer is to minimize the mass of aircraft while ensuring compliance with safety and regulatory requirements. Numerous scenarios are driven by the pursuit of reducing structural mass. These scenarios include increased range resulting from lowered structural mass, allowing for additional payload capacity or reduced fuel consumption for the original range and payload. Moreover, reduced structural mass enables the use of smaller and more fuel-efficient engines, with a larger bypass ratio, creating a snowball effect on aircraft mass as structural mass changes.

14.1 A case in point, the motivation for DD

The range of an aircraft is an important measure of its performance and, the range, is directly influenced by a reduction in mass of an aircraft. According to Breguet's Equation (Torenbeek, 1981, pp. 157-159), an aircraft's range, R, is proportional to ln (W_i/W_f), where W_i=Initial weight, W_f=final weight of an aircraft's flight. It is to be noted that Breguet's equation is a first order approximation and gives a conservative range. The final weight, W_f, for maximum range would be the mass with minimum fuel but with maximum payload and operational equipment. Considering the impact of mass, the range in Breguet's equation is:

$$R \propto \ln \left({}^{W_i}\!/_{W_f} \right)$$

Let us consider a cruise efficient, commercial, the-state-of-the-art, aircraft at the time of this writing, the Airbus 350-1000[1], a modern passenger jet whose structural mass consists primarily of carbon composites. Maximum fuel capacity is 159,000 liters = 159,000*0.804 kg = 127,836 kg

Maximum Takeoff Mass	390,000 kg
Maximum Landing Mass	236,000 kg
Maximum Zero fuel Mass	223,000 kg
Maximum Fuel Mass	127,836 kg
Manufacturers Empty Mass	129,000 kg

Table 14.1: Airbus 350-1000 masses

The mass of two Rolls Royce Trent XWB engines at 7,277 Kgs each, the landing gear, systems, furnishings, paint etc., would be taken out of the declared empty mass. Therefore, let us approximate the net structural mass of the aircraft, to be around 100,000 kgs. Based on prior work (Srivastava, Sharma, Tsai, & Mohite, 2020), where we have proven that using DD can potentially reduce 50-55% of the structural mass. Let us assume that DD and its associated technologies can enable a structural mass reduction of 45% from 100,000 to 55,000 kgs. This gives the aircraft empty mass of around (55,000 + 29,000 kgs) or 84,000 kgs which is a reduction of approximately 35% of the empty mass. Assuming that all the mass's for the aircraft are the same as the original aircraft, and that the reduction in structural mass would be the gain in fuel capacity. Therefore, the same numbers for our fictitious A350-1000-DD would now be:

Maximum Takeoff Mass	390,000 kg
Maximum Landing Mass	236,000 kg
Maximum Zero fuel Mass	178,000 kg
Maximum Fuel Mass	172,836 kg
Manufacturers Empty Mass	84,000 kg

Table 14.2: Airbus 350-1000-DD masses

Let us say R_{NonDD} is the range of A350-1000 and R_{DD} us that of our fictitious aircraft built using DD technologies, A350-1000-DD, according to Breguet's equation would be:

$$R_{DD} = R_{NonDD}\left(\ln\left(\frac{390}{390-172.836}\right)\Big/\ln\left(\frac{390}{390-127.836}\right)\right) = 1.474\,R_{NonDD}$$

Therefore, the A350-1000-DD will have an increase in range of approximately 47.41%. Another parameter that can be viewed here is the reduction in carbon emissions of each passenger of the aircraft. Let us assume that the aircraft in its entire phase of flight is at its optimum fuel consumption, and therefore, using the same argument the fuel/seat/km of an A350-1000-DD reduces by 1/1.4032, to 67.84% of A350-1000, implying a reduction in the carbon footprint of each passenger by 32.16% for each km of flight in an aircraft made using A350-100-DD. Given that net carbon emissions are directly proportional to fuel consumption. Extrapolating what we have evaluated to just one area, that is, air travel, we see the mammoth proportions of the advantages of the discovery of DD. The net carbon emissions from air-transport industry (McKinsey & Company, 2022)are 13% of global

[1] https://aircraft.airbus.com/en/aircraft/a350/a350-1000 and https://en.wikipedia.org/wiki/Airbus_A350

emissions, imagine that if all aircraft were designed and made using DD that number could be reduced by 32.16%, given the pre-covid19 pandemic air-travel statistics.

We have not even begun to scratch the surface of these advantages, they are in the cost of production, with reduced material requirements, reduced wastage of materials, faster layup, reduced time and labor, and the ability to play with finer qualities of aeroelastic tailoring to give better ride quality, more damage resistant aircraft, more crash resistant, and aircraft less vulnerable to corrosion, with respect to what is achievable using conventional technologies today.

14.2 General Aviation Aircraft

General Aviation (GA) aircraft are those aircraft that, classically, were certified under the norms of Code of Federal Regulations CFR Pt23 in the USA; CS23 of European Aviation Safety Agency EASA; or generally called Part 23. GA aircraft can, today, be considered to fall in the class of aircraft that range from single seat aerobatic and pleasure flying aircraft, all the way to those aircraft that carry less than 19 passengers, multiple engines and a maximum take-off mass of less than, 8,618kgs or 19,000 lbs.

FAA, EASA, and aviation regulators of other countries gave the mandate to ASTM International to develop new, modern, industry consensus standards for GA aircraft with the goal of increasing the safety of GA aircraft and halving the certification burden. This has resulted in several standards developed by the F44 committee that are currently used by FAA and EASA, among others, for meeting certification requirements. Existing Airworthiness Directives (AD), Airworthiness Circulars (AC), Acceptable Means of Compliance (AMC), and other airworthiness authority documents are rich with information for specific aspects of airworthiness and safety, they are called out and used in certification, even with the F44's developed standards.

In this chapter we will discuss the development of GA aircraft structures using DD.

14.2.1 ***Effect of weight reduction***

In the case of GA aircraft, the advantage is even more pronounced, with the structural mass taking up around 60-90% of empty mass. Due to varied missions of each type of GA aircraft it is very difficult to ascertain that advantage other than saying that it is greater than that for passenger airliners and, therefore, worth every effort to undertake as an objective. Breguet's equation, as described the previous sections, is still valid and is going to be the primary driver of any function to optimize structural mass.

14.3 Optimum DD for General Aircraft Structures

As has been mentioned in the previous chapters, DD are double angled layup sequences of laminae of composite materials, which have classically consisted of a combination of 0°, 90°, and ±45° laminae, with a few thumb rules thrown in for good measure. Analysis based design and optimization revealed that DD was a more appropriate layup for structurally optimized composites. Once an optimized DD has been found for an application the search for test data that can be used for material characterization becomes relevant. For the DD based carbon composite aircraft structures, a need for test data is frequently cited as a limitation. Analysis was done in the previous chapters has shown that $[\pm23/\pm75]_{Rpt10}$ is the best laminate to test if one is limited to a single test. This laminate had the least coefficient of variation for the stiffness components of all DD (91 of them using an angle increment of 7.5 degree) and 15 CFRP materials. This implies that coupon tests of a $[\pm23/\pm75]_{Rpt10}$ laminate and its Tsai's Modulus, when extrapolated to unknown laminate sequences will result in an accurate stiffness tensor for that material. Thus, let us say that the DD that interests us the most in our aircraft is [±0/±50], we could take the data from the test results of $[\pm23/\pm75]_{Rpt10}$ to arrive at that laminate's Tsai's Modulus, which will be then used to estimate its Master Ply and then derive our laminate properties for [±0/±50]. This process can be used to estimate the properties of almost any laminate that would be of interest to us.

14.4 Composite Aircraft Certification

Due to their material properties and behavior, composite materials, esp. carbon fiber reinforced plastics (CFRP) are highly suited for use in applications where low structural mass is desirable. Carbon composites provide:

1. High specific strength
2. High specific stiffness
3. High corrosion resistance
4. Shape conformance to a structural load spectrum
5. Aerodynamic cleanliness
6. High fatigue strength and toughness
7. Capability to design high energy dissipation structures for crash worthiness

Aircraft certification is a means to prove that a structure, designed, built, operated, and maintained using the certified processes would be safe in its entire mission profile for the lifetime of its certification. Historically, the Federal Aviation Administration (FAA) in the USA and European Aviation Safety Agency (EASA), in Europe and many other countries are the two main agencies in the world that have set forth the standards for aircraft certification that are currently used world-wide. Once an aircraft is certificated, its fleet's operational experience is recorded, and such records have been analyzed by the FAA and EASA generating a vast database of the aircraft's operational and maintenance life.

The experience with metal aircraft certification has pointed to some challenges in the certification of composite aircraft, due to which starting in the 1970's to 1990's significant work was carried to understand the predictability of composite aircraft's safety. Early composite aircraft designs used some thumb-rules, guided by engineering opinion as opposed to analytical understanding one of which is the 10% rule, which says that at least 10% of fibers in any structure should be in one of the 0/90/+-45 deg angles. Many of these thumb-rules are still followed in the industry. Variability in composite constituents such as epoxies, their chemical composition and structural behavior, fibers and the production techniques of composites has historically ebbed on the advantage of composite materials. Lack of standardized data and capability of reproducing that data in production conditions is another issue marring the work till late 1990's.

In the 2000's, significant investment by companies in production techniques, standardization of material systems, development of new fibers etc., has helped this aspect. Today, there is a public database by NCAMP at Wichita State University, standardized materials which are accepted by FAA and EASA. The old composite design philosophies, in absence of their theoretical basis, such as the 10% rule, still are a norm.

The recent work by Prof. Tsai and his group, has analytically proven that composite orthotropic materials have an elastic property, a constant, Tsai's Modulus. Tsai's modulus is unique for a material system and fiber combination and can, therefore, be used to generate the material behavior of any orthotropic carbon composite from a single test data.

In this chapter we'll discuss the differences in the behavior of carbon composite structures compared to those of metal. The current certification mechanisms for aircraft made in metal as compared to composites. We'll also touch on the exciting new Double Double technologies developed by Prof. Tsai's group and their impact on aircraft design, certification, optimization, production, and maintenance.

Additionally, the DD technologies described in this book can significantly reduce layup costs and give a 10-fold boost to production efficiency by reducing the time and wastage during part production. Therefore, carbon composites are suited for aircraft applications.

Legacy certification of aircraft built using composites is based on statistical tools and methods which have not kept pace with the development of new analytical theories. The recent pivotal discovery of master-ply and trace, an invariant, enables an analytically precise behavior of carbon composite laminates used in most of the new aircraft today. The discovery of trace shows that CFRP aircraft certification should be viewed from a new perspective. After a short review of the current regulations, we present a methodology in this chapter for the certification of CFRP aircraft using trace. The novel methodology shows that the theoretical trace of a CFRP material follows experimental results accurately and consistently. Trace based material properties replicate the behavior of the real test data. We will show how the newer certification methodology can reduce certification burden compared to existing regulations.

14.5 Introduction to the Certification of Aircraft Structures

Certification of aircraft structures is done by government regulators, such as the Federal Aviation Administration (FAA) of the USA, the European Aviation Safety Agency (EASA) in Europe and the International Civil Aviation Organization (ICAO) worldwide. The structural certification usually begins with an analysis of the design requirements and the mission profile of an aircraft, based on regulations such as CFR Pt 23 (FAA, 2011), LSA, e-CFT Pt 25 (FAA, 2019) for the FAA and similar numbered regulations in most other ICAO signatory countries, since they have agreed to uphold certain minimum common standards for aircraft certification. Based on the aircraft mission, and the regulations that we've cited above, the analysis includes the generation of loads that the aircraft will be subject to. These loads are a combination of the aerodynamic, inertia, static and dynamic loads. Furthermore, a detailed test, or analysis supported by test, of the aircraft components is done in a systematic manner to ascertain that the structure will meet its mission objectives in a safe manner. The result is a certificate of airworthiness for the aircraft from the regulator which defines the restrictions under which the aircraft would be allowed to operate. To produce the aircraft, the aircraft manufacturer should prove to the regulators that they can build the aircraft with a consistent quality equal to or better than the certified aircraft.

Composite materials are significantly different in their structural behavior and hence the certification of composite structures requires a different approach as compared to the certification of metal aircraft structures. An aircraft must withstand static, dynamic, handling, and operational loads, thermal gradients, solar irradiation, and hygroscopic environment. Therefore, it is imperative that the material data used by the designer guarantees the structural performance up to the design life of an aircraft.

To arrive at consistent and reliable composite material properties is one of the biggest challenges in the design of composite aircraft structures. The Composite Materials Handbook (CMH17), (DOD, 2002), has historically called out the requirements of material testing by defining the reliability of the properties used in design and analysis, and the manner to arrive at those properties. In the case of aircraft these are A-basis and B-basis properties as called out in CMH17. Material coupons are tested in accordance to accepted standards by ASTM International and analyzed using statistical methods defined in CMH17 to arrive at the properties that can then be used in the design, validation, and eventual certification of composite materials. These properties are then called the material allowables.

Till recently, only the CMH17's statistical processes, which do not consider the complete physical behavior of the material itself, were used in the development of material allowables. The recent discovery by Prof. Tsai and his team that the trace, a summation of the diagonal elements of a stress or strain tensor, of an orthotropic system of plies enables an invariant-based approach to stiffness, gives us additional insight onto the behavior of composites (Tsai & Melo, 2015). It shows that statistical analysis of coupons of composite materials can be significantly reduced, since there is an inherent and an exact relationship between the trace of a stress/strain tensor and the remaining material properties. Comparing 30 different carbon composite materials we see that the correlation has a standard deviation of less than 1.3% (Tsai & Melo, 2015). The variation resulting from physical testing of materials is significantly higher than 1.3%, it, therefore, proves the efficacy and the exactness of the discovery of trace.

Due to the potential of significantly reducing the regulatory burden of statistical analysis and testing of carbon composites, there have been some experiments by material testing professionals to derive material properties from trace. However, some of these have failed to properly apply the theory of trace, thus incorporating inexplicable errors (Freed, 2016).

In this chapter we will also try to show how trace has been used to generate reliable material properties using test data. This data is generated from orthotropic coupons by us as well as correlated with existing test data. These results, if used in the specification of material allowables, have the potential to reduce the certification burden and enhance safety compared to existing legacy processes.

14.5.1 Material Allowables

Complete replication of the behavior of any aircraft structure in the lab or in test scenarios is an impossible task due to infinite flight, thermal, mechanical, and chemical conditions in which an aircraft operates. Therefore, analysis and modeling of the structures to simulate the operational conditions and structural response of an aircraft are fundamental to the certification.

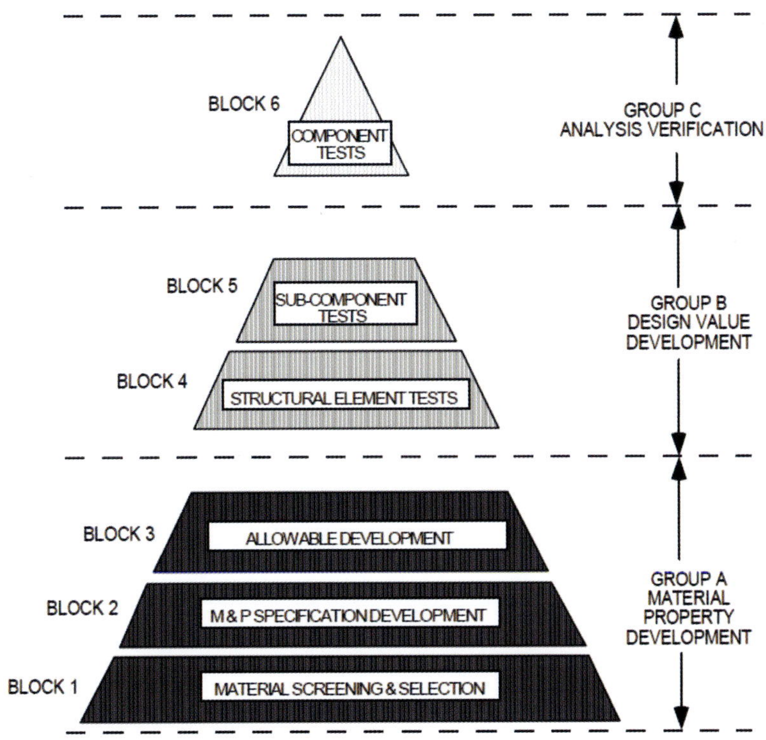

Figure 14.1: Building Block Approach (DOD, 2002)

Regulators such as the FAA and the EASA accept certification of aircraft by test, or analysis supported by test, of progressively complex composite structures. This can be aptly illustrated by the building block approach, where the bottom blocks represent test results of coupons of composite materials and at the top of the pyramid are structural components of a complete aircraft made using the materials developed in the lower blocks of the pyramid [Figure 14.1].

This allows the progressive increase in confidence that the structure would behave in real life as designed, modeled, analyzed on the computer, and tested to support the analysis.

In the case of CFRP composites, using Tsai's Modulus, the bottom three blocks can be reduced to a single test (Steve Tsai, 2019). Such allowables have been shown to be the foundation for virtual material characterization.

CONDITION		COMPOSITE BEHAVIOR W.R.T. METALS
LOAD-STRAIN RELATIONSHIP		MORE LINEAR STRAIN TO FAILURE
NOTCH SENSITIVITY	STATIC	GREATER SENSITIVITY
	FATIGUE	LESS SENSITIVITY
TRANSVERSE PROPERTIES		WEAKER
MECHANICAL PROPERTIES		HIGHER
SENSITIVITY TO AIRCRAFT HYGROTHERMAL ENVIRONMENT		GREATER
DAMAGE GROWTH MECHANISM		IN-PLANE DELAMINATION INSTEAD OF THROUGH-THICKNESS CRACKS

Table 14.3:Differences between Composite and Metal behavior

14.5.2 Fatigue and Durability

Long fiber reinforced materials are inherently fatigue resistant and damage tolerant. However, the matrix material which keeps the fibers in place and transfers the loads between them using shear is prone to fatigue and environmental degradation. The behavior of composites under fatigue and with built-in damage that is below **T**he limit **O**f **D**etectability (TOD) and their modeling is complex.

Metals, on the other hand, are fatigue prone. This dichotomy in behavior of composites w.r.t. metal has been lucidly illustrated in (Whitehead, Cordero, & Saether, 1986) [Table 14.3]. Whitehead showed that low cycle fatigue behavior of composites is controlled by static strength, while high cycle life fatigue behavior is controlled by the net section stress level. According to Whitehead, fatigue notch insensitivity, weak transverse properties, matrix dominated failures, high failure variability, hygrothermal sensitivity and delamination fatigue growth mechanism must be addressed in structural design and certification.

Whitehead outlined four approaches and their advantages and disadvantages for estimation of the minimum acceptable design allowables, B-basis, and the certification of composites as follows:

1. Scatter factor approach.
2. Load Enhancement Factor (LEF) approach, applies a multiplier to the fatigue test spectrum, called the LEF. Some composite structures are required to be tested for 13

lifetimes, to reduce this test duration is fundamental to the LEF approach. The LEF once generated gives guidance to reduce the fatigue test cycles while maintaining equal reliability to that of metals.

3. Ultimate Strength approach.
4. Change in spectrum approach.

Whitehead's approach towards composite material fatigue was further enhanced upon by Tomblin and Seneviratne, the key difference in their approach is the reduction of the LEF as compared to Whitehead's approach (Tomblin & Seneviratne, AR 10/6, 2011). Reducing LEF would allow certification applicants to show equivalent fatigue life with lower number of cycles at a slightly higher load level than normal. However almost all other parameters were kept the same, including the assumption that: the application of either of the above approaches required a component level fatigue test. Recent work from Freed and Elmalich evaluated life scatter parameters based on A-basis and B-basis fatigue tests for hundreds of test coupons and found that the assumptions of both Whitehead and Tomblin were overly conservative (Freed & Elmalich, 2017). Freed and Elmalich report their scatter factor in the table below divided into Material Systems and Specimen Type.

Material System	β_L	N_F
Carbon	4.78	1.85
Kevlar	4.25	2.0
Whitehead et al. [1]	1.25	13.30
Ratier	2.31	3.78

Table 14.4: Material System based Weibull shape parameter β_L and Life factor N_F

Specimen Type	β_L	N_F
Open-hole (solid laminate)	5.75	1.66
Disbonding flaw (sandwich)	3.03	2.70
Impact flaw (sandwich)	2.16	4.18
Whitehead et al. [1]	1.25	13.30
Ratier	2.31	3.78

Table 14.5: Specimen Type based Weibull shape parameter β_L and Life factor N_F

The LEF is greatly dependent on the scatter factor of the material properties. Composites show larger scatter in their fatigue behavior compared to metals, but the scatter factors suggested by Whitehead were over four times of what were discovered from the two independent recent studies, one by Freed and other by Ratier. The scatter of damaged test coupons, essentially open-hole or notched coupons, on the other hand is extremely limited. In aircraft, wherein, the damage tolerance criterion requires the monitoring of the damage progression, scatter of material data for open-hole or notched coupons should be used in certification (Tomblin & Seneviratne, AR 10/6, 2011).

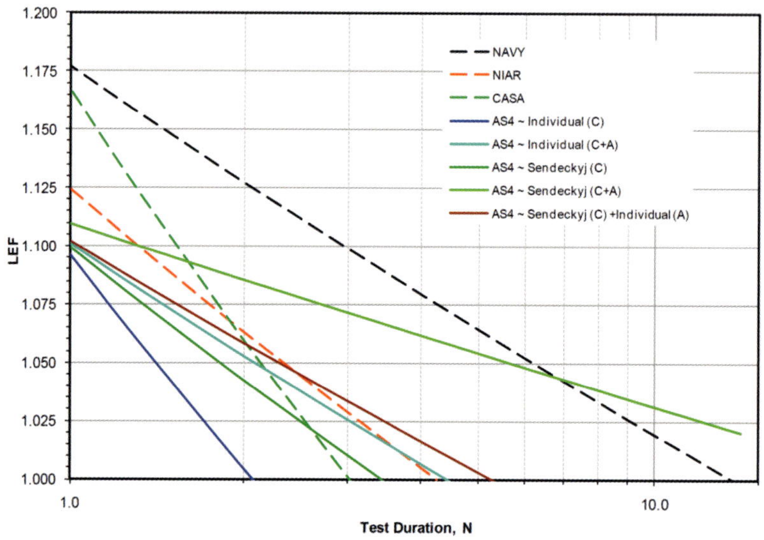

Figure 14.2: Influence of Test Duration on B-basis LEF of AS4-PW from Different Analytical Techniques (Tomblin & Seneviratne, AR 10/6, 2011, p. 81).

Furthermore, the LEF approach provides an efficient way to assure the structural life reliability (DOD, 2002, p. 442). However, other effects may also require a load enhancement and result in an undesirably high load factor. For example, an environmental compensation factor is usually applied in order to account for service environment effects, and a spectrum severity factor is usually applied for military aircraft. Thus, an LEF of 1.18, an environmental compensation factor of 1.06, and a spectrum severity factor of 1.20 would result in an overall fatigue test factor of 1.50. This could change the fatigue failure mode of hybrid metal and composite parts or reach the static strength of the structure. Therefore, in the application of the LEF approach, it is very important to ensure that the fatigue failure mode is preserved.

According to Tomblin, fatigue of sandwich structures with damage is not a key concern because of design allowable at 80% of mean minus 3σ of stress (Tomblin, et al., 1999). In all cases, specimens reached the runout limit of typically 2.5 million cycles in a hot-wet environment thought to be conservative. The ability to use compressive strength for shear requirements was confirmed by testing 24"x24" sandwich panels in a picture frame fixture. However, post-buckled thin walled composite structures stiffened by stringers require a complete gamut of fatigue and damage tolerance studies.

Most Pt23 composite light aircraft are made from sandwich stiffened stressed skin construction where skin buckling is a failure criterion. Additionally, the main structural load carrying structures, such as spars and ribs are protected by the sandwich skin and even external damage that is at the limit of detectability (TOD) does not travel to the spars. Any damage of sandwich skinned structures can be very easily repaired to bring the structure back to its pre-damage strength level.

14.5.3 <u>Environmental degradation</u>

Environmental factors such as UV radiation, temperature, moisture and humidity and salt-water spray influence structural performance of composite aircraft. Most of these have been well researched and there is guidance available in the norms that are currently in vigor. Recent research on hysteresis in the transverse residual stresses due to hygrothermal cycling shows that environment affects both the stiffness and strength of the composite.

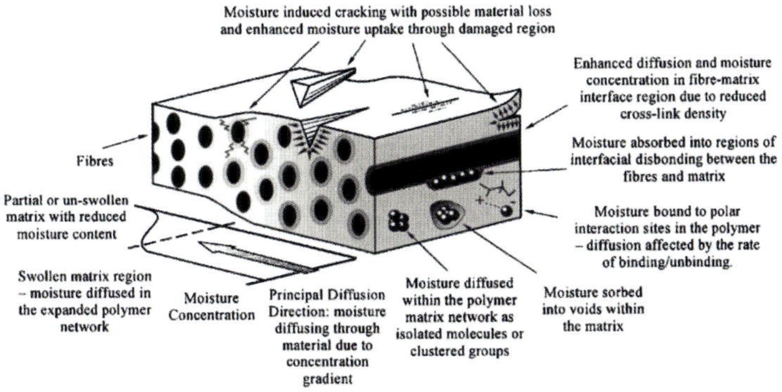

Figure 14.3: Moisture Absorption Mechanisms in Composites (Bond & Smith, 2006)

The change in flexural strength of carbon fiber composites was researched by Komorowski as shown in Figure 14.4 (Komorowski, 1983). King Jye Wong tested moisture absorption behavior and validated a numerical model of the behavior for the purpose of structural design, and a new residual property empirical model is proposed (Wong, 2013).

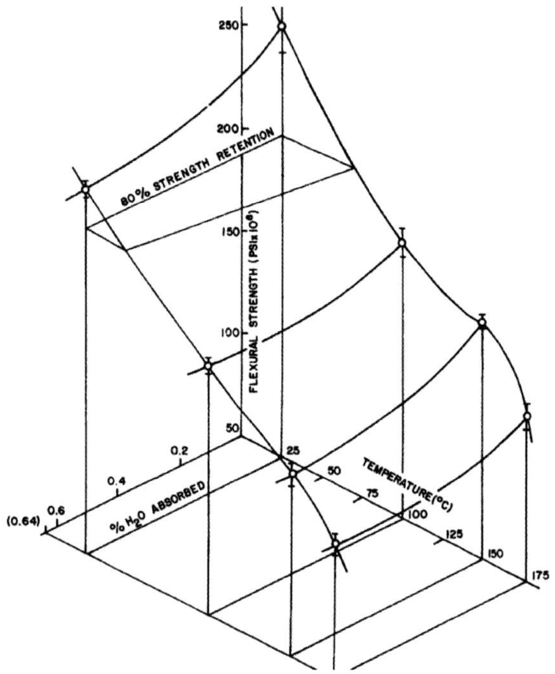

Figure 14.4: Flexural Strength of Unidirectional Carbon Fiber Composites as a Function of Temperature and Percent of Moisture Absorbed (Komorowski, 1983)

14.6 The Certification of Composite Aircraft Structures in Practice

Certification of structures is mission and risk profile dependent for different types of aircraft. Regulators, such as FAA & EASA, specify the limits of each of these factors, first in the norms and then during the certification process. Since 2004, there has been an activity taken up by ASTM International, wherein regulators, industry, users, and general interest groups have participated to develop the standards for future aircraft certification. This began with the development of Light Sport Aircraft standards, and now has developed to General Aviation standards, for aircraft up-to 19 seats of capacity. Structural certification is a crucial element of these standards. FAA and EASA have both accepted the ASTM International standards since they provide equivalent level of safety compared to the current FAA and EASA norms. In the future, many FAA and EASA norms may just refer to the ASTM International Standards.

Here we discuss the requirements of certification to meet the standards in vigor today.

14.6.1 Material Allowables

The procedures to design and certify composite structures defined in the CMH 17 are based on the building-block approach (DOD, 2002). Here, different Structural Complexity Levels ranging from the lamina to the component level can be identified. Although the relative emphasis placed on each level depends upon many factors, such as manufacturing process, structural application, corporate/organizational practices, and/or the procurement or certification agency, implementation of the building-block approach invariably leads to lengthy test campaigns that hinder the fast introduction of novel, optimized materials technologies. Therefore, being able to reduce such testing effort based on the improved knowledge of the physical behavior of the material would accelerate material integration and certification and facilitate the development of more efficient aircraft.

Material allowables are defined at the lower Structural Complexity Levels. These allowables are statistically based parameters that account for the variability of the material properties, manufacturing process, structural details and loading conditions. Based on extensive campaigns of several standardized tests on several coupons for different material configurations and process conditions, A-basis, and B-basis values, defined respectively as the 95% lower confidence bound on the first and tenth percentile of a specified population of measurements, are determined to drive the safe design of structures at larger scales.

To account for the different sources of variability at the material level, the experimental determination of composite material allowables requires testing of different standardized coupons from several laminate configurations (unidirectional, quasi-isotropic, hard, and soft laminates) and environmental conditions (as-received/room temperature (AR/RT), hot-wet (HW) and cold-dry (CD)) obtained from at least three different material batches. Consequently, qualification test campaigns required to determine such allowables are costly and time-consuming, easily ascending to thousands of test coupons.

14.6.2 Fatigue and Durability

In this section we'll present one of the current processes recommended by FAA for demonstration of safe service life. As has been shown in section 14.5.2, there is still a significant amount of controversy surrounding the best method of demonstrating that the structure has a service life that it was designed for. We believe that the progress in this matter over time will result in more robust understanding of the underlying phenomenon of fatigue of composites and result in an easier and less onerous test technique.

For any given new material, process, and structure, built using the state-of-the-art technology, significant amount of testing and analysis is needed to prove to the regulators that it would meet all the fatigue and durability requirements that are currently in-force. To characterize such material, process and structure, a complete test matrix for static, environmental, and dynamic behavior is needed. This would begin with the testing of coupons. The results of these tests would be used in analytical models to simulate a complex component of the aircraft such as the wing or wing spar or horizontal stabilizer. This component will be tested for fatigue and durability, and the test results corelated to structural simulation. A typical test program would need to derive the fatigue test spectrum for the aircraft in a systematic manner as described below [**Error! Reference source not found.**].

At a minimum, the designer would need to identify 6-8 worst case areas of interest, also called fatigue-modes, in the structure and statically test at least six coupons each for Open Hole Tension (OHT), Open Hole Compression (OHC), Uniaxial Tension (UNT), and Uniaxial Compression (UNC), of those areas. A minimum of five points are required to build a Weibull distribution [Figure 14.5], therefore, one would need at least six coupons to have a good representative variability for a Weibull curve. The static testing would have to include hygroscopic testing resulting in 8×24×2 = 384 coupon test matrix. This testing would result in a Weibull distribution of the static test results of B-basis properties of the material (Tomblin & Seneviratne, AR 10/6, 2011). This can then be used to generate the strength shape parameter α_R. The B-basis and A-basis material properties generated from the testing should also be used in design, analysis of the aircraft.

Fatigue test coupons will be required to generate a Weibull distribution of fatigue tests to generate the SN curves for notched or open-hole coupons from the above samples. This will result in the shape life parameter α_L and life factor N_L, from these three the LEF is evaluated. LEF allows one to reduce component test by enhanced loading, i.e. for instance a LEF of 1.12 will require 1.2 times the fatigue spectrum loading for 1 life-time test with a confidence level of B-basis strength.

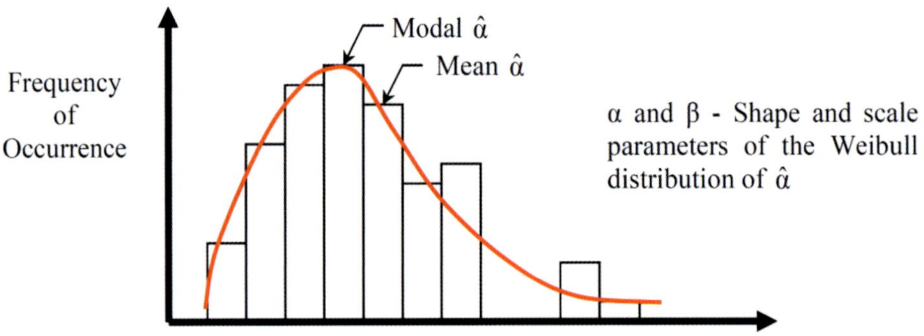

Figure 14.5: Scatter Analysis of Weibull Distribution of Shape Parameters (Tomblin & Seneviratne, AR 10/6, 2011).

Once fatigue testing of coupons is completed, and if it meets the design life cycle of the product, the values generated as B-basis from the fatigue test can then be used in the actual certification of the design including that for static design. These are durability derived allowables.

Such testing is, however, resource intensive and beyond the means of all except the largest of the aircraft producers.

14.7 Proposal for Certification of Composite Aircraft Structures

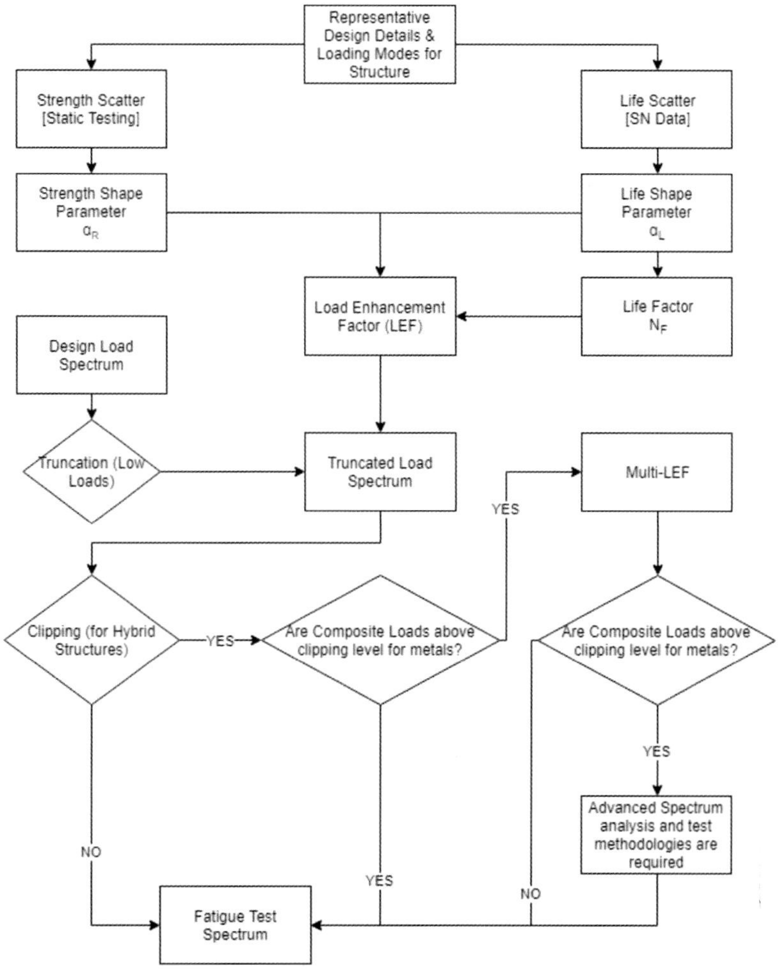

Figure 14.6: Fatigue Test Spectrum development for Composite Structural Testing (Tomblin & Seneviratne, AR10/6, 2011).

The inefficiency of current design and certification procedures for composite structures, along with their reliance on legacy laminate configurations, provides ample motivation for the development of new qualification methods in aircraft design. These procedures severely limit the introduction of new materials and design concepts, such as DD technologies. By improving the understanding of the physical behavior of carbon fiber-reinforced polymers (CFRPs) and establishing precise relationships between their material properties, the design and certification processes can become more cost-effective and open up possibilities for advanced materials and innovative laminate configurations. Trace and omni envelopes, combined with fracture mechanics concepts, establish the necessary relationships for efficient and reliable certification procedures for CFRP structures, independent of the material form and laminate configuration.

Even today, there is little agreement on the correct fatigue and durability behavior of composite materials. The lack of understanding of the fatigue and durability behavior of composites has forced researchers to focus on the statistical behavior of the elements in the building block pyramid. Statistics, although it gives us a safe basis, it cannot completely explain the physics of the phenomenon of fatigue. Fatigue is a mechanism often described by a stress level and the number of cycles that a metallic material can withstand cyclic loading before failure. Metals are amorphous isotropic and the mere presence of an impurity, inclusion or defect, in a loaded part, causes a microscopic anisotropy and added stresses giving rise to a crack and eventual fatigue related failure. However, composites are anisotropic and therefore using a material property like uniaxial stress is inappropriate in that it does not represent the true behavior of the composite.

Our goal here is to present what we believe is the most efficient and least burdensome approach to obtaining certification for a composite aircraft. In this section, we will suggest design considerations for a composite structure that could meet reduced certification requirements at the expense of a slight over-design.

Aircraft parts made from orthotropic carbon composites exhibit high in-plane resistance and low out-of-plane and bearing resistance. Strength is significantly reduced in the presence of holes and notches, thereby lowering the endurance limit. Therefore, aircraft parts should be designed to minimize load transfer points and avoid the use of drilled holes in the composite laminate for load transfer between parts. We propose that metallic inserts can elegantly accomplish load transfer by transferring in-plane loads to a high bearing strength hard point embedded within the composite. This approach reduces stress concentrations and enhances the service life of composite parts.

The buckling strength or stability of flat and curved composite skin panels is strongly influenced by factors such as geometry, stacking sequence, boundary conditions, and loading conditions. A post-buckled design is susceptible to crippling failure of the skins, where significant out-of-plane and out-of-phase deformation of the top and bottom skins, or skin and stiffeners, leads to core delamination or delamination along the skin-stiffener bond line, resulting in reduced structural integrity.

Therefore, it is advisable to avoid post-buckled designs for sandwich structures with thin face sheets and relatively thick cores, as commonly found in simpler General Aviation aircraft. Although this type of structure may incur a weight penalty, it will be easier to certify due to its inherent strength.

14.8 Manufacture of DD Composites

Test result consistency is one of the issues with composite materials which is usually attributed to variability in the properties of materials made using various production techniques. In our experience, such variability is primarily related to the fiber volume fraction, the material sizing, the curing behavior, and the void content in the material. To accelerate the layup and to ease handling, most layup requires the conversion of fiber tows into a layered fabric. These can be characterized into woven or 3D stitched materials. So, when we compare cured composite material properties, we should compare the variability of coupons keeping the carbon material, the fabric, the epoxy and the curing process the same.

Fiber production today in the year 2022, has become a very controlled process where fiber and sizing are highly consistent and variability, is in an extremely narrow band.

14.8.1 Manufacturing process consistency

The need for a process that is reproduceable and whose parameters can be closely controlled is highly desirable. Among the automated layup processes, such as Automated Tape Layup (ATL) or Automated Fiber Placement (AFP), the placement of layup of pre-impregnated (prepreg) materials over a mold can be done accurately and better than the processes done by hand. Added to this is a limitation of the ATL and AFP processes to follow complex mold contours. However, among other issues, prepreg materials cannot remove embedded voids, since pressure and heat can reduce the size of embedded voids formed during the layup process, but they cannot eliminate them. Prepreg materials in the aerospace industry usually contain a pre-calibrated amount of epoxy, which is related to an eventual required fiber volume fraction of the produced part. In most other industries, the prepregs are resin rich to address the issues of proper debulking of materials. Resin rich prepregs are usually covered with a perforated plastic release film which is then covered with a resin absorbing felt and then covered with a bag. The idea of such a configuration is to allow excess resin to travel across the perforations to be absorbed by the felt. Such a configuration does give a very good aesthetic look to the laminate but adds to the inconsistency of the results. The above highlighted issues result in inconsistent laminate quality which requires extensive statistical material qualification for certification. Finally, prepreg's during application on the mold and during cure, release harmful volatile emissions that are in constant contact with workers making this process subject to work-related long-term health issues. Prepreg curing requires extremely high temperatures in an inert environment, which is usually nitrogen at 8-9 atm of pressure and 180deg C. This makes each cure cycle extremely onerous on environment adding to the carbon footprint of the manufactured item.

There are other promising composite fabrication processes that are getting attention such as, vacuum infusion in multitude of its forms. Some of these have been patented by large aircraft manufacturers such as Boeing and Airbus. A patent search on these will unveil a myriad of technologies. Vacuum infusion process can potentially minimize worker health related risks from the exposure of volatiles since infusion is carried out in a vacuum environment where the workers are not subjected to volatile emissions. However, even now, there is no patented process that can guarantee total elimination of voids. NASHERO has developed the ZeroVoid® proprietary vacuum infusion process that can eliminate voids. The ZeroVoid® infusion processes can be carried out in an autoclave and the process parameters can be controlled by sensors in the infusion system, which once set, provide consistent and repeatable composite parts with predetermined fiber volume fractions. Such composites result in extremely low structural variability and a high fatigue resistance as has been witnessed by the tests carried out at the National Institute for Aviation Research (NIAR), at the Wichita State University.

14.8.2 Production molds and parts for composite aircraft structures

To ensure that an aircraft part meets the required shape tolerances, the mold used for manufacturing must guarantee precise shaping. Composite parts, especially thin-walled components like the winglet of an NASHERO aircraft, are prone to out-of-plane deflections caused by curing and thermal stresses. These deflections need to be considered during mold design, as well as the thermal cycling experienced by the mold itself. It is a common misconception among engineers that creating a carbon composite mold for a carbon composite part with an identical thermal coefficient will eliminate any deformation relative to the part. However, achieving perfect matching stiffness tensors between the mold and part is practically unfeasible. Molds have different requirements for surface quality, strength accuracy, cost, and repeatability, whereas the part is typically much thinner than the mold.

Given these considerations, one may wonder about the recommended approach for constructing a high-end mold that ensures perfect shape conformity and tolerance adherence. Here, we will present a simplified version of NASHERO's procedure for directly machined molds (Figure 7), which has been employed to manufacture certain parts (Figure 9). For the sake of brevity and due to the proprietary nature of the procedure, some aerospace-specific elements have been omitted.

In this procedure, the mold design commences with a CAD model of the part, followed by analyzing thermal deformation at the initial in-mold cure temperature. Subsequently, the metal mold is developed, considering the thermal deformation of the mold from the cure temperature to room temperature. By utilizing a split mold, extremely tight tolerance requirements were achieved for a production part. The part is predominantly produced using DD layup, apart from a single outer layer made from a hybrid carbon-kevlar woven material (Figure 9).

On the other hand, soft molds are employed in conjunction with low-temperature curing resins, minimizing deformation due to the limited temperature excursions and low exotherm of the epoxy. These molds can be economically created from highly polished master models or patterns using a mold-making system that does not shrink during the curing process. A thin-walled part manufactured using DD [±0/±50] layup, with a minimum of 2 DD layups at the edges and a ply-drop strategy employing a maximum of 10 DD layups in the middle, exhibited no out-of-plane deformation when placed on the mold surface (Figure 8).

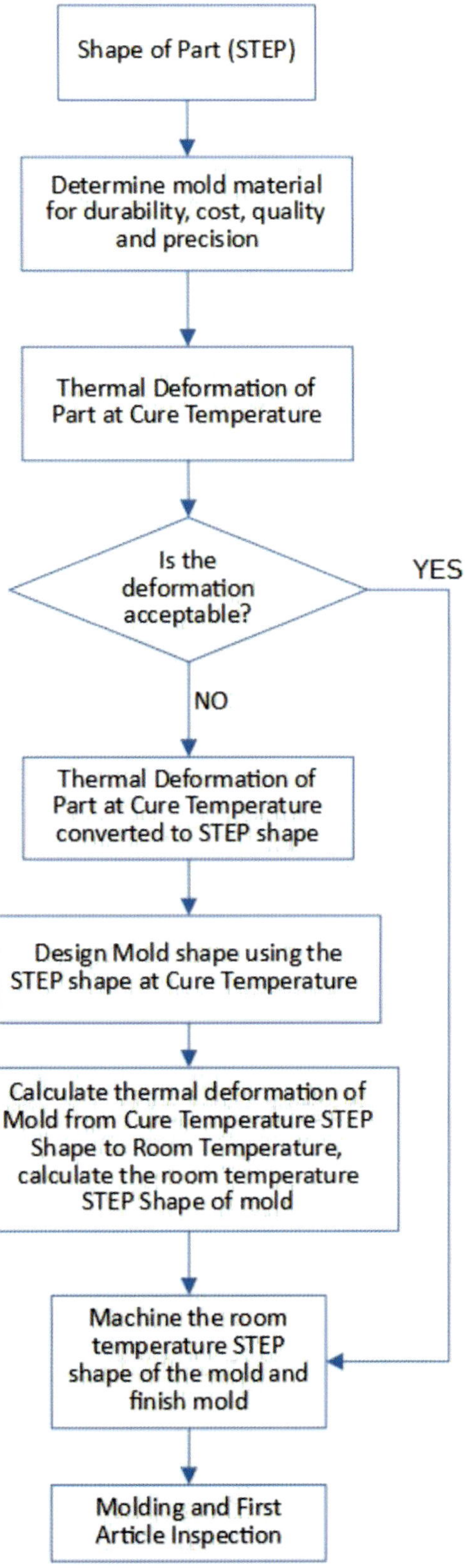

Figure 14.7: simplified version of NASHERO's procedure for directly machined molds.

Figure 14.9: A DD based landing gear fork made from a directly machined metal mold, with co-cured metal hardpoints

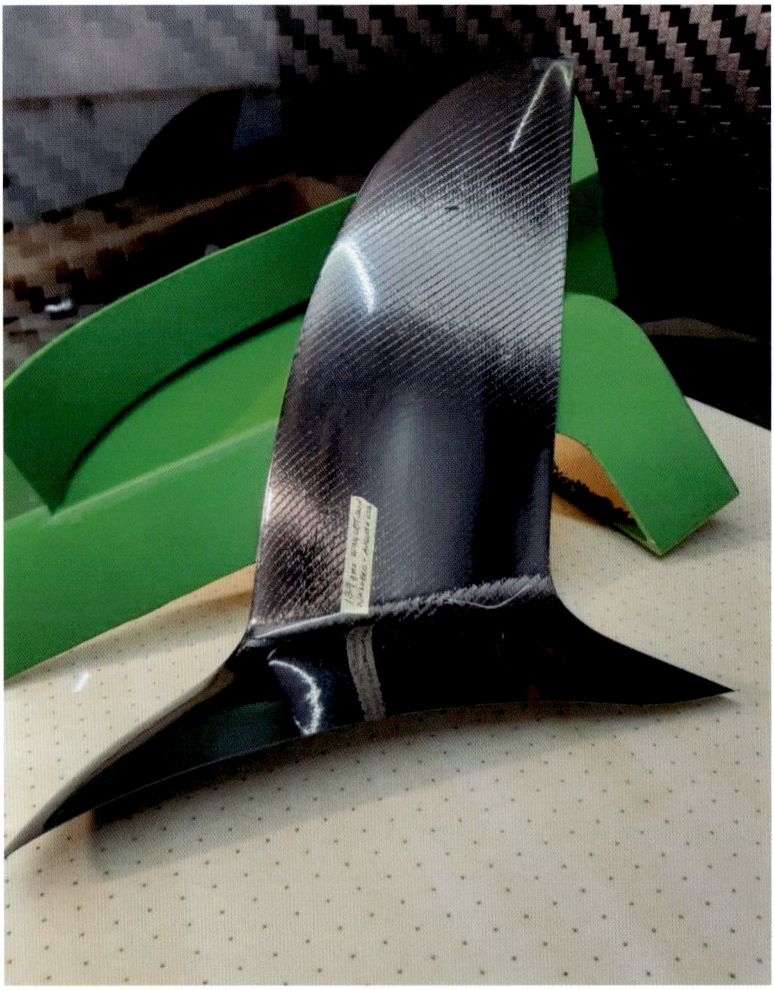

Figure 14.8: DD [±0,±50] part made in a soft mold. Mold is green and part is black.

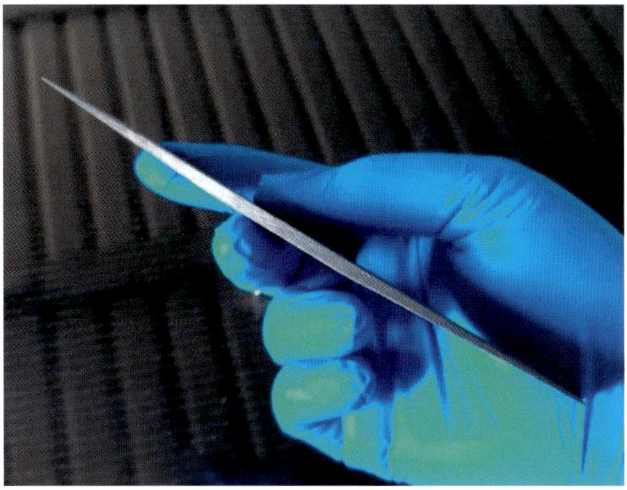

Figure 14.10: Tapered beam coupon made in DD [±0,±50], as tested by NIAR in the USA

Bond, D. A., & Smith, P. A. (2006). Modeling the transport of low-molecular-weight penetrants. *Applied Mechanics Reviews*, 59(5): 249-268.

DOD. (2002). *COMPOSITE MATERIALS HANDBOOK, VOLUME 3. POLYMER MATRIX COMPOSITES* (Vols. MIL-HDBK-17-3F). DEPARTMENT OF DEFENSE.

FAA. (2011). *PART 23 AIRWORTHINESS STANDARDS: NORMAL, UTILITY, ACROBATIC, AND COMMUTER CATEGORY AIRPLANES.* Code of Federal Regulations (annual edition).

FAA. (2019). *ART 25—AIRWORTHINESS STANDARDS: TRANSPORT CATEGORY AIRPLANES.* US Government Publishing Office.

Freed, Y. (2016). Implementation of Invariant-Based Approaches for Determination of Composite Materials Allowables. Tel Aviv, Israel: 56th Israel Annual Conference on Aerospace Sciences.

Freed, Y., & Elmalich, D. (2017). A New Study on Scatter Factors in Fatigue Testing of. *29th ICAF symposium, 2017.* Nagoya, Japan: ICAF.

Komorowski, J. P. (1983). *Hygrothermal Effects in Conitnuous Fiber Reinforced Composites, Part III, Mechanical Properties.* Aeronautical Note, Conseil national de recherches Canada, National Aeronautical Establishment.

McKinsey & Company. (2022). *The net-zero transition.* McKinsey & Company.

Srivastava, S., Sharma, N., Tsai, S. W., & Mohite, P. (2020). D and DD-drop Layup Optimization of Aricraft Wing Panels under Multi-Load Case Design Environment. *Composite Structures*, 0263-8223.

Tomblin, J., & Seneviratne, W. (2011). *DETERMINING THE FATIGUE LIFE OF COMPOSITE AIRCRAFT STRUCTURES USING LIFE AND LOAD-ENHANCEMENT FACTORS.* Washington DC: U.S. Department of Transportation, Federal Aviation Administration.

Tomblin, J., Lacy, T., Smith, B., Hooper, S., Vizzini, A., & Lee, S. (1999). *REVIEW OF DAMAGE TOLERANCE FOR COMPOSITE SANDWICH AIRFRAME STRUCTURES.* US Department of Transportation, Federal Aviation Administration.

Torenbeek, E. (1981). *Synthesis of Subsonic Airplane Design.* Delft: Delft University Press.

Tsai, S. W., & Melo, J. D. (2015). *Composite Materials Design and Testing, unlocking the mystery with invarients.* Stanford, CA, USA: Composites Design Group, Department of Aeronautics and Astronautics, Stanford University.

Whitehead, H. P., Cordero, R., & Saether, E. S. (1986). *Certification Testing Methodology for Composite Structure, Vol I-III.* Northrop Corporation, Aircraft Division.

Wong, K. J. (2013). *MOISTURE ABSORPTION CHARACTERISTICS AND EFFECTS ON MECHANICAL BEHAVIOUR OF CARBON/EPOXY COMPOSITE: APPLICATION TO BONDED PATCH REPAIRS OF COMPOSITE STRUCTURES.* Université de Bourgogne, Mechanics. Université de Bourgogne.

CHAPTER 15

Case Studies Using Lamsearch and MicMac to Find the Best DD

Thierry Massard – Think Composites - France

Introduction : How to find an efficient double-double solution ?

There are at least two ways of finding a double-double (DD) laminate for a composite construction. One is the replacement of an existing laminate, the other is the direct design of a double-double laminate.

15.1 – Replacement : If a general laminate is already known from previous studies, it is possible to find a replacement by a double-double with the same stiffness, and similar strength. The advantage of switching to double-double lies in the simplification of the stacking sequence, the homogenization for easy ply-drop and tapering, the ease of manufacturing and repairing, thus leading to lower cost and faster production. In this case it is recommended to use the **MicMac** software which can handle up to 24 different ply orientations, symmetric or not. The equivalent double-double is given providing that the amount of 0°plies or 90° plies does not exceed 50%. In most practical cases the solution is an exact replacement having the same stiffness. (see chap 2). For the other cases, an approximate solution can be also found having a stiffness close to the initial laminate.

2 – **Direct design** of the laminate : when no preliminary design exists, the problem to solve is given a set of loads applied to the structure what would be the best double-double solution? In this case **Lamsearch** is the perfect tool to find it.

Both software is provided free of charge, unlimited and ready to download (see § 3 in this chapter).

In this chapter we show some practical cases studies with these two methods using MicMac or Lamsearch.

1 – Replacement of an existing laminate

Case study 1 - How to find an equivalent double-double solution to replace an existing laminate using MicMac ?

In this case, a general quad laminates already exists from a previous design.

$$[0_3/45/90_2/\text{-}45_3/0/\text{-}45/45_2/90/45/\text{-}45/0/45]_{2S}$$

This laminate has 18 plies for the elementary block and is a symmetric stacking of 2 blocks with a total of 72 plies and a thickness of 9mm. The material is IM6/epoxy.

The stiffness and strength of the laminate are given in the figure below. They are calculated using MicMac or any software solving the classical laminate theory (CLT).

GPa	[A*]	[D*]	[B*]
11	94,2	104,6	0,0
22	72,7	70,3	0,0
21	27,9	23,9	0,0
66	32,7	28,7	0,0
61	0,0	-1,8	0,0
62	0,0	-1,8	0,0
Tsai-Wu(MPa)	404,8 MPa		

The table shows the stiffness matrix of the laminate and the strength in tension.
The laminate is not homogenized since A* and D* matrices are different with also shear coupling for the bending stiffness. If ply drops are necessary through the structure, the drop step would be 4,5 mm or removing only a few plies would lead to different material properties from location to location.

Replacement method

We use MicMac spreadsheet. It is based on Microsoft Excel. MicMac offers a number of practical solutions for general laminates. In this case, the GenLam option is the most appropriate choice since it can accommodate up to 24 different orientations, each orientation having any number of plies.

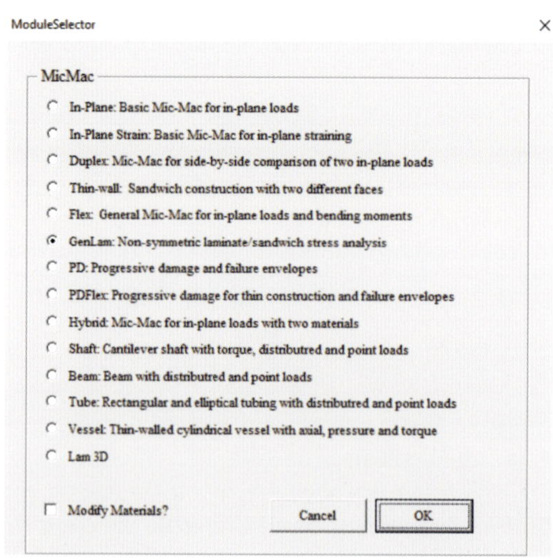

In the Genlam, the input is the following (cell B4 and following) :

theta/24	theta/23	theta/22	theta/21	theta/20	theta/19	theta/18	theta/17	theta/16	theta/15	theta/14	theta/13		Ply mat:	IM6/ep[SI]		
0,0	45,0	90,0	-45,0	0,0	-45,0	45,0	90,0	45,0	-45,0	0,0	45,0	[repeat]	h, #	T	h+, E-3	x
3	1	2	3	1	1	2	1	1	1	1	1	2,00	36		4,50	
theta/1	theta/2	theta/2	theta/2	theta/5	theta/6	theta/7	theta/8	theta/9	theta/10	theta/11	theta/12		Ply mat:	IM6/ep[SI]		
0,0	45,0	90,0	-45,0	0,0	-45,0	45,0	90,0	45,0	-45,0	0,0	45,0	[repeat]	h, #	T	h-, E-3	x
3	1	2	3	1	1	2	1	1	1	1	1	2,00	36		4,50	

The replacement solution is given in the cell A61 and beyond :

The replacement of the quad laminate is $[\pm20/\pm62]_{18T}$ to match the inplane stiffness :

PSI (+/-)		20
PHI (+/-)		62
		A*
	11	94,17
	22	72,74
	12	27,86
	66	32,65
	16	0,00
	26	0,00

Inplane stiffness matrix of the replacement DD solution

In this case the inplane tension strength of the laminate is 404.4 MPa for the First Ply Failure (FPF) and 1528 MPa for the Last Ply Failure (LPF).

To match the flexural stiffness of the quad the best DD would be $[\pm14/\pm61]_{18T}$ given in cells G62-G63. The A* and D* stiffness matrices are :

PSI (+/-)		14
PHI (+/-)		61
		D*
	11	104,58
	22	70,33
	12	23,85
	66	28,65
	16	0
	26	0

Flexural stiffness of the replacement double-double solution

The strength of the DD is 468 MPa for the First Ply Failure (FPF) and 1670 MPa for the Last Ply Failure (LPF) which is slightly higher than the original quad laminate.

Conclusion

The double-double laminate is an exact replacement of the initial quad. The main advantage is that it is homogenized which makes ply drop easy. The step will be 4 plies by 4 plies that is 0.5mm instead of 4.5 mm for the initial quad laminate. The stacking is simpler and can be achieved with a one-axis ATL machine if a pre-built double-double prepreg is used. The strength of the double-double laminate is at least equal or higher than the quad laminate.

Case 2 – Replacement of an unconventional laminate by a double-double laminate

In the literature, many examples of optimum stacking sequence for general laminates can be found. In all the cases that has been surveyed, MicMac was able to find an exact DD solution to replace the unconventional laminate.

We have studied one of the solutions proposed by Moritz Sprengholz and al. in their article in Composite Structures, 2021 *(https://doi.org/10.1016/j.compstruct.2021.114514)*

	θ_n											
Sol.	1	2	3	4	5	6	7	8	9	10	11	12
7	73.8	−34.4	42.8	−54.3	−43.7	−34.8	32.5	83.0	−9.4	−9.1	7.7	32.6

The laminate is thus [73,8/-34,4/42,8/-54,3/-43,7/-34,8/32,5/83,0/-9,4/-9,1/7,7/32,6]$_{2S}$ (24 plies)
with IM7/977 material from our database.

Using MicMac Genlam we input the above stacking sequence:

theta/24	theta/23	theta/22	theta/21	theta/20	theta/19	theta/18	theta/17	theta/16	theta/15	theta/14	theta/13		Ply mat:	IM7/977[SI]	
73,8	-34,4	42,8	-54,3	-43,7	-34,8	32,5	83,0	-9,4	-9,1	7,7	32,6	[repeat]	h, # T	h+, E-3	x
1	1	1	1	1	1	1	1	1	1	1	1	2	24	3,00	
theta/1	theta/2	theta/2	theta/2	theta/5	theta/6	theta/7	theta/8	theta/9	theta/10	theta/11	theta/12		Ply mat:	IM6/ep[SI]	
73,8	-34,4	42,8	-54,3	-43,7	-34,8	32,5	83,0	-9,4	-9,1	7,7	32,6	[repeat]	h, # T	h-, E-3	x
1	1	1	1	1	1	1	1	1	1	1	1	2	24	3,00	
											Repeat	[core] th	Total ply#	h,total E-3	
												0,0	48	6,00	

The replacement is found cells B62-B63 : [± 17 / ±57]$_{6T}$ to match the inplane stiffness or [±25/±61]$_{6T}$ to match the flexural stiffness.

Original laminate stiffness is :

GPa	[A*]	[D*]	[B*]
11	95,8	79,1	0,0
22	59,3	70,0	0,0
21	27,1	30,0	0,0
66	31,4	34,3	0,0
61	-4,3	-8,9	0,0
62	-0,2	-1,0	0,0
Tsai-Wu(MPa)	487,4 MPa		

Stiffness matrices of the original laminate

The strength in tension of this laminate is 487 MPa (FPF) / 1029 MPa (LPF)

The double-double replacement stiffness matrices are given in the table below :

PSI (+/-)	17
PHI (+/-)	57
	A*
11	95,8
22	59,3
12	27,1
66	31,4
16	0,0
26	0,0

FPF 492 MPa

LPF 1366 MPa

Matching the inplane stiffness

PSI (+/-)	25
PHI (+/-)	61
	D*
11	79,15
22	70,01
12	30,04
66	34,33
16	0
26	0

FPF 394 MPa

LPF 1197

Matching the flexural stiffness

Conclusion

An exact replacement solution is found both for inplane or flexural stiffness. In both cases we have A* = D* and B* < 1% of A* which shows that both DD are homogenized laminates and can be tapered if needed for thickness optimization which is impossible to do with the initial unconventional laminate.

Case study 3 - Approximate replacement of a hard laminate

The replacement double-double **[±Ψ/±Φ]** solution can be expressed as a function of V1* and V2* of the initial laminate

$$\mathbf{COS\ 2\Psi = V_1^* + [-V_1^{*2} + V_2^*/2 + 1/2]^{1/2}}$$

$$\mathbf{COS\ 2\Phi = 2V_1^* - COS\ 2\Psi}$$

where Vi* are function of the material properties :

$$V_1^* = (A_{11}^* - A_{22}^*)/2U_2$$

$$V_2^* = (A_{11}^* + A_{22}^* - 2U_1)/2U_3$$

$$V_3^* = (A_{61}^* + A_{62}^*)/U_2$$

$$V_4^* = (A_{61}^* - A_{62}^*)/2U_3$$

(See also chapter 2)

The relations above give an exact solution for double-double replacement of any laminate. The value of the cosine function must be between [-1 and + 1] to have a real solution. There are no Double-double in the region marked in violet in the figure below, and so there is no exact replacement for a laminate which falls in these two regions. For a quad laminate, it appends when the fraction of 0 or the fraction of 90 are greater than 50% as shown in the graph below

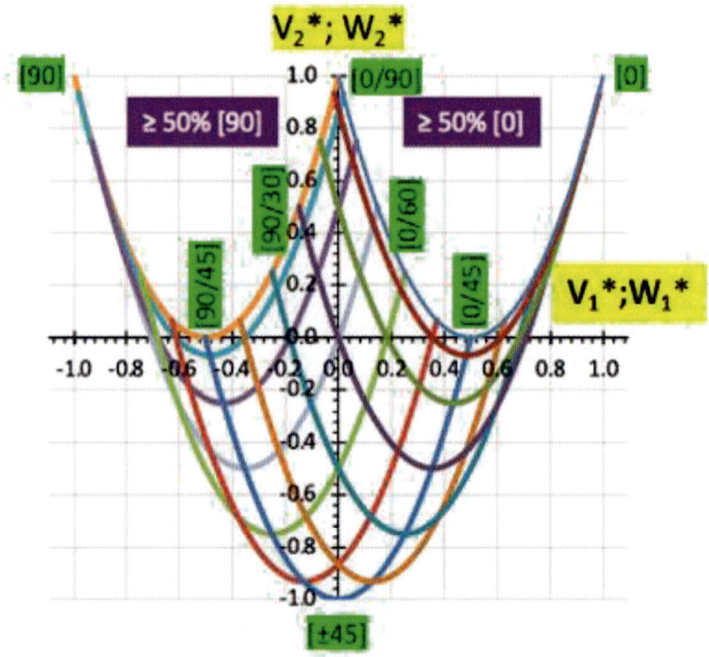

The domain of double-double in the (V1*,V2*) space

However, an approximate solution can be found in the general form of [0/±Φ] for the upper right and [90/±Φ] for the upper left of the diagram which are the border lines of the two regions. For example, we choose the following laminate :

$[5_5/45/-45_3/12]_s$ in IM7/977

With the following inplane stiffness matrix :

	A* GPa
11	140,88
22	31,40
12	21,83
66	26,37
16 = 61	1,79
26 = 62	-8,59

In this case, MicMac Genlam gives COS2Ψ =1,12 (cell B94). So, there is no exact replacement solution. To find the best approximate Double-Double we use MicMac inplane and the graphic macro Chart Quick.
The Chart Quick will make the following graph giving the A*11, A*22, A*66 variations for a laminate in a general form such as [0/±Φ] where Φ varies from 0 to 90. The best replacement will be the one matching the value of the stiffness matrix components of the original laminate which are also given in dash lines on the chart below.

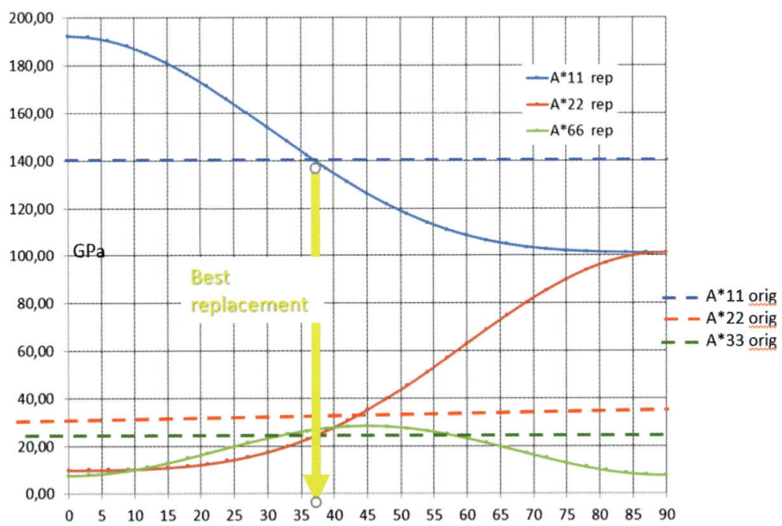

From the graph above there is a region of variation of ±Φ where the Double-double stiffness can match the stiffness of the original laminate.
If we math A11, the best approximate replacement is found here to be $[0_2/\pm 36.5]_T$

The A matrix of the replacement is :

	A* GPa	original
11	141,23	0%
22	23,48	34%
12	22,26	-1%
66	26,55	0%
16 = 61	0,00	
26 = 62	0,00	

the table gives also the percentage of difference compared to the original.

From the graph, it would also be possible to match the A^*_{22} or A^*_{66} or any combination of the 3 main components.

Conclusion

In some cases, there are no exact replacement, but an approximate solution can be found on the border line of the existing domain of Double-double using the Chart Quick graphic tool of the MicMac.

2 - Find the best Double-Double laminate using Lamsearch

Lamsearch is an Excel-base program. It solves the following problem: given a set of loads and having chosen a material from the database, which is the best laminate in the general form of $[\pm \Phi/\pm\Psi]n$ to sustain the loads?

Case 4 - Find the best Double-Double laminate for seven different loads at a point.

2.1 Input data

The seven loads are input in cell U20 and following. Each load has three components sigma1, sigma2, and sigma6

Load	All 7 loads		
#	sig1	sig2	sig6
1	0,43	0,98	0,21
2	0,24	1,00	0,02
3	-0,06	0,88	0,046
4	0,49	0,45	0,17
5	0,00	0,07	0,19
6	0,06	0,54	0,00
7	0,52	-0,01	0,01

Input data for the 7 loads

Material selection: pull-down						Anchors FPF		Nettles circle	
IM6/ep	Trace	X	X'	Degr. trace	Em°	e/x	e'/x	e/x	e'/x
IM6/ep	231	3500	1540	203	1,00	5,0	7,6	4,0	4,0

The material selection refers to a database of 25 materials which can be extended by users to their own material systems. A pull-down menu is used to select the material.

2.2 Overlooking the possible solutions

The program will calculate all the possible solutions in a 13 x 13 table where Φ and Ψ varies from 0 to 90 with 7.5-degree increment. The solution appears in term of strength for 4 different criteria Tsai Wu, Max Strain, Unit circle and Nettles.

Each table will return the value of the strength ratio R for each criterion. The table shows in bold type which DD has the highest strength. If the loads are normalized to 1 then the R value is the strength of the material under these loads. If not, the R value is the safety factor to failure.

R/smooth	[±Ψ]	2	3	4	5	6	7	8	9	10	11	12	13	
[±Φ]	0,0	7,5	15,0	22,5	30,0	37,5	45,0	52,5	60,0	67,5	75,0	82,5	90,0	max
0,0	56	57	60	67	84	117	169	243	332	404	411	368	338	411
7,5	57	58	60	68	84	116	168	243	333	388	394	359	336	394
15,0	60	60	64	69	84	114	167	244	336	395	410	392	378	410
22,5	67	68	69	76	87	114	165	240	337	421	446	439	431	446
30,0	84	84	84	87	99	118	160	233	334	447	**495**	491	485	495
37,5	117	116	114	114	118	136	166	231	332	422	430	436	439	439
45,0	169	168	167	165	160	166	197	244	283	288	298	305	308	308
52,5	243	243	244	240	233	231	244	240	207	202	207	212	214	244
60,0	332	333	336	337	334	332	283	207	176	157	155	156	157	337
67,5	404	388	395	421	447	422	288	202	157	140	130	127	127	447
75,0	411	394	410	446	**495**	430	298	207	155	130	121	115	114	495
82,5	368	359	392	439	491	436	305	212	156	127	115	111	109	491
90,0	338	336	378	431	485	420	289	201	150	125	114	109	108	485
max	411	394	410	446	495	436	305	244	337	447	495	491	485	495

Table for the Tsai WU failure criteria. (Cells C20-O32)

The optimal Double-double laminate is [±30/±75]T. It appears in the table above in bold face and rounded in red is given the maximum strength (495 MPa) of the laminate.

A very practical feature of Lamsearch is the search for alternate solution. If the optimal laminate is not suitable, one can look for other solutions which do not degrade the strength. The colorful zone is where the solution does not differ too much from the optimum. The clear zone goes away from the optimum.
In this case going from [±30/±75] to [±30/90_2] solution degrades only by 2% from the optimum while [±30/±60] will degrade by over 30% as shown in table below.

	9	10	11	12	13
	60,0	67,5	75,0	82,5	90,0
3	332	404	411	368	338
3	333	388	394	359	336
4	336	395	410	392	378
0	337	421	446	439	431
3	334	[illegible]	**495**	[illegible]	485
1	332	422	430	436	439
4	283	288	298	305	308
0	207	202	207	212	214

2.3 Using the zoom option for more precise value of the optimum DD

As the double-double field of solution is a continuous field it is possible to generate a table with a smaller increment by defining new values of variation for the angles Φ and Ψ. With this zoom option a very fine optimum can be found if necessary.

	ZOOM: DD		
	Min	Max	Delta
[±Φ]	28	37	0,75
[±Ψ]	73	78	0,42

R/smooth	[±Ψ]	2	3	4	5	6
[±Φ]	73,0	74,0	75,0	76,0	77,0	78,0
28,0	478	480	481	481	482	481
29,0	485	487	488	489	489	488
30,0	492	494	495	496	496	496
31,0	500	502	503	503	503	503
32,0	508	510	511	511	511	510
33,0	516	518	518	**519**	518	518
34,0	503	504	504	505	506	506
35,0	481	482	483	483	484	485
36,0	459	460	461	462	463	464
37,0	438	439	440	441	442	443

Increment selection and results from the zoom option

In this case, with a 1 degree increment, the optimum double-double is found : $[\pm 76/\pm 33]_T$ with a strength of 519 MPa and over 20% better than a quasi-isotropic laminate.

The same tables are given for other failure criteria: Max strain, Unit Circle and Nettles.

2.4 – Finding the critical load

Lamsearch will also give an important information which is the critical load among the 7. It can be useful to know which load is critical to eventually mitigate the risk on a given load having a large uncertainty.

Load	All 7 loads			
#	sig1	sig2	sig6	R
1	0,43	0,98	0,21	495
2	0,24	1,00	0,02	600
3	-0,06	0,88	0,046	548
4	0,49	0,45	0,17	684
5	0,00	0,07	0,19	1283
6	0,06	0,54	0,00	1038
7	0,52	-0,01	0,01	595
		Critical load is load#		1

2.5 – Tapering the laminate to find a lighter solution

In the case where the 7 loads are given for 7 different points in the structure, Lamsearch will give the tapering capacity at each location. In the case of DD tapering is easy as in metal since the material is homogenized.

Load #	R	Taper
1	495	100%
2	600	83%
3	548	90%
4	684	72%
5	1283	39%
6	1038	48%
7	595	83%

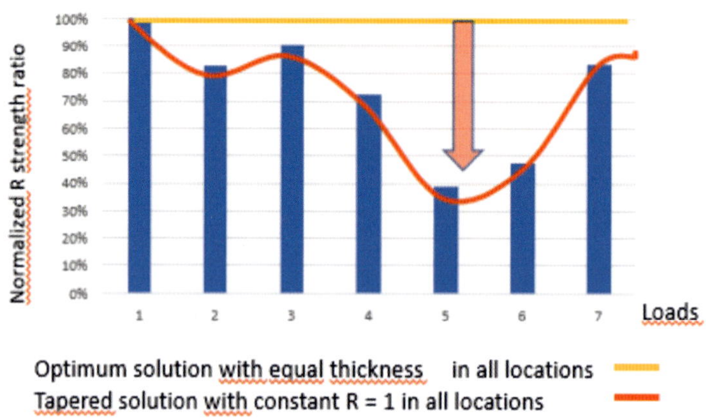

Tapering the thickness in all 7 locations will bring more weight savings.

Other Lamsearch tools

1 - Lamsearch with 49 loads

In a more advanced version, Lamsearch can input 49 loads which makes structural optimization easier on 49 locations of the structure or in the case of complex load cases with many different loadings. This version il also suitable for FEM optimization input.
In this version, the effect of secondary loads can be studied by generating random loads from a set of seven primary loads.

Lam	All 49 loads		
#	sig1	sig2	sig6
1	0,29	0,45	0,61
2	0,09	0,34	1,08
3	0,04	0,27	1,140
4	0,22	0,30	0,91
5	0,51	-0,24	1,16
6	0,49	0,29	0,33
43	0,05	-0,17	0,75
44	0,58	0,30	0,29
45	0,00	0,00	1,00
46	0,20	0,00	1,00
47	0,20	-0,20	1,00
48	0,00	0,50	0,00
49	0,50	0,00	0,50

Lam search 49 loads V1.0 revised 0301202	output in blue
Steve Tsai, Thierry Massard, Pranav D. Shah	input in red
stsai@stanford.edu, tmassard@stanford.edu	For strength prediction
	use degraded U's

Material selection: pull-down						Anchors FPF		Nettles circle	
IM7/MTM	Trace	X	X'	ded	Em°	e/x	e'/x	e/x	e'/x
IM7/MTM	199	2500	1700	142	1,00	8,4	9,7	4,0	4,0

2 – Lamsearch with 7 loads and 7 materials

This version of Lamsearch handle 7 loads and also shows the optimum DD for 5 materials at the same time. It is very useful to understand how the optimum DD depends on the choice of the material.

Lamseaarch 5 Material 7 loads v2.1	2/16/2022
Steve Tsai, Thierry Massard, Pranav D. Shah	
stsai@stanford.edu, tmassard@stanford.edu	

Input of the data : the 7 loads are given as in any other Lamsearch versions :

Load	Input all 7 loads		
#	sig1	sig2	sig6
1	0,43	0,98	0,21
2	0,24	1,00	0,02
3	-0,06	0,88	0,046
4	0,49	0,45	0,17
5	0,00	0,07	0,19
6	0,06	0,54	0,00
7	0,52	-0,01	0,01

The material selection is by a pull-down menu for 5 different materials from the database :

Input Material selection: pull-down menu				
Material1	Material2	Material3	Material4	Material5
IM6/ep	IM7/977	T7/2510	T8S/3900	C-Ply 64

Results are listed below. The best double-double laminate for each material is gen in the general form of DD1 and also DD2.

R FPF is given for the 4 main failure criteria: Tsai-Wu, Max strain, Unit Circle and Nettles. If loads are normalized then the R is the strength of the laminate under the 7 loads. This Lamsearch also gives the R ratio for the case of the laminate with an open-hole. Then the corresponding knock-down factor to the intact R is given in the last line. The critical load is indicated for each material selection.

In this case, the optimum Double-double are very similar for all 5 materials. The higher R value is obtained for the IM7 material. If we also consider open-hole strength ratio the IM6 materials exhibit a lower knock down factor and in final a higher strength with open-hole. So the analyst designer have the option between choosing the higher strength intact (IM7) or the higher open-hole strength for better damage tolerance (IM6).

Results					
	Input Material selection: pull-down menu				
	Material1	Material2	Material3	Material4	Material5
	IM6/ep	IM7/977	T7/2510	T8S/3900	C-Ply 64
Best smooth DD1					
±Φ	33	30	37,5	30	30
±Ψ	76	83	60	75	75
Best smooth DD2					
±ϕ	54,58	56,25	48,75	52,50	52,50
±ψ	21,33	26,25	11,25	22,50	22,50
Strength (MPa)					
R Tsai Wu	521	574	360	957	481
R Max strain	510	455	300	353	338
R Unit	502	451	298	351	336
R Nettles	391	359	237	278	267
Critical load	1	1	1	1	1
R Open Hole	188	169	112	132	126
Knock Down Factor to Unit circle	2,8	3,4	3,2	7,3	3,8

3 - Free download

All the software described in this book can be obtained free of charge and unlimited. A free on-line tutorial can be required. It will be organized at the best of our availability.

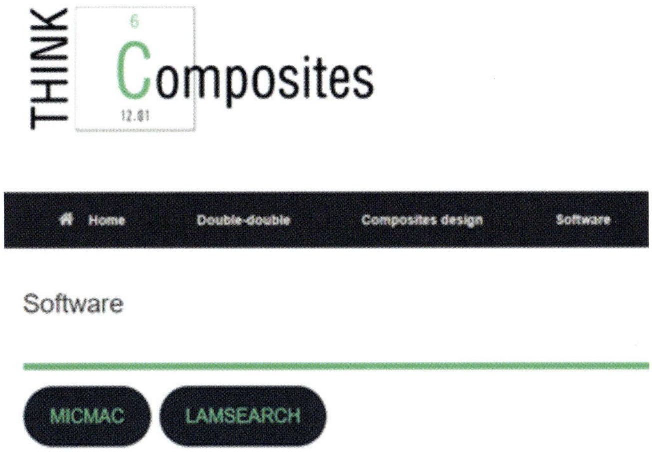

Web site to download MicMac and/or Lamsearch

Request of software and/or on-line tutorial : Stsai@stanford.edu
Download any of the software : https://www.think-composites.com/software/

MicMac 6.2	02/20/2022
Lam search Light 7 loads	02/16/2022
Lamseaarch 5 Material 7 loads	02/26/2022
Lamsearch 49 loads	on request

MicMac/Duplex for master envelope plotting

Easy to use Excel-based MicMac for generating of failure envelopes for DD and quad. Free copy will be sent upon request. Material macro is not working. So material files must be copied and pasted. Input and output (at GetVal) are shown in two slides below.

Operation: 1. Select material; 2. Enter laminate angles and ply numbers; 3. Push Chart-quick

For master envelope, select T4708/MR as a good approximation.

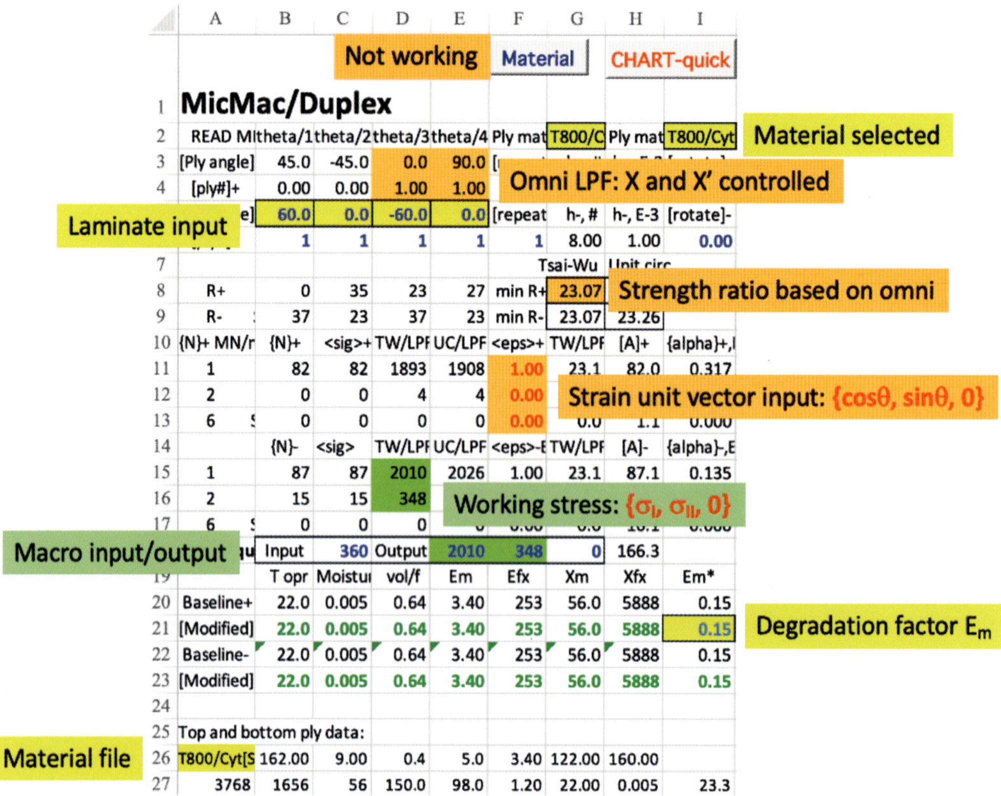

Mic-Mac input: Laminate definition followed by Chart-Quick for calculating omni LPF envelopes

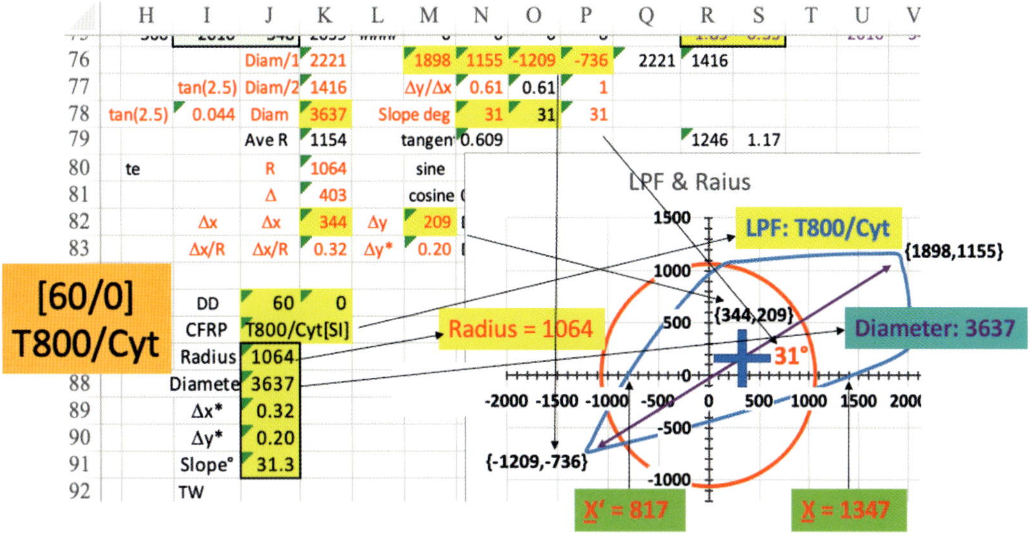

Typical MicMac output for failure envelopes. X and X' can be found directly from envelope

Material **CHART-quick**

MicMac/Duplex

	A	B	C	D	E	F	G	H	I
2	READ MI	theta/1	theta/2	theta/3	theta/4	Ply mat	T800/C	Ply mat	T800/Cyt
3	[Ply angle]	45.0	-45.0	0.0	90.0	[repeat	h+, #	h+, E-3	[rotate]+
4	[ply#]+	0.00	0.00	1.00	1.00	2	8.00	1.00	0.00
5	[Ply angle]	60.0	0.0	-60.0	0.0	[repeat	h-, #	h-, E-3	[rotate]-
6	[ply#]-	1	1	1	1	1	8.00	1.00	45.00
7							Tsai-Wu	Unit circ	
8	R+	0	61	34	12	min R+	12.03	8.50	
9	R-	14	61	39	61	min R-	38.56	13.24	
10	{N}+ MN/r	{N}+	<sig>+	TW/LPF	UC/LPF	<eps>+	TW/LPF	[A]+	{alpha}+,I
11	1	58	58	2234	767	0.71	8.5	82.0	0.317
12	2	-58	-58	-2232	-766	-0.71	-8.5	82.0	0.317
13	6	0	0	0	0	0.00	0.0	1.1	0.000
14		{N}-	<sig>	TW/LPF	UC/LPF	<eps>-E	TW/LPF	[A]-	{alpha}-,E
15	1	23	23	274					
16	2	-23	-23	-273					
17	6	0	0	0	0	0.00	0.0	26.0	-0.481
18	CHART-qu	Input	315	Output	274	-273	0	166.3	

Rigid-body rotation

Laminate shear strength S: 274 MPa

Angle of rotation for $\sigma_1 = -\sigma_2$

Determination of laminate shear strength S

Ch.15

This double-double came before ours and is very popular in California and Arizona. Should be as easy to swallow as our DD